Microorganisms for Sustainability

Volume 15

Series Editor
Naveen Kumar Arora
Environmental Microbiology
School for Environmental Science, Babasaheb Bhimrao Ambedkar University
Lucknow, Uttar Pradesh, India

More information about this series at http://www.springer.com/series/14379

Dilfuza Egamberdieva • Antonio Tiezzi
Editors

Medically Important Plant Biomes: Source of Secondary Metabolites

Springer

Editors
Dilfuza Egamberdieva
Faculty of Biology
National University of Uzbekistan
Tashkent, Uzbekistan

Key Laboratory of Biogeography and Bioresource in Arid Land
Xinjiang Institute of Ecology and Geography, CAS
Urumqi, People's Republic of China

Antonio Tiezzi
Department for the Innovation on Biological, Agrofood and Forestal systems (DIBAF)
Laboratoy of Plant Cytology and Biotechnology, Tuscia University
Viterbo, Vatican City State

ISSN 2512-1901 ISSN 2512-1898 (electronic)
Microorganisms for Sustainability
ISBN 978-981-13-9565-9 ISBN 978-981-13-9566-6 (eBook)
https://doi.org/10.1007/978-981-13-9566-6

This Springer imprint is published by the registered company Springer Nature Singapore Pte Ltd.
The registered company address is: 152 Beach Road, #21-01/04 Gateway East, Singapore 189721, Singapore

Foreword

Plants have been used for centuries as one of the main sources of natural drugs, and this tradition is well documented. The role of plants as health agents in multiple cultures of the world, transmitted through generations is widely known. Their compounds have important ecological functions, providing protection against pests, diseases, ultraviolet-B damage, and other environmental stresses. Plants are also producers of pharmaceutical drugs, such as antibiotics, anticancer, and antifungals, food supplements, and agrochemicals and have a wide variety of other industrial biotechnology applications, such as the steroid production. The new advances in genome sequencing technology enhance the progress in the study of metabolic pathways to be able to understand the role of some enzymes, improving human health and industrial uses.

In recent time, the study of the plant biome opened new perspectives for the identification and production of medically important secondary metabolites, and this book, an impressive collection of chapters, is an important contribution and provides an up-to-date review of each topic of this interesting field.

Each chapter intends to present relevant information about the state of the art and the basis for new research and application of plants. The book yields information about the compounds produced by plants, their relation with endophytes, and the role of these microbes in nature, working on the protection against diseases and abiotic stress, growth promotion, and production of useful secondary metabolites as antimicrobial and anticancer drugs, mycotoxins, insecticides, and agrochemical compounds useful for industrial crops. Ferns have relevance as nutraceuticals due to their rich content of protein, fiber, minerals, vitamins, essential amino acids, and fatty acids. Besides bioherbicide potential, ferns are endowed with chemopreventive, hepatoprotective, cytotoxic, antihyperglicemic, leishmanicidal, trypanocidal, antimicrobial, antinociceptive, and immunomodulatory metabolites. Lately, the application of recombinant techniques and tissue culture involves the in vitro production of plants cells, allowing the strain improvement, selection of high-producing cell lines, and medium optimization that can lead to an enhancement in secondary metabolite production.

The editors of this book, Dilfuza Egamberdieva and Antonio Tiezzi, in putting together this excellent volume which highlights plant biomes and their relevant importance as source of secondary metabolites offered a real contribution to the enhancement of this research field. As a final result, this book will be of great value to students and researchers interested in the study of new sources of secondary metabolites.

University of Concepcion
Concepcion, Chile

Mario J. Silva Osorio,

Contents

About the Series Editor

Naveen Kumar Arora, PhD, Microbiology, Professor in the Department of Environmental Science, Babasaheb Bhimrao Ambedkar University (A Central University), Lucknow, Uttar Pradesh, India, is a renowned researcher in the field of Environmental Microbiology and Biotechnology. His specific area of research is rhizosphere biology and PGPRs. He has more than 50 research papers published in premium international journals and several articles published in magazines and dailies. He is an editor of 10 books, published by Springer. He is a member of several national and international societies, Secretary General of Society for Environmental Sustainability, in editorial board of four journals, and reviewer of several international journals. He is also the Editor in Chief of the journal *Environmental Sustainability* published by Springer Nature. He has delivered lectures in conferences and seminars around the globe. He has a long-standing interest in teaching at the PG level and is involved in taking courses in bacteriology, microbial physiology, environmental microbiology, agriculture microbiology, and industrial microbiology. He has been an advisor to 118 postgraduate and 8 doctoral students. Recently, he was awarded for excellence in research by the Honorable Governor of Uttar Pradesh. Although an academician and researcher by profession, he has a huge obsession for the wildlife and its conservation and has authored a book *Splendid Wilds*. He is the President of Society for Conservation of Wildlife and has a dedicated website www.naveenarora.co.in for the cause of wildlife and environment conservation.

About the Editors

Dilfuza Egamberdieva graduated in Biology from the National University of Uzbekistan in 1993 and received her PhD in Agricultural Sciences from the Humboldt University of Berlin, Germany, in 2000. She conducted her postdoctoral studies at the Helsinki University of Finland, University of Florence, Italy; Manchester Metropolitan University, UK; Leiden University in the Netherlands; and Leibniz Centre for Agricultural Landscape Research, Germany. She established the research laboratory "Plant Microbe Interactions" at the National University of Uzbekistan. She has received numerous fellowship and awards for her academic achievements, including UNESCO-L'Oréal Fellowship for Women in Science, UNESCO-MAB Award, and Alexander Von Humboldt Fellowship. In 2012, she was awarded the TWAS Prize in Agricultural Sciences for her innovative contributions to the study of plant-microbe interactions in stressed environments. She has published over 100 peer-reviewed papers in international journals and 40 chapters, including 4 books published by Springer.

Antonio Tiezzi graduated in Biology from the University of Siena, Italy, in 1979. He then worked at Siena University and was a Visiting Scholar at the Biozentrum, Basel, Switzerland, under an EMBO fellowship. In 1982, he gained experience in the vaccine industry at Sclavo (now GSK) in Siena, and in 1983, he undertook research at the Einstein College of Medicine, NY, USA. In 1985, he worked at the Department of Environmental Sciences, Siena University, and since 1992, he has been a Professor of Plant Cytology at Tuscia University, Viterbo, Italy. His main research areas include plant cytoskeleton – he discovered the motor protein kinesin in plants and became a board member of the European Cytoskeletal Forum – and bioactive plant substances. He has published 95 papers in peer-reviewed journals and 25 book chapters; has edited 3 books, including 1 book published by Springer; has participated in numerous international meetings; and has coordinated national and international educational/research networks and projects.

Chapter 1
Ethnobotanical Aspects of Some Traditional Medicinal Plants

Iftikhar Ahmad, Muhammad Sajid Aqeel Ahmad, Mumtaz Hussain, and Mansoor Hameed

Abstract Ethnobotany (study of usage of plant parts for human health) is considered to be a part of Economic Botany, which emphasizes on the economic utilization of plants for human welfare. Biological diversity is universally recognized as an important part of the world's natural heritage and an essential component for the sustainability of global ecosystems. In the current era, modern allopathic medicines are very fast effective and have over-ridden the traditional herbal remedies. Additionally, the diversity of traditional medicinal plants is facing a continuous decline due to a number of natural and anthropogenic activities including the clear-cutting of forests, conversion of grasslands into cultivated lands, industrialization, overgrazing, soil erosion, desertification, etc. Similarly, overexploitation also poses a severe threat to diversity of medicinal plants and has led to decline severely a number of species. It should be recognized that plant diversity has a commendable importance as a source of pharmaceutically active substances. In this chapter, the medicinal value and usage of various medicinal plants typically used in traditional medicine have been discussed.

Keywords Medicinal plants · Diversity · Active ingredients · Soon Valley · Salt Range

1.1 Introduction

Indigenous knowledge of plants of different areas is as old as human civilization. However, the term "ethnobotany" was first used by an American botanist Johan W. Harshberger in 1896, "to the study of plants used by primitive and aboriginal people." In modern ecological terms ethnobotany was described as "The study of

I. Ahmad (✉)
Department of Botany Sciences, University of Sargodha, Sargodha, Pakistan

M. S. A. Ahmad · M. Hussain · M. Hameed
Department of Botany, University of Agriculture, Faisalabad, Pakistan

D. Egamberdieva, A. Tiezzi (eds.), *Medically Important Plant Biomes: Source of Secondary Metabolites*, Microorganisms for Sustainability 15,
https://doi.org/10.1007/978-981-13-9566-6_1

direct interactions between human and plant populations" (Plotkin 1991; Heiser 1993). Today ethnobotany is widely accepted as a science of human interactions with plants and related ecosystems. Plants have been used as medicine since ancient times. The use of plants to improve economy is an old tradition of human history.

An ethnobotanical use of plants is more common in many parts of the world especially developing countries like Pakistan due to lack of medical facilities in the far lying areas from cities. For example, 40 medicinally important plant species of 21 families were reported in Kala Chitta Hills (Salt Range) of District Attock which were under common use of indigenous people. Due to increase in population, demands of people increase causing great pressure on the products of the area. The region is very rich in having medicinal plants. To understand the indigenous knowledge of the local people, ethnomedicinal study is very important. This helps a lot for creating awareness among them regarding sustainable natural resource management. Local people, hakims, and medicinal businessmen are very important in this regard (Mahmood et al. 2004).

In Africa, people use different plant extracts for treating trypanosomiasis. Methanolic extracts from 23 plants collected from the Savannah vegetation belt of Nigeria were analyzed in vitro for trypanocidal activity against *Trypanosoma brucei brucei* and *Trypanosoma congolense* at concentrations of 4 mg/ml, 0.4 mg/ml, and 0.04 mg/ml. Extracts of *Khaya senegalensis*, *Piliostigma reticulatum*, *Securidaca longipedunculata*, and *Terminalia avicennioides* were strongly trypanocidal to both organisms, while extracts of *Anchomanes difformis*, *Cassytha* spp., *Lannea kerstingii*, *Parkia clappertoniana*, *Striga* spp., *Adansonia digitata*, and *Prosopis africana* were trypanocidal to either *T. brucei brucei* or *T. congolense*. This provided scientific basis for the use of some plants in the traditional management of trypanosomiasis (Atawodi et al. 2003). Some of the indigenous plants are very important in the diets of post-postpartum women during which time it is claimed that these spices and herbs aid the contraction of the uterus. Spices and herbs are generally known to possess antibacterial and antioxidant properties (Iwu 1989).

1.2 Some Traditional Medicinal Plants

A large number of medicinal plant species have been reported growing in various parts of the world. For example, Leporatti and Lattanzi (1994) studied the ethnobotanical use of 27 medicinal plants species. They reported and discussed their traditional medicinal uses. The inhabitants use the medicinal plants for various purposes and are dependent on surrounding plant resources for their food, shelter, and health. A total of 25 species of herbs belonging to 18 families and their medicinal uses by indigenous people were recorded from the area. Some of these species were used for the treatment of cholera, dyspepsia, fever, herpes, eczema, jaundice, and liver complaints (Qureshi and Khan 2001).

The vegetation of Lawat in the District of Muzaffarabad, Azad Jammu and Kashmir, for ethnobotanical purposes was investigated. He recorded 52 species out of which 3 species were of 2 gymnospermic families while 49 species were of 35 angiospermic families. Most of the plants were used medicinally. The investigation indicated that the medicinal plants were either used singly or with mixtures by local inhabitants. The area under investigation, due to unplanned utilization, had resulted in loss of medicinally important plant species (Dar 2003).

Euphorbiaceae is an important plant family especially recognized for its anticancer components, anti-hepatitis B components, and carcinogenic factors. In the literature of ancient traditional Chinese medicine, 33 species of plants from 17 genera of Euphorbiaceae have been mentioned as medicines. Presently 111 species within 35 genera of medicinal euphorbiaceous plants have been reported. Among them, 17 species were used to treat snakebites. It was observed that most of the species within the Euphorbiaceae family contained toxic components. Only a few species were employed as widespread medicines. Most species were recognized only as local minority tribe medicines (Lai et al. 2005).

Sambucus nigra bush of family Caprifoliaceae is one of the plants which are most commonly used for medicinal and various other purposes by the inhabitants of Catalonia and in many Mediterranean regions. It is a most versatile plant, being used for food and medicine. In addition, almost every part of the plant, including the bark, roots, leaves, flowers, and fruit, has some uses (Valles et al. 2004). Similarly, leaves and roots of *Justicia adhatoda* L. (Acanthaceae) are used for coughs, bronchitis, asthma, and rheumatism. Leaf buds are also used in diabetes and for joints and as antiseptic. Green leaves of *Withania somnifera* (L.) Dunal (Solonaceae) are used to relieve the pain from joints and painful swelling. Roots are used as diuretic and tonic. Juice of the whole plant is useful in rheumatism, whereas seeds have been reported to be used as to coagulate milk (Figs. 1.1 and 1.2).

Whole plant of *Buxus papillosa* is used as diaphoretic, purgative, and antirheumatic. Different species of *Dicliptera* shoots are used as tonic. *Peganum harmala*, a herbal whole plant, is used as an analgesic, aphrodisiac, emmenagogue, hypnotic, and antispasmodic. *Salvia virgata* leaves are applied to tumors and ulcers. *Solanum indicum* roots, leaves, and fruits are used as expectorant, carminative analgesic, and febrifuge. *Solanum surattense* whole plant is used as vasodilator, astringent, and expectorant. *Sophora mollis* or Khumbi seeds are used as anthelmintic (Ahmad et al. 2002; Khan 1951). *Adiantum* species are used for chest complaints, cough, expectorant, increasing lactation, colds, emmenagogue, aiding kidney function, antiparasitic, dandruff, and general cure-all. The fresh or dried leafy fronds are antidandruff, antitussive, astringent, demulcent, depurative, emetic, weakly emmenagogue, emollient, weakly expectorant, febrifuge, galactogogue, laxative, pectoral, refrigerant, stimulant, sudorific, and tonic (Grieve 1984). In Nepal, a paste made from the fronds is applied to the forehead to relieve headaches and to the chest to relieve chest pains. The plant is best used fresh, though it can also be harvested in the summer and dried for later use (Chiej 1984; Launert 1981).

Fig. 1.1 Some important medicinal plants of Acanthaceae, Adiantaceae, and Solanaceae families found commonly growing in Salt Range of Pakistan

Fig. 1.2 Some important medicinal plants of Buxaceae, Leguminosae, Lamiaceae, and Zygophyllaceae families found commonly growing in Salt Range of Pakistan

1.3 Proximate Composition

Composition described in terms of main classes of substances is called proximate composition, for example, proteins or minerals that first arrived in the process of analysis. In proximate analysis the groups are measured as such, instead of individual proteins or specific minerals (FAO 2001). It mostly includes proteins, fats, minerals, and carbohydrates. Almost all plants contain these substances and are initially analyzed proximately.

Morinda citrifolia is an important medical plant in Southeast Asian countries. It was analyzed proximately and biochemically to make a more modern drug from a traditional product. In order to obtain better understanding of the medicinal characteristics of the *M. citrifolia* a fruit cultivated in Cambodia, fatty acids, proteins, amino acids, and sugars of juices were analyzed (Chunhieng et al. 2005).

Kochhar et al. (2006) analyzed three traditional medicinal plants known for their hypoglycemic action, namely, bitter gourd, fenugreek seeds, and jambu seeds, for proximate composition, available carbohydrates, dietary fibers, and anti-nutritional factors. Protein, fat, ash, crude fiber, carbohydrate, and energy content of these medicinal plants ranged from 4.16% to 25.80%, 0.49% to 6.53%, 2.16 to 9.89%, 1.28 to 10.92%, 58.13 to 90.85%, and 319.11% to 394.46% kcal, respectively. Total soluble sugars, reducing sugars, nonreducing sugars, and starch content varied from 2.03% to 11.76%, 0.78 to 4.43%, 1.03 to 8.0%, and 29.20 to 33.63%, respectively. Dietary fiber constituents like hemicellulose, cellulose, lignin, pectin, and total dietary fiber varied from 8.44% to 34.75%, 1.46 to 8.23%, 0.38 to 5.18%, 0.36 to 2.95%, and 22.4 to 40.38%, respectively. It was found that these plants were a good source of protein, fat, minerals, crude fiber, and energy and a rich source of available carbohydrates and dietary fiber. It was concluded that these hypoglycemic traditional medicinal plants provide various nutrients which are not provided by allopathic medicine, and these plants have no adverse effects. Therefore, the diabetic patients should be encouraged to include these medicinal plants in their daily diet to control blood sugar level.

Ripe fruits of *Dennettia tripetala* were analyzed for proximate composition. *Dennettia tripetala* contained crude protein (15.31%), total carbohydrate (62%), crude fibers (9.84%), crude lipids (3.47%), and moisture (8.0%). It had an energy value of 480.24 g cal·100 g^{-1} of fresh fruit (Okwu and Morah 2004). This justifies the use of *Dennettia tripetala* fruits as food and a drug in herbal medicine in Southeastern Nigeria.

Cymbopogon jwarancusa is a useful plant in diseases of blood, skin, vomiting, abdominal tumors, unconsciousness, and fever (Kirtikar and Basu 1982). This plant was proximately studied (Mahmud et al. 2002) and was found to contain moisture contents, 67.02%; ash contents, 9.52%; carbohydrates, 1.8%; reducing sugar, 1.07%; nitrogen, 0.67%; crude proteins, 5.02%; and crude fiber, 9.50%.

Four medicinal plants belonging to the family Lamiaceae were chemically screened (Edeoga et al. 2006) for their chemical constituents and nutritional value. The medicinal plants contained crude protein (9.19–17.94%), crude fiber (4.88–9.04%), ash

(5.68–6.88%), carbohydrate (66.24–75.87%), crude lipid (3.48–4.90%), and food energy (357.68–373.26 mg/cal). These plants play an important role not only in nutrition but also in traditional medicine and in pharmaceutical industry.

Fagoina arabica is among the widely used medicinal plants all over the world including Pakistan. Generally the plant is located on dry calcareous rocks, distributed in most parts of the Mediterranean region to South Africa, Afghanistan, India, and Pakistan especially Sindh, Punjab, and NWFP (Rizvi et al. 1996). Proximate analysis of this plant revealed that leaves and seeds have maximum moisture content (58.51 ± 0.50) followed by shoots and roots (43.29 ± 0.42, 29.45 ± 0.28, respectively). Ash and protein (1.85 ± 0.12, 0.64 ± 0.01, respectively) increased in different parts in descending order, i.e., roots < shoots < leaves and seeds, whereas fat and fiber contents (1.33 ± 0.05, 56.80 ± 0.23, respectively) decreased in ascending order, i.e., roots>shoots>leaves and seeds (Shad et al. 2002).

Now interest has been developed in wild species for their possible medicinal values in diets. Wild plant species provide minerals, fibers, vitamins, and essential fatty acids and enhance taste and color in diets. In addition, they have antibacterial, hepatoprotective, and anticarcinogenic properties and therefore have medicinal value (Green 1992; Bianco et al. 1998). Yildirim et al. (2001) analyzed eight plant species in Turkey for dry matter, ascorbic acid, nitrogen, and protein which are important nutritionally as well as for medicinal value.

Piliostigma thonningii is a leguminous medicinal plant belonging to the family Caesalpiniacea used for the treatment of dysentery, fever, infections, respiratory ailments, snake bites, hookworm, and skin diseases (Jimoh and Oladiji 2005). Proximate composition of *Piliostigma thonningii* seeds showed that seeds contained moisture contents 6.71%, ash 3.50%, crude proteins 30.33%, crude fibers 35.03%, lipids 1.42%, and carbohydrates 23.00%.

Two rural settled Fulani villages, northeastern Nigeria, were surveyed for the use of wild plants as food or medicine (Lockett et al. 2000). Different parts of commonly used plant species were proximately analyzed for protein, fat, carbohydrate, and mineral contents. Kuka bark (*Adansonia digitata*) given to infants to increase weight gain contained high fat, calcium, copper, iron, and zinc contents. Cediya (*Ficus thonningii*), dorowa (*Parkia biglobosa*), and zogale (*Moringa oleifera*) were found to be the good sources of protein and fat and excellent sources of calcium and iron or copper and zinc. Fruits, leaves, and nuts of aduwa (*Balanites aegyptiaca*) were widely used during the dry season and drought. Edible wild species available during the wet season generally were inferior in energy and mineral content as compared to dry season plants. Fruits commonly eaten by children were poor sources of protein and minerals but rich in carbohydrate and fibers. Shiwaka leaves (*Veronia colorate*) that were mostly consumed by pregnant women to increase breast milk production and to expel intestinal worms contained high-fiber contents.

In Nigeria four medicinal plants belonging to the family Lamiaceae were chemically screened and found to contain high percentage of phytochemicals. The medicinal plants investigated were *Hyptis suaveolens* and three putative hybrids of *Ocimum gratissimum* (Hybrid A, B, and C). The plants contained crude protein (9.19–17.94%), crude fiber (4.88–9.04%), ash (5.68–6.88%), carbohydrate (66.24–75.87%), crude lipid (3.48–4.90%), and food energy (357.68–373.26 mg/cal). This showed the significance of these plants not only in traditional medicine but also in food and in pharmaceutical industries (Edeoga et al. 2006).

Carica papaya belonging to the family Caricaceae is an important and common medicinal plant in tropical Africa. Proximate analysis of the unripe pulp of *Carica papaya* was analyzed for the presence of different phytochemicals and minerals (Oloyede 2005). It showed that the pulp contained starch (43.28%), sugars (15.15%), crude protein (13.63%), crude fat (1.29%), moisture (10.65%), and fiber contents up to 1.88%. These results indicated that the pulp of mature unripe *Carica papaya* contained nutrients and mineral elements useful in nutrition. The presence of some phytochemicals like saponins and cardenolides explained the astringent action of the plant encountered in the numerous therapeutic uses.

Arubi (2003) analyzed papaya kernel flour for proximate composition and functional properties. The flour was high in protein (32.4%) but moderate in available carbohydrates (49.9%) and low in moisture (7.5%) contents. The total minerals and fiber contents were 5.3% and 4.2%, respectively. Oil and water absorption capacities of the flour sample were high. The flour had very good foaming and emulsifying properties. These results suggested that papaya kernel flour can be used in a number of food formulations.

The underground caudex of the cycad, *Stangeria eriopus*, is used extensively by several communities in South Africa, mainly as an emetic. It was found that only in the month of July during 1992, 3410 plants were sold, which threatened the remaining plant populations. Proximate analysis of the caudex material gives high carbohydrate content with only small percentages of fat, protein, fiber, and ash (Osborne et al. 1994).

Hassan and Umar (2006) studied the nutritive value of *Momordica balsamina* L. leaves by analyzing their proximate composition, amino acids, and mineral constituents. The results showed that the plant leaves had high moisture contents (71.00 ± 0.95% fresh weights). The concentration of estimated crude protein and available carbohydrates on dry weight (DW) basis was 11.29 ± 0.07% and 39.05 ± 2.01%, respectively. The leaves also have high mineral (18.00 ± 0.56% DW) and crude fiber (29.00 ± 1.23% DW) contents, while crude lipid contents (2.66 ± 0.13% DW) and energy value (191.16 kcal/100 g DW) were low. The results indicated that the *Momordica balsamina* leaves could be a good supplement for mineral, protein, carbohydrate, and fiber contents.

Wild edible plants form an important part of traditional diets in the Himalaya. In the Sikkim Himalaya, 190 species were screened as edible species, out of which nearly 47 species came to the market. Twenty-seven plant species were analyzed proximately for their nutritive values, 22 were edible for their fruits and 5 for leaves and shoots. Among different plant parts, generally higher nutrient concentration was recorded for leaves, followed by new shoots and fruits (Sundriyal and Sundriyal 2004). For different species the crude fiber contents ranged between 2.15% and 39.90%, total soluble salts between 4.66% and 21.00%, and vitamin C contents from 6 to 286 mg/100 g. The fat contents were determined high in the fruits of

Castanopsis hystrix, *Machilus edulis*, and *Cinnamomum* species, while the protein contents were highest in *Hippophae rhamnoides*, *Cucumis melo*, and *Elaeagnus latifolia*. The total carbohydrate contents ranged from 32% to 88% in the fruits of various wild plants, whereas the reducing sugar ranged from 1.25% to 12.42% and total sugar from 2.10% to 25.09% and the lignin contents varied from 9.05% to 39.51%, the hemicellulose between 25.63% and 55.71%, and cellulose contents from 9.57% to 33.19% in different species. It was suggested that a few wild edible species were needed to be grown for commercial cultivation and included in the traditional agro-forestry systems, which will reduce pressure on them in natural forest stands as well as producing economic benefits for poor farmers.

1.4 Minerals

The attention toward the inorganic constituents of medicinal plants was drawn by Hakim Abdul Hameed, President of Hamdard National Foundation, India, who is the originator of the discipline "Elementology" (Arora et al. 1984).

Health depends upon the organized state of elements in the body, and their imbalance causes disease (Golden 1988), and restoration of balance by drugs can cure diseases.

Medicinal plants show therapeutic effects for the treatment of different diseases due to the presence of certain chemical compounds in these plants. These are mostly organic compounds which have biological activities, but none of these act independently and they cannot perform the functions of the medicinal plant as a whole (Mutaftchiev 2003).

Analyses have showed that medicinal plants are rich in many trace elements, and it was suggested that this was an important factor in the curative effect of these plants (Olabanji et al. 1997; Pereira and Felcman 1998). The trace elements can be found in free-states or organically bound in a complex. It is a well-established fact that different states and forms can have different functions in its physiological activities such as biotoxicity and percent absorption in the body (Svendsen and Lund 2000). Trace elements coexist with various organic compounds in medicinal plants (Remington 1995), and mostly they are bound to organic compounds. So the concentration of the free trace elements will be very low.

Medicinal properties of most of the medicinal plants are attributed mainly due to their cultivation in different parts of the world (Rajput et al. 1996), and the active constituents, especially inorganic elements present in plants, are in very variable quantities (Gauch 1972) if grown in different environmental conditions and different types of varieties used for cultivation. It is very much clear now that inorganic trace elements are very active in very small concentrations, and the analysis of different parts of both plants and decoction has shown the presence of many essential and important elements such as Ca, Mg, Zn, Fe, Co, Mn, etc. Zn is very effective in killing virus (Randal 1984).

Sahito et al. (2001) investigated two varieties of medicinal herb *Catharanthus roseus* for elemental composition as Ca, Na, K, Mg, Zn, Fe, Cu, Co, Mn, Ni, Cd, Pb, Ba, and Al. The level of essential elements such as Zn, Fe, Mn, and Cu was present in considerable amount. In decoction the level of essential elements was high as compared to toxic elements.

Most of the medicinal plants qualify as nonprescription drugs, and some of them are taken in low doses as food drugs in these days (Obiajunwa et al. 2002), for example, Se, Zn, vitamin E, and other antioxidants of plant origin are proving to be reliable weapons in the effort against premature aging and the postponement of degenerative diseases. There are at least 50 elements which are vital for the well-being of humans (Tolonen 1990). Now the people are very much interested in trace elements in the area of medical science.

Obiajunwa et al. (2002) investigated different major and minor elements in different plants. Fourteen different elements, namely, K, Ca, Ti, V, Cr, Mn, Fe, Ni, Cu, Zn, Se, Br, Rb, and Sr were detected in varied amounts in different plants. The concentrations of Ca and K were the highest, whereas Br and Se were the least. All the essential elements were present in required dose, but some plants such as *C. procera*, *A. indica*, and *A. wilkesiana* showed toxicity due to their high Cr level.

The high concentration of Ca is very important as Ca enhances the qualities of bones and teeth and also of neuromuscular systemic and cardiac functions. Iron is another important element present in all the specimens which plays a role in oxygen and electron transport in human beings. The high amount of Fe and Ca in *C. alata* showed that it could be especially useful in the treatment of constipation in nursing and pregnant mothers at 5 g/dose, as its main medicinal activity is as a mild laxative. *E. hirta*, *A. melegueta*, *M. indica*, and *G. kola* contained high concentration of Zn and could be used in cases of Zn deficiency which includes impairment in healing, taste, and growth (Obiajunwa et al. 2002). In addition to identifying the active secondary metabolites of these medicinal plants, the knowledge of their elemental composition is very important in determining their toxicity or safety for use.

More than 25 naturally occurring elements perform essential functions in the human body. Some of them such as zinc, copper, selenium, cobalt, chromium, molybdenum, manganese, and iodine are required in small amounts, and each comprises less than 0.01% of the body weight, and they are called as essential trace elements. They work in a similar way in the body. Most of them are at the active site of enzymes or of physiologically active substances of the body. Dietary deficiency of these elements causes various problems, which are consistent with the decreased activity of these active substances (Wada and Yanagisawa 1996).

Zinc is one of the most important trace elements in the body as it performs various biological activities. It is an essential catalytic component for more than 200 enzymes and also acts as a structural component of many proteins, hormones, neuropeptides, hormone receptors, and most likely polynucleotide (Fabris and Mocchegiani 1995). Due to its role in cell division, differentiation, programmed cell death, gene transcription, biomembrane functioning, and obviously many enzymatic activities, zinc is considered a most important element in the accurate working of an organism, from the very first embryonic stages to the last periods of life

(Fabris and Mocchegiani 1995). The zinc supplementation is efficacious in most of the problems. It is said to be a therapeutic support instead of a simple dietary supplement. The relevance of zinc to many age-associated diseases and the aging itself of the major homeostatic mechanisms of the body, i.e., the nervous, neuroendocrine, and immune systems, places zinc in an essential position in the economy of the aging organism.

Rajurkar and Damame (1998) studied elemental composition of some Ayurvedic medicinal plants used for healing urinary tract disorders. Fourteen elements were estimated in different plants: among these Cu, Cr, Co, and Cd were found to be present at the trace level; Mn, Pb, Zn, Ni, Na, Fe, and Hg at minor level; and K, Ca, and Cl at major level. The differences in the concentration of the elements are attributed to soil composition and the climate in which the plant grows. It was found that these elements play an important role in treatment of urinary tract disorders.

Some inorganic trace elements such as V, Zn, Cr, Cu, Fe, K, Na, and Ni play an important role in maintaining normoglycemia by activating the beta-cells of the pancreas. Leaves of four traditional medicinal *plants* (*Murraya koenigii*, *Mentha piperitae*, *Ocimum sanctum*, and *Aegle marmelos*) widely used in the treatment of diabetes and related metabolic disorders were analyzed for different inorganic elements. The levels of Cu, Ni, Zn, K, and Na were found to be in trace amounts, and Fe, Cr, and V levels were found in minor levels (Narendhirakannan et al. 2005).

According to Anke (1986) 7 quantitative elements and perhaps 18 trace elements are of vital importance for the animals. Their metabolism is antagonistically or synergistically influenced by the inorganic and organic constituents of the food of different kinds. More than 30 elements (Cu, Zn, Mg, Mn, Cr, V, and so on) are involved in the treatment in the process of arteriosclerosis.

Singh and Garg (1997) analyzed specific plant parts of several plants (fruits, leaves, or roots) often used as medicines in the Indian Ayurvedic system for 20 elements (As, Ba, Br, Ca, Cl, Co, Cr, Cu, Fe, K, Mn, Mo, Na, P, Rb, Sb, Sc, Se, Sr, and Zn). Most of the medicinal herbs were found to be rich in one or more of the elements under study.

Dennettia tripetala or pepper fruit plant is a well-known Nigerian spicy medicinal plant. *Dennettia tripetala*, besides protein, carbohydrate, fibers, and lipids, also contains important mineral contents as calcium (1.80%), phosphorus (0.33%), potassium (2.50%), and magnesium (0.42%). Trace elements included Fe, Cu, Zn, and Cd; however, Cr was not detected. This justifies the use of *Dennettia tripetala* fruits as food and a drug in herbal medicine in Southeastern Nigeria (Donatus and Morah 2004).

Mineral analysis of *Piliostigma thonningii* showed that seeds were good source of antioxidant micronutrients such as Fe, Ca, Se, Zn, and Mn. So it could serve as a cheap source of antioxidant micronutrients supplements in both man and animal.

The level of iron among all minerals analyzed was found to be the highest (782 ppm). This might be of nutritional importance especially in the part of the world where anemia and iron deficiency is more common (Jimoh and Oladiji 2005). *P. thonningii* seeds are also good sources of calcium (43.11 ppm), while zinc (0.016 ppm), manganese (1.00 ppm), and phosphorus (0.02 ppm) levels were quite low when compared with iron and calcium but comparable with values for some legumes (Elegbede 1998). Iron, selenium, zinc, and manganese are antioxidant micronutrients (Talwar et al. 1989), and their presence could thus boost the immune system.

Fagoina arabica is among the widely used medicinal plants in Pakistan. Its mineral composition showed that Zn and Na were maximum in roots and minimum in leaves and seeds. Concentrations of Fe, P, K, and Ca decreased in order of leaves, seeds, shoots, and then roots. *Fagonia arabica* contained Ca, Na, P, Cu, Fe, Mn, and Zn. Zn plays an important role as an antioxidant in animals (Bray and Betteger 1990) as well as in plant membranes (Cakmak and Marschner 1988). *Fagonia arabica* contains lower amounts of heavy metals. The possible reason for low concentrations of heavy metal could be due to the fact that this herb is found mostly in desert and dry calcareous rocks where industrial pollutions are not found, which might have resulted in least amount of heavy metal. However, the macro elements (P, K, Na, Ca) were found to be maximum. Phosphorus is mainly involved in RNA, DNA sugar-phosphate backbone, in the process of energy transfer, and it is the inorganic phosphate that appears as an intermediate product during photosynthesis and respiration pathways of metabolism (Shad et al. 2002).

Fagonia arabica contains a fair amount of K and Na. Due to this reason, it is mostly used in diseases like diarrhea, stomatitis, and deobstruent (Dey et al. 1980) where mostly fluid losses take place (Whitney and Hamilton 1984).

Cu, Zn, Mn, and Fe are considered as trace elements due to their relatively minute quantity that is essential to the body. Copper is important for red blood cell formation, mitochondria function, and a component of ribonucleic acid, whereas Zn, Mn, and Fe are necessary for the development of bones and connective tissues (Nielsen 1987).

Unlike other compounds, living organisms cannot synthesize mineral elements. Only small fraction of the Ca, Mg, and P and most of the Na, K, and Cl are present as electrolytes in the body fluids and soft tissues. Electrolytes present in blood or cerebrospinal fluid maintain acid-base and water balance and adjust osmotic pressure. They regulate membrane permeability and exert characteristic effects on the excitability of muscles and nerves (Nielsen 1987; Bukhari et al. 1987).

The distribution of the elements in various genera and species of plants will be highlighted in the knowledge of the distribution of certain valuable trace elements and their availability from medicinal plants. The uptake of mineral constituents depends on many factors such as the amount of mineral elements present in soil, their availability, moisture contents of soil, and the botanical factor. The variation in mineral composition was observed in different varieties of the same species (Sahito et al. 2001).

1.5 Bioactive Substances

Medicinal plants are considered very important for the health of individuals and communities. These plants mostly display a wide range of biological and pharmacological activities such as anti-inflammatory, antibacterial, and antifungal properties (Okwu and Ekeke 2003; Okogun 1985). The medicinal value of such plants is due to the presence of some chemical substances which create or enhance a definite physiological function in the human body. Phytochemicals perform many ecological and physiological functions and are widely distributed in plant kingdom. Medicinal plants can synthesize and accumulate a great variety of phytochemicals including alkaloids, flavonoids, tannins, cyanogenic glycosides, phenolic compounds, and saponins (Pandey 1980; Edeoga et al. 2005; Okwu 2004). Phytochemicals are present in different plants and are used as important components of both human and animal diets. Diets containing an abundance of fruits and vegetables are protective against a variety of diseases, particularly cardiovascular diseases (Uruquiaga and Leighton 2000). Herbs and spices are harmless sources for obtaining natural antioxidants (Okwu 2004; Kim et al. 1994). Most of these are potent bioactive compounds found in different parts of medicinal plant that can be used for therapeutic purpose or which are precursors for the synthesis of useful drugs (Sofowara 1993). The active principles differ from plant to plant due to their biodiversity, and they produce a definite physiological action on the human body (Edeoga et al. 2006).

Leaves and stems of most of the plants were found rich in alkaloids, flavonoids, tannins, and phenolic compounds. They had already been examined to show medicinal activity as well as exhibiting physiological activity (Sofowara 1993). Natural products have been an important source of drugs for centuries, and about half of the pharmaceutical market presently depends on natural products (Clark 1996).

Ten medicinally important plants belonging to different families were analyzed and compared for alkaloids, tannins, saponins, terpenoid, flavonoids, and phenolics. The medicinal plants investigated were *Cleome rutidosperma*, *Emilia coccinea*, *Euphorbia heterophylla*, *Physalis angulata*, *Richardia brasiliensis*, *Scoparia dulcis*, *Sida acuta*, *Spigelia anthelmia*, *Stachytarpheta cayennensis*, and *Tridax procumbens*. All the plants were found to contain alkaloids, tannins, and flavonoids except for the absence of tannins in *S. acuta* and flavonoids in *S. cayennensis*, respectively. These plants were found very important in traditional medicine (Edeoga et al. 2005).

Phenolic compounds are widely distributed in the plant kingdom, and the presence of phenols is considered to be potentially toxic to the growth and development of pathogens (Singh and Sawhney 1988). Phenolic compounds act as electron donors and are readily oxidized to form phenolate ion or quinine which is an electron acceptor. Protonated phenol is used as a cleaning agent (Uruquiaga and

Leighton 2000) and acts as anti-inflammatory, anticlotting, antioxidant, immune enhancer, and hormone modulator agents. Phenols have been the subjects of extensive research as disease preventives (Duke 1992). Phenols have been shown to have the ability to block specific enzymes that cause inflammation. They also modify the prostaglandin synthesis pathways and thereby protect platelets from clumping.

Phytochemicals show a wide range of biological effects due to their antioxidant properties. Several types of polyphenols (phenolic acid and flavonoids) show anticarcinogenic and antimutagenic effects (Uruquiaga and Leighton 2000). Polyphenols are considered to interfere several steps in the development of malignant tumors, inactivating carcinogens and inhibiting the expression of mutagens. Several studies have shown that in addition to their antioxidant protective effect, polyphenols, particularly flavonoids, also inhibit the initiation, promotion, and progression of tumors (Okwu 2004; Uruquiaga and Leighton 2000). Recently plant flavonoids have attracted attention as potentially important dietary cancer chemo-protective agents (Okwu and Okwu 2004). In addition, the possible antitumor action of certain flavonoids has also generated interest (Kandaswami et al. 1991). Moreover, naturally occurring phytochemicals are potential anti-allergic, anticarcinogenic, antiviral, and antioxidant agents (Okwu 2004; Uruquiaga and Leighton 2000). Flavonoids represent the most common and widely distributed groups of plant phenolics that are potent water-soluble super antioxidants and free radical scavengers which prevent oxidative cell damage, have strong anticancer activity, and protect against all stages of carcinogenesis (Okwu 2004). Flavonoids in intestinal tract lower the risk of heart disease (Cook and Samman 1996). Most of these effects of flavonoids have been linked to their known functions as strong antioxidants, free radical scavengers, and metal chelators (Torel et al. 1986; Nakayama et al. 1993). Flavonoids are mostly 15-carbon compounds and are distributed throughout the plant kingdom (Harborne 1973). Some isoflavones act as allelochemicals widely used in insecticides. They are also important in disease resistance (Salisbury and Ross 1992).

Some other biological functions of flavonoids include protection against allergies, inflammation, free radicals, platelet aggregation, microbes, ulcers, hepatoxins, viruses, and tumors (Okwu 2004; Okwu and Omodamiro 2005).

The presence of phenolic compounds in certain plants makes them very good antimicrobial and antibacterial agent in different infections. This is the reason that *B. pinnatum*, with higher phenolic compounds, is effective in the treatment of typhoid fever and other bacterial infections, particularly those caused by *Staphylococcus aureus*, *Escherichia coli*, *Bacillus subtilis*, *Pseudomonas aeruginosa*, *Klebsiella aerogenes*, *Klebsiella pneumoniae*, and *Salmonella typhi* (Ofokansi et al. 2005). These results supported the use of *B. pinnatum* in treating the placenta and navel of a newborn baby, which not only heals fast but also prevents the formation of infections (Okwu 2001, 2003).

Alkaloids were also detected in these plants. Alkaloids and their synthetic derivatives are used as basic medicinal agents for their analgesic, antispasmodic, and bactericidal effects (Stary 1998). They show marked physiological activity when given to animals (Okwu and Okwu 2004).

The phytochemical analyses of *Piliostigma thonningii* seeds showed that seeds contain saponins, flavonoids, phenolics, glycosides, as well as cardiac glycosides (Jimoh and Oladiji 2005). Some of these chemical compounds have been reported to have inhibitory effects on some gram-negative bacteria such as *Escherichia coli* and *Bacillus subtilis* among others. They also have prominent effects on animal systems and microbial cells (Liu et al. 1990; Oyagade et al. 1999). The presence of these chemical compounds therefore suggests the pharmacological activities of *P. thonningii*.

Four medicinal plants of family Lamiaceae were analyzed chemically for certain phytochemical constituents including alkaloids, tannins, saponins, flavonoids, and phenols. The plants investigated were *Hyptis suaveolens* and three putative hybrids of *Ocimum gratissimum*. All the plants contain high percentage yield of crude alkaloids and flavonoids ranging from 10.44–14.32% to 9.28–12.54%, respectively. Tannins and phenols were present in all plants; however, saponins were absent in these plants. This gives significance of these plants in traditional medicine and in the pharmaceutical industries (Edeoga et al. 2006).

Two Nigerian medicinal plants (*Garcinia kola* Heckel and *Aframomum melegueta*) were analyzed for their phytochemicals (Okwu 2005). These plants were found to contain bioactive constituents as flavonoids (5.76–1.98 mg/100 g), phenols (0.09–0.11 mg/100 g), saponins (1.24–11.48 mg/100 g), and tannins (0.26–0.38 mg/100 g). These constituents are considered responsible for the health-related properties of these plants, which are based on their antioxidant, anticancer, antitumor, antiviral, anti-inflammatory, and anti-allergic activities. These facts justify the popular use of *G. kola* and *A. melegueta* in herbal medicine in Nigeria.

1.6 Conclusion

According to the World Health Organization (WHO), more than 80% of the world's population relies on traditional herbal medicine for their primary health-care needs. These traditional systems are culturally and psychologically more tolerable in most of the societies as compared to western allopathic medicines. In addition, being the natural plant products, they are considered to be the safest way of treating diseases with least side effects on human health as compared to allopathic or homeopathic medicines. Medicinal plants not only serve as important source of raw materials for the manufacture of traditional medicines but also used for the preparation of a number of modern allopathic medicines. The use of herbal medicines is increasing day by day, and efforts are underway to examine the medicinal plant resources and their active ingredients. However, the research in medicinal plants requires a considerable interaction of researchers with indigenous communities. In addition, successful research must involve peoples from other disciplines such as ethnobotanists, natural product chemists, pharmacologists, taxonomists, traditional healers, and/or user communities, and if useful compounds are isolated that have need of development, then synthetic chemists are compulsory.

This chapter on medicinal flora of the Soon Valley in Salt Range shows that more than 98 angiosperms of 45 families are traditionally locally used as healing agents. However, a large number of plant species belonging to different plant groups still need a thorough pharmacognostic assessment. Local people collect and sell a large number of species of plants to merchants in the market or to larger pharmaceutical trading houses. This, however, is not done on a scientific basis, and species may be mixed, or not collected at the time of maximum potency because the synthesis of various nutritional and medicinal components varies considerably during different seasons and at different localities. Different plants are used for different medicinal purposes throughout the country. During the last quarter century, environmental and cultural changes and market-based economics have seriously influenced all aspects of traditional medicine systems by affecting environment and resources of traditional medicine. Over harvesting of medicinal plants and animal species has resulted in resource degradation, loss of biodiversity, and the loss of indigenous and traditional medicinal knowledge. Such extensive uses are common threats to most of the plant species.

Acknowledgment This book chapter has been extracted form "Review of Literature" section presented in the Ph.D. thesis of Iftikhar Ahmad (89-ag-1513) submitted to University of Agriculture, Faisalabad in 2008. A partial material has also been drawn from the article "WORLD IS TURNING BACK TO NATURAL MEDICINES: Past, present and future of medicinal plants" published as a new article by the same author(s).

References

Ahmad H, Ahmad A, Jan MM (2002) The medicinal plants of salt range. Online J Biol Sci 2:175–177

Anke M (1986) Role of trace elements in the dynamics of arteriosclerosis. Z Gesamte Inn Med 41(4):105–111

Arora RB, Vohora SB, Khan MSY (1984) Proceedings of the first international conference on elements in health and disease. World Health Organization and Institute of History of Medicine and Medical Research, New Delhi

Arubi PA (2003) Proximate composition and selected functional properties of defatted papaya (*Carica papaya* L.) kernel flour. Biomed Life Sci 58(3):1–7

Atawodi SE, Bulus T, Ibrahim S, Ameh DA, Nok AJ, Mamman M, Galadima M (2003) In vitro trypanocidal effect of methanolic extract of some Nigerian savannah plants. Afr J Biotechnol 2(9):317–321

Bianco VV, Santamaria P, Elia A (1998) Nutritional value and nitrate contents in edible wild species used in Southern Italy. In: Proceedings 3rd international summit on diversification of vegetable crops. Acta Horticult 467:71–87

Bray TM, Betteger WJ (1990) The physiological role of zinc as an antioxidant. Free Radic Biol Med 8:281–291

Bukhari AQ, Ahmad S, Mirza M (1987) The role of trace elements in health and disease. In: Elements in health and diseases. In: 2nd international conference held in Karachi, Pakistan

Cakmak I, Marschner H (1988) Increase in membrane permeability and exudation in roots of zinc deficient plants. J Plant Physiol 132:356–361

Chiej R (1984) Encyclopedia of medicinal plants. MacDonald 1984. ISBN:0-356-10541-5. Covers plants growing in Europe. Good photographs

Chunhieng T, Hay L, Montet D (2005) Detailed study of the juice composition of noni (*Morinda citrifolia*) fruits from Cambodia. Fruits 60:13–24

Clark AM (1996) Natural products as a resource for new drugs. Pharma Res 13:1133–1141

Cook NS, Samman S (1996) Flavonoids – chemistry, metabolism, cardioprotective effect and dietary sources. J Nutr Biochem 7:66–76

Dar ME (2003) Ethnobotanical uses of plants of Lawat District Muzaffarabad, Azad Jammu and Kashmir. Asian J Plant Sci 2(9):680–682

Dey AC, Singh B, Singh MP (1980) Indian medicinal plants used is ayurvedic preparation. Bishen Singh Mahendrapal Singh publishers, Dehra Dun

Donatus EO, Morah FNI (2004) Mineral and nutritive value of Dennettia tripetala fruits. Fruits 59:437–442

Duke J (1992) Handbook of biological active phytochemicals and their activities. CRC Press, Boca Raton, pp 99–131

Edeoga HO, Okwu DE, Mbaebie BO (2005) Phytochemical constituents of some Nigerian medicinal plants. Afr J Biotechnol 4(7):685–688

Edeoga HO, Omosun G, Uche LC (2006) Chemical composition of *Hyptis suaveolens* and *Ocimum gratissimum* hybrids from Nigeria. Afr J Biotech 5(10):892–895

Elegbede JA (1998) Legumes. In: Osagie AU, Eka OU (eds) Nutritional quality of plant foods. Post Harvest Research Unit, University of Benin, pp 53–83

Fabris N, Mocchegiani E (1995) Zinc, human diseases and aging. Aging (Milano) 7(2):77–93

FAO (2001) FAO in partnership with support unit for International Fisheries and Aquatic Research, SIFAR. Aberdeen

Gauch HG (1972) Inorganic plant nutrition. Dowden, Hutchinson and Ross, Stroudsburg

Golden MH (1988) Trace elements in human nutrition. Hum Clin Nutr 6:448–455

Green C (1992) An overview of production and supply trends in the U.S. specialty vegetable market. Acta Horticult 318:41–45

Grieve (1984) A modern herbal. Penguin, London. 919 p

Harborne JB (1973) Phytochemical methods. Chapman and Hall, Ltd, London, pp 49–188

Hassan LG, Umar KJ (2006) Nutritional value of balsam apple (*Momordica balsamina* L.) leaves. Pak J Nutr 5(6):522–529

Heiser CB (1993) Ethnobotany and economic botany. In: Flora of North America Editorial Committee (ed) Flora of North America. Oxford University Press, New York, pp 199–206

Iwu M (1989) Food for medicine. In: Iwu M (ed) Dietary plants and masticetories as sources of biologically active substances. University of IFE Press, pp 303–310

Jimoh FO, Oladiji AT (2005) Preliminary studies on *Piliostigma thonningii* seeds: proximate analysis, mineral composition and phytochemical screening. Afr J Biotechnol 4(12):1439–1442

Kandaswami C, Perkins E, Soloniuk DS, Arzewiecki G, Middle E (1991) Anti-preventative effects of citrus flavonoids on a human squamous cell carcinoma in vitro. Cancer Lett 56:147–152

Khan AH (1951) The medicinal plants, their past and present (with special reference to the work being done in Pakistan). Pak J For 1:353–367

Kim SY, Kim JH, Kim SK, Oh MJ, Jung MY (1994) Antioxidant activities of selected oriental herb extracts. J Am Oil Chem Soc 71:633–640

Kirtikar KR, Basu BD (1982) Indian medicinal plants, 2nd edn. Vol I and II International Book distributor, Dera Dune, India

Kochhar A, Nagi M, Sachdeva R (2006) Proximate composition, available carbohydrates, dietary fibre and anti nutritional factors of selected traditional medicinal plants. J Hum Ecol 19(3):195–199

Lai XZ, Yang YB, Shan XL (2005) The investigation of euphorbiaceous medicinal plants in Southern China. Econ Bot 58(1):307–320

Launert E (1981) Edible and medicinal plants. The Hamlyn Publishing Group Ltd., London

Leporatti ML, Lattanzi E (1994) Traditional physiotherapy periodic grazing on coastal areas of Makran (Southern Pakistan). Fitotherapia 65:158–161

Liu S, Babajide O, Charles DH, Alvie M (1990) 3-methoxysampangine, a Noval antifungal copyrine alkaloids fungi deistopholis pattern. Antimicrob Agents Chemother 34(4):529–533

Lockett CT, Calvert CC, Grivetti LE (2000) Energy and micronutrient composition of dietary and medicinal wild plants consumed during drought. Study of rural Fulani, Northeastern Nigeria. Int J Food Sci Nutr 51(3):195–208

Mahmood T, Khan MA, Ahmad J, Ahmad M (2004) Ethnomedicinal studies of Kala Chitta Hills of District Attock, Pakistan. Asian J Plant Sci 3(3):335–339

Mahmud K, Naseer R, Shahid M, Rashid S (2002) Biochemical studies and trace elements profiles of *Cymbopogon Jwarancusa*. Asian J Plant Sci 1(1):57–58

Mutaftchiev KL (2003) Catalytic spectrophotometric determination of manganese in some medicinal plants and their infusions. Turk J Chem 27:619–626

Nakayama NG, Lindsey ML, Michael LH (1993) Inhibition of the infectivity of influenza virus by tea polyphenols. Antivir Res 21:289–299

Narendhirakannan RT, Subramanian S, Kandaswamy M (2005) Mineral content of some medicinal plants used in the treatment of diabetes mellitus. Biol Trace Elem Res 103(2):109–115

Nielsen FH (1987) In: Mertz W (ed) Trace elements in human and animal nutrition, 5th edn. Academic, New York, p 2

Obiajunwa EI, Adebajo AC, Omobuwajo OR (2002) Essential and trace element contents of some Nigerian medicinal plants. J Radioanal Nucl Chem 252(3):473–476

Ofokansi KC, Esimone CO, Anele CK (2005) Evaluation of the in vitro combined antibacterial effects of the leaf extracts of *Bryophyllum pinnatum*.(Fam: crassulaceae) and *Ocimum gratissimum* (Fam: Labiate). Plant Prod Res J 9:23–27

Okogun JI (1985) Drug production efforts in Nigeria. Chemistry Research and missing link. Being the text of a lecture given to the Nigeria. Acad Sci 29–52

Okwu DE (2001) Evaluation of the chemical composition of indigenous spices and flavouring agents. Glob J Pure Appl Sci 7:455–459

Okwu DE (2003) The potentials of *Ocimum gratissimum, Pengluria extensa* and *Tetrapleura tetraptera* as spice and flavouring agents. Nig Agric J 34:143–148

Okwu DE (2004) Phytochemicals and vitamin content of indigenous spices of South Eastern Nigeria. J Sustain Agric Environ 6:30–34

Okwu DE (2005) Phytochemicals, vitamins and mineral contents of two Nigerian medicinal plants. Int J Mol Med Adv Sci 1(4):375–381

Okwu DE, Okwu ME (2004) Chemical composition of *Spondias mombin* Linn plant parts. J Sustain Agric Environ 6:140–147

Okwu DE, Omodamiro OD (2005) Effects of hexane extract and phytochemical content of *Xylopia acthiopica* and *Ocimum gratissimum* on the uterus of Guinea pig. Bio-Research 3(2):40–44

Okwu DE, Ekeke O (2003) Phytochemical screening and mineral composition of chewing sticks in south eastern Nigeria. Global J Pure Appl Sci 9(2):235–238

Okwu DE, Morah FN (2004) Mineral and nutritive value of Dennettia tripetala fruits. Fruits 59(6):437–442

Olabanji SO, Makanji OV, Ceccato D, Buoso MC, Haque AM, Cherubini R, Moschini G (1997) Biol Trace Elem Res 58:223–236

Oloyede OI (2005) Chemical profile of unripe pulp of *Carica papaya*. Pak J Nutr 4(6):379–381

Osborne R, Grove A, Oh P, Mabry TJ, Ng JC, Seawright AA (1994) The magical and medicinal usage of *Stangeria eriopus* in South Africa. J Ethnopharmacol 43(2):67–72

Oyagade JO, Awotoye OO, Adewumi TJ, Thorpe HT (1999) Antibacterial activity of some Nigerian medicinal plants. I. Screening for antibacterial activity. Biosci Res Comm 11(3):193–197

Pandey BP (1980) Economic botany for degree honours and post-graduate students. S. Chand and Company Ltd., Ram Nagar

Pereira CE, Felcman J (1998) Correlation between five minerals and the healing effect of Brazilian medicinal plants. Biol Trace Elem Res 65(3):251–259

Plotkin MJ (1991) Traditional knowledge of medicinal plants – the search for new jungle medicines. In: Akerele O, Heywood V, Synge H (eds) The conservation of medicinal plants. In: Proceedings of an International Consultation, 21–27 March 1988, Chiang Mai, Thailand. Cambridge University Press, Cambridge, pp 53–64

Qureshi SJ, Khan MA (2001) Ethnobotanical study of Kahuta from Rawalpindi District Pakistan. Online J Biol Sci 1(1):27–30

Rajput MT, Hassney SS, Khan KM (1996) Plant taxonomy. Oxford Publisher Ltd., Pakistan

Rajurkar NS, Damame MM (1998) Mineral contents of medicinal plants used in the treatment of diseases resulting from urinary tract disorders. Appl Radiat Isot 49(7):773–776

Randal J (1984) Am Health 3:37. cited in cur. Sci 54:364

Remington JP (1995) Solutions, emulsions, suspensions, and extracts. In: The science and practice of pharmacy, vol 2. Mack Publishing Company, Easton. 19th 1495–1523

Rizvi MA, Ahmad L, Sarwar GR (1996) Wild medicinal plants of Madinatul Hikmah and its adjacent areas. Hamdard Medicus 39:8–10

Sahito SR, Kazi TG, Kazi GH, Jakhrani MA, Shaikh MS (2001) Trace elements in two varieties of indigenous medicinal plant *Catharanthus roseus* (*Vinca rosea*). Sciences 1(2):74–77

Salisbury FB, Ross CW (1992) Plant physiology. Wadsworth. Benjamin/Cummings Publishing company, Redwood City

Shad AA, Shah H, Khattak FK, Dar NG, Bakht J (2002) Proximate and mineral constituents of medicinal herb *Fagonia Arabica*. Asian J Plant Sci 1(6):710–711

Singh V, Garg AN (1997) Availability of essential trace elements in Ayurvedic Indian medicinal herbs using instrumental neutron activation analysis. Appl Radiat Isot 48(1):97–101

Singh R, Sawhney SK (1988) Advances in frontier areas of plant biochemistry. Prentice Hall in India Private Ltd., New Delhi, p 487

Sofowara A (1993) Medicinal plants and traditional medicine in Africa. Spectrum Books Ltd., Ibadan, p 289

Stary F (1998) The natural guide to medicinal herbs and plants. Tiger Books International, London, pp 12–16

Sundriyal M, Sundriyal RC (2004) Wild edible plants of the Sikkim Himalaya: nutritive values of selected species. Econ Bot 58(2):286–299

Svendsen R, Lund W (2000) Speciation of Cu, Fe and Mn in beer using ion exchange separation and size-exclusion chromatography in combination with electrothermal atomic absorption spectrometry. Analyst 125(11):1933–1937

Talwar GP, Srivastava LM, Mudgil KD (1989) Textbook of biochemistry and human biology, 2nd edn. Prentice Hall of India, Private Ltd., New Delhi

Tolonen M (1990) Vitamins and minerals in health and nutrition. Ellis Horwood Limited, Chichester

Torel JB, Scontia S, Blevo ZO, Lapis MV (1986) Inhibition of lipid peroxidation by flavonoids. Superoxide anions, hydroxyl ions. BBA 759:38–41

Uruquiaga I, Leighton F (2000) Plant polyphenol antioxidants and oxidative stress. Biol Res 33:159–165

Valles J, Angels Bonet M, Agelet A (2004) Ethnobotany of in Catalonia (Iberian Peninsula): the integral exploitation of a natural resource in mountain regions. Econ Bot 58(3):456–469

Wada O, Yanagisawa H (1996) Trace elements and their physiological roles. Nippon Rinsho 54(1):5–11

Whitney EN, Hamilton EMN (1984) The trace minerals. In: Understanding nutrition, 3rd edn. West Publishing Company, St. Paul

Yildirim E, Dursun A, Turan M (2001) Determination of the nutrition contents of the wild plants used as vegetables in Upper Çoruh Valley. Turk J Bot 25:367–371

Chapter 2
Medicinal Plant: Environment Interaction and Mitigation to Abiotic Stress

Murtaza Abid and M. M. Abid Ali Khan

Abstract Herbal/traditional plant medicine is the most antioxidant-rich category. Abiotic stresses including climatic factors, plant species, extreme temperatures, light intensity, soil and air pollution, drought, flooding, salinity and osmotic changes, and other environmental factors affected both the enzymatic and nonenzymatic antioxidant defense system in plants. The activities of the antioxidant enzymes such as polyphenol oxidase (PPO), catalase (CAT), superoxide dismutase (SOD), peroxidase (POD), phenylalanine ammonia-lyase (PAL), ascorbate peroxidase (APX), and lipoxygenase (LOX) are altered in stressed conditions which lead to the changes in malondialdehyde (MDA), superoxide radical, and hydrogen peroxide content of the cells. The different components of nonenzymatic defense system such as glycine betaine (GB), proline, glutathione (GSH), ascorbic acid (AsA), tocopherols, carotenoids, flavonoids, and phenolic compounds also play a crucial role as they interact at cellular level. A common factor between most stresses is the active production of reactive oxygen species (ROS). They are actively produced and used as signaling molecules by cells in response to most abiotic stresses. Due to the highly reactive nature of ROS, their production and detoxification need to be strictly controlled. Studies on transformed plants expressing increased activities of single enzymes of the antioxidant defense system indicate that it is possible to confer a degree of tolerance to stress by these means. The advent of plant transformation has placed within our grasp the possibility of engineering greater stress tolerance in plants by enhancements of the antioxidant defense system.

Keywords Medical plant · Environment · Interaction · Abiotic stress · Antioxidant defense system

M. Abid
Ex-Research Assistant, Department of Biochemistry, K. G. M. University, Lucknow, UP, India

M. M. Abid Ali Khan (✉)
Department of Botany, Shia P. G. College, Lucknow, Uttar Pradesh, India

D. Egamberdieva, A. Tiezzi (eds.), *Medically Important Plant Biomes: Source of Secondary Metabolites*, Microorganisms for Sustainability 15,
https://doi.org/10.1007/978-981-13-9566-6_2

2.1 Introduction

2.1.1 Reactive Oxygen Species and Antioxidant System

Plants have integrated antioxidant systems, which include enzymatic and nonenzymatic antioxidants that are usually effective in blocking harmful effects of ROS. Ethylene biosynthesis and membrane breakdown involving lipid peroxidation seem to involve free radicals. Since plants have less evolved mechanisms of stress avoidance, they require important means of adaptation to changing environmental conditions. A cyanide-insensitive respiratory pathway in chloroplasts competes for electrons with photosynthetic electron transport (Bennoun 1994) and may also reduce oxygen. Furthermore, some important sites, such as the reaction center protein of PSII (DI) and the apoplastic space, appear to have very little protection against oxidative damage (Castillo and Greppin 1988; Luwe et al. 1993). To save themselves from these lethal oxygen intermediates, plant cells and its organelles like chloroplast, mitochondria, and peroxisomes employ antioxidant defense systems. A great deal of research has established that the induction of the cellular antioxidant machinery is important for protection against various stresses (Tuteja 2007; Khan and Singh 2008; Gill et al. 2011; Singh et al. 2008). The components of antioxidant defense system are enzymatic and nonenzymatic antioxidants. Enzymatic antioxidants include SOD, CAT, POD, APX, monodehydroascorbate reductase (MDHAR), dehydroascorbate reductase (DHAR), and glutathione reductase (GR), and nonenzymatic antioxidants are GSH, AsA (both water soluble), carotenoids and tocopherols (lipid soluble), proline, glycine betaine, flavonoids, and phenols (Gill et al. 2011; Mittler et al. 2004; Singh et al. 2008).

2.1.2 Abiotic Stresses

Plants are often subjected to hostile environmental conditions which cause abiotic stress conditions that play a major role in determining productivity and yields (Boyer 1982) and also the differential ecological distribution of the plants species (Chaves et al. 2003). A significant feature of plant adaptation to abiotic stresses is the activation of multiple responses involving complex gene interactions and cross-talk with many molecular pathways (Basu 2012; Umezawa et al. 2006). Abiotic stresses elicit complex cellular responses that have been elucidated by studying plant abiotic responses at the whole-plant, morphological, physiological, biochemical, cellular, and molecular levels (Grover et al. 2001). The development of stress-tolerant plants either by genetic engineering or through conventional breeding has been done. The elucidation of the different components and molecules playing important role in abiotic stress responses of a broad range of species in both model and crops plant is in progress. Now, efforts are being made to expand our knowledge on plant response to abiotic stresses using holistic system biology approaches, taking advantage of available high-throughput tools such as transcriptomics, proteomics, and metabolomics. The objective of this chapter is to provide an insight of abiotic stress biology in medicinal plants. In the present chapter, we present some details about the enzymatic and nonenzymatic responses of plants to various abiotic stresses for the better adaptation to face the environmental constraints.

2.1.2.1 Types of Abiotic Stresses

Stress is usually defined as an external factor that exerts a disadvantageous influence on the plant (Taiz and Zeiger 1991). Alternatively, stress could be defined as a significant deviation of the optimal condition of life (Larcher 2003). The effects of the following types of abiotic stresses have been studied in quite detail (Fig. 2.1).

Temperature

Among stile conditions, temperature stress is considered to be one of the most damaging because of the ever-changing components of the environment. Heat stress has several impacts on the life processes of organisms, and plants, in particular, are most affected since they are as sessile and cannot move to more favorable environments.

The changing climate and global warming have made the study related to temperature stress as the major concerns for plant scientists worldwide. High temperature (HT) is now considered to be one of the major abiotic stresses for decreasing crop production and yield (Hasanuzzaman et al. 2012). As there is an optimum temperature limit in every plant species for plant growth and metabolism, there may be devastating effects of temperature on the growth and survival of plants. The US Environmental Protection Agency (EPA) indicates that global temperatures have risen during the last 30 years (EPA Student's Guide to Global Climate Change. www.epa.gov 2011) and also said that the decade from 2000 to 2009 was the hottest period ever recorded. High temperature stress is defined as the rise in temperature beyond a critical threshold for a period of time sufficient to cause irreversible damage to plant growth and development (Wahid 2007). The growth and development of plants involves numerous biochemical reactions, all of which operates at a particular temperature (Zróbek 2012).

Low temperature (LT) or cold stress also affects plant growth and crop productivity and leads to substantial crop losses (Xin and Browse 2000; Sanghera et al. 2011). Chilling stress results from cool temperatures low enough to produce injury without forming ice crystals in the cells, whereas freezing stress results in ice formation within plant tissues. Plants differ in their tolerance to chilling (0–15 °C) and freezing (<0 °C) temperatures. Both chilling and freezing stresses are together termed low temperature or cold stress: the damage due to cold stress can range from chilling injury and freezing injury to suffocation and heaving. In general, plants from temperate climatic regions are considered to be chilling tolerant to variable degrees, and their freezing tolerance can be increased by exposing to cold, but

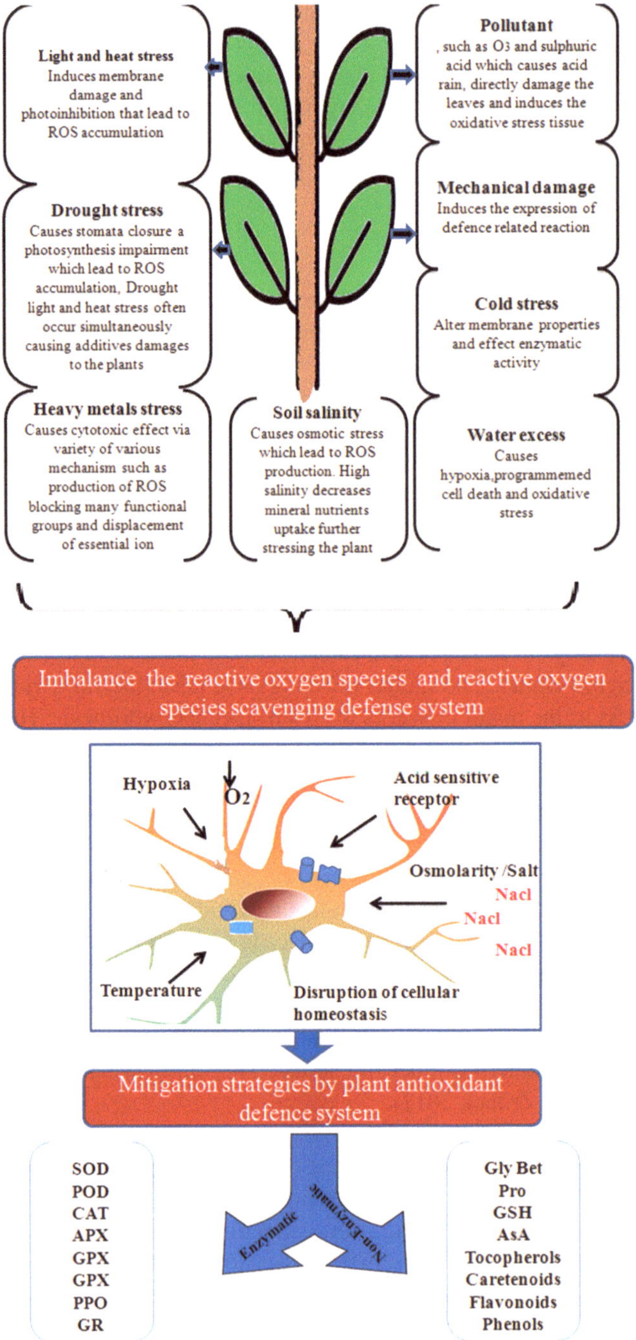

Fig. 2.1 Different types of abiotic stresses and their enzymatic and nonenzymatic responses

non-freezing, temperatures; this process is known as cold acclimation. However, generally, the plants of tropical and subtropical origins are sensitive to chilling stress and lack this mechanism of cold acclimation (Sanghera et al. 2011). Low temperature may affect several aspects of crop growth, viz., survival, cell growth and division, photosynthesis, water transport, growth, and finally crop yield.

The alterations at the cellular level due to HT or LT lead to the excess accumulation of toxic compounds, especially reactive oxygen species (ROS). The end result of ROS accumulation is oxidative stress (Mittler 2002; Yin et al. 2008; Suzuki and Mittler 2006). In response to HT, the reaction catalyzed by ribulose-1,5-bisphosphate carboxylase oxygenase (RuBisCO) results into the production of H_2O_2 due to increases in its oxygenase reactions (Kim and Portis 2004). LT conditions can also inhibit the activity of the Calvin-Benson cycle and create an imbalance between light absorption and light use. Enhanced photosynthetic electron flux to O_2 and over-reduction of the respiratory electron transport chain (ETC) can also result in ROS accumulation during chilling which causes oxidative stress (Hu et al. 2008). Plants have developed a variety of responses to extreme temperatures that reduce injury and ensure the maintenance of cellular homeostasis (Kotak et al. 2007). Researchers have explored the direct link between ROS scavenging and plant stress tolerance under extreme temperature (Suzuki and Mittler 2006). There is a relation between temperature stress tolerance and enhanced activities of enzymes involved in antioxidant systems of plants. Plants cope with the deleterious effects of oxidative damage caused by extreme temperatures by making use of several nonenzymatic and enzymatic antioxidants. Numerous studies on plants have shown that enhancing antioxidant defense confers stress tolerance to either HT or LT stress (Almeselmani et al. 2006, 2009; Nagesh Babu and Devaraj 2008; Huang and Guo 2005).

Water

Water is a universal solvent and 70% of living cell comprises of water. Water stress limits the growth and productivity of crops particularly in arid and semi-arid areas causing heavy economic losses in agriculture. Inoculation of plants with native beneficial microorganisms may increase rough tolerance of plants growing in arid or semi-arid areas (Marulanda et al. 2007). A large number of microorganisms exist in the rhizosphere, and plants select those bacteria which secrete organic compounds through exudates and contribute to their well-being (Bazin et al. 1990). Water (drought) stress impairs electron transport system leading to the formation of activated oxygen (Chandra et al. 1998). Activated oxygen compound such as H_2O_2, $O\text{-}^{2}$, and OH- may accumulate during water deficit stress and damage the photosynthetic apparatus. Superoxide dismutase (SOD) and ascorbate peroxidase along with the antioxidant ascorbic acid and glutathione prevent oxidative damage in plants (Allen 1995). Oxidative molecules initiate damage in the chloroplast and cause a series of damaging effects including chlorophyll (Chl) degradation, lipid peroxidation, and loss of protein activity (Zhang and Kirkham 1994). In aromatic plants, growth and essential oil production are influenced by various environmental factors, such as water stress (Burbott and Loomis 1969).

Drought

The response of plants toward drought stress is a complex phenomenon, and it seems to involve the synthesis of polyamines, some new proteins whose function is still not clear (Caplan et al. 1990). Abscisic acid is an important component since it stimulates stomatal guard cells to close, reducing water loss. This process also reduces the availability of CO_2 for photosynthesis, which can lead to the formation of reactive oxygen species from plants, which direct the electrons in the photo system. In tomato, cytosolic Cu/Zn-SOD was induced strongly by drought, while the chloroplastic Cu/Zn-SOD remained unchanged. An increase in glutathione reductase activity was reported in wheat and cotton plants under drought stress (Burke et al. 1985). It was assumed that in addition to removing H_2O_2, NADP was made available that can accept electrons from ferredoxin, thereby minimizing chances of superoxide formation. In drought-tolerant *Hordeum* species, levels of glutathione reductase and ascorbate peroxidase increased, but SOD activity was not examined (Smirnoff and Colombe 1988). Drought-stressed cotton was found to be resistant to paraquat (Burke et al. 1985), which proves the existence of a common protective mechanism against these stresses. Drought-induced changes in lipid peroxidation and the activities of SOD and catalase were compared in two mosses, the drought-tolerant *Tortula ruralis* and the drought-sensitive *Cratoneuron filicinum* (Dhindsa and Matowe 1981). In the presence of stress, the drought-tolerant moss showed lower levels of lipid peroxidation and increased levels of enzymes. The opposite occurred in the sensitive moss. Oxidized glutathione (GSSG) was shown to be a good indicator of drought stress (Dhindsa 1991). Drought-tolerant and intolerant maize inbred were analyzed by Malan et al. (1990), and resistance was found to correlate with Cu/Zn-SOD and glutathione reductase activities. Sairam et al. (1997, 1998) showed that H_2O_2 scavenging systems comprising of ascorbate peroxidase, glutathione reductase, and catalase are more important in conferring tolerance against drought-induced oxidative stress than superoxide dismutase alone.

Flood or Anorexia

Stress on plants imposed by water logging and deeper submergence (flooding) of the soil is one of the major abiotic constraints on growth, species' distribution and agricultural productivity (Jackson 2004), and grain yields (Setter and Waters 2003). Flooding stress also plays a role in adaptive strategies and evolution. A major constraint resulting from excess water is an inadequate supply of oxygen to submerged tissues (Armstrong and Drew 2002) and other changes in the soil that influence plants; levels of the plant hormone ethylene (Smith and Russell 1969; Jackson 1982) and products of anaerobic metabolism by soil microorganisms (e.g., Mn^{2+},

Fe^{2+}, S_2, H_2S, and carboxylic acids) can accumulate (Ponnamperuma 1983; McKee and McKevlin 1993). Moreover, availability of carbon dioxide, light, and oxygen to the shoots is reduced (Jackson and Ram 2003).

Oxidative stress and increased ROS production are an important part of many stress situations, including hypoxia. Post-hypoxic hydrogen peroxide accumulation has been shown in the roots and leaves of *Hordeum vulgare* (Kalashnikov et al. 1994) and in wheat roots (Biemelt et al. 2000). The presence of H_2O_2 in the apoplast and in association with the plasma membrane under hypoxic conditions in four plant species has been shown (Blokhina et al. 2001). Indirect evidence of ROS formation such as TBARS contents (i.e., lipid peroxidation products) under low oxygen have been detected (Yan et al. 1996; Chirkova et al. 1998; Blokhina et al. 1999). Flooding in *Zea mays* resulted in a significant increase in TBARS content, production of superoxide radical and hydrogen peroxide, and membrane permeability in the leaves (Yan et al. 1996). An excessive accumulation of superoxide due to the reduced activity of SOD under flooding stress has also been reported (Yan et al. 1996).

Light

Plants require adequate light in order to grow and survive. Light is the primary source of energy, and plants convert light to chemical energy through photosynthesis. Light is an essential prerequisite for chlorophyll (Chl) biosynthesis and chloroplast development. Light reactions of photosynthesis require light and are essential for the synthesis of carbohydrates.

Intense light has long been known to disrupt metabolic processes in plants, including photosynthesis, glucose assimilation, electron transport chain, and phosphorylation (Egneus et al. 1975). However, excessive exposure to high light intensities can cause considerable damage to plants. The effects of H_2O_2 on gene expression have also been reported to be different when it was induced in response to high light (Golemiec et al. 2014).The flux of ROS generated in cells is activated under high light and is capable of leaving the thylakoid membrane and reaching the cytoplasm or even the nucleus (Fischer et al. 2007), which makes its role as a signaling molecule feasible. In a recent study (Zhao et al. 2011), it was shown that under high temperature treatments, large amounts of $O^{\bullet 2-}$and H_2O_2 were generated and accumulated in cucumber leaves, leading to premature senescence, which is indicated by the changes in protein, lipid peroxidation (LPO), and chlorophyll content. Following high light illumination, $^1O_2$ accumulates and modifies the expression of a group of genes encoding chloroplast proteins, leading to a significant change in chloroplast structure and functional modifications.

Low light induced oxidative stress by modulating the activity of antioxidant enzymes. It has been shown that exogenous H_2O_2 can have a beneficial effect on low light-induced oxidative stress (Zhang et al. 2011). Low light induces oxidative stress (Sielewiesiuk 2002), which increases ROS and causes lipid peroxidation. H_2O_2 pretreatment of cucumber leaves resulted in decreased levels of $O^{2}-$, endogenous H_2O_2, and malonaldehyde by moderating the activities of antioxidant enzymes, thus reducing lipid peroxidation and stress intensity at low light.

Salinity

Salt stress causes oxidative damage as a result of water deficit and triggers generation of reactive oxygen species, which disrupts biological membranes and thus results in either the death of the plant or reduction in productivity. High salt concentration might interfere with the electron transport chain in different organelles and generates ROS such as singlet oxygen, superoxide, hydrogen peroxide, and hydroxyl radicals (Elstner 1982; Hernandez et al. 1993, 1995). Excess of ROS triggers phytotoxic reactions such as lipid peroxidation, protein degradation, and DNA mutation (Stewart and Bewley 1980; Fridovich 1986; Davies 1987). Mannitol is accumulated by a wide range of species in response to salinity (Stoop et al. 1996).

pH

Soil pH is one of the important factors determining plant growth as it may affect the availability of nutrients to the plants. Nutrients are most available to plants in the optimum 5.5–7.0 range. In some cases, aluminum (Al) becomes soluble at pH levels below 5.0 becoming toxic to plant growth.

Soil acidification is becoming a very serious environmental problem affecting plant growth and yield since the use of acidic and physiologically acidic nitrogen fertilizers is increased and the ever-increasing environmental pollution causes acid rain (Russell et al. 2006; Zhang et al. 2009; Guo et al. 2010). The pH value of most acidic soil is highly reduced from the 1980s to the 2000s, and the pH is under 4.0 in some highly acidic soils (Guo et al. 2010). Proton toxicity (low pH stress) is considered to be one of the major stresses limiting plant growth in acid soils (Kochian et al. 2005). Low pH levels directly inhibited plant growth via high H activity (Schubert et al. 1990; Koyama et al. 2001). A high concentration of H triggers typical oxidative stress on plants by inducing the accumulation of excess reactive oxygen species (ROS), such as superoxide radicals (O) and hydrogen peroxide (HO) in plant tissues (Shi et al. 2006; Liu et al. 2011). To counteract oxidative damage, plants have evolved complex antioxidant systems including antioxidant enzymes such as superoxide dismutase (SOD), catalase (CAT), peroxidases (POD), ascorbate peroxidase (APX), glutathione reductase (GR), dehydroascorbate reductase (DR), and antioxidants such as α-tocopherol, ascorbate, and reduced glutathione (Asada 1999; Mittler 2002). Studies have indicated that higher activity levels of antioxidant enzymes may contribute to better H tolerance by increasing the protective capacity against oxidative damage (Liu and Liu 2011; Chen et al. 2013).

Alkaline stress is defined as the presence of alkaline salts (Na_2CO_3 or $NaHCO_3$) in the soil (Paz et al. 2012). It is one of the most critical abiotic stresses which plants face in the era of climate change. A number of researches have shown that alkaline stress is more hazardous than saline stress because of its additional high pH stress (Campbell and Nishio 2000; Hartung et al. 2002; Chen et al. 2012; Radi et al. 2012). High pH value may lead to reduction in seed germination, damage to the root cell structure, alterations in the nutrient availability, and disorder in nutrient uptake, thus resulting in a decreased yield of agricultural plants (Peng et al. 2008). The effects of alkalinity was studied on maize plants, and it was shown that alkaline-stressed plants showed a decrease in growth parameters, leaf relative water content (LRWC), and the contents of photosynthetic pigments, soluble sugars, total phenols, and potassium ion (K^+), as well as potassium/sodium ion (K^+/Na^+) ratio. By contrast, alkaline stress increased the contents of soluble proteins, total free amino acids, proline, Na^+, and malondialdehyde (MDA), as well as the activities of superoxide dismutase (SOD), catalase (CAT), and peroxidase (POD) in stressed plants (Latef and Tran 2016).

The effects of NaCl and $NaHCO_3$ stresses were investigated in tomato roots. The relative growth rate and respiration of tomato plants were considerably decreased in both NaCl and $NaHCO_3$ treatments, especially under $NaHCO_3$ stress. As the concentration of NaCl and $NaHCO_3$ increased, Na accumulation in roots was increased, while accumulation of N, P, K, Fe, and Mg was less. Among both the treatments, $NaHCO_3$ treatments induced much higher levels of reactive oxygen species (ROS) and lipid peroxidation as well as higher activities of antioxidant enzymes and higher concentrations of ascorbate-glutathione. However, after few days of treatment, $NaHCO_3$ stress led to decreased accumulation of ROS, antioxidant enzyme activities, and ascorbate-glutathione content (Gong et al. 2014).

Pollutants

Atmospheric pollutants such as ozone (O_3) and sulfur dioxide (SO_2) have been shown to generate free radical (Mehlhorn 1990; Pell et al. 1977) and influence the growth of forest. Ozone seems to be a greater threat to plants than SO_2 (Heagle and Annu 1989). Mehlhorn (1990) suggested that the damaging effects of O_3 are due to its oxidizing potential and the consequent formation of free radicals followed by initiation of chain reactions. The O_3 concentration in the intercellular air spaces of leaves is close to zero (Laisk et al. 1989). Thus, ozone is not likely to reach the chloroplast, but it causes pigment bleaching and lipid peroxidation (Heath 1987). Stimulation of synthesis and degradation of the PSII-DI protein occurs in spruce trees following O_3 treatment (Lutz et al. 1957) and a decrease in the activity and quantity of RuBisCO has been found in poplar following exposure to O_3 (Landry and Pell 1993).

SO_2 exposure results in tissue damage and release of stress ethylene from both photosynthetic and non-photosynthetic tissues (Peiser and Yang 1985), and fumigation with SO_2 results into a change in cytoplasmic pH. The concentration of proton in the cytoplasm is one of the most important factors regulating cellular activity. SO_2 causes considerable acidification of the cytoplasm it reacts with water to form sulfurous acid which may then be converted into sulfuric acid (Laisk et al. 1988;

Veljovic-Jovanovic et al. 1993). The oxidation of sulfite to sulfate in the chloroplast also gives rise to the formation of free radicals (Polle et al. 1992). The oxidation of sulfite is initiated by light and is mediated by photosynthetic electron transport. This results in loss of photosynthetic function caused by inhibition of the activity of SH-containing, light activated enzymes of the chloroplast (Shimazaki and Sugahara 1980; Covello et al. 1989).

Radiation

Sunlight contains energetic short wavelength ultraviolet (UV) photons which are highly injurious because of their destructive interactions with amino acids, nucleic acids, or membrane lipids (1). The intensity of UV radiation reaching the earth's surface varies greatly with season, time of day, latitude, ozone layer thickness, altitude, and cloud cover. Sometimes, differential responses are studied in UV-A (400–320 nm) and the UV-B (320–290 nm) regions. In both cases, however, the basic mechanisms of photochemical damage remain similar although different receptor molecules (chromophores) may be involved.

A depletion of ozone layer led to a significant increase in ultraviolet-B (UV-B) radiation (280–320 nm). UV-B can manipulate plant processes either causing direct damage or via different regulatory effects (Rozema et al. 1999; Potters et al. 2009). It can cause either direct injury to DNA leading to mutations which could be heritable or direct or indirect damage to physiological functions of plant (Ormrod and Hale 1995; Lidon 2012). UV-B results in altered plant growth and productivity. UV-B injury also causes membrane changes and protein denaturation.

A wide range of morphological, growth, biochemical, and physiological responses of plant have been reported in the presence of UV-B radiation (Caldwell et al. 1998; Zhang et al. 2009; He et al. 2003). Flavonoids play a major role in protecting plants from UV-B damage (Liang et al. 2006). These flavonoids generally absorb the light in the region of 280–320 nm and thus are capable of acting as a UV filter, thereby protecting the photosynthetic tissues from damage (Siefermann and Harms 1987). Flavonoids stabilize and protect the lipid phase of the thylakoid membrane and are quenchers of the excited triplet state of chlorophyll and singlet oxygen (Agrawal and Rathore 2007). In addition, carotenoids also have antioxidant properties which act as an internal filter against UV-B radiation. Plants scavenge reactive oxygen species by detoxification mechanism produced by enzymatic antioxidant such as catalase, peroxidase, superoxide dismutase, phenylalanine ammonia-lyase, etc. (Moran and Porath 1980).

2.1.3 *Effects of Abiotic Stresses on Antioxidant Status of Medicinal Plants (Table 2.1)*

***Ocimum basilicum* L.** Water stress caused the following physiological and biological changes in basil plants. It resulted in the accumulation of reactive oxygen species in the cell. It resulted in higher antioxidant activity, and the highest concentration of CAT and GPX activity was reported. Increased water stress caused increased chlorophyll content in leaves, whereas APX activity decreased. Inoculation with rhizobacteria could be efficiently used to improve growth, antioxidant status, and photosynthetic pigments in basil under water stress. *Pseudomonades* sp. under water stress considerably improved CAT enzyme activity in the leaves and increased it. Combination of three bacterial species caused the highest GPX and APX activity and chlorophyll content in leaves under water stress (Heidari and Golpayegani 2012).

***Withania somnifera* L.** Iron stress was induced by adding higher quantity of $FeSO_4$. It caused disturbance in balancing of nutrients and induces oxidative stress in plants (root and leaf tissues analyzed for catalase, superoxide dismutase, and guaiacol peroxidase (GPX) have shown, an increase in content with respect to exposure of time) (Rout et al. 2015).

***Camellia sinensis* L.** Drought stress caused increased water loss rate (WLR) and decrease in relative water content (RWC), dry mass, chlorophyll, carotenoid, and total phenolic contents of leaf and antioxidants like ascorbate and glutathione in tea. Leaf antioxidant enzymes SOD, CAT, and GR showed differential activities, whereas there was an increase in reactive oxygen species (ROS) and lipid peroxidation resposible for gradual decrease in POD activities. An increased activity of POD, GR, CAT, and higher phenol content was reported. Drought stress altered antioxidant response with apparent decrease in mineral nutrient (Zn, Ca, Na, Fe, Mg, and K) contents of leaves suggesting that mineral deficiency-mediated drought stress-induced oxidative damage in tea. Tea plants exposed to heavy metals (HM) (e.g., Cd, Cu, Al) also showed reduction in growth and antioxidant responses (Upadhyaya and Panda 2013).

The antioxidant responses to increasing concentrations of copper were investigated in the leaves of two cultivars (TS-462 and TS-520) of tea commonly grown in the Darjeeling hills. Exposure to excess Cu resulted in increased lipid peroxidation, reduced chlorophyll content, higher levels of phenolic compounds, and an increase in peroxidase enzyme levels. Two new peroxidase isozymes (POD1 and POD2) were detected in plants exposed to Cu. TS-520 was found to be more sensitive to increasing concentrations of Cu. Superoxide dismutase activity increased in TS-462 but declined in TS-520 when exposed to higher Cu concentrations. A sharp increase in the activity of ascorbate peroxidase was noticed at the 10 days of exposure in the more tolerant cultivar. On the other hand, catalase levels increased only marginally

Table 2.1 Enzymatic and nonenzymatic response of some medicinal plants toward abiotic stresses

No.	Plant	Stress	Site	Enzymatic effects	Non-enzymatic effects	References
1.	*Ocimum basilicum* L.	Water	Leaf	CAT ↑ GPX ↑ APX ↓	–	Heidari and Golpayegani (2012)
3.	*Urtica dioica* L.	Pb	Leaf and root	–	–	Gjorgieva et al. (2013)
6.	*Olea europaea* L.	Temperature + radiation	Leaf and root	SOD ↓ CAT ↓ APX ↓ POD ↓ LOX ↓	Phenols	Sofo et al. (2004)
7.	*Plantago ovata* Forsk.	Salt	Leaf	SOD ↑ CAT ↑ POD ↑	–	Kala (2015)
8.	*Plantago maritima*	Salt	Leaf	SOD ↑ CAT ↑ GR ↑ APX ↑	–	Sekmen et al. (2007)
9.	*Plantago media*	Salt		SOD ↓ CAT ↓ GR ↓ APX ↑	–	Sekmen et al. (2007)
10.	*Plantago crassifolia*	Salt	Leaf	–	Proline ↑ GB ↑ Sorbitol ↑	Al Hassan et al. (2016)
11.	*Plantago coronopus*	Salt	Leaf	–	Proline ↑ GB ↑ Sorbitol ↑	Al Hassan et al. (2016)
12.	*Plantago major*	Salt	Leaf	–	Proline – GB ↑ Sorbitol ↑	Al Hassan et al. (2016)
13.	*Scutellaria baicalensis* L.	Temperature	Leaf	PAL ↓ CAT ↓ SOD ↓ POD ↓	Baicalin ↓ Baicalein ↓	Yuan et al. (2011)

(continued)

Table 2.1 (continued)

No.	Plant	Stress	Site	Enzymatic effects	Non-enzymatic effects	References
		Hg	Leaf and root	SOD_L ↑	$NPSH_L$ ↑	Calgaroto et al. (2010)
				SOD_R ↑	$NPSH_R$ ↑	
				CAT_L ↑	AsA_L ↑	
				CAT_R –	AsA_R ↑	
				APX_L ↑	Carotenoids–	
				APX_R ↓	$Proline_L$ ↑	
					$Proline_R$ ↓	
19.	*Trifolium resupinatum* L.	SO_2	Leaf	SOD	–	Bayat et al. (2014)
				CAT ↑		
				GPX ↑		
20.	*Thymus vulgaris* L.	Drought	Leaf	–	Phenols ↓	Khosh-Khui et al. (2012)
		Temperature + extraction	Leaf	–	–	Hossain et al. (2013)
25.	*Catharanthus roseus* (L.) G. Don.	Salt	Leaf	POD ↓	AsA ↑	Jaleel et al. (2007, 2008) and Amirjani (2015)
				SOD ↓	Glutathione ↑	
				PPO ↓		
				CAT ↑		
				GR ↑		
				POD ↑		

"↑" – increased activity
"↓" – decreased activity

in both the cultivars. Exposure to Cu resulted in accumulation of products of lipid peroxidation in the leaves. The level of thiobarbituric acid reactive substances (TBARS) increased steadily with Cu concentration and time of exposure in both cultivars up to 7 days beyond which the rate of increase declined. Aluminum exposure also caused an increase in SOD activity in cultured tea cells (Ghanati et al. 2005; Saha et al. 2012).

Seedlings of *Camellia sinensis* were grown hydroponically in order to study the effect of fluorine (F) on growth parameters, antioxidant defense system, photosynthesis, and leaf ultrastructure. Fresh and dry mass, chlorophyll (Chl) content, and net photosynthetic rate (PN) decreased with increasing F concentration. Superoxide dismutase activity decreased significantly, and catalase and guaiacol peroxidase activities reached maximum under F stress. Proline, malondialdehyde, and hydrogen peroxide contents increased significantly. These results suggested that antioxidant defense system of leaves did not sufficiently scavenge excessive reactive oxygen species. The cell ultrastructure was not changed under low F stress; however, it was destroyed at high F stress (Li et al. 2011).

Seedlings of *Camellia sinensis* (L.) were studied for the effect of aluminum (Al) on leaf antioxidant defense system and cell ultrastructure. It was seen that malondialdehyde content decreased at low Al concentration but increased at high Al concentration. Hydrogen peroxide content increased at high Al dose, and no differences were observed at low Al dose. Superoxide dismutase activity remained practically constant at low Al concentration but increased sharply at high Al concentration. Catalase and guaiacol peroxidase activities decreased following an initial increase, reaching their peaks at low Al dose. Ascorbate peroxidase activity increased and glutathione level fluctuated with increasing Al concentrations. Transmission electron microscope analysis of Al-treated leaves showed that although cell ultrastructural integrity was maintained at low concentration of Al, significant membrane damage was observed at high concentration (Li et al. 2011).

A possible connection between the effects of aluminum (Al) on the growth of tea plants and the active oxygen species scavenging system in root tips of intact tea plants and suspension-cultured tea cells was examined. Compared with the control, the activities of superoxide dismutase, catalase, and ascorbate peroxidase increased by Al both in roots of intact plants and cultured cells. The level of lipid peroxidation of membrane, the activity of membrane bound peroxidase, and the content of lignin and cell wall-bound phenols were reduced by the treatment with Al in cultured tea cells (Ghanati et al. 2005).

Olea europaea **L.** The effects of water recovery on the activities of superoxide dismutase, catalase, ascorbate peroxidase, guaiacol peroxidase, polyphenol oxidase, and lipoxygenase and on malondialdehyde levels were investigated in 2-year-old *Olea europaea* L. (cv. "Coratina") plants grown in environmental conditions characterized by high temperatures and irradiance levels and gradually subjected to a controlled water deficit. After reaching the maximum level of water stress, plants were subjected to a rewatering treatment for 30 days, under both environmental irradiance and semi-shade conditions.

The activities of SOD, CAT, APX, POD and LOX, and MDA levels decreased during the rewatering period in both leaves and roots, and these decrements were faster in plants rewatered in semi-shade conditions (SHP) than in plants under environmental light (NSHP). In contrast, PPO activity increased during rewatering in both leaf and root tissues. Thus, the lower expression of the enzymatic antioxidant system in SHP with respect to NSHP may be due to a reduced need of activated oxygen species removal (Sofo et al. 2004).

Since SOD and APX are the main antioxidant enzymes of chloroplasts (Alscher et al. 2002; Mehlhorn et al. 1996), the more marked reduction of SOD and APX activities in SHP with respect to that in NSHP at the same rewatering level suggested that lower PPFD levels induce a different response of olive tree to oxidative stress because in semi-shade conditions, the need of antioxidant defenses is reduced. Different POD isoforms have a higher affinity for H_2O_2 if compared with CAT but require some phenolic compounds (e.g., guaiacol) as substrates (Mehlhorn et al. 1996; Sofo et al. 2004).

***Plantago* spp.** Activity of antioxidant enzymes such as superoxide dismutase, catalase, and peroxidase in leaves of isabgol (*Plantago ovata* Forsk.) was studied under salt stress. Salt stress caused significant increase in the activity of superoxide dismutase, catalase, and peroxidase in the leaves of isabgol. All the isabgol genotypes responded differently with respect to CAT and POD activity under salt stress. At biochemical level, SOD, CAT, and POD are major antioxidant enzymes associated with scavenging reactive oxygen species (ROS), and SOD is likely to be central in the defense against toxic ROS (Marschner 1995). However, SOD detoxifies superoxide anion free radicals accompanying the formation of H_2O_2 which is very damaging to the chloroplasts, nucleic acids, and proteins and can be eliminated by CAT and POD (Marschner 1995). An increase in the antioxidant enzymes under salt stresses could be indicative of an increased production of ROS and buildup of a protective mechanism to reduce oxidative damage triggered by stress in plants. Catalase in peroxisomes breaks down H_2O_2. Peroxidase in cytosol and chloroplast can perfectly scavenge H_2O_2 (Kala 2015). Increase of peroxidase activity by salt treatment in plants has also been reported by Kahrizi et al. (2012).

Effect of drought and salinity on growth, development, and yield of isabgol HI-5 has been investigated by Surekha (1997) and Varshney and Surekha (2001). Vandana (2003) screened five isabgol genotypes, viz., HI-5, HI-34, HI-96, GI-2, and PB-80, for salt tolerance. Among these genotypes GI-2 and HI-96 were found salt tolerant while PB-80 and HI-5 salt-sensitive on the basis of growth, development, and yield parameters (Kala 2015).

Salt-tolerant *Plantago maritima* and salt-sensitive *Plantago media* were studied for plant growth, relative water content, stomatal conductance, lipid peroxidation, and antioxidant system in relation under salt stress. Reduction in shoot length was higher in *P. media* than in *P. maritima*, and shoot dry weight decreased in *P. media* and did not change in *P. maritima*. There was reduction in RWC and stomatal conductance in *P. media*, whereas no effect was seen on leaf RWC in *P. maritima*, and negligible reduction in stomatal conductance was observed. Activities of superoxide dismutase, catalase, and glutathione reductase decreased in *P. media* with increasing salinity. Ascorbate peroxidase activity in leaves of *P. media* was increased. However, activities of CAT, APX, and GR increased at high concentration of NaCl, while their activities did not change at low concentrations in *P. maritima*. SOD activity in leaves of *P. maritima* increased with increasing salinity. Concomitant with this, four SOD activity bands were identified in leaves of *P. maritima*; two bands only were observed in *P. media*. Peroxidase activity increased under both salt concentrations in *P. maritima* but only at lower concentrations in *P. media*. Confirming this, five POX activity bands were identified in leaves of *P. maritima*, but only two bands were determined in *P. media*. Malondialdehyde levels in the leaves increased under salt stress in *P. media* but showed no change and decreased in *P. maritima* (Sekmen et al. 2007).

Three *Plantago* species, *P. crassifolia* and *P. coronopus* both halophytes and *P. major* (salt-sensitive), were studied for salt stress. It was seen that *P. major* was quite resistant to salt stress on the basis of growth parameters. Salt-treated plants of the three taxa accumulated Na^+ and Cl^- in response to increasing NaCl concentrations to a lesser extent in *P. major* than in the halophytes. In the halophytes, K^+

concentration decreased at moderate salinity levels but increased at high salt conditions, whereas in *P. major* K^+ contents were reduced at NaCl stress.

The content of the common osmolytes in plants – proline (Pro), glycine betaine (GB), and total soluble sugars (TSS) – was evaluated in leaves treatments with salt. Pro contents showed no significant increase in controls, but stronger salt stress conditions induced the accumulation of this osmolyte in *P. crassifolia*; the induction of Pro biosynthesis was even stronger in *P. coronopus*, while no increase over the control was detected in *P. major* under the same conditions. In the case of GB, it accumulated in the leaves of *Plantago* plants to maximum concentrations in *P. crassifolia* and *P. major*, and it did not increase significantly at higher salt concentrations in *P. coronopus*. Total soluble sugar levels decreased moderately in *P. coronopus* and *P. crassifolia* with increasing NaCl concentrations, while no significant change was detected in *P. major*. In the presence of NaCl, sorbitol levels increased in leaves of all tested *Plantago* species in a concentration-dependent manner (Al Hassan et al. 2016).

***Scutellaria baicalensis* L.** *Scutellaria baicalensis* is a traditional Chinese medicinal plant, but increasing average annual temperatures have made plants unsuitable for medicinal use. Two flavones, baicalin and baicalein, are the major active ingredients of *S. baicalensis*. It was demonstrated that protracted heat treatment inhibited the accumulation of baicalin and baicalein as well as the activity of phenylalanine ammonia-lyase (PAL). PAL is involved in the phenylpropanoid pathway, which produces baicalin in the plant.

Heat treatment also affected the activities of the antioxidant enzymes such as catalase, superoxide dismutase, and peroxidase. Cells continued growing during the protracted heat stress. Long-term exposure to high temperatures did not affect *S. baicalensis* cell growth but inhibited flavonoid biosynthesis and reduced the content of baicalin and baicalein. These two compounds play important roles in the balance between ROS and antioxidant enzyme activities in adaptive responses to high heat (Yuan et al. 2011).

***Glycyrrhiza uralensis* L.** *Glycyrrhiza uralensis* seeds were germinated and grown with different concentrations of cadmium acetate, in order to investigate the effects of cadmium on the growth, uptake, SOD, POD, CAT, PPO, and PAL activities in *Glycyrrhiza uralensis* seedlings. Results suggested that increased cadmium concentrations lead to decreased shoot elongation and seedling biomass. SOD activity in the cotyledons, hypocotyls, and radicles increased gradually. POD activity in the cotyledons, hypocotyls, and radicles concentrations increased continuously with rising cadmium concentrations. CAT activity in the cotyledons, hypocotyls, and radicles increased gradually with increasing cadmium concentrations. PPO activity showed significant increases in the cotyledons, hypocotyls, and radicles. A significant change of PAL activity in the cotyledons, hypocotyls, and radicles was observed with increasing cadmium concentrations (Zheng et al. 2010).

***Pfaffia glomerata* L.** The role of the antioxidant enzymes in adaptive responses of the accumulator *P. glomerata* species under cadmium (Cd) stress was studied. The lipid peroxidation rates in leaves and roots were smaller at the start of the experiment for all Cd levels. SOD activity increased in leaves and in roots as Cd levels increased. Cd stress induced an increase in the activity of APX in leaves, whereas in roots APX activity was reduced at high concentration of Cd. At the end of the experiment, CAT activity in leaves was reduced as Cd concentration increased. Nevertheless, the GR and GPX activities increased. In roots, GR activity was reduced (Marques and Soares 2011).

Oxidative stress caused by mercury (Hg) was investigated in *Pfaffia glomerata* plantlets. Accumulation of Hg in tissue increased with increasing concentration of Hg. Root and shoot fresh weight and delta-ALA-D activity were significantly decreased, and chlorophyll and carotenoid concentrations were not affected at high concentration of Hg. H_2O_2 concentration in shoot increased curvilin early with higher level of mercury, whereas lipid peroxidation increased in roots and shoots. SOD activity showed a straight correlation with H_2O_2 concentration, whereas CAT activity increased in shoots. Shoot APX activity was either decreased at low Hg concentration or increased at high Hg concentration. Conversely, root APX activity was only increased at low Hg concentration. In general, AsA, non-protein thiols (NPSH), and proline concentrations increased upon addition of Hg, with the exception of proline in roots, which decreased (Calgaroto et al. 2010).

***Bacopa monnieri* L.** *Bacopa monnieri* L. plants were exposed to cadmium (Cd) stress and analyzed for the accumulation of metal and its influence on various enzymatic and nonenzymatic antioxidants, TBARS, photosynthetic pigments, and protein content. The accumulation of Cd was found to be increased with increasing concentration and time duration, and more Cd was accumulated in the root. TBARS content of the roots and leaves increased with increase in Cd concentration and exposure time. Enhancement in the activities of SOD, APX, and GPX was recorded in stressed roots and leaves of *B. monnieri*. A considerable decrease in CAT activity in Cd-treated *B. monnieri* was seen. There is an increase in the cysteine amino acid and non-protein thiol contents of the roots in *B. monnieri* followed by a decline but in leaves, cysteine and non-protein thiol contents were found to be enhanced at all the Cd concentrations and exposure periods. A significant reduction in the level of ascorbic acid, total chlorophyll, and protein contents was observed. *B. monnieri* was able to combat metal-induced oxidative injury involving a mechanism of activation of various enzymatic and nonenzymatic antioxidants (Singh 2006).

***Salicornia brachiate* L.** The effect of salinity stress was studied on the activities of antioxidant enzymes and polyphenol content in *S. brachiata*. Polyphenol content was found to increase in *S. brachiata* treated with waste water of tannery. PPO activity of *S. brachiata* was increased and then significantly declined under stress. CAT activity of the control and the experimental plants of *S. brachiata* showed an increase initially and then drastically reduced when compared to the control. The results indicate a decline in CAT activity under extreme salinity, which suggests that

CAT appears not to be an efficient scavenger of H_2O_2 in *S. brachiata*. The activity of SOD increased with increasing concentrations of NaCl during the growth period (Santhanakrishnan et al. 2014).

***Zygophyllum* Species** Two *Zygophyllum* species (*Z. album* and *Z. coccineum*) were grown, and the effects of soil heavy metal stress on shoot heavy metal concentrations, lipid peroxidation, antioxidant enzyme activities, and the root plasma membrane (PM) lipid composition were analyzed. Heavy metal concentrations and lipid peroxidation increased in the shoot of both species grown in the polluted area. The activities of ascorbate oxidase (ASO), guaiacol peroxidase (GPX), ascorbate peroxidase (APX), and superoxide dismutase (SOD) were increased, whereas those of catalase (CAT) were decreased in both species under the polluted conditions. PM total lipids, phospholipids, glycolipids, and sterols were decreased in *Z. album* and *Z. coccineum* as a result of the polluted soil. Heavy metal stress increased phosphatidylethanolamine (PE) and decreased phosphatidylinositol (PI) and phosphatidylglycerol (PG), with no significant change in phosphatidylcholine (PC) in the root PM of both species. Phosphatidylserine (PS) decreased in the PM of *Z. album*, whereas it increased in the PM of *Z. coccineum* under the pollution conditions. Heavy metal stress changed the composition and concentration of fatty acids of the root PM, resulting in increased saturated/unsaturated ratio of both species (Morsy et al. 2012).

***Trifolium resupinatum* L.** Different concentrations of SO_2 had a significant effect on Persian clover root weight and antioxidant system. Increasing SO_2 stress decreased root fresh and dry weight and antioxidant capacities (IC_{50}) and increased antioxidant activities (I%) of Persian clover leaves significantly in comparison to the control plants (under 0 ppm) and increased SOD, CAT, and GPX activity. Inoculation of Persian clover plants with native and standard *Rhizobium* increased root weight and did not show a significant effect on antioxidants activity and capacity, but interaction between *Rhizobium* inoculation and SO_2 treatment reduced significantly the stress effects of high concentration of SO_2 on root growth and antioxidants activity and capacity. In fact, level of this change of root growth and antioxidant system under SO_2 pollution stress in inoculated plants was lower than in the non-inoculated plants. As a result, an increase in SO_2 concentration caused a decrease in root weight and increase in antioxidants activity and capacity of Persian clover. Inoculation with *Rhizobium* strains could alleviate the effect of SO_2 pollution on antioxidant system by effects on root growth (Bayat et al. 2014).

***Thymus vulgaris* L.** In this study, a pot experiment was conducted to assess the effect of drought on the antioxidant activity of thyme. The FRAP (ferric reducing antioxidant power) and DPPH (2,2-diphenylpicrylhydrazyl) scavenging assays. Thus results indicated that severe water stress significantly decreased the antioxidant activity of thyme. A higher value of IC_{50} showed a lower antioxidant activity, which indicated that severe water stress significantly decreased the antioxidant activity of thyme. FRAP values showed a trend to reduction by increasing the irrigation intervals, but this trend was not significant. FRAP value and total phenolic contents of samples were decreased with increase in the duration of irrigation intervals, but the differences were not significant. Control showed higher values and 8-day interval had lower values. The results of the DPPH assay showed that with water deficit conditions, IC50 values of samples were increased significantly. IC50 values were increased from control to severe stress condition (Khosh-Khui et al. 2012).

The effect of temperature and extraction process on the antioxidant activity of various organic crude extracts from the leaves of *Thymus vulgaris* species native to Sultanate of Oman was evaluated. The antioxidant activity of different crude extracts from both extraction methods was measured by DPPH with modification. By Soxhlet extraction method, the activity result found in butanol crude extracts was highest and the lowest in hexane crude extract shown in the following order: butanol>methanol>ethyl acetate extract>chloroform>hexane extract. However, by maceration method, the activity was highest in ethyl acetate and lowest in chloroform: ethyl acetate>methanol extract>butanol>hexane >chloroform (Rahman 2013).

***Elodea* (Egeria) *densa* Planch** *Elodea* plants were incubated in the presence of individual and mixed sulfate salts of Ni, Cd, Cu, Zn, and Mn to study the influence of heavy metals (HM) on shoot growth, structural and functional parameters of the photosynthetic apparatus, lipid peroxidation, enzymatic activities of the antioxidant defense system (superoxide dismutase and catalase), and the content of non-protein (NPSH) and protein thiols (PSH) in leaves.

The accumulation of HM in leaves decreased in a row: Mn>Cu>Cd>Zn>Ni. The largest reduction in chlorophyll content was caused by Mn and Cu, whereas the strongest reduction in carotenoid content was induced by Cu. The presence of Cu produced the largest decrease in the maximal quantum efficiency of photosystem II (PSII) (Fv/Fm). The presence of Cd elevated the content of chlorophyll and carotenoids without altering the photochemical efficiency of PSII; Cd retarded the shoot growth but had no appreciable effect on leaf mesostructure. The addition of the second metal to the growth medium alleviated in most treatments the detrimental action of individual ions owing to the enhanced activities of SOD and catalase and because of the significant increase in the content of NPSH. It is supposed that the observed antagonism of metal ions is related to their competitive interactions restricting the entry of HM into the cell.

The chloroplast dimensions in elodea cells showed no uniform change under the action of HM. The addition of Ni caused a significant reduction of chloroplast volume (more than a 2.5-fold decrease compared to control values). A similar effect was noted under combined application of all metals examined (the reduction by 1.4 times). On the other hand, the long-term exposure (68 days) of plants to Cd or Cd + Ni induced the reliable increase in the chloroplast volume, which was likely caused by the chloroplast swelling. It is not excluded that the presence of Cd resulted in partial destruction of chlorophyll, which was evident from the pale leaf color.

The addition of Ni, Zn, and Mn to the growth medium elevated SOD activity by 20% on the average compared to control values. Under combined application of most metals, SOD activity was substantially higher, especially in the presence of Ni + Cd, Ni + Zn, and Ni + Mn combinations. The heavy metals examined had little influence on CAT activity when applied individually. However, the combined application of two metals enhanced CAT activity by 1.5–2.0 times, with an exception of Mn + Cd and Mn + Cu combinations.

Synthesis of NPSH and PSH is a known means of plant protection against deleterious action of HM. The content of non-protein thiols increased significantly in the presence of individual HM and their combinations. The largest increase was observed in the presence of metal pair Mn + Cu. Non-protein thiols, reduced glutathione (GSH) in particular, play an active role in membrane protection against free radical damage (Maleva et al. 2012).

***Jatropha curcas* L.** In the present study, the effects of aluminum (Al) concentrations on growth, superoxide dismutase, peroxidase, catalase, and phenylalanine ammonia-lyase activities in *Jatropha curcas* L. seedlings were investigated. It was seen that with the increasing Al concentrations, the biomass of cotyledons increased initially and then decreased, but the biomass of hypocotyls and radicles decreased gradually. Compared to the control, SOD activity in the cotyledons, hypocotyls, and radicles was all enhanced by Al stress. SOD activity in the hypocotyls increased significantly with increasing Al concentrations. The pattern of SOD isoforms was analyzed by native PAGE, and activity staining revealed that at least four SOD isoenzyme bands in the cotyledons, hypocotyls, and radicles were detected, respectively. Al stress significantly affected the POD activity in the cotyledons showing significant increase. On the activity gels, at least six bands in the cotyledons, hypocotyls, and radicles were observed. POD isoenzyme (II and III) in the cotyledons showed an increase in the staining intensities with the increasing of Al concentration. In the hypocotyls and radicles, the main increase in the staining intensities was isoenzyme IV and III, respectively. Compared to control, CAT activities in hypocotyls and radicles were all increased, while in cotyledons, CAT activities were increased first and then decreased with the increasing Al concentration. Compared to the control, the PAL activities were all increased, but the change trends were different. In the cotyledons and radicles, PAL activities were increased first and then decreased with the increasing Al concentration (Ou-yang et al. 2014).

***Ctenanthe setosa* (Rosc.) Eichler** The relationship of the antioxidant enzyme to drought stress tolerance was studied during leaf rolling in the leaf, petiole, and root of *Ctenanthe setosa*. Chlorophyll and carotenoid content and the chlorophyll stability index decreased in the early period of drought stress but increased in later periods, approaching the control level as leaf rolling increased. Relative water content decreased, while the root/shoot ratio increased during drought stress. LPO measured as MDA content also increased and then declined in the same drought period, contrary to photosynthetic pigment content. SOD activity did not significantly change in leaves. In the petiole and root, however, it decreased in the early drought period but increased later. GR activity did not significantly change in the leaf and petiole versus the control but increased in root. POD activity increased in the leaf and petiole but decreased in the root. A peroxidase isoenzyme activity band present in the control leaves did not appear in leaves exposed to drought, but in the latter periods, that activity increased. Tolerance of drought stress apparently is closely associated with the antioxidant enzyme system as well as leaf rolling in *C. setosa* (Terzi and Kadioglu 2006).

***Cleome gynandra* L.** The effects of heavy metals on antioxidant defense system were studied in *Cleome gynandra* plants. The decreased value of phenolics with increased concentration of heavy metal copper and cadmium shows that phenol form chelation with metal and thereby reduce the toxicity of the plant during accumulation of the metal. Antioxidant activity was found to be maximum in the plant exposed to control soil, while free radical scavenging activity was reduced much in the plant exposed to heavy metal-contaminated soils. The proline value was considerably increased with increased concentration of copper and cadmium. Superoxide dismutase, catalase, and glutathione were increased significantly in the plant sample exposed to heavy metal-contaminated soils. Among the two metals, cadmium affects the plant to a greater extent than copper (Haribabu and Sudha 2011).

***Catharanthus roseus* (L.) G. Don.** *Catharanthus roseus* (L.) G. Don. was studied for salinity stress, and the ability of triadimefon (TDM), a triazole group of fungicide, to ameliorate the stress was also studied. There was decreased overall growth of this plant and reduced chlorophyll content, protein, and antioxidant enzymes such as POX, SOD, and PPO. The root alkaloid ajmalicine increased under salt treatment. When these stressed plants were treated with TDM, it minimized the injurious effects of NaCl stress by increasing the root and shoot growth, leaf area, dry weight (DW), chlorophyll and protein contents, and the activities of antioxidant enzymes like POD, SOD, and PPO. The quantity of ajmalicine was also increased with the TDM treatment when compared to both control and NaCl-treated plants (Jaleel et al. 2008).

Antioxidant responses were analyzed in *Catharanthus roseus* under salt stress in order to investigate the plant's protective mechanisms against long-term salt-induced oxidative stress. High salinity caused a decrease in reduced glutathione and an enhancement in total ascorbate content and the antioxidant enzyme and ascorbate peroxidase activities. Moreover, salinity induced a significant decline in superoxide dismutase and peroxidase activities. The changes found in catalase activities may be of great importance in the H_2O_2 detoxification mechanism under oxidative stress (Jaleel et al. 2007).

The effect of responses of *Catharanthus roseus* to NaCl stress has been explored. The plants were exposed to different concentrations of salt and the effect of treatment on germination, growth parameter, and antioxidant defense system investigated. Increasing the NaCl concentration reduced germination percentage, and the fresh and dry weights of treated plants also showed a decrease. Ascorbic acid content increased in the presence of stress, and glutathione concentration showed a significant increase. NaCl caused a significant decrease of SOD activity and enhanced the activities of catalase, peroxidase, and glutathione reductase. The MDA content increased with the increasing concentrations of NaCl. MDA content of samples treated by NaCl also increased (Amirjani 2015).

2.1.4 Medicinal Plants and Their Antioxidant Properties

In addition to providing defense, plants have long been a source of exogenous (i.e., dietary) antioxidants. It is believed that two-thirds of the world's plant species have medicinal importance, and almost all of these have excellent antioxidant potential (Krishnaiah et al. 2011). The in vitro evaluation of antioxidant activity of medicinal plants or their phytochemicals several biochemical tests have been used. In ethanopharmacological and nutraceutical investigations, these assays are done to understand the probable mechanism of action of plant antioxidants (Antolovich et al. 2002) in minimizing the oxidative stress linked pathophysiology of diseases. There are several in vitro assays used to measure and confer antioxidant activity to plants; however, each of these has its own limitations regarding applicability. In these assays, plants are generally assessed for their function as reducing agents, hydrogen donors, singlet oxygen quenchers, or metal chelators, after which they are classified as primary (chain-breaking) and secondary (preventive) antioxidants. Primary antioxidants act by donating a hydrogen atom, while secondary antioxidants function via binding of metal ions capable of catalyzing oxidative processes and scavenging oxygen, absorbing UV radiation, inhibiting enzymes, or decomposing hydroperoxides (Kasote 2013).

Since time immemorial, plants have been a source of food and medicines, either in the form of traditional preparations or as pure active principles (Hegde et al. 2014). Most of the medicinal effects of plants have been attributed to their potent antioxidant activity. It has been suggested that free radicals are involved in the pathology of more than 50 human diseases, including aging (Halliwell 1991). Plants are rich storehouse of secondary metabolites, and the complex diversity of these metabolites makes them fascinating candidates for study. Plant antioxidants such as ascorbic acid and flavonoids have been shown to be the best exogenous antioxidants. Indeed, these compounds not only restrain ROS production by scavenging free radicals but also help boost endogenous antioxidant defenses of the body (Halliwell 2006).

The chemical structure of polyphenols is responsible for its antioxidant potential as they determine the conjugation reactions with methyl, sulfate, or glucuronide groups (Scalbert and Williamson 2000). Flavonoids are the most important and abundant dietary polyphenols, with over 5000 reported to date (Ross and Kasum 2002; Dai and Mumper 2010). In addition to their remarkable antioxidant property, polyphenols have pro-oxidant properties also.

2.2 Conclusion

Plants are exposed to harsh climatic and environmental conditions which lead to stress. Abiotic stresses overproduce ROS in plants which are highly unstable and toxic to the cells and leads to oxidative damage. Manageable amounts of ROS are produced during normal metabolic processes, but excessive amounts damage nucleic acids, lipids, and proteins, causing them to lose their activity. Since plants are sessile, they need to be equipped with excellent antioxidant defense mechanisms to detoxify the harmful effects of ROS. The antioxidant defenses could be either enzymatic (e.g., superoxide dismutase, catalase, peroxidases, and glutathione reductase) or nonenzymatic (e.g., glutathione, glycine betaine, proline, α-tocopherols, phenols, carotenoids, and flavonoids).

References

Agrawal SB, Rathore D (2007) Changes in oxidative stress defense system in wheat (*Triticum aestivum* L.) and mung bean (*Vigna radiata* L.) cultivars grown with and without mineral nutrients and irradiated by supplemental ultraviolet-B. Environ Exp Bot 59(1):21–33

Al Hassan M, Pacurar A, López-Gresa MP, Donat-Torres MP, Llinares JV, Boscaiu M, Vicente O (2016) Effects of salt stress on three ecologically distinct *Plantago* species. PLoS One 11(8):e0160236

Allen RD (1995) Dissection of oxidative stress tolerance using transgenic plants. Plant Physiol 107(4):1049

Almeselmani M, Deshmukh PS, Sairam RK, Kushwaha SR, Singh TP (2006) Protective role of antioxidant enzymes under high temperature stress. Plant Sci 171(3):382–388

Almeselmani M, Deshmukh P, Sairam R (2009) High temperature stress tolerance in wheat genotypes: role of antioxidant defense enzymes. Acta Agron Hung 57(1):1–14

Alscher RG, Erturk N, Heath LS (2002 May 15) Role of superoxide dismutases (SODs) in controlling oxidative stress in plants. J Exp Bot 53(372):1331–1341

Amirjani MR (2015) Effect of salinity stress on seed germination and antioxidative defense system of *Catharanthus roseus*. ARPN J Agric Biol Sci 10:163–171

Antolovich M, Prenzler PD, Patsalides E, McDonald S, Robards K (2002) Methods for testing antioxidant activity. Analyst 127(1):183–198

Armstrong W, Drew MC (2002) Root growth and metabolism under oxygen deficiency. In: Plant roots: the hidden half, 3rd edn. Marcel Dekker, New York, pp 729–761

Asada K (1999) The water-water cycle in chloroplasts: scavenging of active oxygens and dissipation of excess photons. Ann Rev Plant Biol 50(1):601–639

Basu U (2012) Identification of molecular processes underlying abiotic stress plants adaptation using, in: Benkeblia, N. (Ed) "Omics" technologies. In: Sustainable agriculture and new technologies. CRC, Boca Raton, pp 149–172

Bayat L, Askari M, Fariba A, Morteza Z (2014) Effects of Rhizobium inoculation on *Trifolium resupinatum* antioxidant system under sulfur dioxide pollution. Biol J Microorg 2(8):37–50

Bazin MJ, Markham P, Scott EM, Lynch JM (1990) Population dynamics and rhizosphere interactions. In: Lynch JM (ed) The rhizosphere, pp 99–127

Bennoun P (1994) Chlororespiration revisited: mitochondrial-plastid interactions in *Chlamydomonas*. Biochim Biophys Acta (BBA) Bioenerg 1186(1):59–66

Biemelt S, Keetman U, Mock HP, Grimm B (2000) Expression and activity of isoenzymes of superoxide dismutase in wheat roots in response to hypoxia and anoxia. Plant Cell Environ 23(2):135–144

Blokhina OB, Fagerstedt KV, Chirkova TV (1999) Relationships between lipid peroxidation and anoxia tolerance in a range of species during post-anoxic reaeration. Physiol Plant 105(4):625–632

Blokhina OB, Chirkova TV, Fagerstedt KV (2001) Anoxic stress leads to hydrogen peroxide formation in plant cells. J Exp Bot 52(359):1179–1190

Boyer JS (1982) Plant productivity and environment. Science 218(4571):443–448

Burbott AJ, Loomis WD (1969) Evidence for metabolic turnover of monoterpenes in peppermint. Plant Physiol 44(2):173–179

Burke JJ, Gamble PE, Hatfield JL, Quisenberry JE (1985) Plant morphological and biochemical responses to field water deficits I. Responses of glutathione reductase activity and paraquat sensitivity. Plant Physiol 79(2):415–419

Caldwell MM, Björn LO, Bornman JF, Flint SD, Kulandaivelu G, Teramura AH, Tevini M (1998) Effects of increased solar ultraviolet radiation on terrestrial ecosystems. J Photochem Photobiol B Biol 46(1):40–52

Calgaroto NS, Castro GY, Cargnelutti D, Pereira LB, Gonçalves JF, Rossato LV, Nicoloso FT (2010) Antioxidant system activation by mercury in Pfaffia glomerata plantlets. Biometals 23(2):295–305

Campbell SA, Nishio JN (2000) Iron deficiency studies of sugar beet using an improved sodium bicarbonate-buffered hydroponic growth system. J Plant Nutr 23(6):741–757

Caplan A, Claes B, Dekeyser R, Van Montagu M (1990) Salinity and drought stress in rice. R.S. Sangwan and B.S. Sangwan-Norreel (eds.). The impact of biotechnology in agriculture. © 1990 Kluwer Academic Publishers, pp 391–402

Castillo FJ, Greppin H (1988) Extracellular ascorbic acid and enzyme activities related to ascorbic acid metabolism in *Sedum album* L. leaves after ozone exposure. Environ Exp Bot 28(3):231–238

Chandra A, Bhatt RK, Misra LP (1998) Effect of water stress on biochemical and physiological characteristics of oat genotypes. J Agron Crop Sci 181(1):45–48

Chaves MM, Maroco JP, Pereira JS (2003) Understanding plant responses to drought—from genes to the whole plant. Funct Plant Biol 30(3):239–264

Chen S, Xing J, Lan H (2012) Comparative effects of neutral salt and alkaline salt stress on seed germination, early seedling growth and physiological response of a halophyte species *Chenopodium glaucum*. Afr J Biotechnol 11(40):9572–9581

Chen J, Wang WH, Liu TW, Wu FH, Zheng HL (2013) Photosynthetic and antioxidant responses of *Liquidambar formosana* and *Schima superba* seedlings to sulfuric-rich and nitric-rich simulated acid rain. Plant Physiol Biochem 64:41–51

Chirkova TV, Novitskaya LO, Blokhina OB (1998) Lipid peroxidation and antioxidant systems under anoxia in plants differing in their tolerance to oxygen deficiency. Russ J Plant Physiol 45(1):55–62

Covello PS, Chang A, Dumbroff EB, Thompson JE (1989) Inhibition of photosystem II precedes thylakoid membrane lipid peroxidation in bisulfite-treated leaves of *Phaseolus vulgaris*. Plant Physiol 90(4):1492–1497

Dai J, Mumper RJ (2010) Plant phenolics: extraction, analysis and their antioxidant and anticancer properties. Molecules 15(10):7313–7352

Davies KJ (1987) Protein damage and degradation by oxygen radicals. I. General aspects. J Biol Chem 262(20):9895–9901

Dhindsa RS (1991) Drought stress, enzymes of glutathione metabolism, oxidation injury, and protein synthesis in *Tortula ruralis*. Plant Physiol 95(2):648–651

Dhindsa RS, Matowe W (1981) Drought tolerance in two mosses: correlated with enzymatic defense against lipid peroxidation. J Exp Bot 32(1):79–91

Egneus H, Heber U, Matthiesen U, Kirk M (1975) Reduction of oxygen by the electron transport chain of chloroplasts during assimilation of carbon dioxide. Biochim Biophys Acta (BBA) Bioenerg 408(3):252–268
Elstner EF (1982) Oxygen activation and oxygen toxicity. Ann Rev Plant Physiol 33(1):73–96
Fischer BB, Krieger-Liszkay A, Hideg É, Šnyrychová I, Wiesendanger M, Eggen RI (2007) Role of singlet oxygen in chloroplast to nucleus retrograde signaling in *Chlamydomonas reinhardtii*. FEBS Lett 581(29):5555–5560
Fridovich I (1986) Biological effects of the superoxide radical. Arch Biochem Biophys 247(1):1–11
Ghanati F, Morita A, Yokota H (2005) Effects of aluminum on the growth of tea plant and activation of antioxidant system. Plant Soil 276(1–2):133–141
Gill SS, Khan NA, Anjum NA, Tuteja N (2011) Amelioration of cadmium stress in crop plants by nutrients management: morphological, physiological and biochemical aspects. Plant Stress 5(1):1–23
Gjorgieva D, Panovska TK, Ruskovska T, Bačeva K, Stafilov T (2013) Influence of heavy metal stress on antioxidant status and DNA damage in *Urtica dioica*. Biomed Res Int 2013. Article Id 276417:6
Golemiec E, Tokarz K, Wielanek M, Niewiadomska E (2014) A dissection of the effects of ethylene, H2O2 and high irradiance on antioxidants and several genes associated with stress and senescence in tobacco leaves. J Plant Physiol 171(3):269–275
Gong B, Wen D, Bloszies S, Li X, Wei M, Yang F, Wang X (2014) Comparative effects of NaCl and NaHCO3 stresses on respiratory metabolism, antioxidant system, nutritional status, and organic acid metabolism in tomato roots. Acta Physiol Plant 36(8):2167–2181
Grover A, Kapoor A, Lakshmi OS, Agarwal S, Sahi C, Katiyar-Agarwal S, Dubey H (2001) Understanding molecular alphabets of the plant abiotic stress responses. Curr Sci 80(2):206–216
Guo JH, Liu XJ, Zhang Y, Shen JL, Han WX, Zhang WF, Zhang FS (2010) Significant acidification in major Chinese croplands. Science 327(5968):1008–1010
Halliwell B (1991) Drug antioxidant effects. Drugs 42(4):569–605
Halliwell B (2006) Reactive species and antioxidants. Redox biology is a fundamental theme of aerobic life. Plant Physiol 141(2):312–322
Haribabu TE, Sudha PN (2011) Effect of heavy metals copper and cadmium exposure on the antioxidant properties of the plant *Cleome gynandra*. Int J Plant Anim Environ Sci 1(2):80–87
Hartung W, Leport L, Ratcliffe RG, Sauter A, Duda R, Turner NC (2002) Abscisic acid concentration, root pH and anatomy do not explain growth differences of chickpea (*Cicer arietinum* L.) and lupin (*Lupinus angustifolius* L.) on acid and alkaline soils. Plant Soil 240(1):191–199
Hasanuzzaman M, Hossain MA, da Silva JAT, Fujita M (2012) Plant response and tolerance to abiotic oxidative stress: antioxidant defense is a key factor. In: Bandi V, Shanker AK, Shanker C, Mandapaka M (eds) Crop stress and its management: perspectives and strategies. Springer, Dordrecht, pp 261–315
He H, Ding Y, Bartlam M, Sun F, Le Y, Qin X, Tang H, Zhang R, Joachimiak A, Liu J, Zhao N, Rao Z (2003) Crystal structure of tabtoxin resistance protein complexed with acetyl coenzyme a reveals the mechanism for beta-lactam acetylation. J Mol Biol 325(5):1019–1030
Heagle AS, Annu R (1989) Ozone and crop yield. Phytopathology 27:397–423
Heath RL (1987) The biochemistry of ozone attack on the plasma membrane of plant cells. Recent Adv Phytochem 21:29–54
Hegde PK, Rao HA, Rao PN (2014) A review on Insulin plant (*Costus igneus* Nak). Pharmacogn Rev 8(15):67
Heidari M, Golpayegani A (2012) Effects of water stress and inoculation with plant growth promoting rhizobacteria (PGPR) on antioxidant status and photosynthetic pigments in basil (*Ocimum basilicum* L.). J Saudi Soc Agric Sci 11(1):57–61
Hernandez JA, Corpas FJ, Gomez M, Río LA, Sevilla F (1993) Salt-induced oxidative stress mediated by activated oxygen species in pea leaf mitochondria. Physiol Plant 89(1):103–110
Hernandez JA, Olmos E, Corpas FJ, Sevilla F, Del Rio LA (1995) Salt-induced oxidative stress in chloroplasts of pea plants. Plant Sci 105(2):151–167

Hossain MA, AL-Raqmi KAS, AL-Mijizy ZH, Weli AM, Al-Riyami Q (2013 Sep) Study of total phenol, flavonoids contents and phytochemical screening of various leaves crude extracts of locally grown *Thymus vulgaris*. Asian Pac J Trop Biomed 3(9):705–710

Hu WH, Song XS, Shi K, Xia XJ, Zhou YH, Yu JQ (2008) Changes in electron transport, superoxide dismutase and ascorbate peroxidase isoenzymes in chloroplasts and mitochondria of cucumber leaves as influenced by chilling. Photosynthetica 46(4):581–588

Huang M, Guo Z (2005) Responses of antioxidative system to chilling stress in two rice cultivars differing in sensitivity. Biol Plant 49(1):81–84

Jackson MB (1982) Ethylene as a growth promoting hormone under flooded conditions. In: Wareing PF (ed) Plant growth substances, 1982: the proceedings of the 11th international conference on plant growth substances, Aberystwyth, 12–16 July, 1982. Academic, London

Jackson MB (2004) The impact of flooding stress on plants and crops. Available at: http://www.plantstress.com/Articles/waterlogging_i/waterlog_i.htm

Jackson MB, Ram PC (2003) Physiological and molecular basis of susceptibility and tolerance of rice plants to complete submergence. Ann Bot 91(2):227–241

Jaleel CA, Gopi R, Manivannan P, Panneerselvam R (2007) Responses of antioxidant defense system of *Catharanthus roseus* (L.) G. Don. to paclobutrazol treatment under salinity. Acta Physiol Plant 29(3):205–209

Jaleel CA, Gopi R, Kishorekumar A, Manivannan P, Sankar B, Panneerselvam R (2008) Interactive effects of triadimefon and salt stress on antioxidative status and ajmalicine accumulation in *Catharanthus roseus*. Acta Physiol Plant 30(3):287

Kahrizi S, Sedighi M, Sofalian O (2012) Effect of salt stress on proline and activity of antioxidant enzymes in ten durum wheat cultivars. Ann Biol Res 3(8):3870–3874

Kala S (2015) Effect of Nacl salt stress on antioxidant enzymes of Isabgol (*Plantago ovata* Forsk.) genotypes. Int J Food Sci Technol 4(2):40–43

Kalashnikov JE, Balakhnina TI, Zakrzhevsky DA (1994) Effect of soil hypoxia on activation of oxygen and the system of protection from oxidative destruction in roots and leaves of *Hordeum vulgare*. Russ J Plant Physiol 41:583–588

Kasote DM (2013) Flaxseed phenolics as natural antioxidants. Int Food Res J 20(1):27–34

Khan NA, Singh S (2008) Abiotic stress and plant responses. IK International, New Delhi

Khosh-Khui M, Ashiri F, Saharkhiz MJ (2012) Effects of irrigation regimes on antioxidant activity and total phenolic content of Thyme (*Thymus vulgaris* L.). Med Aromat Plants 1:114

Kim K, Portis AR (2004) Oxygen-dependent H2O2 production by Rubisco. FEBS Lett 571(1–3):124–128

Kochian LV, Pineros MA, Hoekenga OA (2005) The physiology, genetics and molecular biology of plant aluminum resistance and toxicity. Plant Soil 274(1–2):175–195

Kotak S, Larkindale J, Lee U, von Koskull-Döring P, Vierling E, Scharf KD (2007) Complexity of the heat stress response in plants. Curr Opin Plant Biol 10(3):310–316

Koyama H, Toda T, Hara T (2001) Brief exposure to low-pH stress causes irreversible damage to the growing root in *Arabidopsis thaliana*: pectin–Ca interaction may play an important role in proton rhizotoxicity. J Exp Bot 52(355):361–368

Krishnaiah D, Sarbatly R, Nithyanandam R (2011) A review of the antioxidant potential of medicinal plant species. Food Bioprod Process 89(3):217–233

Laisk A, Pfanz H, Schramm MJ, Heber U (1988) Sulfur-dioxide fluxes into different cellular compartments of leaves photosynthesizing in a polluted atmosphere. Planta 173(2):230–240

Laisk A, Kull O, Moldau H (1989) Ozone concentration in leaf intercellular air spaces is close to zero. Plant Physiol 90(3):1163–1167

Landry LG, Pell EJ (1993) Modification of Rubisco and altered proteolytic activity in O_3-stressed hybrid poplar (*Populus maximowizii* x trichocarpa). Plant Physiol 101(4):1355–1362

Larcher W (2003) Physiological plant ecology: ecophysiology and stress physiology of functional groups, 4th edn. Springer, Berlin, pp 345–437

Latef AAA, Tran LSP (2016) Impacts of priming with silicon on the growth and tolerance of maize plants to alkaline stress. Front Plant Sci 7:1–10

Li C, Xu H, Xu J, Chun X, Ni D (2011) Effects of aluminium on ultrastructure and antioxidant activity in leaves of tea plant. Acta Physiol Plant 33(3):973–978
Liang B, Huang X, Zhang G, Zhang F, Zhou Q (2006) Effect of lanthanum on plants under supplementary ultraviolet-B radiation: effect of lanthanum on flavonoid contents in soybean seedlings exposed to supplementary ultraviolet-B radiation. J Rare Earths 24(5):613–616
Lidon FJC (2012) Micronutrients' accumulation in rice after supplemental UV-B irradiation. J Plant Interact 7(1):19–28
Liu EU, Liu CP (2011) Effects of simulated acid rain on the antioxidative system in *Cinnamomum philippinense* seedlings. Water Air Soil Pollut 215(1–4):127–135
Liu TT, Wu P, Wang LH, Zhou Q (2011) Response of soybean seed germination to cadmium and acid rain. Biol Trace Elem Res 144(1–3):1186–1196
Lutz HJ, Costello DF, Hickie P, Stone EL, Bates M, Ellison L (1957) Symposium on applications of ecology. Ecology 38(1):46–64
Luwe MW, Takahama U, Heber U (1993) Role of ascorbate in detoxifying ozone in the apoplast of spinach (*Spinacia oleracea* L.) leaves. Plant Physiol 101(3):969–976
Malan C, Greyling MM, Gresse J (1990) Correlation between CuZn superoxide dismutase and glutathione reductase, and environmental and xenobiotic stress tolerance in maize inbreds. Plant Sci 69(2):157–166
Maleva MG, Nekrasova GF, Borisova GG, Chukina NV, Ushakova OS (2012) Effect of heavy metals on photosynthetic apparatus and antioxidant status of elodea. Russ J Plant Physiol 59(2):190–197
Marques TCLLD e MS, Soares AM (2011) Antioxidant system of ginseng under stress by cadmium. Sci Agric 68(4):482–488
Marschner H (1995) Mineral nutrition of higher plants, 2nd edn. Academic, San Diego, Kindly, p 889
Marulanda A, Porcel R, Barea JM, Azcón R (2007) Drought tolerance and antioxidant activities in lavender plants colonized by native drought-tolerant or drought-sensitive Glomus species. Microb Ecol 54(3):543
McKee WH, McKevlin MR (1993) Geochemical processes and nutrient uptake by plants in hydric soils. Environ Toxicol Chem 12(12):2197–2207
Mehlhorn H (1990) Ethylene-promoted ascorbate peroxidase activity protects plants against hydrogen peroxide, ozone and paraquat. Plant Cell Environ 13(9):971–976
Mehlhorn H, Lelandais M, Korth HG, Foyer CH (1996) Ascorbate is the natural substrate for plant peroxidases. FEBS Lett 378(3):203–206
Mittler R (2002) Oxidative stress, antioxidants and stress tolerance. Trends Plant Sci 7(9):405–410
Mittler R, Vanderauwera S, Gollery M, Van Breusegem F (2004) Reactive oxygen gene network of plants. Trends Plant Sci 9(10):490–498
Moran R, Porath D (1980) Chlorophyll determination in intact tissues using N,NDimethylformamide. Plant Physiol 65:478–479
Morsy AA, Salama A, Hamid K, Kamel HA, Fahim Mansour MM (2012) Effect of heavy metals on plasma membrane lipids and antioxidant enzymes of *Zygophyllum* species. Eur Asian J Biosci 6:1–10
Nagesh Babu R, Devaraj VR (2008) High temperature and salt stress response in French bean (*Phaseolus vulgaris*). Aust J Crop Sci 2(2):40–48
Ormrod DP, Hale BA (1995) Physiological response of plants and crops to ultravoilet - B radiation stress. In: Pessarakli M (ed) Handbook of plant and crop physiology. Marcel Dekker Inc, New York, pp 761–770
Ou-yang C, Gao S, Mei LJ, Chung TW, Tang L, Wang SH, Chen F (2014) Effects of aluminum toxicity on the growth and antioxidant status in *Jatropha curcas* seedlings. J Med Plants Res 8(3):178–185
Paz RC, Rocco RA, Reinoso H, Menéndez AB, Pieckenstain FL, Ruiz OA (2012) Comparative study of alkaline, saline, and mixed saline–alkaline stresses with regard to their effects on

growth, nutrient accumulation, and root morphology of *Lotus tenuis*. J Plant Growth Regul 31(3):448–459
Peiser G, Yang SF (1985) Biochemical and physiological effects of SO2 on non-photosynthetic processes in plants. AGRIS, Publisher: Stanford University Press, CA
Pell EJ, Lukezic FL, Levine RG, Weissberger WC (1977) Response of soybean foliage to reciprocal challenges by ozone and a hypersensitive-response-inducing Pseudomonad. Phytopathology 67:1342–1345
Peng YL, Gao ZW, Gao Y, Liu GF, Sheng LX, Wang DL (2008) Eco-physiological characteristics of alfalfa seedlings in response to various mixed salt-alkaline stresses. J Integr Plant Biol 50(1):29–39
Polle A, Chakrabarti K, Chakrabarti S, Seifert F, Schramel P, Rennenberg H (1992) Antioxidants and manganese deficiency in needles of Norway spruce (*Picea abies* L.) trees. Plant Physiol 99(3):1084–1089
Ponnamperuma FN (1983) Effects of flooding on soils. In: Kozlowski TT (ed) Flooding and plant growth, vol 24. Academic, New York, pp 29–96
Potters G, Pasternak TP, Guisez Y, Jansen MA (2009) Different stresses, similar morphogenic responses: integrating a plethora of pathways. Plant Cell Environ 32(2):158–169
Radi AA, Abdel-Wahab DA, Hamada AM (2012) Evaluation of some bean lines tolerance to alkaline soil. J Biol Earth Sci 2(1):18–27
Rahman SM (2013) Effect of temperature and extraction process on antioxidant activity of various leaves crude extracts of *Thymus vulgaris*. J Coast Life Med 1(2):130–134
Ross JA, Kasum CM (2002) Dietary flavonoids: bioavailability, metabolic effects, and safety. Ann Rev Nutr 22(1):19–34
Rout JR, Behera S, Keshari N, Ram SS, Bhar S, Chakraborty A, Sahoo SL (2015) Effect of iron stress on *Withania somnifera* L.: antioxidant enzyme response and nutrient elemental uptake of in vitro grown plants. Ecotoxicology 24(2):401–413
Rozema J, van de Staaij JWM, Bjorn LO, Bakker N (1999) Depletion of stratospheric ozone and solar UV-B radiation: evolution of land plants, UV-screens and function of phenolis. Editors J. Rozema, In: Stratopheric ozone depletion: the effects of enhanced UV-B radiation on terrestrial ecosystems, Leiden, Publisher Backhuys Publishers, pp 1–19
Russell AE, Laird DA, Mallarino AP (2006) Nitrogen fertilization and cropping system impacts on soil quality in Midwestern Mollisols. Soil Sci Soc Am J 70(1):249–255
Saha D, Mandal S, Saha A (2012) Copper induced oxidative stress in tea (*Camellia sinensis*) leaves. J Environ Biol 33(5):861
Sairam RK, Shukla DS, Saxena DC (1997) Stress induced injury and antioxidant enzymes in relation to drought tolerance in wheat genotypes. Biol Plant 40(3):357–364
Sairam RK, Deshmukh PS, Saxena DC (1998) Role of antioxidant systems in wheat genotypes tolerance to water stress. Biol Plant 41(3):387–394
Sanghera GS, Wani SH, Hussain W, Singh NB (2011) Engineering cold stress tolerance in crop plants. Curr Genomics 12(1):30
Santhanakrishnan D, Perumal RK, Chandrasekaran B (2014) Effect of tannery soaking water on antioxidant enzymes of *Salicornia brachiata*. Int J Curr Microb Appl Sci 3(2):359–367
Scalbert A, Williamson G (2000) Dietary intake and bioavailability of polyphenols. J Nutr 130(8):2073S–2085S
Schubert S, Schubert E, Mengel K (1990) Effect of low pH of the root medium on proton release, growth, and nutrient uptake of field beans (*Vicia faba*). Plant Soil 124(2):239–244
Sekmen HA, Türkan I, Takio S (2007) Differential responses of antioxidative enzymes and lipid peroxidation to salt stress in salt-tolerant *Plantago maritima* and salt-sensitive *Plantago media*. Physiol Plant 131(3):399–411
Setter TL, Waters I (2003) Review of prospects for germplasm improvement for waterlogging tolerance in wheat, barley and oats. Plant Soil 253(1):1–34
Shi QH, Zhu ZJ, Juan LI, Qian QQ (2006) Combined effects of excess Mn and low pH on oxidative stress and antioxidant enzymes in cucumber roots. Agric Sci China 5(10):767–772

Shimazaki KI, Sugahara K (1980) Inhibition site of the electron transport system in lettuce chloroplasts by fumigation of leaves with SO2. Plant Cell Physiol 21(1):125–135
Siefermann D, Harms H (1987) The last harvesting and protective function of carotenoids in photosynthetic membranes. Physiol Plant 69:51–568
Sielewiesiuk J (2002) Why there are photodamages to photosystem II at low light intensities. Acta Physiol Plant 24(4):399–406
Singh TP (2006) Protective role of antioxidant enzymes under high temperature stress. Plant Sci 171(3):382–388
Singh S, Anjum NA, Khan NA, Nazar R (2008) Metal-binding peptides and antioxidant defence system in plants: significance in cadmium tolerance. Abiotic stress and plant responses. IK International, New Delhi, pp 159–189
Smirnoff N, Colombé SV (1988) Drought influences the activity of enzymes of the chloroplast hydrogen peroxide scavenging system. J Exp Bot 39(8):1097–1108
Smith SR, Russell KA (1969) Occurrence of ethylene and its significance in anaerobic soil. Nature 222:769–771
Sofo A, Dichio B, Xiloyannis C, Masia A (2004) Effects of different irradiance levels on some antioxidant enzymes and on malondialdehyde content during rewatering in olive tree. Plant Sci 166(2):293–302
Stewart RR, Bewley JD (1980) Lipid peroxidation associated with accelerated aging of soybean axes. Plant Physiol 65(2):245–248
Stoop JM, Williamson JD, Pharr DM (1996) Mannitol metabolism in plants: a method for coping with stress. Trends Plant Sci 1(5):139–144
Surekha K (1997) Effect of drought and salinity on growth and development of isabgol (*Platago ovate* Forsk.). MSc thesis, CCS Haryana Agricultural University, Hisar, India
Suzuki N, Mittler R (2006) Reactive oxygen species and temperature stresses: a delicate balance between signaling and destruction. Physiol Plant 126(1):45–51
Taiz L, Zeiger E (1991) Plant physiology. The Benjamin/Cummings Publishing Company, Redwood City, pp 346–368
Terzi R, Kadioglu A (2006) Drought stress tolerance and the antioxidant enzyme system. Acta Biol Cracov Ser Bot 48:89–96
Tuteja N (2007) Mechanisms of high salinity tolerance in plants. Methods Enzymol 428:419–438
Umezawa T, Fujita M, Fujita Y, Yamaguchi-Shinozaki K, Shinozaki K (2006) Engineering drought tolerance in plants: discovering and tailoring genes to unlock the future. Curr Opin Biotechnol 17(2):113–122
Upadhyaya H, Panda SK (2013) Abiotic stress responses in tea [*Camellia sinensis* L. (O) Kuntze]: an overview. Rev Agric Sci 1:1–10
Vandana N (2003) Physiological response of Isabgol (*Plantago ovata* Forsk.) genotypes to salt stress. Doctoral dissertation, MSc thesis, CCS Haryana Agricultural University, Hisar, India
Varshney UK, Surekha K (2001) Differential response of chloride and sulphate dominated salinity on growth and developmental aspects of isabgol (*Plantago ovate* Forsk.). Abstract number: 1121, National symposium on plant biodiversity and its conservation, Patiala, p 56
Veljovic-Jovanovic S, Bilger W, Heber U (1993) Inhibition of photosynthesis, acidification and stimulation of zeaxanthin formation in leaves by sulfur dioxide and reversal of these effects. Planta 191(3):365–376
Wahid A (2007) Physiological implications of metabolite biosynthesis for net assimilation and heat-stress tolerance of sugarcane (*Saccharum officinarum*) sprouts. J Plant Res 120(2):219–228
Xin Z, Browse J (2000) Cold comfort farm: the acclimation of plants to freezing temperatures. Plant Cell Environ 23(9):893–902
Yan B, Dai Q, Liu X, Huang S, Wang Z (1996) Flooding-induced membrane damage, lipid oxidation and activated oxygen generation in corn leaves. Plant Soil 179(2):261–268
Yin H, Chen Q, Yi M (2008) Effects of short-term heat stress on oxidative damage and responses of antioxidant system in *Lilium longiflorum*. Plant Growth Regul 54(1):45–54

Yuan Y, Liu Y, Luo Y, Huang L, Chen S, Yang Z, Qin S (2011) High temperature effects on flavones accumulation and antioxidant system in *Scutellaria baicalensis* Georgi cells. Afr J Biotechnol 10(26):5182–5192

Zhang J, Kirkham MB (1994) Drought-stress-induced changes in activities of superoxide dismutase, catalase, and peroxidase in wheat species. Plant Cell Physiol 35(5):785–791

Zhang HM, Wang BR, Xu MG, Fan TL (2009) Crop yield and soil responses to long-term fertilization on a red soil in southern China. Pedosphere 19(2):199–207

Zhang XL, Jia XF, Yu B, Gao Y, Bai JG (2011) Exogenous hydrogen peroxide influences antioxidant enzyme activity and lipid peroxidation in cucumber leaves at low light. Sci Hortic 129(4):656–662

Zhao X, Nishimura Y, Fukumoto Y, Li J (2011) Effect of high temperature on active oxygen species, senescence and photosynthetic properties in cucumber leaves. Environ Exp Bot 70(2):212–216

Zheng G, Lv HP, Gao S, Wang SR (2010) Effects of cadmium on growth and antioxidant responses in *Glycyrrhiza uralensis* seedlings. Plant Soil Environ 56(11):508–515

Zróbek SA (2012) Temperature stress and responses of plants. In: Ahmad P, Prasad MNV (eds) Environmental adaptations and stress tolerance of plants in the era of climate change. Springer, New York, pp 113–134

Chapter 3
Antibacterial, Antifungal, and Antiviral Properties of Medical Plants

Dilfuza Jabborova, Kakhramon Davranov, and Dilfuza Egamberdieva

Abstract There is evidence of medicinal plants having been used in the treatment of human disease caused by various pathogenic microorganisms in many countries of the world. Plants with known antimicrobial activities were used for therapeutic treatments. They contain various biological compounds which could be used in the development of novel drugs for human well-being. Their phytochemical constituents include alkaloids, saponins, tannins, flavonoids, and glycosides, which serve as defense mechanisms against various microbes including insects. These compounds may include antibacterial, antifungal, and anticancer activities. The search for new antimicrobial compounds from medicinal plants from many continents is an important line of research because of the increased number of multidrug resistance pathogenic microorganisms. However, the therapeutic ability of a number of medicinal important plants is still unknown. Considering the importance of medicinal plants as sources for antimicrobial drugs, in this review, we report on progress to date in antimicrobial activities of medicinal plants.

Keywords Medicinal plants · Antibacterial · Antifungal · Bioactive compounds

D. Jabborova
Laboratory of Medical Plants Genetics and Biotechnology, Institute of Genetics and Plant Experimental Biology, Academy of Sciences of Uzbekistan, Tashkent, Uzbekistan

K. Davranov
Faculty of Biology, National University of Uzbekistan, Tashkent, Uzbekistan

D. Egamberdieva (✉)
Faculty of Biology, National University of Uzbekistan, Tashkent, Uzbekistan

Key Laboratory of Biogeography and Bioresource in Arid Land, Xinjiang Institute of Ecology and Geography, CAS, Urumqi, People's Republic of China

D. Egamberdieva, A. Tiezzi (eds.), *Medically Important Plant Biomes: Source of Secondary Metabolites*, Microorganisms for Sustainability 15,
https://doi.org/10.1007/978-981-13-9566-6_3

3.1 Introduction

The increasing incidence of multidrug resistance microorganisms has constantly become a scientific community concern (Compean and Ynalvez 2014). The members of gram-negative and gram-positive bacteria such as *Escherichia coli*, *Salmonella typhi*, *Pseudomonas aeruginosa*, *Staphylococcus aureus*, and *Bacillus cereus* were known as the causal agents of food-borne diseases (Pandey and Singh 2011; Braga et al. 2005). The dermatophytes and also *Candida* spp. are considered an important group of skin pathogens which cause many skin disorders. Many of the medicinal plant species are used for the treatment of various diseases (Bussmann et al. 2010; Duraipandiyan and Ignacimuthu 2011; Mamedov and Egamberdieva 2018). To date many plant secondary metabolites known to contain various antimicrobial compounds were screened against human pathogenic microbes (Egamberdieva and Teixeira da Silva 2015). Several scientists studied the biological activity of medicinal plants and their metabolites with antimicrobial activity against food spoilage bacteria (Gnat et al. 2017; David et al. 2010; Egamberdieva and Jabborova 2018). The phytochemical constituents of medicinal plants play a major role in plant biological activity, e.g., saponins (Lacaille-Dubois and Wagner 1996), flavonoids (David et al. 2010), and alkaloids (Omulokoli et al. 1997) were reported for their antiviral and antibacterial properties (Egamberdieva et al. 2017). The screening of medicinal plants for their biological active metabolites might lead to the isolation of compounds that are effective as antifungal, antiviral, or antibacterial agents (Cushnie and Lamb 2005; Shrivastava et al. 2015). In previous work it has been observed that alkaloids and phenolic compounds have strong interaction with microbial cells through enzymes and proteins (Burt 2004; Gill and Holley 2006). Antimicrobial activity of Indian medicinal plants broadly reported based on folklore knowledge (Duraipandiyan and Ignacimuthu 2011). The Middle East has thousands of year's history in traditional medicine, which has been used for treatment of various ailments. The flora of Uzbekistan covers more than 4500 species of vascular plants, of them around 20% has showed positive effect on various ailments (Mamedov et al. 2005; Shurigin et al. 2018).

3.2 Antimicrobial Activities of Medicinal Plants

The antimicrobial activities of medicinal and aromatic plants from various countries were described, and some results (Ahmad and Beg 2001; Kokoska et al. 2002; Alzoreky and Nakahara 2003; Rios and Recio 2005; Sher 2009; Pirbalouti et al. 2010; Verma et al. 2012; Akinpelu et al. 2015) were listed in Table 3.1. Tajkarimi et al. (2010) described antimicrobial activities of aromatic plants. In another study, Gupta et al. (2010) reported antibacterial activity of *Achyranthes aspera*, *Tagetes patula*, and *Lantana camara* plant extracts against *Staphylococcus aureus*, *Pseudomonas aeruginosa*, and *Bacillus subtilis*.

Table 3.1 Antimicrobial activity of medicinal plants

Plant species	Antimicrobial properties	References
Cinnamomum cassia, *Rumex nervosus*, *Ruta graveolens*, *Thymus serpyllum*	Antibacterial	Alzoreky and Nakahara (2003)
Allium sativum	Antibacterial	Verma et al. (2012)
Punica granatum		
Persea americana	Antibacterial	Akinpelu et al. (2015)
Achyranthes aspera	Antibacterial	Gupta et al. (2010) and Beaulah et al. (2011)
Lantana camara	Antibacterial	Gupta et al. (2010)
Tagetes patula		
Lavandula multifida	Antibacterial	Guesmi et al. (2017)
Annona squamosa	Antibacterial	Patel and Kumar (2008) and Padhi et al. (2011)
Punica granatum	Antibacterial, antifungal	Silva et al. (2008a, b), Alzoreky (2009), Mangang and Chhetry (2012), Mangang and Chhetry (2012), Mahboubi et al. (2015), Guesmi et al. (2017), Mishra et al. (2017), and Mostafa et al. (2018)
Ocimum gratissimum, *Eugenia uniflora*, *Murraya koenigii*, *Cynodon dactylon*, *Lawsonia inermis*, *Adha-thoda vasica*	Antibacterial	Fadeyi and Alcapan (1989)
Cuminum cyminum	Antibacterial	Arora and Kaur (1999). Shan et al. (2007), Shan et al. (2007), Chaudry and Tariq (2008), Dua et al. (2013), and Mostafa et al. (2018)
Zingiber officinale	Antibacterial	Alzoreky and Nakahara (2003), Betoni et al. (2006), Ushimaru et al. (2007), Sapkota et al. (2012), Qader et al. (2013), and Mostafa et al. (2018)
Syzygium aromaticum	Antibacterial	Mostafa et al. (2018)
Thymus vulgaris		
Psidium guajava	Antibacterial	Farjana et al. (2014)
Calendula officinalis	Antibacterial	Chakraborthy (2008) and Farjana et al. (2014)
Azadirachta indica	Antibacterial, antifungal	Alzoreky and Nakahara (2003), El-Mahmood et al. (2010), Koona and Budida (2011), Sapkota et al. (2012), Jabeen et al. (2013), Farjana et al. (2014), Rakholiya et al. (2014), and Mishra et al. (2017)
Camellia sinensis	Antibacterial	Farjana et al. (2014)
Tussilago farfara	Antibacterial	Hleba et al. (2014)
Aesculus hippocastanum		
Equisetum arvense		

(continued)

Table 3.1 (continued)

Plant species	Antimicrobial properties	References
Terminalia arjuna	Antimicrobial	Gupta et al. (2016)
Polyalthia longifolia		
Momordica charantia	Antifungal	Wang et al. (2016)
Alstonia boonei	Antibacterial	Ogueke et al. (2014)
Solanum coagulans	Antifungal	Qin et al. (2016)
Pituranthos tortuosus	Antibacterial	Mighri et al. (2015)
Anogeissus acuminata	Antibacterial	Mishra et al. (2017)
Boerhavia diffusa	Antibacterial	Mishra et al. (2017)
Bauhinia variegata	Antibacterial	Mishra et al. (2017)
Soymida febrifuga	Antibacterial	Mishra et al. (2017)
Aristolochia indica	Antibacterial	Kumar et al. (2011)
Terminalia chebula	Antibacterial	Mishra et al. (2017)
Tinospora cordifolia	Antibacterial	Mishra et al. (2017)
Tribulus terrestris	Antibacterial	Mishra et al. (2017)
Annona squamosa	Antifungal	Kalidindi et al. (2015)
Rhanterium epapposum	Antibacterial, antifungal	Adam et al. (2011), Akbar and Al-Yahya (2011), and Demirci et al. (2017)
Lumnitzera littorea	Antibacterial, antifungal	Saad et al. (2011)
Alternanthera sessilis	Antibacterial	Johnson et al. (2010)
Cinnamomum zeylanicum	Antifungal	Ajay et al. (2009)
Dahlia pinnata	Antibacterial	Bissa et al. (2011)
Piper nigrum	Antibacterial	Karsha and Bhagyalakshmi et al. (2010)
Plumeria rubra	Antibacterial	Baghel et al. (2010)
Achillea millefolium, *Ipomoea pandurata*, *Hieracium pilosella*, and *Solidago canadensis*	Antibacterial	Frey and Meyers (2010)
Glycyrrhiza glabra	Antibacterial, antifungal	Patil et al.(2009)
Allium sativum	Antibacterial	Betoni et al. (2006), Ushimaru et al. (2007), and Sapkota et al. (2012)
Phyllanthus niruri	Antibacterial	Selvamohan et al. (2012)
Baccharis dracunculifolia	Antibacterial	Ferronato et al. (2007)
Chamaecyparis obtuse, *Chrysanthemum boreale*, *Cryptomeria japonica*	Antibacterial, antiviral	Lee and Choi (2015)
Cynara scolymus,	Antibacterial	Asolini et al. (2006)
Achyrocline satureioides		
Dennettia tripetala	Antibacterial, antifungal	Ejechi and Akpomedaye (2005) and Oyemitan et al. (2019)
Rosmarinus officinalis	Antibacterial	Silva et al. (2008a, b) and Adam et al. (2014)

(continued)

Table 3.1 (continued)

Plant species	Antimicrobial properties	References
Cyclocarya paliurus	Antibacterial, antifungal	(Xie et al. 2012)
Malva aegyptiaca	Antibacterial	Fakhfakh et al. (2017)
Blepharis cuspidata, Boswellia ogadensis, Thymus schimperi	Antibacterial	Gadisa et al. (2019)
Periploca laevigata	Antibacterial	Hajji et al. (2019)
Tridax procumbens	Antibacterial	Bharati et al. (2012) and Andriana et al. (2019)
Prunus domestica	Antibacterial	Islam et al. (2017) and El-Beltagi et al. (2019)
Artemisia nilagirica, Artocarpus integrifolia, Citrus maxima, Coix lacryma-jobi, Hedychium coronarium, Lantana camera, Michelia champaca, Passiflora foetida, Strobilanthes flaccidifolius	Antifungal	Mangang and Chhetry (2012)
Helicteres hirsuta	Antibacterial	Pham et al. (2018)
Syzygium aromaticum	Antibacterial	Vizhi et al. (2016)
Anagallis arvensis	Antifungal	Soberón et al. (2017)
Cichorium intybus	Antibacterial, antifungal	Mares et al. (2005), Nandagopal and Kumari (2007), Verma et al. (2013), Rehman et al. (2014), and Shaikh et al. (2016)
Polygonum hydropiper	Antibacterial, antifungal	Hasan et al. (2009)
Kigelia africana	Antibacterial, antifungal	Owolabi et al. (2007)
Cnicus benedictus	Antibacterial	Szabó et al. (2009)
Seriphidium kurramense	Antibacterial, antifungal	Ahmad et al. (2018) and Mahmoud et al. (2011)
Rosmarinus officinalis	Antifungal	Adam et al. (2014)
Salvia bicolor	Antifungal	Taghreed (2012)

In another study Guesmi et al. (2017) reported that *Lavandula multifida* showed the most powerful activity against *Bacillus cereus* strain. The extract of *Punica granatum* showed antibacterial activity against *Staphylococcus aureus*, *Bacillus cereus*, *Escherichia coli*, and *Salmonella typhi*, which cause borne diseases (Alzoreky 2009; Mahboubi et al. 2015; Guesmi et al. 2017; Mishra et al. 2017). In other reports cumin seed (*Cuminum cyminum*) extract exhibited antimicrobial activity against gram-positive and gram-negative bacteria (Shan et al. 2007; Chaudry and Tariq 2008). Dua et al. (2013) reported that extracts of cumin effective against *E. coli*, *P. aeruginosa*, *S. aureus*, and *B. pumilus* were ranged between 6.25 and 25 mg/ml. Qader et al. (2013) studied *Zingiber officinale* and *Thymus kotschyana* for their effect on human pathogenic bacteria *S. aureus* and *E. coli*, and they found antimicrobial activity of plant extracts. Similar reports were published by other authors, where *Zingiber officinale* and *Allium sativum* extracts inhibited growth of *S. aureus* (Betoni et al. 2006; Ushimaru et al. 2007; Sapkota et al. 2012).

Mostafa et al. (2018) observed an antimicrobial activity of plant extracts of *Zingiber officinale*, *Punica granatum*, *Syzygium aromaticum*, and *Thymus vulgaris* against *Escherichia coli*, *Bacillus cereus*, *Pseudomonas aeruginosa*, *Staphylococcus aureus*, and *Salmonella typhi* at concentration of 10 mg/ml. Silva et al. (2008a, b) reported that extracts of *Punica granatum* fruit (pomegranate) were inhibitory against *Staphylococcus aureus.*

The plant extracts of guava (*Psidium guajava*), neem (*Azadirachta indica*), and marigold (*Calendula officinalis*) also inhibited growth of bacteria belonging to *Pseudomonas*, *Vibrio*, *Klebsiella*, *Escherichia*, *Salmonella*, and *Staphylococcus* genera (Farjana et al. 2014). Plants belonging to *T. farfara* and *Equisetum arvense* also showed antimicrobial properties against human pathogenic bacteria (Hleba et al. 2014). Gupta et al. (2016) reported that human pathogenic bacteria *Escherichia coli*, *Pseudomonas aeruginosa*, *Candida albicans*, and *Staphylococcus aureus* were inhibited by methanol extracts of *Terminalia arjuna*, *Camellia sinensis*, and *Polyalthia longifolia*. The ethanol extract of *Alstonia boonei* inhibited growth of *E. coli* with inhibition zone of 23.73 mm (Ogueke et al. 2014).

Several crop extracts also showed antifungal activity against plant pathogenic fungi such as *Fusarium*, *Rhizoctonia*, and *Verticillium*. For example, vegetable crop extract *Momordica charantia* inhibited the mycelial growth of *Fusarium solani*, a plant pathogen which causes root rot disease (Wang et al. 2016). The extract of *Solanum coagulans* showed remarkable antifungal activity against *T. mentagrophytes*, *M. gypseum*, and *E. floccosum* (Qin et al. 2016). In another report *Annona squamosa* Linn. leaf extract showed antifungal activity against *Alternaria alternata*, *Fusarium solani*, *Microsporum canis*, and *Aspergillus niger* (Kalidindi et al. 2015). Following other reports we found that *Artemisia nilagirica*, *Artocarpus integrifolia*, *Citrus maxima*, *Hedychium coronarium*, *Lantana camera*, *Passiflora foetida*, and *Strobilanthes flaccidifolius* showed also antifungal activity against *R. solani* (Mangang and Chhetry 2012). Similar results were obtained by Mahmoud et al. (2011) where ethanol extract of *S. kurramense* was effective against *A. flavus*.

Mighri et al. (2015) reported the antibacterial activity of *P. tortuosus* on *E. coli* and *Klebsiella pneumoniae*, moderate activity against S. *aureus*, and high activity against *Streptococcus pyogenes* and *Enterobacter aerogenes*. Methanol extract of plants such as *Anogeissus acuminata*, *Boerhavia diffusa*, *Soymida febrifuga*, and *Tribulus terrestris* showed antimicrobial activity against *Enterococcus faecalis*, *Staphylococcus aureus*, *Escherichia coli*, *Klebsiella oxytoca*, *Klebsiella pneumoniae*, and *Pseudomonas aeruginosa* (Mishra et al. 2017).

Annona squamosa Linn. is cultivated throughout America, Brazil, and India and is used as traditional medicine in treatment of various diseases (Kaleem et al. 2008; Raj et al. 2009).

The antimicrobial properties of *Rhanterium epapposum* were positive against *B. cereus*, *S. aureus*, and *P. vulgaris* (Adam et al. 2011; Akbar and Al-Yahya 2011). Furthermore several biological active compounds with antimicrobial properties such as flavonoids, tannins, sterols, triterpenes, and essential oils were found (Al-Yahya et al. 1990; Akbar and Al-Yahya 2011).

Demirci et al. (2017) evaluated the antimicrobial potential of *R. epapposum* essential oil against *Bacillus subtilis*, *Enterobacter aerogenes*, *Proteus vulgaris*, *Salmonella typhimurium*, *Staphylococcus aureus*, *Staphylococcus epidermidis*, and the yeast *Candida parapsilosis*. The essential oil was able to inhibit growth of microbial strains. In another study the extracts from different mangrove plants have been reported to possess inhibition action against human and plant pathogens (Chandrasekaran et al. 2009; Sivaperumal et al. 2010; Ravikumar et al. 2010; Hu et al. 2010; Khajure and Rathod 2010). Saad et al. (2011) investigated the antimicrobial properties of ethyl acetate and methanol extracts of *Lumnitzera littorea* leaves against *Staphylococcus aureus*, *Bacillus cereus*, *Pseudomonas aeruginosa*, *Escherichia coli*, and two fungal strains *Candida albicans* and *Cryptococcus neoformans*.

Mathabe et al. (2006) reported that methanol, ethanol, and acetone extracts from *Indigofera daleoides*, *Punica granatum*, *Elephantorrhiza burkei*, *Ximenia caffra*, *Schotia brachypetala*, and *Spirostachys africana* showed antimicrobial activity against *Vibrio cholerae*, *Escherichia coli*, *Staphylococcus aureus*, *Shigella* species, and *Salmonella typhi*. Some plants such as *Ocimum gratissimum* and *Eugenia uniflora* have been reported to be rich in volatile oils, which have antimicrobial effect against *Staphylococcus* sp., *Escherichia coli*, and *Shigella* sp. and are mainly used in the treatment of diarrhea and ear infection in human beings. However, the ethanol and aqueous extracts of *Murraya koenigii*, *Cynodon dactylon*, *Lawsonia inermis*, and *Adha-thoda vasica* showed least inhibitory activity (Fadeyi and Alcapan 1989). Frey and Meyers (2010) reported antimicrobial properties of *Achillea millefolium*, *Ipomoea pandurata*, *Hieracium pilosella*, and *Solidago canadensis* against *Salmonella typhimurium*. Similarly, Patil et al. (2009) reported a significant antifungal and antibacterial activity against *Candida albicans*, *Staphylococcus aureus*, *Pseudomonas aeruginosa*, and *Escherichia coli* by the diethyl ether fraction of ethanolic extract of *Glycyrrhiza glabra*. The methanolic extract *Phyllanthus niruri* (stone breaker) showed the maximum activity against *Staphylococcus* sp. (Selvamohan et al. 2012). In another study *Baccharis dracunculifolia* oil at a 10-μL dose prevented microbial growth of *E. coli*, *S. aureus*, and *P. aeruginosa* in antimicrobial assays (Ferronato et al. 2007). The methanolic extracts of *Chamaecyparis obtusa* and *Cryptomeria japonica* possessed strong antiviral activity against HRV3 at a concentration of 100 μg/mL with no cytotoxicity. Similarly, methanolic extract of *Chrysanthemum boreale* possesses strong antimicrobial activity against *Staphylococcus aureus*, *Bacillus cereus*, *Escherichia coli*, and *Yersinia enterocolitica* (Lee and Choi 2015). Asolini et al. (2006) reported that ethanol extracts of artichoke (*Cynara scolymus*) inhibited the growth of *Bacillus cereus*, *B. subtilis*,

Pseudomonas aeruginosa, and *S. aureus*. The essential oil of *Dennettia tripetala* fruit possesses antimicrobial activities against bacterial and fungal isolates (Ejechi and Akpomedaye 2005). The hydroalcoholic extract of *Rosmarinus officinalis* Linn. showed antibacterial activity against *Streptococcus* spp. and *Lactobacillus casei* (Silva et al. 2008a, b). Adam et al. (2014) reported high antifungal activity of aqueous extract of *Rosmarinus officinalis* toward *Candida albicans* and *Aspergillus niger.* In another study Fakhfakh et al. (2017) reported the highest inhibitory effect of polysaccharide extract of *Malva aegyptiaca* against gram-negative bacteria. Polysaccharides derived from plants *Cyclocarya paliurus* (Batal.) showed antifungal activity against *Saccharomyces cerevisiae* and *Candida* sp. and antibacterial activity against *E. coli*, *S. aureus*, and *B. subtilis* (Xie et al. 2012).

The essential oils of medicinal plants that contain phenols also possess antimicrobial activities. For example, the essential oils from *Blepharis cuspidata*, *Boswellia ogadensis*, and *Thymus schimperi* showed antimicrobial activity against multidrug *resistance E. coli*, *K. pneumoniae*, and *S. aureus* (Gadisa et al. 2019).Essential oil extracted from *B. cuspidate* had elicited high antibacterial effect on tested *Enterobacteriaceae*. A novel water-soluble polysaccharide isolated from root barks of *Periploca laevigata* demonstrated antioxidant potential and high antibacterial activity against several gram-positive and gram-negative bacteria (Hajji et al. 2019). *Tridax procumbens* L. showed effective inhibition on the growth of *Escherichia coli*, *Staphylococcus aureus*, *Bacillus subtilis*, and *Proteus mirabilis* (Bharati et al. 2012; Andriana et al. 2019). El-Beltagi et al. (2019) evaluated the phytochemical composition of *Prunus domestica* fruit and their antimicrobial activity. They found that ethanol extract of fruit exhibited antibacterial activity against *Staphylococcus aureus* (ZI = 18.51 mm). Islam et al. (2017) reported antimicrobial potential, gram-positive and gram-negative bacteria have been found susceptible to the *P. domestica* extract, for example, strain of *S. aureus* (19.7 ± 0.4 mm) and *E. coli* (14.4 ± 0.7 mm). There are other plants with antimicrobial potential; however, they were not fully studied yet. For example, *Helicteres hirsuta* Lour. known with wide pharmacological properties showed antimicrobial activity against *E. coli* (MIC values of 2.5 and 5.0 mg/mL) and *S. lugdunensis* (MIC values of 0.35 and 0.50 mg/mL) (Pham et al. 2018).

In another study Vizhi et al. (2016) tested the antibacterial activity of methanol, ethyl acetate, and acetone extracts of *Syzygium aromaticum* medicinal plant against *Bacillus subtilis*, *Pseudomonas aeruginosa*, and *Staphylococcus aureus.* Methanol extract of *S. aromaticum* showed good antimicrobial activity against *Bacillus subtilis*, *Pseudomonas aeruginosa*, and *Staphylococcus aureus.* The antifungal compounds derived from plant *Anagallis arvensis* L. showed higher inhibitory activity against human pathogenic yeast *Candida albicans* (Soberón et al. 2017). Mares et al. (2005) reported antifungal activity of *C. intybus* against anthropophilic fungi *Trichophyton tonsurans*, *T. rubrum*, and *T. violaceum*. *Cichorium intybus* leaf extracts showed antimicrobial activity against *S. aureus*, *P. aeruginosa*, *E. coli*, and *C. albicans.* Root extracts had pronounced effects on *B. subtilis*, *S. aureus*, *Salmonella typhi*, *Micrococcus luteus*, and *E. coli* (Nandagopal and Kumari 2007).

Cichorium intybus crude extract exhibited wide range of antimicrobial activity against *E. coli*, *K. pneumoniae*, *P. aeruginosa*, *S. epidermidis*, *S. aureus*, and *B. subtilis* (Rehman et al. 2014). Moreover, the growth of fungi such as *Aspergillus flavus*, *Fusarium solani*, *Aspergillus fumigatus*, and *Aspergillus niger* was inhibited by plant extract.

Shaikh et al. (2016) tested seed extract of *Cichorium intybus* showed antimicrobial activity against several human pathogenic bacteria such as *Staphylococcus aureus*. Ethyl acetate and ethanol extract were found to be significant against *P. aeruginosa*. The biological active compounds such as lactucin and lactucopicrin derived from *C. intybus* exhibited antibacterial activity (Verma et al. 2013). *Polygonum hydropiper* (L.) root extract showed significant antibacterial activities against four gram-positive (*Bacillus subtilis*, *Bacillus megaterium*, *Staphylococcus aureus*, and *Enterobacter aerogenes*) and four gram-negative (*Escherichia coli*, *Pseudomonas aeruginosa*, *Salmonella typhi*, and *Shigella sonnei*) bacteria (Hasan et al. 2009). The ethanolic and aqueous extract of *Kigelia africana* showed antimicrobial activity against both bacteria and fungi (Owolabi et al. 2007). Other plants such as *Cnicus benedictus* L. showed antibacterial activity against ten pathogens such as *Salmonella typhimurium*, *Salmonella enteritidis*, *Staphylococcus aureus* ssp., *Escherichia coli*, *Streptococcus pyogenes*, *Pseudomonas aeruginosa*, *Enterococcus faecalis*, and *Shigella sonnei* (Szabó et al. 2009). Ahmad et al. (2018) investigated the antimicrobial activity of crude ethanolic and aqueous extracts of *Seriphidium kurramense* by agar well diffusion assays against five bacterial species such as *Staphylococcus aureus*, *Escherichia coli*, *Klebsiella pneumoniae*, *Bacillus subtilis*, and *Salmonella typhi*, and six fungal species such as *Aspergillus niger*, *Aspergillus flavus*, *Alternaria solani*, *Rhizoctonia solani*, *Fusarium solani*, and *Pleurotus florida*. The ethanol extract showed its highest growth inhibition (74.4%) toward *B. subtilis* and its lowest inhibition (32.2%) toward *K. pneumoniae*. A petroleum ether extract and a methanolic extract of aerial parts of *Salvia bicolor* against *Staphylococcus epidermidis* and *Candida albicans*.

3.3 Conclusion

From published reports, it is evident that antimicrobial properties of medicinal plants were reported based on folklore information. They synthesize various biological active compounds that possess antimicrobial properties. The compounds contain alkaloids, saponins, coumarins, steroids, flavonoids, glycosides, phenols, and tannins. A number of essential oils that contain aldehydes or phenols were also used as antimicrobial agents. These reports provide an insight into the antibacterial properties of medicinal plants used in traditional medicine and justification for the use of medicinal plants in medicine to treat infectious diseases. It will also lead to the development of some new biologically active compounds which can be formulated as antimicrobial agents.

References

Adam SIY, Ahmed AYA, Omer AKM, Bashir AMA, Abdel Rahman OAM, Abdelgadir WS (2014) In vitro antimicrobial activity of *Rosmarinus officinalis* leave extracts. J Agri-Food Appl Sci 2(1):15–21

Adam SIY, El-Kamali HH, Adama SEI (2011) Phytochemical screening and antibacterial activity of two Sudanese wild plants, *Rhanterium epapposum* and *Trichodesma africanum*. J Fac Sci Technol 2:83–96

Ahmad I, Beg AZ (2001) Antimicrobial and photochemical studies on 45 Indian medicinal plants against multi-drug resistant human pathogens. J Ethnopharmocol 74:113–123

Ahmad K, Ali A, Afridi WA, Somayya R, Ullah MJ (2018) Antimicrobial, hemagglutination and phytotoxic activity of crude ethanolic and aqueous extracts of *Seriphidium kurramense*. J Tradit Chin Med 38(3):433–438

Ajay KM, Mishra A, Kehri HK, Pandey BSAK (2009) Inhibitory activity of Indian spice plant *Cinnamomum zeylanicum* extracts against *Alternaria solani* and *Curvularia lunata*, the pathogenic dematiaceous moulds. Ann Cli Micro Antimicrob 8:1–9

Akbar S, Al-Yahya MA (2011) Screening of Saudi plants for phytoconstituents, pharmacological and antimicrobial properties. Aust J Med Herbalism 23:76–87

Akinpelu DA, Aiyegoro OA, Akinpelu OF, Okah AI (2015) Stem bark extract and fraction of *Persea americana* (Mill) exhibits bactericidal activities against strains of *Bacillus cereus* associated with food poisoning. Molecules 20:416–429

Al-Yahya MA, Al-Meshal IA, Mossa JS, Al-Badr AA, Tariq M (1990) Saudi plants, a phytochemical and biological approach. KACST, Riyadh, pp 75–80

Al-Zoreky NS (2009) Antimicrobial activity of pomegranate (*Punica granatum* L.) fruit peels. Int J Food Microbiol 134(3):244–248

Alzoreky NS, Nakahara K (2003) Antibacterial activity of extracts from some edible plants commonly consumed in Asia. Int J Food Microbiol 80:223–230

Andriana Y, Xuan TD, Quy TN, Minh TN, Van TM, Viet TD (2019) Antihyperuricemia, antioxidant, and antibacterial activities of Tridax procumbens. L Foods 8(21):1–12

Arora DS, Kaur J (1999) Antimicrobial activity of spices. Int J Antimicrob Agents 12:257–262

Asolini FC, Tedesco AM, Ferraz C, Alencar SM, Carpes ST (2006) Antioxidant and antibacterial activities of phenolic compounds in extracts of plants used as tea. Braz J Food Technol 9(6):209–215

Baghel AS, Mishra CK, Rani A, Sasmal D, Nema RK (2010) Antibacterial activity of Plumeria rubra Linn. plant extract. J Chem Pharma Res 2(6):435–440

Beaulah AG, Sadiq AM, Santhi RJ (2011) Antioxidant and antibacterial activity of Achyranthes aspera: an in vitro study. Pharma Chem 3(5):255–262

Betoni JE, Mantovani RP, Barbosa LN, Di Stasi LC, Fernandes JA (2006) Synergism between plant extract and antimicrobial drugs used on *Staphylococcus aureus* diseases. Mem Inst Oswaldo Cruz 101(4):387–390

Bharati V, Varalakshmi B, Gomathu S, Shanmugapriya A, Karpagam T (2012) Antibacterial activity of *Tridax procumbens* Linn. Int J Pharm Sci Res 3:364–367

Bissa S, Bohra A, Bohra A (2011) Screening of *Dahlia pinnata* for its antimicrobial activity. J Res Biol 1:51–55

Braga LC, Shupp JW, Cummings C, Jett M, Takahashi JA, Carmo LS (2005) Pomegranate extract inhibits *Staphylococcus aureus* growth and subsequent enterotoxin production. J Ethnopharmacol 96:335–339

Burt S (2004) Essential oils: their antibacterial properties and potential application in foods: a review. Int J Food Microbiol 94:223–253

Bussmann RW, Glenn A, Meyer K, Kuhlman A, Townesmith A (2010) Herbal mixtures in traditional medicine in Northern Peru. J Ethnobiol Ethnomed 6:10

Chakraborthy GS (2008) Antimicrobial activity of the leaf extracts of Calendula officinalis (LINN.). J Herb Med Toxicol 2(2):65–66

Chandrasekaran M, Kannathasan K, Venkatesalu V, Prabhakar K (2009) Antibacterial activity of some salt marsh halophytes and mangrove plants against methicillin resistant *Staphylococcus aureus*. World J Microbiol Biotechnol 2:155–160

Chaudry NMA, Tariq P (2008) In vitro antibacterial activities of Kalonji, cumin and poppy seeds. Pak J Bot 40(1):461–467

Compean KL, Ynalvez RA (2014) Antimicrobial activity of plant secondary metabolites: a review. Res J Medicinal Plant 8(5):204–213

Cushnie TP, Lamb AJ (2005) Antimicrobial activity of flavonoids. Int J Antimicrob Agents 26:343–356

David E, Elumalai EK, Sivakumar C, Therasa SV, Thirumalai T (2010) Evaluation of antifungal activity and phytochemical screening of Solanum surattense seeds. J Pharm Res 3:684–687

Demirci B, Yusufoglu HS, Tabanca N, Temel HE, Bernier UR, Agramonte NM, Alqasoumi SI, Al-Rehaily AJ, Can Bas KH, Demirci F (2017) *Rhanterium epapposum* Oliv. essential oil: chemical composition and antimicrobial, insect-repellent and anticholinesterase activities. Saudi Pharm J 25:703–708

Dua A, Gaurav G, Balkar S, Mahajan R (2013) Antimicrobial properties of methanolic extract of cumin (*Cuminum cyminum*) seeds. Int J Res Ayurveda Pharm 4(1):104–107

Duraipandiyan V, Ignacimuthu S (2011) Antifungal activity of traditional medicinal plants from Tamil Nadu, India. Asian Pac J Trop Biomed 1:204–215

Egamberdieva D, Jabborova D (2018) Medicinal plants of Uzbekistan and their traditional use. In: Egamberdieva D, Ozturk M (eds) Vegetation of Central Asia and environs. Springer, New York. https://doi.org/10.1007/978-3-319-99728-5

Egamberdieva D, Teixeira da Silva JA (2015) Medicinal plant and PGPR: a *new* frontier for-*phytochemicals*. In: Egamberdieva D, Shrivastava S, Varma A (eds) Plant Growth-Promoting Rhizobacteria (PGPR) and medicinal plants. Springer, Berlin, pp 287–303. https://doi.org/10.1007/978-3-319-13401-7_14

Egamberdieva D, Wirth S, Behrendt U, Parvaiz A, Berg G (2017) Antimicrobial activity of medicinal plants correlates with the proportion of antagonistic endophytes. Front Microbiol 8:199. https://doi.org/10.3389/fmicb.2017.00199

Ejechi BO, Akpomedaye DE (2005) Activity of essential oil and phenolic extract of pepperfruit (Dennettia tripetala G. Baker); Anonaceae against some food borne microorganisms. Afr J Biotechnol 4(3):258–261

El-Beltagi HS, El-Ansary AE, Mostafa MA, Kamel TA, Safwat G (2019) Evaluation of the phytochemical, antioxidant, antibacterial and anticancer activity of Prunus domestica. Not Bot Horti Agrobo 47(2):395–404

El-Mahmood AM, Ogbonna OB, Raji M (2010) The antibacterial activity of Azadarichta indica (neem) seeds extracts against bacterial pathogens associated with eye and ear infections. J Med Plants Res 4(14):1414–1421

Fadeyi MO, Alcapan UE (1989) W Afr J Pharmaco Drug Res 9:29–30

Fakhfakh N, Abdelhedi O, Jdir H, Nasri M, Zouari N (2017) Isolation of polysaccharides from Malva aegyptiaca and evaluation of their antioxidant and antibacterial properties. Int J Biol Macromol 105:1519–1525

Farjana A, Zerin N, Kabir M (2014) Antimicrobial activity of medicinal plant leaf extracts against pathogenic bacteria. Asian Pac J Trop Dis 4:920–923

Ferronato R, Marchesan ED, Pezenti E, Bednarski F, Onofre SB (2007) Antimicrobial activity of essential oils produced by Baccharis dracunculifolia D.C. and Baccharis uncinella D.C. (Asteraceae). Rev Bras Pharmacogn 17(2):224–230

Frey FM, Meyers R (2010) Antibacterial activity of traditional medicinal plants used by Haudenosaunee peoples of New York State. BMC Complement Altern Med 10:64

Gadisa E, Weldearegay G, Desta K, Tsegaye G, Hailu S, Jote K, Takele A (2019) Combined antibacterial effect of essential oils from three most commonly used Ethiopian traditional medicinal plants on multidrug resistant bacteria. BMC Complement Altern Med 19:24

Gill AO, Holley RA (2006) Disruption of Escherichia coli, Listeria monocytogenes and Lactobacillus sakei cellular membranes by plant oil aromatics. Int J Food Microbiol 108:1–9
Gnat S, Majer-Dziedzic B, Nowakiewicz A, Trościańczyk A, Ziolkowska G, Jesionek W, Choma I, Dziedzic R, Zięba P (2017) Antimicrobial activity of some plant extracts against bacterial pathogens isolated from feces of red deer (Cervus elaphus). Pol J Vet Sci 20(4):697–706
Guesmi F, Ben Hadj A, Landoulsi A (2017) Investigation of extracts from Tunisian ethnomedicinal plants as antioxidants, cytotoxins, and antimicrobials. Biomed Environ Sci 30(11):811–824
Gupta D, Dubey J, Kumar M (2016) Phytochemical analysis and antimicrobial activity of some medicinal plants against selected common human pathogenic microorganisms. Asian Pac J Trop Dis 6(1):15–20
Gupta RN, Kartik V, Manoj P, Singh PS, Alka G (2010) Antibacterial activities of ethanolic extracts of plants used in flok medicine. Int J Res Ayurveda Pharm 1(2):529–535
Hajji M, Hamdi M, Sellimi S, Ksouda G, Laouer H, Li S, Nasri M (2019) Structural characterization, antioxidant and antibacterial activities of a novel polysaccharide from Periploca laevigata root barks. Carbohydr Polym 206:380–388
Hasan MF, Das R, Khan A, Hossain MS, Rahman M (2009) The determination of antibacterial and antifungal activities of Polygonum hydropiper (L.) root extract. Adv Biol Res 3(1–2):53–56
Hleba L, Vukovic N, Horskac E, Petrova J, Sukdolak S, Kacaniova M (2014) Phenolic profile and antimicrobial activities to selected microorganisms of some wild medical plant from Slovakia. Asian Pac J Trop Dis 4(4):269–274
Hu HQ, Li ZS, He H (2010) Characterization of an antimicrobial material from a newly isolated Bacillus amyloliquefaciens from mangrove for biocontrol of Capsicum bacterial wilt. Biol Control 54:359–365
Islam NU, Amin R, Shahid M, Amin M, Zaib S, Iqbal J (2017) A multitarget therapeutic potential of Prunus domestica gum stabilized nanoparticles exhibited prospective anticancer, antibacterial, urease inhibition, anti-inflammatory and analgesic properties. BMC Complement Altern Med 17(1):276
Jabeen K, Hanif S, Naz S, Iqbal S (2013) Antifungal activity of Azadirachta indica against Alternaria solani. J Life Sci Technol 1(1):89–93
Johnson M, Wesely EG, Selvan N, Kavitha MS (2010) In vivo and in vitro anti-bacterial efficacy of Alternanthera sessilis (Linn.). Int J Pharma Res Dev 2(10):72–79
Kaleem M, Medha P, Ahmed QU, Asif M, Bano B (2008) Beneficial effects of Annona squamosa extract in streptozotocin-induced diabetic rats. Singap Med J 49:800–804
Kalidindi N, Thimmaiah NV, Jagadeesh NV, Nandeep R, Swetha S, Kalidindi B (2015) Antifungal and antioxidant activities of organic and aqueous extracts of Annona squamosa Linn. leaves. J Food Drug Anal 23:795–802
Karsha PV, Bhagyalakshmi O (2010) Antibacterial activity of black pepper (Piper nigrum Linn.) with special reference to its mode of action on bacteria. Ind J Nat Prod Res 1(2):213–215
Khajure PV, Rathod JL (2010) Antimicrobial activity of extracts of Acanthus ilicifolius extracted from the mangroves of Karwar coast Karnataka. Rec Res Sci Technol 2(6):98–99
Kokoska L, Polesny Z, Rada V, Nepovim A, Vanek T (2002) Screening of some Siberian medicinal plants for antimicrobial activity. J Ethnopharmacol 82:51–53
Koona S, Budida S (2011) Antibacterial potential of the extracts of the leaves of Azadirachta indica Linn. Not Sci Biol 3(1):65–69
Kumar SM, Rajeswari, Astalakshmi M (2011) Evaluation of antimicrobial activities of Aristolochia Indica (Linn). Int J Pharm Pharma Sci 3(4):271–272
Lacaille-Dubois MA, Wagner H (1996) A review of the biological and pharmacological activities of saponins. Phytomedicine 2:363–386
Lee JH, Choi HJ (2015) Antiviral and antimicrobial activity of medicinal plant extracts. J Microb Biochem Technol 7:286–288
Mahboubi A, Asgarpanah J, Sadaghiqani PN, Faizi M (2015) Total phenolic and flavonoid content and antibacterial activity of Punica granatum L. Var. pleniflora flower (Golnar) against bacterial strains causing food borne diseases. BMC Complem Altern Med 15:366–373

Mahmoud DA, Hassanein NM, Youssef KA, Abou Zeid MA (2011) Antifungal activity of different neem leaf extracts and the nimonol against some important human pathogens. Braz J Microbiol 42:1007–1016
Mamedov N, Gardner Z, Craker LE (2005) Medicinal plants used in Russia and Central Asia for the treatment of selected skin conditions. J Herbs Spices Med Plants 11(1–2):191–222
Mamedov, Egamberdieva (2018) Phytochemical constituents and pharmacological effects of liquorice: a review. In: Ozturk M, Hakeem KR (eds) Plant and human health, Volume 3
Mangang HC, Chhetry GKN (2012) Antifungal properties of certain plant extracts against Rhizoctonia solani causing root rot of French bean in organic soil of Manipur. Int J Sci Res Publ 2(5):001
Mares D, Romagnoli C, Tosi B, Andreotti E, Chillemi G, Poli F (2005) Chicory extracts from Cichorium intybus L. as potential antifungals. Mycopathologia 160(1):85–91
Mathabe MC, Nikolova RV, Lall N, Nyazema NZ (2006) Antibacterial activities of medicinal plants used for the treatment of diarrhoea in Limpopo Province, South Africa. J Ethnopharmacol 105:286–293
Mighri H, Sabri K, Eljeni H, Neffati M, Akrout A (2015) Chemical composition and antimicrobial activity of Pituranthos chloranthus (Benth.) Hook and Pituranthos tortuosus (Coss.) Maire essential oils from Southern Tunisia. Adv Biol Chem 5:273–278
Mishra MP, Rath S, Swain SS, Ghosh G, Das D, Padhy RN (2017) In vitro antibacterial activity of crude extracts of 9 selected medicinal plants against UTI causing MDR bacteria. J King Saud Univ Sci 29:84–95
Mostafa AA, Al-Askar AA, Almaary KS, Dawoud TM, Sholkamy EN, Bakri MM (2018) Antimicrobial activity of some plant extracts against bacterial strains causing food poisoning diseases. Saudi J Biol Sci 25:361–366
Nandagopal S, Kumari RBD (2007) Phytochemical and antibacterial studies of chicory (Cichorium intybus L.)—a multipurpose medicinal plant. Adv Biol Res 1:17–21
Ogueke CC, Uwaleke J, Owuamanama CI, Okolue B (2014) Antimicrobial activities of Alstonia boonei stem bark, a Nigerian traditional medicinal plant. Asian Pac J Trop Dis 4:S957–S962
Omulokoli E, Khan B, Chhabra S (1997) Antiplasmodial activity of four Kenyan medicinal plants. J Ethnopharmacol 56:133–137
Owolabi J, Omogbai EKI, Obasuyi O (2007) Antifungal and antibacterial activities of ethanolic and aqueous extract of Kigelia african (Bignoniaceae) stem bark. Afr J Biotechnol 6(14):882–885
Oyemitan IA, Elusiyan CA, Akinkunmi EO, Obuotor EM, Akanmu MA, Olugbade TA (2019) Memory enhancing, anticholinesterase and antimicrobial activities of βphenylnitroethane and essential oil of Dennettia tripetala Baker f. J Ethnopharmacol 229:256–261
Padhi LP, Panda SK, Satapathy SN, Dutta SK (2011) In vitro evaluation of antibacterial potential of Annona squamosa Linn. and Annona reticulata L. from Similipal Biosphere Reserve, Orissa, India. J Agric Technol 7(1):133–142
Pandey A, Singh P (2011) Antibacterial activity of Syzygium aromaticum (Clove) with metal ion effect against food borne pathogens. Asian J Plant Sci Res 1(2):69–80
Patel DJ, Kumar V (2008) Annona squamosal L.: phytochemical analysis and antimicrobial screening. J Pharm Res 1(1):34–38
Patil SM, Patil MB, Sapkale GN (2009) Antimicrobial activity of Glycyrrhiza glabra Linn. roots. Int J Chem Sci 7(1):585–591
Pham HNT, Sakoff JA, Bond DR, Vuong QV, Bowyer MC, Scarlett CJ (2018) In vitro antibacterial and anticancer properties of Helicteres hirsuta Lour. leaf and stem extracts and their fractions. Mol Biol Rep 45:2125–2133
Pirbalouti AG, Jahanbazi P, Enteshari S, Malekpoor F, Hamedi B (2010) Antimicrobial activity of some Iranian medicinal plants. Arch Biol Sci Belgrade 62(3):633–642
Qader MK, Khalid NS, Abdullah AM (2013) Antibacterial activity of some plant extracts against clinical pathogens. Int J Microbiol Immunol Res 1(5):53–56
Qin XL, Lunga PK, Zhao YL, Liu YP, Luo XD (2016) Chemical constituents of Solanum coagulans and their antimicrobial activities. Chin J Nat Med 14(4):308–312

Raj DS, Vennila JJ, Aiyavu C, Panneerselvam K (2009) The hepatoprotective effect of alcoholic extract of Annona squamosa leaves on experimentally induced liver injury in Swiss albino mice. Int J Integr Biol 5:182–186

Rakholiya K, Kaneria M, Chanda S (2014) Inhibition of microbial pathogens using fruit and vegetable peel extracts. Int J Food Sci Nutr 1–7

Ravikumar S, Muthuraja M, Sivaperumal P, Gnanadesign M (2010) Antibacterial activity of the mangrove leaves Exoecaria agallocha against selected fish pathogens. Asian J Med Sci 2(5):211–213

Rehman A, Ullah N, Ullah H, Ahmad I (2014) Antibacterial and antifungal study of Cichorium intybus. Asian Pac J Trop Dis 4(2):S943–S945

Rios JL, Recio MC (2005) Medicinal plants and antimicrobial activity. J Ethnopharmacol 4:80–100

Saad S, Taher M, Susanti D, Qaralleh H, Abdul Rahim NA (2011) Antimicrobial activity of mangrove plant (Lumnitzera littorea). Asian Pac J Trop Med 523–525

Sapkota R, Dasgupta R, Nancy, Rawat DS (2012) Antibacterial effects of plants extracts on human microbial pathogens & microbial limit tests. Int J Res Pharm Chem 2(4):926–936

Selvamohan T, Ramadas V, Shibila Selva Kishore S (2012) Antimicrobial activity of selected medicinal plants against some selected human pathogenic bacteria. Adv Appl Sci Res 3(5):3374–3381

Shaikh T, Rub RA, Sasikumar S (2016) Antimicrobial screening of Cichorium intybus seed extracts. Arab J Chem 9:S1569–S1573

Shan B, Cai Y, Brooks JD, Corke H (2007) The in vitro antibacterial activity of dietary spice and medicinal herb extracts. Int J Food Microbiol 117:112–119

Sher A (2009) Antimicrobial activity of natural products from medicinal plants. Gomal J Med Sci 7(1):72–78

Shrivastava S, Egamberdieva D, Varma A (2015) PGPRs and medicinal plants- the state of arts. In: Egamberdieva D, Shrivastava S, Varma A (eds) Plant Growth-Promoting Rhizobacteria (PGPR) and medicinal plants, Springer, pp 1–16. https://doi.org/10.1007/978-3-319-13401-7_1

Shurigin V, Davranov K, Wirth S, Egamberdieva D, Bellingrath-Kimura SD (2018) Medicinal plants with phytotoxic activity harbour endophytic bacteria with plant growth inhibitory properties. Environ Sustain 1(2):209–215

Silva MAR, Higino JS, Pereira JV, Siqueira-Júnior JP, Pereira MSV (2008a) Antibiotic activity of the extract of Punica granatum Linn. over bovine strains of Staphylococcus aureus. Rev Bras Pharmacogn 18(2):209–212

Silva MSA, Silva MAR, Higino JS, Pereira MSV, Carvalho AAT (2008b) In vitro antimicrobial activity and antiadherence of Rosmarinus officinalis Linn. against oral planktonic bacteria. Rev Bras Pharmacogn 18(2):236–240

Sivaperumal P, Ramasamy P, Inbaneson S, Ravikumar S (2010) Screening of antibacterial activity of mangrove leaf bioactive compounds against antibiotic resistant clinical isolates. World J Fish Mar Sci 2(5):348–353

Soberón JR, Sgariglia MA, Pastoriza AC, Soruco EM, Jager SN, Labadie GR, Sampietro DA, Vattuone MA (2017) Antifungal activity and cytotoxicity of extracts and triterpenoid saponins obtained from the aerial parts of Anagallis arvensis L. J Ethnopharmacol 203:233–240

Szabó I, Pallag A, Cristian- Blidar F (2009) The antimicrobial activity of the Cnicus benedictus L. extracts. Analele Universităţii din Oradea, Fascicula Biol Tom 1:126–128

Taghreed AI (2012) Chemical composition and biological activity of extracts from Salvia bicolor Desf. growing in Egypt. Molecules 17:11315–11334

Tajkarimi MM, Ibrahim SA, Cliver DO (2010) Antimicrobial herb and spice compounds in food. Food Control 21:1199–1218

Ushimaru PI, Silva MTN, Di Stasi LC, Barbosa L, Fernandes JA (2007) Antibacterial activity of medicinal plant extracts. Braz J Microbiol 38(1):717–719

Verma R, Rawat A, Ganie SA, Agnihotri RK, Sharma R, Mahajan S (2013) In vitro antibacterial activity of Cichorium intybus against some pathogenic bacteria. Br J Pharm Res 3(4):767–775

Verma V, Singh R, Tiwari RK, Srivastava N, Verma S (2012) Antibacterial activity of extracts of Citrus, Allium and Punica against food borne spoilage. Asian J Plant Sci Res 2(4):503–509
Vizhi DK, Irulandi K, Mehalingam P, Kumar NN (2016) In vitro antimicrobial activity and phytochemical analysis of fruits of Syzygium aromaticum (L.) Merr. & L.M.Perry – an important medicinal plant. J Phytopharmacol 5(4):137–140
Wang S, Zheng Y, Xiang F, Li S, Yang G (2016) Antifungal activity of Momordica charantia seed extracts toward the pathogenic fungus Fusarium solani L. J Food Drug Anal 24(4):881–887
Xie JH, Shen MY, Xie MY, Nie SP, Chen Y (2012) Ultrasonic-assisted extraction, antimicrobial and antioxidant activities of Cyclocarya paliurus (Batal.) Iljinskaja polysaccharides. Carbohydr Polym 89:177–184

Chapter 4
Biologically Active Components of the Western Ghats Medicinal Fern *Diplazium esculentum*

Ammatanda A. Greeshma, Kandikere R. Sridhar, and Mundamoole Pavithra

Abstract The riparian fern *Diplazium esculentum* is nutritionally and medicinally valuable in the ethnic population of the Western Ghats of India. Fiddle heads of this fern are nutraceutically versatile and consumed similar to other leafy vegetables. The present study has addressed biologically active compounds (total phenolics, tannins, flavonoids, vitamin C, phytic acid, L-DOPA, trypsin inhibition and haemagglutination) and antioxidant potential (total antioxidant activity, ferrous ion-chelating capacity, reducing power, DPPH and ABTS radical-scavenging activities) in uncooked and cooked fiddle heads. Fiddle heads were devoid of L-DOPA as well as haemagglutinin activity. Total phenolics and flavonoids contents were not influenced by cooking, while tannins, vitamin C, phytic acid and trypsin inhibition activity were higher in uncooked than cooked fiddle heads. Among the antioxidant properties, total antioxidant activity and ferrous ion-chelating capacity were not influenced by cooking, whereas reducing power, DPPH and ABTS radical-scavenging activities were higher in uncooked than cooked fiddle heads. The principal component analysis was performed to ascertain the link between bioactive components and antioxidant potential of uncooked and cooked fiddle heads. Vitamin C and trypsin inhibition activity of uncooked fiddle heads influenced the ABTS radical-scavenging activity, while total phenolics, flavonoids and tannins of cooked samples influenced the total antioxidant activity, ferrous ion-chelating capacity and reducing power. Cooking has differentially influenced the bioactive components as well as antioxidant potential of fiddle heads. There also seems to be geographical difference in quantity of bioactive components (phenolics, flavonoids and vitamin C) as well as antioxidant potential (reducing power). Further insights are warranted to utilize different parts of the ethnically valued fern *D. esculentum* for nutritional and therapeutic advantages.

Keywords Antioxidant potential · Bioactive compounds · Ethnic value · Leafy vegetable · Non-conventional food · Nutraceutical potential · Riparian fern

---

A. A. Greeshma · K. R. Sridhar (✉) · M. Pavithra
Department of Biosciences, Mangalore University, Mangalore, Karnataka, India

D. Egamberdieva, A. Tiezzi (eds.), *Medically Important Plant Biomes: Source of Secondary Metabolites*, Microorganisms for Sustainability 15,
https://doi.org/10.1007/978-981-13-9566-6_4

4.1 Introduction

Consumption of nutrient-rich vegetables is one of the alternatives to overcome the malnutrition as well as nutrition-dependent human ailments (Ruma 2016; Sridhar and Karun 2017). Various ethnic groups are formulating and utilizing plant-based nutrients as well as medicines for several generations worldwide. In the recent past, ethnobotanical studies are advancing rapidly towards evaluation of traditional knowledge of wild plant sources as food and medicine (Jain 1987a, b, 1991; Sridhar et al. 2016; Sridhar and Karun 2017). Up to 80% of world's foods are derived from plants belonging to 17 families, and the most important families include Brassicaceae, Fabaceae and Poaceae (Fici 2016). However, pteridophytes (ferns) are largely ignored and untapped natural resource for food as well as medicine (Singh and Singh 2012). Ferns are well known for food value (nutritional), medicine (homeopathic, ayurvedic and unani), insecticide (anti-herbivory) and antibiotic properties (Benniamin 2011).

Pteridophytes and their allies of the Western Ghats include up to 256 illustrated forms (Dudani et al. 2012). One of the nonconventional edible and medicinal ferns of the Western Ghats is *Diplazium esculentum*, which is riparian and commonly occurs on the banks of streams and rivers. It is an important delicacy of multiethnic population of the Western Ghats (Archana et al. 2013; Greeshma and Sridhar 2016; Greeshma et al. 2018). Rinsed fiddle heads of *D. esculentum* are pan-frayed by seasoning with edible oil, spices and grated coconut to serve as a starter dish. Tender leaves of *D. esculentum* are consumed with hot sauce in Uttarkhand, India (Alderwerelt 1989). It also serves as an ingredient in culinary dishes in the Philippines (Tongco et al. 2014).

Apart from nutritional qualities, *D. esculentum* is also known for biologically active constituents (Kaushik et al. 2012; Archana et al. 2013; Tongco et al. 2014; Greeshma et al. 2018). Its foliage is traditionally used to cure headache, pain, fever, wounds, dysentery, glandular swelling, diarrhoea and skin infections (Akter et al. 2014). Roy et al. (2013) demonstrated through in vitro assays that the extract of *D. esculentum* possesses significant quantity of natural antioxidants, which prevents progression of various oxidative stress-associated diseases. The tribal communities and ethnic groups of the Western Ghats are utilizing different parts of this fern (e.g. rhizome, stem, fronds, pinna and spores) in treatment of many human ailments (Akter et al. 2014). Dried rhizomes of *D. esculentum* also serve as insecticides, and its decoction is useful in curing haemoptysis as well as cough (Anderson et al. 2003). In Albino mice model, aqueous extract of fresh leaves of *D. esculentum* at low doses (100 mg/kg) served as significant CNS stimulant against standard caffeine (Kaushik et al. 2012). Fiddle heads of *D. esculentum* being known for nutritional and medicinal values, the present study envisaged to emphasize some of the biologically active components, antioxidant potential and their interrelationships.

Fig. 4.1 Side view of grown-up *Diplazium esculentum* (**a**), close-up view of fiddle heads with tender pinna (**b**) and different swirling patterns of fiddle heads (**c**–**f**)

4.2 The Fern

The fiddle heads of fern *Diplazium esculentum* (Retz.) Sw. (family, Athyriaceae) were sampled from five different locations of the Western Ghats of Karnataka (Bethri, Kiggal, Mekeri, Murnad and Nelji) during southwest monsoon season (June–August, 2014) and brought to the laboratory in cold packs. The identity of the fern was confirmed by taxonomic descriptions by (Beddome 1865; Manickam and Irudayaraj 1992) (Fig. 4.1). Five fiddle head samples were independently processed within 6–8 h of sampling by rinsing in distilled water to remove the debris followed by pressing with paper towel to remove surface water. Each sample was divided into two groups, the first group was dried in an oven (50–55 °C), while the second group was pressure cooked without addition of more water followed by oven drying. The dried samples were milled (Wiley Mill, mesh # 30), and powder samples were preserved in refrigerator for further analysis.

4.3 Assessment

4.3.1 Bioactive Components

The fern samples were assessed for eight bioactive components like total phenolics, tannins, flavonoids, vitamin C, phytic acid, L-DOPA, trypsin inhibition and haemagglutination.

Total Phenolics Total phenolics content of fern flour was assessed by the method outlined by Rosset et al. (1982). To fern flour (100 mg) methanol (50%, 10 ml) was added, mixed, incubated in water bath (95 °C, 10 min), cooled and centrifuged (2000 rpm, 20 min) and the supernatant recovered. The process was repeated, and the pooled final volume of supernatant was made to 20 ml. Flour extract (0.5 ml) was mixed with equal volume of distilled water, incubated (10 min, room temperature) on adding sodium carbonate (prepared in 0.1 N NaOH, 5 ml). On addition of Folin-Ciocalteu reagent (dilution 1:2, 0.5 ml), the absorbance was read (725 nm; UV-VIS Spectrophotometer-118, Systronics, Ahmedabad, Gujarat, India). The content of total phenolic was expressed as standard mg tannic acid equivalents/gram fern power (mg TAEs/g).

Tannins Tannin content of fern flour was evaluated based on the procedure by Burns (1971). To fern powder (1 g) methanol (50 ml) was mixed to extract tannins and shaken on a rotary shaker (28 °C, 24 h) followed by centrifugation (1500 rpm) to sample the supernatant. To the supernatant (1 ml) vanillin hydrochloride was added (5 ml: 4% in methanol +8% concentrated HCl in methanol; 1:1) followed by incubation (20 min, room temperature), and the absorbance was measured at 500 nm. The catechin dissolved in methanol served as standard, and tannin content was expressed in mg catechin equivalents (mg CEs/g).

Flavonoids Total flavonoids content in fern flour was detected by the method outlined by Chang et al. (2002). Fern flour (1 mg) was extracted with methanol (1.5 ml), aliquots of extract (0.5 ml each) were mixed with aluminium chloride (10%, 0.1 ml) + potassium acetate (1M, 0.1 ml), and the final volume was made to 3 ml in distilled water followed by incubation (30 min, room temperature). The standard used was quercetin dihydrate, and absorbance was measured (415 nm) and expressed the flavonoids in mg equivalents/gram fern powder (mg QEs/g).

Vitamin C Vitamin C content of fern powder was evaluated based on the procedure by Roe (1954). Powder flour (1 g) was extracted using trichloroacetic acid (TCA, 5%, 10 ml), and aliquots of extract (0.2 ml) were made up to 1 ml using TCA (5%) and mixed followed by addition of chromogen (1 ml) (dinitrophenyl hydrazine thiourea copper sulphate solution: 5 parts of 5% thiourea +5 parts of 0.6% copper sulphate + 90 parts of 2% 2,4-dinitrophenylhydrazine in H_2SO_4). The mixture was incubated (boiling water bath, 10 min), cooled, and on addition of H_2SO_4 (65%, 4 ml) further incubated (room temperature, 10 min), and absorbance was measured (540 nm). The standard used was ascorbic acid for quantification of vitamin C and represented in mg ascorbic acid equivalents/gram fern powder (mg AAEs/g).

Phytic Acid Phytic acid in fern powder was determined based on the method by Deshpande et al. (1982) and Sathe et al. (1983). Fern powder (2 g) was extracted with sodium sulphate (10 ml, 10% in 1.2% HCl) and stirred (room temperature,

2 h). On centrifugation (3000 rpm, 10 min), the supernatant was made up to 10 ml in sodium sulphate. The extract (5 ml) was blended with ferric chloride (2 g in 16.3 ml of 12N HCl, diluted to 1 L) and vortexed followed by centrifugation (3000 rpm, 10 min). Filtered the supernatant (Whatman # 1), and the filtrate was made to 10 ml using distilled water. Free soluble phosphorous was determined by vandomolybdophosphoric acid method by potassium dihydrogen phosphate as reference to express phytic acid in percentage.

$$\text{Total phosphorus}\left(\text{g}/100\,\text{g}\right)=\frac{M\times V\times F}{10{,}000\times W\times V}\times\Delta A_{\text{p}} \quad (4.1)$$

[M, average of phosphorus standard (μg/ΔA_{p}); V, original sample in ml; F, dilution factor; ΔA_{p}, absorbance; W, weight of sample (g); V, volume of sample (ml)]

$$\text{Phytic acid}\left(\%\right)=\frac{\text{Phosphorus}\left(\text{g}/100\,\text{g}\right)}{0.282} \quad (4.2)$$

(0.282, factor used to convert phosphorus into phytic acid as it contains 28.2% of phosphorus).

L-DOPA The L-DOPA (L-3,4-dihydroxyphenylalaninne) of fern powder was determined by method proposed by Fujii et al. (1991). Fern samples were mixed with distilled water (1 ml), incubated (room temperature) for 2 h and centrifuged (1500 rpm, 10 min), and the supernatant is allowed to concentrate to dryness using a rota evaporator. To eliminate high molecular weight compounds, the extract was dissolved in distilled water followed by filtering through ultrafilter overnight. The fraction was purified using ODS extraction mini column (C18 Sep-Pak Cartridge, Waters) with water followed by evaporation to dryness. The L-DOPA was determined in HPLC (Tosoh system DP-8020; UV-8020, 280 nm; Column, Aqua 180 Mightsil; Kanto chemical Co. Inc., Japan) as well as LC-ESI/MS (Positive mode; Waters 181 Associates Inc., Milford, MA).

Trypsin Inhibition Trypsin activity was evaluated according to the method by Kakade et al. (1974). Fern powder (1 g) was stirred constantly with NaOH (0.01N, 50 ml) up to 10 min. The extract (1 ml) was diluted with distilled water (1:1), followed by addition of enzyme standard (2 ml) (2 mg trypsin/100 ml 0.001 M NaOH) and incubated in water bath (37 °C, 10 min), and 5 ml BAPNA (40 mg N_{α}-Benzoyl-L-arginine 4-nitroanilide hydrochloride dissolved in dimethyl sulphoxide and made to 100 ml with Tris-buffer at 37 °C) was added and incubated at room temperature (10 min). Acetic acid (30%; 1 ml) was added to stop the reaction followed by measurement of absorbance (410 nm). Control was prepared as per protocol without addition of the fern extract. Trypsin inhibition (TIu)/mg of fern flour was calculated.

$$\text{TIu / mg} = \frac{\left[\left(A_{c410} - A_{s410}\right) \times 100\right] \text{per ml extract}}{\text{Mg sample per ml of extract}} \tag{4.3}$$

(A_c, absorbance of control; A_s, absorbance of sample)

Haemagglutination The haemagglutinin activity in fern powder was determined by the method outlined by Occeña et al. (2007). The fern extract was prepared by mixing defatted powder (1 g) in NaCl (0.9%; 10 ml) and incubated at room temperature (1 h). Centrifuged (2000 rpm, 10 min), supernatant was collected and filtered to use as crude agglutinin. Heparinized human blood samples were centrifuged (2000 rpm, 10 min) to separate erythrocytes. Erythrocytes (A^+, B^+, AB^+, O^+) were washed repeatedly until the clear supernatant was obtained (1:4; chilled saline, 0.9%). Processed erythrocytes (4 ml) were transferred into phosphate buffer (100 ml; 0.0006 M, pH 7.4) and incubated (37 °C, 1 h) by addition of trypsin (2%, 1 ml) on mixing. On incubation, trypsinized solution was repeatedly washed using saline (0.9%) to remove trypsin content. The erythrocytes were suspended in saline (0.9%) and made to 100 ml. Round bottomed 96-well microtitre plate was used for assay. Phosphate buffer (50 μl) was added in the well # 1–11, followed by the addition of crude agglutinin extract (50 μl) to the well # 1, and mixed, and twofold serial dilution was made up to well # 11. Erythrocytes suspension (50 μl) was added to well # 1–11. The well # 12 served as control for sample. Contents in the wells were gently mixed followed by incubation at room temperature (4 h) to observe haemagglutination in each well. Haemagglutination unit/gram (Hu/g) was calculated.

$$\text{Hu / g} = \frac{D_a \times D_b \times S}{V} \tag{4.4}$$

(D_a, dilution factor of extract in well #1; D_b, dilution factor of well containing 1 Hu is the well in which the haemagglutination was observed; S, initial extract/gram fern powder; V, volume of extract in well # 1).

4.3.2 Antioxidant Properties

Evaluation of antioxidant potential of any plant material has no universal method. In almost all methods, a radical has been generated, and the capability of test sample in quenching the radical is assessed (Erel 2004). According to Wong et al. (2006), it is necessary to evaluate at least two methods for a fair assessment of antioxidant potential of a biological material. In our study, five methods of assessment have been followed to get a fair idea of antioxidant potential of uncooked and cooked fiddle heads of *D. esculentum* (total antioxidant activity, ferrous ion-chelating capacity, reducing power, DPPH radical-scavenging activity and ABTS radical-scavenging activity).

The fern powder samples each of 0.5 g were extracted with 30 ml methanol using a rotary shaker (150 rpm, 48 h). After the samples were centifuged, the supernatant was transferred to a preweighed Petri dish and allowed to evaporate at room temperature. The extract weight was assessed gravimetrically and dissolved in methanol at concentration 1 mg/ml to assess antioxidant potential.

Total Antioxidant Activity Total antioxidant activity (TAA) was determined by the method by Prieto et al. (1999). To methanolic extract of fern (1mg/ml; 0.1 ml) added the reagent mixture (28 mM sodium phosphate +4 mM ammonium molybdate in 0.6 M sulphuric acid) followed by incubation (95 °C, 90 min). The absorbance was measured (695 nm), and the TAA was expressed in µM equivalents of ascorbic acid/gram (µM AAEs/g).

Ferrous Ion-Chelating Capacity Ferrous ion-chelating capacity of the methanolic extract of fern was determined by the protocol by Hsu et al. (2003). On mixing methanol extract (1 ml) with 2 mM ferrous chloride (0.1 ml) + ferrozine (5 mM, 0.2 ml), the volume was made to 5 ml (in methanol) followed by incubation (room temperature) for 10 min, and the absorbance was measured (562 nm). Control was prepared similar to the sample without addition of fern extract to calculate percent ferrous ion-chelating capacity.

$$\text{Ferrous ion-chelating activity}\,(\%) = \left(1 \ \frac{A_{s562}}{A_{c562}}\right)100 \tag{4.5}$$

(A_s, absorbance of sample; A_c, absorbance of control).

Reducing Power Reducing power of the fern extract was detected following the method by Oyaizu (1986) with a slight modification. Different concentrations of fern extract (0.2–1.0 mg/ml) were prepared in phosphate buffer (0.2 M, pH 6.6), and potassium ferricyanide (1%, 2.5 ml) was added and incubated (50 °C) up to 20 min. The TCA (10%, 2.5 ml) was added to the mixture followed by centrifugation (3000 rpm, 10 min), and supernatant (2.5 ml) was mixed with double-distilled water (2.5 ml) followed by addition of $FeCl_3$ (0.1%, 0.5 ml), and absorbance was measured (700 nm).

DPPH Radical-Scavenging Activity Radical-scavenging activity of the fern extract was determined according to Singh et al. (2002). Different concentrations of fern extract (0.2–1.0 mg/ml) were made up to 1 ml using methanol, and reagent was added (0.001 M DPPH in methanol, 4 ml). On mixing it was incubated in dark room temperature up to 20 min. The reagent devoid of extract served as control, and the absorbance was measured (517 nm).

$$\text{Free radical-scavenging activity}\,(\%) = \left(\frac{[A_{c517} - A_{s517}]}{A_{c517}}\right) \times 100 \tag{4.6}$$

(where A_c, absorbance of control; A_s, absorbance of sample)

ABTS Radical-Scavenging Activity The ABTS [(2, 2′-azinobis (3-ethylbenzothiazoline-6-sulfonic acid)] cationic radical decolourization assay was performed based on the procedure by Adedapo et al. (2008). Stock solution ($ABTS^+$, 7.4 mM, and potassium persulphate, 2.6 mM) and working solution (mixing stock solutions 1:1; allowed to react at room temperature, for 12 h in dark) were prepared. The working solution was diluted with methanol to get suitable absorbance (1 ± 0.01 units at 734 nm). Different concentrations of fern extract (made up to 62 μl using absolute alcohol) were treated with $ABTS^+$ (188 μl, in dark, 30 min) followed by measurement of absorbance (734 nm) to determine percent inhibition.

$$\text{Inhibition percentage}(\%) = \left(\frac{\text{Assay control} - (\text{Test} - \text{Control})}{\text{Assay control}} \right) 100 \quad (4.7)$$

(Assay control, ethanol + ABTS reagent; control, sample + ethanol + methanol).

4.3.3 Data Analysis

The distinction between uncooked and cooked fern flours in assays was assessed by t-test (StatSoft Inc. 2008). To establish the relationship between the bioactive components (total phenolics, tannins, flavonoids, vitamin C, phytic acid and trypsin inhibition activity) and antioxidant potential (total antioxidant activity, ferrous ion-chelating capacity, reducing power assay, DPPH radical-scavenging activity and ABTS radical-scavenging activity), the principal component analysis (PCA) was employed (SPSS version 16.0: www.spss.com).

4.4 Observations and Discussion

4.4.1 Bioactive Components

Total Phenolics Total phenolics content of the fiddle heads of *D. esculentum* was not influenced by pressure cooking ($p > 0.05$) (Fig. 4.2a). Turkmen et al. (2005) observed increased content of phenolics in cooked vegetables such as green beans, pepper and broccoli. The phenolic contents of fiddle heads of *D. esculentum* in the present study are higher than the fiddle heads from Assam, while opposite for the young pinna reported from Bangladesh, India (Darjeeling, Maharashtra) and the Philippines (Das et al. 2013; Roy et al. 2013; Akter et al. 2014; Tongco et al. 2014; Saha et al. 2015). Moderately high quantity of total phenolics in the study is comparable with an earlier report by Archana et al. (2013). Phenolic compounds are well known for their importance as antioxidants, antimicrobial agents and insecticidal potential (De Britto et al. 2012).

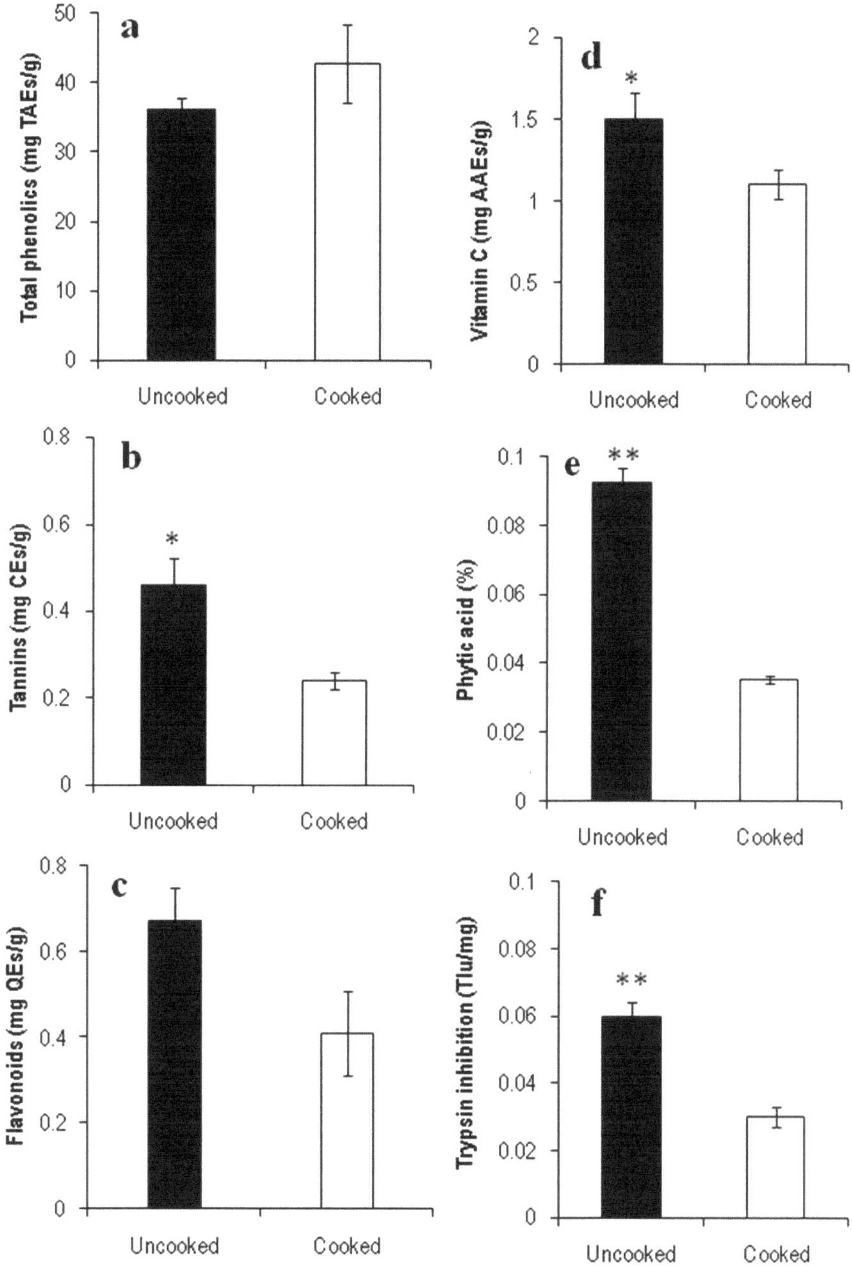

Fig. 4.2 Bioactive principles of fiddle heads of *Diplazium esculentum*: total phenolics (**a**), tannins (**b**), flavonoids (**c**), vitamin C (**d**), phytic acid (**e**) and trypsin inhibition (f) (*t*-test: $*p < 0.05$; $**p < 0.01$)

Tannins Tannins content was significantly higher in uncooked than cooked fiddle heads ($p < 0.05$) (Fig. 4.2b). Its content in uncooked samples in our study is higher than the young pinna from Assam (0.44 vs. 0.1 mg/g) (Saha et al. 2015). Tannins are known for their wide influence on the nutritive values of foodstuffs of humans as well as livestock (Saxena et al. 2013).

Flavonoids Although flavonoids content was higher in uncooked fiddle heads, it was not significantly differed compared to cooked fiddle heads ($p > 0.05$) (Fig. 4.2c). However, Stewart et al. (2000) noted that heat treatment increases the level of free flavonols. Flavonoids content in our study is higher than the fiddle heads sampled from Malaysia, while lower than the young pinna from Bangladesh, India (Darjeeling and Maharashtra) and the Philippines (Miean and Mohamed 2001; Das et al. 2013; Roy et al. 2013; Akter et al. 2014; Tongco et al. 2014). Flavonoids have a wide array of biochemical and pharmacological effects especially anti-oxidation, anti-inflammation, antiplatelet, antithrombotic and anti-allergic effects (Havsteen 1983; Gryglewski et al. 1987; Middleton and Kandaswami 1992; Cook and Samman 1996).

Vitamin C Vitamin C content was significantly higher in uncooked than cooked fiddle heads ($p < 0.05$) (Fig. 4.2d), and its quantity is higher than the fiddle heads from Assam, India (Saha et al. 2015). Archana et al. (2013) also reported presence of higher quantity of vitamin C in mature fronds than young petioles of *D. esculentum*. The vitamin C elicits many functions in the humans especially by boosting the overall health status (Walingo 2005). In addition to vitamin C, presence of β-carotene, tocopherol, thiamine, riboflavin and niacin was also reported from young and mature fronds of *D. esculentum* by Archana et al. (2013).

Phytic Acid Phytic acid content is significantly higher in uncooked than cooked fiddle heads ($p < 0.01$) (Fig. 4.2e), while its content is comparable to the fiddle heads sampled from Assam, India (Saha et al. 2015). Low quantity of phytic acid in *D. esculentum* in our study corroborates with an earlier study by Archana et al. (2013). Phytic acid serves as antioxidant, and it also involves in digestion and absorption of minerals in the intestine (Saha et al. 2015).

L-DOPA In uncooked and cooked fiddle heads, L-DOPA content was below detectable levels. Like L-DOPA, a number of nonprotein amino acids are known to be produced by plants, which possess strong allelopathic properties (Soares et al. 2014). The L-DOPA is also known for its application in treating Parkinson's disease (Hornykiewicz 2002).

Trypsin Inhibition Trypsin inhibition of uncooked fiddle heads was significantly higher than cooked fiddle heads ($p < 0.01$), which is nutritionally advantageous ($p < 0.01$) (Fig. 4.2f). Usually, deficiency of sulphur amino acids has been connected to presence of trypsin inhibitors in food stuffs owing to utilization of sulphur amino acids for synthesis of trypsin and chymotrypsin (Liener and Kakade 1969). Interestingly, in uncooked fiddle heads, the sulphur amino acids methionine and cystine were significantly higher ($p < 0.05$) than cooked fiddle heads (Greeshma et al. 2018).

Haemagglutination Uncooked and cooked fiddle heads were devoid of haemagglutinin activity against human erythrocytes (A^+, B^+, AB^+ and O^+), which signifies that it is safe for human consumption. Lectins are involved in carbohydrate storage, binding symbiotic rhizobia to develop root nodules, carrier for the delivery of chemotherapeutic agents, and also serve as tumour markers (Kumar et al. 2012). According to Hartmann and Meisel (2007), the lectins in food stuffs are known for immunomodulation processes.

4.4.2 Antioxidant Potential

Total Antioxidant Activity Although total antioxidant activity was higher in the uncooked fiddle heads, it was not significantly higher than cooked fiddle heads ($p > 0.05$) (Fig. 4.3a). Therefore, the total antioxidant activity in the fiddle heads was not influenced by cooking method followed. Evaluation of total antioxidant activity will be helpful to determine the additive effect of antioxidant properties of plant food stuffs (Pellegrini et al. 2003).

Ferrous Ion-Chelating Capacity Ferrous ion-chelating capacity also followed a similar pattern as in total antioxidant activity ($p > 0.05$) (Fig. 4.3b) and not influenced by the method of cooking applied. Metal ion-chelating capacity is one of the significant aspects because it reduces the concentration of the transition metal, which catalyses lipid peroxidation (Mohan et al. 2012).

Reducing Power The present study showed higher reducing power in uncooked than cooked fiddle heads ($p < 0.05$) (Fig. 4.3c). The reducing power of fiddle heads in the present study is higher compared to the fiddle heads of *D. esculentum* sampled from Darjeeling, India (Roy et al. 2013).

DPPH and ABTS Radical-Scavenging Activities The DPPH and ABTS assays estimate the capacity of antioxidant present in the fiddle heads to scavenge free radical. As seen in reducing power, DPPH as well as ABTS radical-scavenging activities were higher in uncooked than cooked fiddle heads ($p < 0.05$) (Fig. 4.3d, e). The DPPH radical-scavenging activity of the present study is lower compared to the fiddle heads of *D. esculentum* of Assam, India (Saha et al. 2015). Method of cooking fiddle heads might have affected the radical-scavenging activities and thus warrants to apply alternate methods of cooking (e.g. microwave, extrusion cooking and steaming) maximizing the radical-scavenging potential.

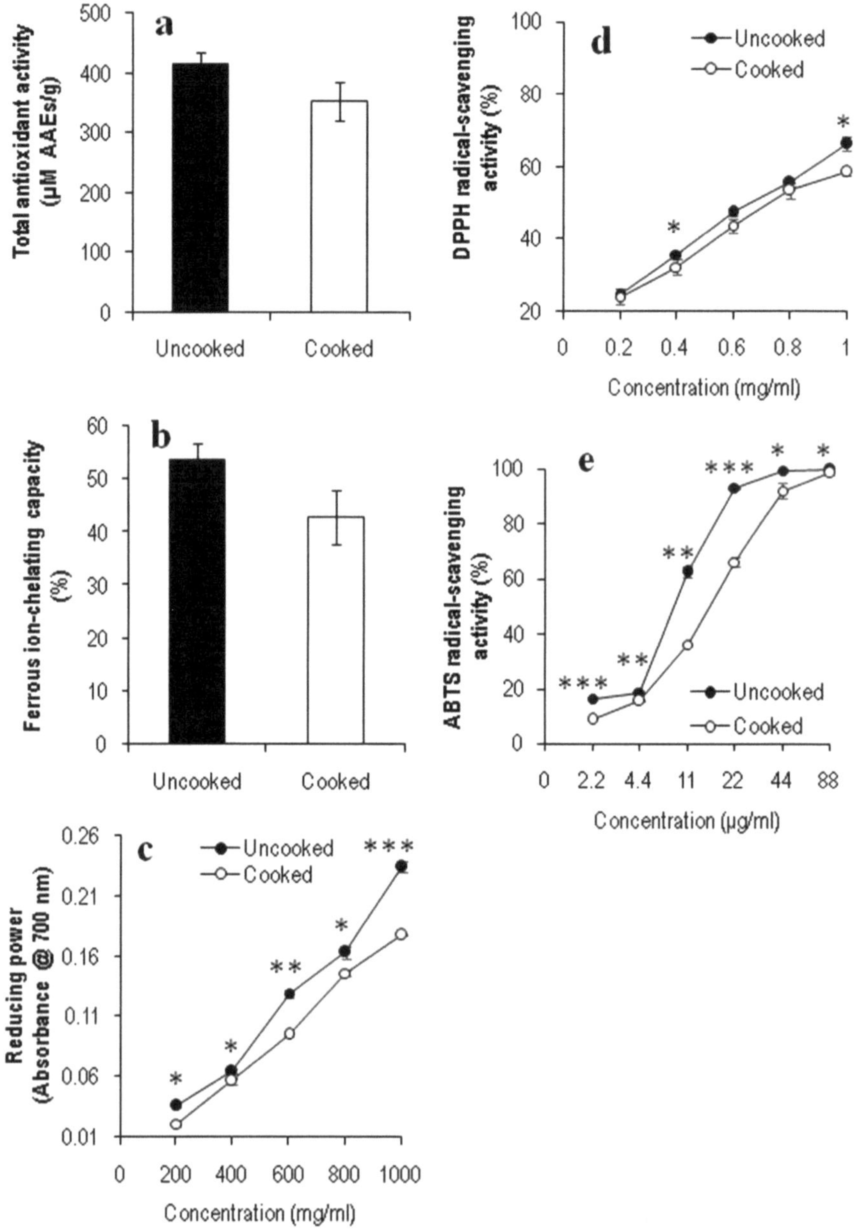

Fig. 4.3 Antioxidant activities of fiddle heads of *Diplazium esculentum*: total antioxidant activity (**a**), ferrous ion-chelating capacity (**b**), reducing power (**c**), DPPH radical-scavenging activity (**d**) and ABTS radical-scavenging activity (**e**) (*t*-test: $*p < 0.05$; $**p < 0.01$; $***p < 0.001$)

4.4.3 *Bioactive Components vs. Antioxidant Potential*

Principal component analysis (PCA) of bioactive components against antioxidant potential of uncooked fiddle heads of *D. esculentum* resulted in two components with 100% variance. The score plot showed variance 54.5% for component 1 and 45.5% for component 2 (Fig. 4.4a). The bioactive principles vitamin C (VCU) and trypsin inhibition (TiU) were clustered with ABTS radical-scavenging activity (ABTSU) at the right hand corner of the plot. Thus, the quantities vitamin C and trypsin inhibition potential of uncooked fiddle heads have a major role to play in radical-scavenging activity.

The PCA for bioactive principles of cooked fiddle heads against antioxidant capacity yielded two components with 100% variance. The score plot showed variance of 68.1% for component 1 and 31.9% for component 2 (Fig. 4.4b). The bioactive components like total phenolics (TpC), flavonoids (FlC) and tannins (TaC) were grouped with total antioxidant activity (TaaC), ferrous ion-chelating capacity (FiC) and reducing power (RpC) at the right hand corner of the plot. This confirms that although the total phenolics and flavonoids contents were not significantly higher in cooked fiddle heads compared to uncooked fiddle heads, they showed major influence on total antioxidant activity, ferrous ion-chelating capacity and reducing power. Thus, the quantities of total phenolics as well as flavonoids retained in the cooked fiddle heads are mainly responsible for ferrous ion-chelating capacity as well as total antioxidant activity.

4.5 Conclusions and Outlook

The present study addressed bioactive components and antioxidant potential of uncooked and cooked fiddle heads of ethnically valued fern *Diplazium esculentum* of the Western Ghats of India. Among the eight bioactive components assessed, the fiddle heads were devoid of L-DOPA and haemagglutinin activity; thus lack of latter activity signifies its nutritional advantage. Cooking has differential impacts on the bioactive components as well as antioxidant potentials of fiddle heads. Total phenolics and flavonoids contents were not influenced by cooking, while tannins, vitamin C, phytic acid and trypsin inhibition was higher in uncooked than cooked fiddle heads. Among the antioxidant properties, total antioxidant activity and ferrous ion-chelating capacity in fiddle heads were not influenced by the cooking, whereas cooking decreased the reducing power, DPPH and ABTS radical-scavenging activities. The principal component analysis indicated that vitamin C and trypsin inhibition activity of uncooked fiddle heads influenced the ABTS radical-scavenging activity, while total phenolics, flavonoids and tannins contents of cooked fiddle heads influenced the total antioxidant activity, ferrous ion-chelating capacity and reducing power. There seems to be geographical difference in phenolics, flavonoids and vitamin C of fiddle heads of the Western Ghats of India against Assam (Northeast

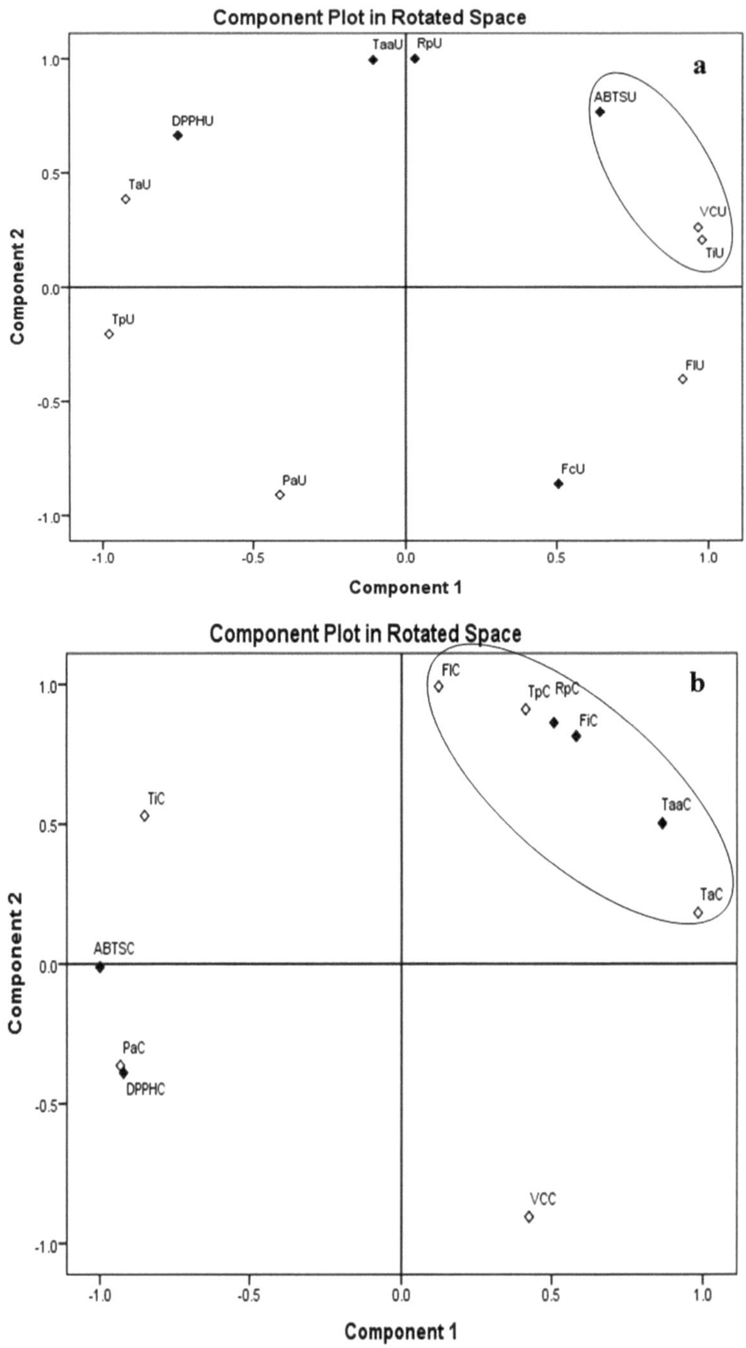

Fig. 4.4 Principal component analysis (PCA) of uncooked (with suffix U) (**a**) and cooked (with suffix C) (**b**) fiddle heads of *Diplazium esculentum*: Bioactive principles (total phenolics, Tp; tannins, Ta; flavonoids, Fl; vitamin C, VC; phytic acid, Pa; trypsin inhibition, Ti) and antioxidant activities (total antioxidant activity, Taa; ferrous ion-chelating capacity, Fc; reducing power, Rp; DPPH radical-scavenging activity, DPPH; ABTS radical-scavenging activity, ABTS)

India) and Malaysia. Similar trend was seen in reducing power of fiddle heads between the Western Ghats and Darjeeling (Northern India). As many vitamins (β-carotene, tocopherol, thiamine, riboflavin and niacin) were assessed qualitatively, future studies are warranted for their quantification. There are ample possibilities to manoeuver the quantity of bioactive components and antioxidant potential of fiddle heads on application of different thermal treatments (microwave, steaming and extrusion cooking) in favour of nutritional and or therapeutic benefits.

Acknowledgements Authors are grateful to Mangalore University to perform this study in the Department of Biosciences. GAA greatly acknowledges the award of INSPIRE Fellowship, Department of Science and Technology, New Delhi, Government of India. KRS is grateful to the University Grants Commission, New Delhi, India, for the award of UGC-BSR Faculty Fellowship during the tenure of this study.

References

Adedapo AA, Jimoh FO, Afolayan AJ, Masika PJ (2008) Antioxidant activities and phenolic contents of the methanol extracts of the stems of *Acokanthera oppositifolia* and *Adenia gummifera*. BMC Complement Altern Med 8:1–7

Akter S, Hossain MM, Ara I, Akhtar P (2014) Investigation of *in vitro* antioxidant, antimicrobial and cytotoxic activity of *Diplazium esculentum* (Retz). Sw. Int J Adv Pharm Biol Chem 3:723–733

Alderwerelt V (1989) Malaysian ferns. Asher and Co, Amsterdam

Anderson DK, Hall ED, Kumar R, Fong V, Endrinil LMS, Sani HA (2003) Determination of total antioxidant activity in three types of local vegetables shoots and the cytotoxic effect of their ethanol extracts against different cancer cell lines. Asia Pac J Clin Nutr 12:292–295

Archana GN, Pradeesh S, Chinmayee ME, Mini I, Swapna TS (2013) *Diplazium esculentum*: a wild nutrient-rich leaf vegetable from Western Ghats. In: Sabu A, Augustine A (eds) Prospects in bioscience: addressing the issues. Springer, New Delhi, pp 293–301

Beddome RH (1865) The ferns of South India. Gantz Bros, Madras. Reprint edition 1970, Today and Tomorrow's Printers & Publishers, New Delhi

Benniamin A (2011) Medicinal ferns of North Eastern India with special reference to Arunachal Pradesh. Ind J Tradit Knowl 10:516–522

Burns R (1971) Methods for estimation of tannins in grain sorghum. Agron J 63:511–512

Chang C, Yang M, Wen H, Chern J (2002) Estimation of total flavonoid content in propolis by two complementary colorimetric methods. J Food Drug Anal 10:178–182

Cook NC, Samman S (1996) Flavonoids – chemistry, metabolism, cardioprotective effects and dietary sources. J Nutr Biochem 7:66–76

Das B, Paul T, Apte KG, Chauhan R, Saxena RC (2013) Evaluation of antioxidant potential & quantification of polyphenols of *Diplazium esculentum* Retz. with emphasis on its HPTLC chromatography. J Pharm Res 6:93–100

De Britto AJ, Gracelin DHS, Kumar PBJR (2012) Phytochemical studies on five medicinal ferns collected from Southern Western Ghats, Tamil Nadu. Asia Pac J Trop Biomed 2:S536–S538

Deshpande SS, Sathe SK, Salunkhe DK, Cornforth DP (1982) Effects of dehulling on phytic acid, polyphenols and enzyme inhibitors of dry beans (*Phaseolus vulgaris* L). J Food Sci 47:1846–1850

Dudani S, Chandran SMD, Ramachandra TV (2012) Pteridophytes of Western Ghats: biodiversity documentation and taxonomy. Biju Kumar Narendra Publishing House, New Delhi, pp 343–351

Erel O (2004) A novel automated method to measure total antioxidant response against potent free radical reactions. Clin Biochem 37:112–119
Fici S (2016) A new narrow-leaved species of *Capparis* (Capparaceae) from central Palawan, Philippines. Phytotaxa 267:146–150
Fujii Y, Shibuya T, Yasuda T (1991) L 3,4-dihydroxyphenylalanine as an allelochemical candidate from *Mucuna pruriens* (L.) DC. var. *utilis*. Agric Biol Chem 55:617–618
Greeshma AA, Sridhar KR (2016) Ethnic plant-based nutraceutical values in Kodagu region of the Western Ghats. In: Pullaiah T, Rani SS (eds) Biodiversity in India, vol 8. Regency Publications, New Delhi, pp 299–317
Greeshma AA, Sridhar KR, Pavithra M, Ashraf M, Ahmad MSA (2018) Nutritional prospects of edible fern of the Western Ghats of India. In: Öztürk M, Hakeem KR (eds) Global perspectives on underutilized crops. Springer, pp 251–164
Gryglewski RJ, Korbut R, Robak J, Swies J (1987) On the mechanism of antithrombotic action of flavonoid. Biochem Pharmacol 36:317–322
Hartmann R, Meisel H (2007) Food-derived peptides with biological activity: from research to food application. Curr Opin Biotechnol 18:163–169
Havsteen B (1983) Flavonoids, a class of natural products of high pharmacological potency. Biochem Pharmacol 32:1141–1148
Hornykiewicz O (2002) L-DOPA: from a biologically inactive amino acid to a successful therapeutic agent. Amino Acids 23:65–70
Hsu CL, Chen W, Weng YM, Tseng CY (2003) Chemical composition, physical properties and antioxidant activities of yam flours as affected by different drying methods. Food Chem 83:85–92
Jain SK (1987a) A manual of ethnobotany. Scientific Publishers, Jodhpur
Jain SK (1987b) Glimpses of Indian ethnobotany. Oxford IBH Publishing Co, New Delhi
Jain SK (1991) Dictionary of Indian folk medicine and ethnobotany. Deep Publishers, New Delhi
Kakade ML, Rackis JJ, McGhee JE, Puski G (1974) Determination of trypsin inhibitor activity of soy products, a collaborative analysis of an improved procedure. Cereal Chem 51:376–382
Kaushik A, Jijta C, Kaushik JJ, Zeray R, Ambesajir A, Beyene L (2012) FRAP (ferric reducing ability of plasma) assay and effect of *Diplazium esculentum* (Retz) Sw. (a green vegetable of North India) on central nervous system. Ind J Nat Prod Res 3:228–231
Kumar KK, Chandra KLP, Sumanthi J, Reddy GS, Shekar PC, Reddy BVR (2012) Biological role of lectins: a review. J Orofac Sci 4:20–25
Liener IE, Kakade ML (1969) Protease inhibitors. In: Liener IE (ed) Toxic-constituents of plant foodstuffs, 2nd edn. Academic, New York, pp 7–71
Manickam VS, Irudayaraj V (1992) Pteridophyte flora of the Western Ghats – South India. B.I. Publications, New Delhi
Middleton E Jr, Kandaswami C (1992) Effect of flavonoids on immune and inflammatory cell function. Biochem Pharmacol 43:1167–1179
Miean KH, Mohamed S (2001) Flavonoid (myricetin, quercetin, kaempferol, luteolin, and apigenin) content of edible tropical plants. J Agric Food Chem 49:3106–3112
Mohan SC, Balamurugan V, Salini ST, Rekha R (2012) Metal ion chelating activity and hydrogen peroxide scavenging activity of medicinal plant *Kalanchoe pinnata*. J Chem Pharma Res 4:197–202
Occenã IV, Majica E-RE, Merca FE (2007) Isolation of partial characterization of a lectin from the seeds of *Artocarpus camansi* Blanco. Asian J Plant Sci 6:757–764
Oyaizu M (1986) Studies on products of browning reactions: antioxidative activities of products of browning reaction prepared from glucosamine. Jap J Nutr 44:307–315
Pellegrini N, Serafini M, Colombi B, Del Rio D, Salvatore S, Bianchi M, Brighenti F (2003) Total antioxidant capacity of plant foods, beverages and oils consumed in Italy assessed by three different in vitro assays. J Nutr 133:2812–2819
Prieto P, Pineda M, Aguilar M (1999) Spectrophotometric quantitation of antioxidant capacity through the formation of a phosphomolybdenum complex: specific application to the determination of vitamin E. Anal Biochem 269:337–341

Roe JH (1954) Chemical determination of ascorbic, dehydroascorbic and diketogluconic acids. In: Glick D (ed) Methods of biochemical analysis, vol 1. InterScience Publishers, New York, pp 115–139
Rosset J, Bärlocher F, Oertli JJ (1982) Decomposition of conifer needles and deciduous leaves in two Black Forest and two Swiss Jura streams. Int Rev Gesamten Hydrobiol 67:695–711
Roy S, Hazra B, Mandal N, Chaudhuri TK (2013) Assessment of the antioxidant and free radical scavenging activities of methanolic extract of *Diplazium esculentum*. Int J Food Prop 16:1351–1370
Ruma OC (2016) Phytochemical screening of selected indigenous edible plants from the towns of Isabela, Philippines. Asian J Nat Appl Sci 5:36–45
Saha J, Biswal AK, Deka SC (2015) Chemical composition of some underutilized green leafy vegetables of Sonitpur District of Assam, India. Int Food Res J 22:1466–1473
Sathe SK, Deshpande SS, Reddy NR, Goll DE, Salunkhe DK (1983) Effects of germination on proteins, raffinose oligosaccharides and antinutritional factors in the Great Northern Beans (*Phaseolus vulgaris* L.). J Food Sci 48:1796–1800
Saxena V, Mishra G, Vishwakarma KK, Saxena A (2013) A comparative study on quantitative estimation of tannins in *Terminalia chebula, Terminalia belerica, Terminalia arjuna* and *Saraca indica* using spectrophotometer. Asian J Pharma Clin Res 6:148–149
Singh S, Singh R (2012) Ethnomedicinal use of pteridophytes in reproductive health of tribal women of Pachmarhi biosphere reserve, Madhya Pradesh, India. Int J Pharma Sci Res 28:4780–4790
Singh RP, Murthy CKN, Jayaprakasha GK (2002) Studies on antioxidant activity of pomegranate (*Punica granatum*) peel and seed extracts using in vitro methods. J Agric Food Chem 50:81–86
Soares AR, Marchiosi R, Siqueira-Soares RDC, Barbosa de Lima R, Santos WD, Ferrarese-Filho O (2014) The role of L-DOPA in plants. Plant Signal Behav 9(3):e28275
Sridhar KR, Karun NC (2017) Plant based ethnic knowledge on food and nutrition in the Western Ghats. In: Pullaiah T, Krishnamurthy KV, Bhadur B (eds) Ethnobotany of India, 2, Western Ghats and west coast of Peninsular India. Apple Academic Press, Oakville, pp 255–276
Sridhar KR, Shreelalitha SJ, Supriya P, Arun AB (2016) Nutraceutical attributes of ripened split beans of three *Canavalia* landraces. J Agric Technol 7:1277–1297
StatSoft (2008) Statistica, Version # 8. StatSoft, Tulsa
Stewart AJ, Bozonnet S, Mullen W, Jenkins GI, Michael EJ, Crozier A (2000) Occurrence of flavonols in tomatoes and tomato-based products. J Agric Food Chem 48:2663–2669
Tongco JVV, Villaber RAP, Aguda RM, Razal RA (2014) Nutritional and phytochemical screening, and total phenolic and flavonoid content of *Diplazium esculentum* (Retz.) Sw. from Philippines. J Chem Pharma Res 6:238–242
Turkmen N, Sari F, Velioglu YS (2005) The effect of cooking methods on total phenolics and antioxidant activity of selected green vegetables. Food Chem 93:713–718
Walingo KM (2005) Role of vitamin C (ascorbic acid) on human health – a review. Afr J Food Agric Nutr Dev 5:1–13
Wong SP, Leong LP, Koh JHW (2006) Antioxidant activities of aqueous extracts of selected plants. Food Chem 99:775–783

Chapter 5
Medicinal Plants as a Source of Alkaloids

Valentina Laghezza Masci, Stefano Bernardini, Lorenzo Modesti, Elisa Ovidi, and Antonio Tiezzi

Abstract Plants are an important source of biomolecules widely used in medical treatments, and among such products, the alkaloids are a very interesting and complex group. The definition of the term "alkaloid" is still a cause of controversy due to the similarity of some of these natural molecules with other secondary compounds. From a biological point of view, alkaloids are a group of chemicals actively involved in different biological processes of plants, animals, and microorganisms at different cellular levels. In medical science alkaloids are nitrogenous compounds, derived from vegetables origin, characterized by high molecular masses and complex structures. A possible classification of alkaloids is based on their biological and ecological activity, chemical structures, and biosynthetic pathway. They are grouped according to the shape and origins, and in this view three main groups of alkaloids can be distinguished: true alkaloids, protoalkaloids, and pseudoalkaloids.

Keywords True alkaloids · Protoalkaloids · Pseudoalkaloids · Pharmaceutical properties

5.1 Introduction

For a long time, plants play an important role in medical treatments, and thanks to their wide biochemical diversity, the use is being extensively studied especially as source of biomolecules. To date, about 40% of modern monomolecular drugs derive directly or indirectly from plants and their secondary metabolites (Bernardini et al. 2017). Some plant molecules are marked as pharmaceutical regulators; however, in most cases, in absence of clinical tests, the empiric experience matured in traditional medicine during hundreds or thousands of years can be considered testimony of their efficacy.

V. Laghezza Masci · S. Bernardini · L. Modesti · E. Ovidi · A. Tiezzi (✉)
Laboratory of Plant Cytology and Biotechnology, Department for the Innovation in Biological, Agro-food and Forestal Systems, Tuscia University, Viterbo, Italy
e-mail: antoniot@unitus.it

D. Egamberdieva, A. Tiezzi (eds.), *Medically Important Plant Biomes: Source of Secondary Metabolites*, Microorganisms for Sustainability 15,
https://doi.org/10.1007/978-981-13-9566-6_5

The plant organisms use the products of secondary metabolism as intermediaries with the surrounding environment. Among such products alkaloids are a very interesting and complex group present in different biotopes, and although relatively common in all the kingdoms of living organisms, four fifths of known secondary metabolites come from the plant kingdom (Kim and Buell 2015). Two hundred years of scientific research has not yet fully explained the connections between alkaloids and life, nor has it explained why these diverse chemicals are produced and degraded by organisms or why they have such a very large spectrum of biological activities. This is probably due to the link among plants, soil, and the environment and contributing to develop numerous adaptation mechanisms (Aniszewski 2007).

Alkaloids appeared for the first time in the worldwide scenery in 1805, when a German apothecary assistant named Friedrich Sertürner isolated an impure form of the molecule of morphine, which is still today one of the most important and widely known alkaloid (Bynum and Porter 1994a; Courtwright 2009; Krishnamurti and Rao 2016), thus allowing a very important progress in the fields of chemistry and pharmacology (Sneader 1990; Aniszewski 1994a; Dias et al. 2012; Veeresham 2012). In the following years, the extraction method developed by Sertürner allowed pharmacists to isolate many other alkaloid molecules, such as brucine, strychnine, febrifuge, quinine, caffeine, and veratrine (Fig. 5.1) (Bynum and Porter 1994a; Bynum and Porter 1994b; Dias et al. 2012; Veeresham 2012).

The term "alkaloid" was coined in 1819 by the German apothecary Wilfred Meißner who noticed that such compounds seemed "like alkali" (Clayden et al. 2001).

However, although several attempts since the time of their discovery have been performed, especially within the academic world, the definition of the term "alkaloid" is still a cause of controversy due to the similarity of some of these natural molecules with other secondary compounds. Biologists, for instance, consider alkaloids as pure and perfect natural products: a very general picture of such molecules describe them as compounds biologically active, with a chemical heterocyclic structure containing nitrogen, which can possibly have pharmacological, medicinal, or ecological use (Aniszewski 1994b). The huge biological and chemical differences between such compounds make it difficult to give a general definition of alkaloid groups without adding exceptions, objections and derogations (Koskine 1993). There have been different ways to define the alkaloid molecules: for Winterstein and Tier (1910) the alkaloids were compounds having a chemical structure with a basic composition, heterocyclic nitrogen atoms and amino acid derivative classified according to the order of a greater or less toxicity on the central nervous system and the source plants; Waller and Nowacki (1978) instead focused their attention on the presence of nitrogen connected to at least two carbon atoms and of at least one ring not necessarily heterocyclic and affirmed that the alkaloids could not be structural units of macromolecular cellular substances, vitamins, or hormones; in more recent times, Sengbush (2003) defines alkaloids simply by emphasizing the presence within the molecules of bases that contain nitrogen and that most of them are drugs.

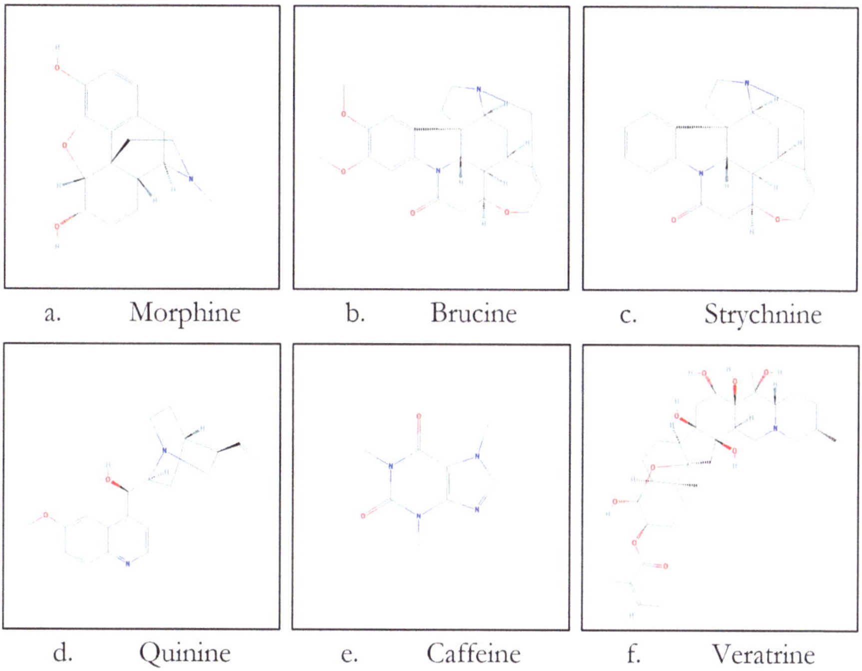

Fig. 5.1 Chemical structures of the earlier discovered alkaloids. (**a**) Morphine, (**b**) brucine, (**c**) strychnine, (**d**) quinine, (**e**) caffeine, (**f**) veratrine

Starting from these assumptions, the alkaloids are defined using the features which better qualify the activity or the structures of such molecules. From a biological point of view, alkaloids are a group of chemicals which are actively involved in different biological processes of plants, animals, and microorganisms at different cellular levels. In medical science substances are considered alkaloids when they are nitrogenous; derived from vegetables, characterized by high molecular masses and complex structures, including heterocycles containing primary, secondary, or tertiary bases or a quaternary ammonium groups (Aniszewski 2007); and soluble in ethanol, benzene, ether and chloroform much better than in water. Moreover, other distinctive features of alkaloids are their power to create intense physiological actions, which allows their application as curative drugs, as well as the possibility to result highly toxic for organisms even when used in very little doses. Different definitions of alkaloids from a medical point of view are also available, and an authoritative definition, among the others, is provided by the National Library of Medicine (NCBI 2005). Another definition of alkaloids has been provided by chemists, who defined such substances as a group of heterocyclic nitrogen compounds which conserve their basic chemical properties and show a strong physiological activity and often result to be toxic, although it has been also asserted that such definition is subject to few exceptions as reported in literature (Jakubke et al. 1994).

Although there are some small differences among them, all such definitions are very similar and sometimes identical.

In any case, whatever the definition, alkaloids are natural compounds synthetized by living organisms as a consequence of their own metabolism and providing biological, chemical, and pharmacological activities which have allowed to develop drugs to fight, and sometimes defeat, many different diseases. Nowadays thousands of natural compounds and derivatives are assessed as alkaloids and new molecules belonging to such group are constantly discovered at the rate of about 100 every year. Alkaloids can be found as acid, salts, amides, and esters and in combination with sugars or also in quaternary salts or tertiary amine oxides (Aniszewski 2007).

5.2 Classification of Alkaloids

A possible classification of alkaloids is based on their biological and ecological activity, chemical structures, and biosynthetic pathway. Considering the biological activity, alkaloids can be divided into different groups: neutral or weakly basic molecules (lactams and certain N-oxides), animal-derived alkaloids (produced by anurans, mammalians, and arthropods), marine alkaloids, moss alkaloids, fungal and bacterial alkaloids, and non-natural alkaloids (structural modified or analogues), the last of which has been continuously growing in recent years as a consequence of bio-organic and stereochemistry research and of the increasing demand for new molecules for possible production and pharmaceutical application.

Anyway, alkaloids classification is generally based on the common molecular precursors, depending on the molecular pathway by which the molecules have been synthetized, and, considering their structures, they are then grouped according to shape and origins. In this view three main groups of alkaloids can be distinguished: true alkaloids, protoalkaloids, and pseudoalkaloids, the only group containing alkaloids not derived from amino acids (Aniszewski 2007).

True alkaloids occur in a limited number of species and families of plants in the free state, as salts and as N-oxides, and are produced as a consequence of the condensation of decarboxylated amino acids with a nonnitrogenous structural moiety. They are highly reactive crystalline substances, generally white solid (with the exception of nicotine that is a brown liquid) able to form water-soluble salts when they are united with acids, bitter in taste, and biologically active even in low doses. Precursors of this kind of molecules are L-ornithine, L-lysine, L-phenylalanine/L-tyrosine, L-tryptophan, and L-histidine (Dewick 2002), and some of the most famous true alkaloids are cocaine, quinine, dopamine, morphine, and usambarensine (Fig. 5.2).

Protoalkaloids are a little group of compounds characterized by very simple chemical structure, consisting in a closed ring, in which the N atom is not included in the heterocyclic (Jakubke et al. 1994). Such kind of alkaloids derived from L-tyrosine and L-tryptophan, and good examples of them are hordenine, mescaline, and yohimbine (Fig. 5.3).

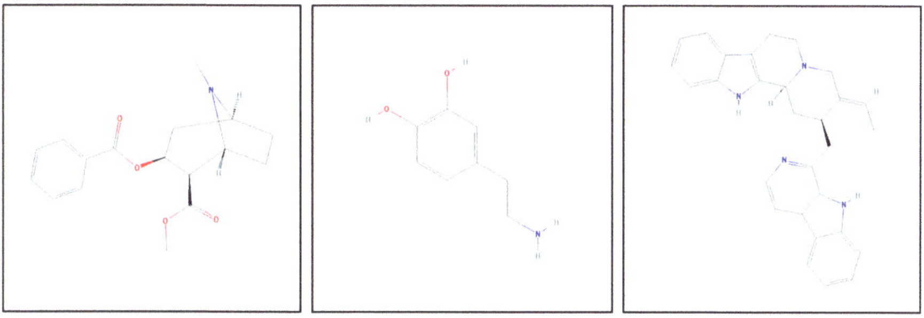

a. Cocaine b. Dopamine c. Usambarensine

Fig. 5.2 Chemical structures of some of the most famous "true alkaloids." (**a**) Cocaine, (**b**) dopamine, (**c**) usambarensine

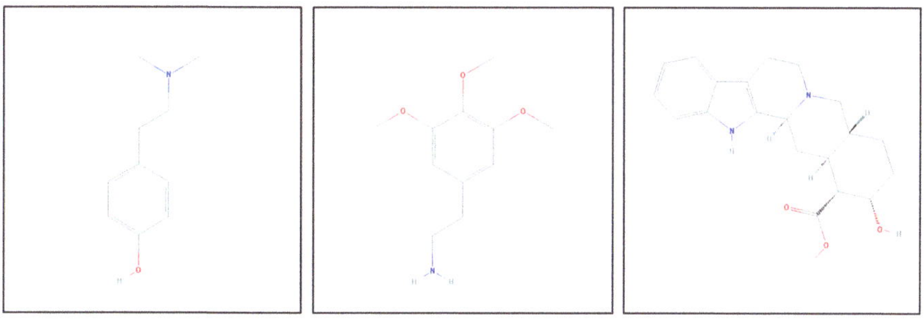

a. Hordenine b. Mescaline c. Yohimbine

Fig. 5.3 Chemical structures of some of the most famous "protoalkaloids." (**a**) Hordenine, (**b**) mescaline, (**c**) yohimbine

Pseudoalkaloids are compounds whose carbon skeleton doesn't derive directly from amino acids (Jakubke et al. 1994) but from their precursors and postcursors (derivatives in the degradation process), from their amination or transamination reactions (Dewick 2002), or from non-amino acid precursor. The N atom can be donated at a relatively late stage in the case of steroidal or terpenoid skeletons or across a transamination reaction from an amino acid source in case of presence of a suitable aldehyde or ketone. Good examples of such kind of alkaloids are coniine, capsaicin, ephedrine, solanidine, caffeine, theobromine, and pinidine (Fig. 5.4).

Alkaloids occur abundantly in higher plants, and at least 25% of them contain alkaloids. Alkaloids are also contained in a larger number of plant species; however since the low alkaloid concentration is unable to influence cellular processes, such plants are not considered alkaloid species; alkaloids plants have been defined by Hegnauer (1966, 1967) as those plants containing more than 0,01% of alkaloids. Considering also the presence of slight traces, alkaloid plant families are the following:

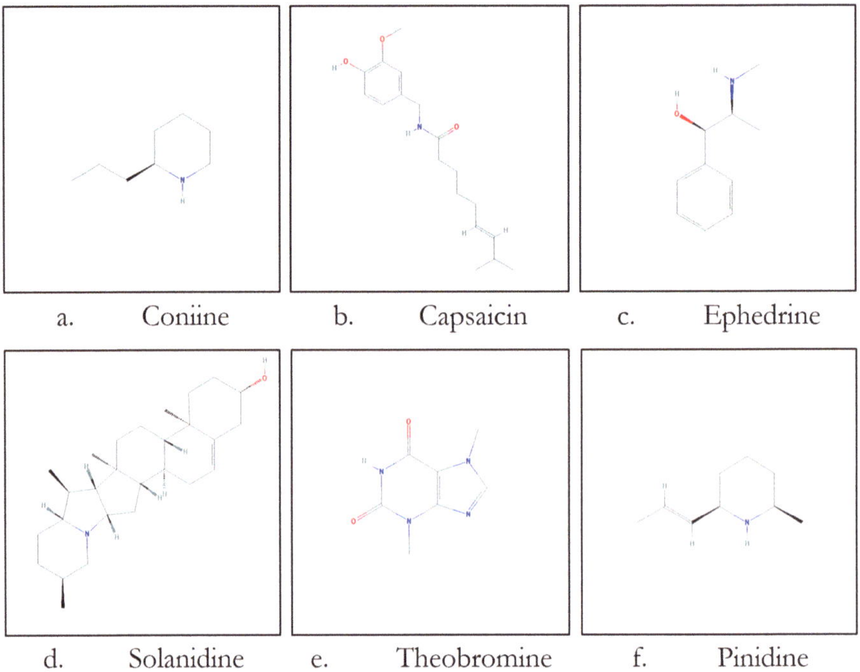

Fig. 5.4 Chemical structures of some of the most famous "pseudoalkaloids." (**a**) Coniine, (**b**) capsaicin, (**c**) ephedrine, (**d**) solanidine, (**e**) theobromine, (**f**) pinidine

- *Apocynaceae*: This family is also known as Dogbane botanical family and contains 150 genera and 1700 species, all particularly rich in L-tryptophan-derived alkaloids having a strong biological and medicinal effect and largely used in cancer chemotherapy (Aniszewski 2007). Typical examples of alkaloids belonging to such species are reserpine and rescinnamine, quinine, ibogaine, deserpine (Varchi et al. 2005), ervatine, tabersonine, coronaridine (Srivastava et al. 2001), and menilamine (Macabeo et al. 2005).
- *Asteraceae*: With 900 genera and more than 2000 species, this family results to be one of the largest and contains many different L-ornithine- and L-tryptophan-derived alkaloids. The most common alkaloids produced by the plants of this family are retrosine, senecionine (Pelser et al. 2005), apigenin (Aniszewski 2007), intergerrimine, and usaramine (Roeder et al. 1996).
- *Loganiaceae*: This family contains 30 genera and more than 500 species containing mainly L-tyrosine-derived biologically and medicinally very important alkaloids such as strychnine, brucine, curare, sungucine, isosungucine (Lansiaux et al. 2002), and isostrychnopentamine (Frédérich et al. 2004).
- *Papaveraceae*: Poppy botanical family contains 26 genera and about 250 plant species which produce many L-tyrosine-derived alkaloids, most of them having

a strong impact on biology and medicine such as morphine, codeine, thebanine, papaverine, narcotine, narceine, isoboldine, salsolinol (Aniszewski 2007), sanguinarine, cholidonine, hydrastine, berberine and chelerythine (Vavrečková et al. 1996a, b).

- *Rutaceae*: This family contains more than 150 genera and 900 species, many of them containing both L-anthranilic acid and L-histidine-derived alkaloids. Alkaloids derived from L-anthranilic acid include alkaloids such as dictamine, skimmianine, acronycine, melicopicine, rutacridone, acutine, acetylfolifidine, bucharidine, dubinidine, dubinine, glycoperine, evoxine, γ- fagarine, folifidine, foliosidine, haplophyline, haplopine, perfamine, skimmianine, helietidine, flindrsine, kokusaginine, and maculasine (Nazrullaev et al. 2001; de Moura et al. 2002). Some of the most famous L-histidine-derived alkaloids are instead represented by pilocarpine, pilosine, fagaronine, isodecarpine, 8-demethyloxychelerythine, 1-hydroxyrutaecarpine, rutaecarpine, 1-methoxyrutaecarpine, hyemaline, melicarpine, samecarpine, galipine, cusparine, cuspareine, demethoxycusparine, galipinine, evocarpine, dihydroevocarpine, and evodiamine (Sheen et al. 1996; Rakotoson et al. 1998; Shin et al. 1998; de Moura et al. 2002; Chen et al. 2003, 2005a).
- *Solanaceae*: It is a family consisting of 90 genera and more than 2000 species abundant in L-ornithine-derived alkaloids such as hyoscyamine, hyoscine, and cuscohygrine, nicotinic acid-derived alkaloids such as nicotine and anabasine, L-phenylalanine-derived alkaloids such as capsaicin, and steroidal-derived alkaloids such as solanidine, solamargine, solasodine, and tomatine (Schwarz et al. 2005).
- *Erythroxylaceae*: This family is also known as the coca plant family and contains a lot of L-ornithine-derived alkaloids commonly used in medicine and also in criminal activity such as cocaine, ecgonine, cinnamylcocaine, α-truxilline, truxilline, methylecgonine, tropine, hygrine, hygroline, and cuscohygrine (Aniszewski 2007).
- *Boraginaceae*: Includes 95 genera and about 2000 species rich in L-ornithine-derived alkaloids such as indicine-*N*-oxide, europine, and ilamine and their N-oxides, which have particularly strong toxic effects (Farsam et al. 2000), and several pyrrolizidine alkaloids used in local folk medicine as a diuretic, analgesic, sedative, and sudorific remedies, against stomach ulcers, and for treatment of skin diseases (Roeder 1995; Al-Douri 2000; Haberer et al. 2002; Said et al. 2002; Bracca et al. 2003; Siciliano et al. 2005).
- *Fabaceae*: This is one of the three largest families, including 650 genera and 18,000 species which contain mostly L-lysine-derived alkaloids such as lupinine, luparine, angustifoline, epi-lupanine, anagyrine, swainsonine, castanospermine, and many others (Przybylak et al. 2005), L-ornithine-derived alkaloids such as senecionine, and L-tryptophan-derived alkaloids such as eserine, eseramine,

physovenine, geneserine, erysovine, wrythraline, erysodine, α-erythroidine, β-erythroidine, dihydro-β-erythroidine, and many others, all of them biologically and ecologically significant (Soto-Hernandez and Jackson 1993; Lou et al. 2001; Tanaka et al. 2001; Wanjala et al. 2002; Kramell et al. 2005).

- *Menispermaceae*: Consists of 70 genera and 450 species which produce many L-tyrosine-derived alkaloids with medicinal effects or other activities that influence cellular processes such as tetrandrine, stephanine, curare, tubocurarine, methylliriodendronine, 2-*O*,*N*-dimetylliriodendronine, liriodenine, dicentronine, corydine, aloe-emodin, liriodenine, corydine, isocorydine, atherospermidine, stephalagine, dehydrostephalagine, stephalonines A–I, norprostephabyssine, isoprostephabyssine, isolonganone and isostephaboline, cepharathine, cepharanoline, isotetrandrine, and berbamine (Nakaoji et al. 1997; Camacho et al. 2000; Gören et al. 2003; Chen et al. 2005b).
- *Berberidaceae*: This family includes 10 genera and 574 species which produce L-tyrosine-derived alkaloids such as berberine, glaucine, hydroxyacanthin, and berbamine widely used in Japanese folk medicine against whooping cough, asthma, pharynx tumors, uterine bleeding, and diabetes and also as antiarrythmic, anti-myocardial, anti-ischemic, and antithrombotic (Khamidov et al. 2003; Orallo 2004; Aniszewski 2007).
- *Ranunculaceae*: A family consisting of about 2000 species distributed in 50 genera and producing L-tyrosine-derived alkaloids such as berberine, hydrastine, fangcholine, and fuzitine (Erdemgil et al. 2000) and terpenoid-derived alkaloids such as aconitine, sinomontanine, karacoline, karakanine, songorine, nepelline, 12-acetylnepelline, cammaconine, secokaraconitine (Tashkhodzhaev et al. 2000; Sultankhodzaev et al. 2002) tangutisine, acorone, acorridine, coryphine, coryphidine (Dzhakhangirov and Bessonova 2002), methyllycaconitine, barbine, delcorinine, delsonine (Salimov 2001), lycoctonine, anthranoyllycoctonine, ajacine, and delpoline (Boronova and Sultankhodzhaev 2000).
- *Liliaceae*: With more than 200 genera and around 3500 species, it is a very large family of plant, most of them producing L-tyrosine-derived alkaloids such as autumnaline, floramultine, kreysigin, and the well-known colchicine (Aniszewski 2007) and steroidal alkaloids such as jervine, cyclopamine, cycloposine, protoveratrine A and protoveratrine B, and *O*-acetyljervine used in Chinese folk medicine as antitussive and expectorant (Suladze and Vachnadze 2002; Jiang et al. 2005).
- *Rubiaceae*: Plants of this family are included in 400 genera and 6000 species producing adenine-/guanine-derived alkaloids, also called purine-derived alkaloids, such as caffeine, theophylline, and theobromine, leading to a positive and prophylactic effect against Parkinson's disease (Aniszewski 2007), L-tryptophan-derived alkaloids such as walterione A (Hoelzel et al. 2005), corynantheidine, corynantheine, dihydrocorynantheine, α-yohimbine, corynanthine, quinine, quinidine, cinchonine, cinchonidine, and many other molecules with biological and ecological important effects (Staerk et al. 2000; Ravishankara et al. 2001; Horie et al. 2005; Matsumoto et al. 2005).

- *Amaryllidaceae*: This family comprises of 50 genera including more than 850 plant species which produce L-tyrosine-derived alkaloids such as lycorine, galanthamine, galanthindole, galanthine, haemanthamine, lycorine, lycorenine, oxomaritidine, maritidine, vittatine, and many other molecules provided the different and important biological activities (Antoun et al. 1993; Bastida et al. 1996a, b; Machocho et al. 1998; Lewis, 1999; Unver et al. 1999; Lewis 2000; Herrera et al. 2001; Abou-Donia et al. 2002; Unver et al. 2003; Szlávik et al. 2004; Forgo and Hohmann 2005).
- *Elaeagnaceae*: This family consists of 3 genera and 50 species which produce among the other the L-tryptophan-derived alkaloid eleagine (Aniszewski 2007).
- *Zygophyllaceae*: With around 30 genera and more than 230 species, this family includes plants containing the L-tryptophan-derived alkaloid known as harman, the anthranilic acid-derived alkaloids called harmine, and the acetate-derived alkaloids dihydroschoberine, nitrabirine N-oxide, komavine, and acetylkomavine (Tulyaganov and Allaberdiev 2001; Tulyaganov et al. 2001).

5.3 Biological and Pharmaceutical Properties of Alkaloids

Alkaloids are the oldest successfully used drugs throughout history in the treatment of many diseases, this for their biological activity that operates not only on the endogenous life processes in the organisms that produce them but also in the organisms that come in contact with them (Wink 1998). These compounds are nontoxic in vacuoles where they are stored but toxic when they get away in different cells and tissues, due to the change in chemical configurations according to pH changes; this implies that alkaloids can have different biological activities in different cell conditions and different receptors (Aniszewski 2007).

Alkaloids play a very important role in an organism's metabolism and functions; their biological activity can be very different and dependent on their chemical structure. These compounds are biologically significant as active stimulators, inhibitors, and terminators of growth or part of endogenous safety and regulation mechanisms. It has been hypothesized that alkaloids, in addition to having this pivotal role in plant metabolism, can also be considered as vegetable waste (Waller and Nowacki 1978).

As a matter of fact, alkaloids are the most important active compounds in natural herbs. The distributing effect on the nervous systems of animals of alkaloids is widely known, among which is the analgesic action of morphine (Benyhe 1994). Some of them have also significance in the hemoglobinizators of leukemia cells, and they can be biologically determined to be estrogenically active molecules (Dupont et al. 2005). They display antimicrobial and antiparasitic properties (Fernandez et al. 2010; Cushnie et al. 2014). Recent research has proved that their biotoxicity is directed only toward foreign organisms or cells and is selective (Aniszewski 2007). Alkaloids can alter DNA, selectively deform cells, and cause locoism, a disease usually of horses, cattle, and sheep caused by chronic poisoning with locoweeds (plant that produces swainsonine, a phytotoxin harmful to

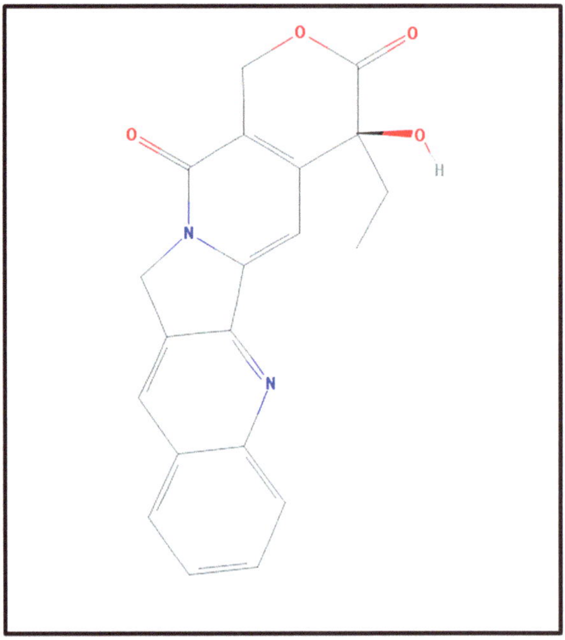

Fig. 5.5 Chemical structure of camptothecin

livestock) (Takanashi et al. 1980; Chenchen et al. 2014). Some alkaloid molecules, both natural and synthetic, can act as narcotics (Das and Ratty, 1987; Collins et al. 2002). Moreover, they play a very important role in the immune systems of animals and plants (Castellino et al. 2006). Several alkaloids demonstrate more important biological activities for treating asthma, such as the relieving action of ephedrine (Lee 2011), and last but not the least the anticancer activities of some compound as camptothecin (CPT), colchicine, and vinblastine (VBL) (Huang et al. 2007; Li et al. 2007; Ghawanmeh et al. 2018).

Some alkaloids have been successfully turned into chemotherapeutic drugs, and an example is camptothecin (CPT) (Fig. 5.5), a topoisomerase I inhibitor (Huang et al. 2007). This compound is a monoterpene indole alkaloid naturally extracted from the bark of *Camptotheca acuminata*, a tree used in traditional Chinese medicine for cancer treatment (Sadre et al. 2016). Wall et al. (1966) in systematic screening of natural products first demonstrated the high anticancer activity of CPT in preliminary clinical trials. The mode of action of CPT is the specific inactivation of topoisomerase I resulting in cell death by apoptosis (Wright et al. 2015).

Due of its low solubility and adverse drug reaction, derivatives to improve pharmacological properties and clinical efficacy have been synthesized, and two semi-synthetic analogues, topotecan and irinotecan (Samuelsson, 2004), have been approved and used in cancer chemotherapy (Takimoto and Calvo 2008).

Another important alkaloid used for cancer therapy is colchicine (Fig. 5.6), a compound that promotes microtubule depolymerization (Ghawanmeh et al. 2018). Colchicine is originally extracted from plants of the genus *Colchicum* (*Colchicum*

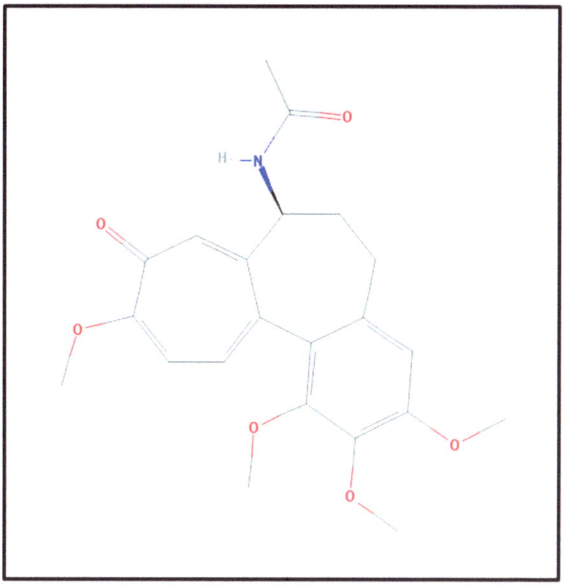

Fig. 5.6 Chemical structure of colchicine

autumnale) and was first isolated in 1820 by two French chemists, Pelletier and Caventou. This compound is known in the treatment of different diseases (Dasgeb et al. 2018), such as gout (Dalbeth et al. 2014) and rheumatism (Slobodnick et al. 2018).

The mechanism of action of colchicine is to bind to tubulin, blocking the microtubule polymerization. The pivotal role of microtubules in various cellular processes classifies colchicine as an antimitotic drug, by forming tubulin-colchicine complexes in a reversible manner and preventing the elongation of the microtubule polymer (Leung et al. 2015).

Probably the most famous alkaloids used for cancer therapy are vinblastine (VBL) and vincristine (VCR) (Fig. 5.7). These two compounds are bisindole alkaloids that consist of two subunits, an upper catharanthine ring system linked to a lower vindoline ring system by a single bond (Kingston 2009), and are the first plant-derived natural products used in the clinical field for cancer treatment (Lukesh et al. 2017). These compounds are extracted from the leaves of the Madagascar periwinkle *Catharanthus roseus* (L.), also called *Vinca rosea* (Kingston 2009), and have been discovered in the 1950s by the Canadian scientists Noble and Beer (Moudi et al. 2013).

The Vinca alkaloids have been also used to treat diseases such as diabetes and high blood pressure (Moudi et al. 2013). These compounds belong to the class of antimicrotubular antiblasts; microtubules are components of the cytoskeleton and play important roles in eukaryotic cellular functions such as intracellular organelle transport, cell migration, cell signalling, and mitosis (Perez 2009). The mechanism of action of vinblastine and vincristine is to prevent the polymerization of microtubules by the interaction with the β-subunit of tubulin during the formation of the mitotic spindle, causing cell metaphase arrest (Himes 1991; Li et al. 2007).

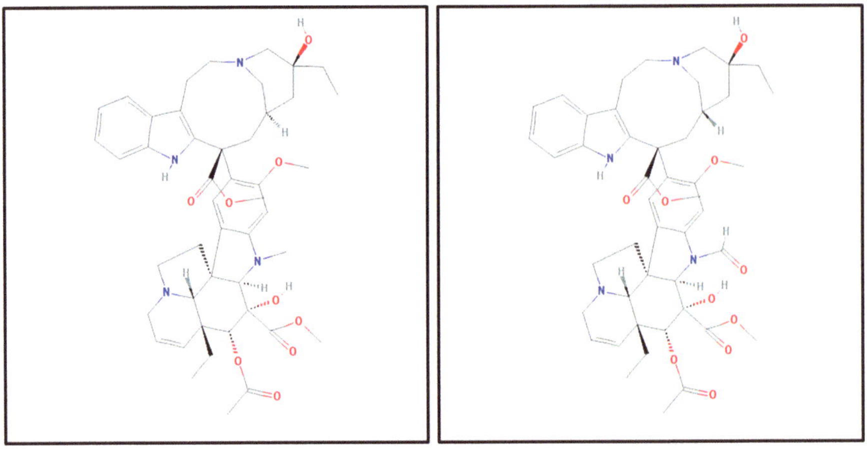

a. Vinblastine b. Vincristine

Fig. 5.7 Chemical structure of vinblastine (**a**) and vincristine (**b**)

This phenomenon contributes to reduce the number of cancer cells. Both VBL and VCR are efficacious clinical drugs used in combination therapies to treat Hodgkin's disease; testicular, ovarian, breast, head, and neck cancer; and non-Hodgkin's lymphoma or in the curative treatment regimens for childhood lymphocytic leukemia. VBL has some side effects to white blood cells and can induce patients' problems such as nausea, vomiting, constipation, dyspnea, chest or tumor pain, wheezing, and fever and is rarely associated with antidiuretic hormone secretion (Rowinsky 2003).

Presently four major Vinca alkaloids are in clinical use: vinblastine (VBL), vincristine (VCR), vinorelbine (VRL), and vindesine (VDS); only VCR, VBL, and VRL are approved for use in the United States (Rowinsky 2003).

Different synthetic Vinca alkaloids were produced in the last few years: vinflunine that is currently approved in Europe for medicinal treatment (Bennouna et al. 2008; Schutz et al. 2011) and vindesine that was the first analogue of VBL entered in clinical use.

Vindesine (Fig. 5.8), unlike VBL, has an amide function rather than a methyl ester on the vindoline ring and does not have an acetyl group on this ring system. It has higher hematological toxicity than vincristine, but it has been incorporated into several effective combination regimens for treatment of leukemia, lymphoma, and non-small cell lung cancer (NSCLC) (Dancey and Steward 1995; Joel 1996; Butler 2005).

Moreover vinorelbine (Fig. 5.9), a semi-synthetic derivative of VBL in which the bridge linking the indole ring to the piperidine nitrogen has been shortened by one carbon and water has been eliminated from the piperidine ring (Clardy and Walsh 2004), was launched in 1989 for the treatment of non-metastatic breast cancer and NSCLC (Mano 2006; Gralla et al. 2007).

Recently, a new synthetic Vinca alkaloid was produced: vinflunine, developed through the addition of two fluor elements by superacidic chemistry (Fig. 5.10). This molecule is the first fluorinated microtubule inhibitor that belongs to the

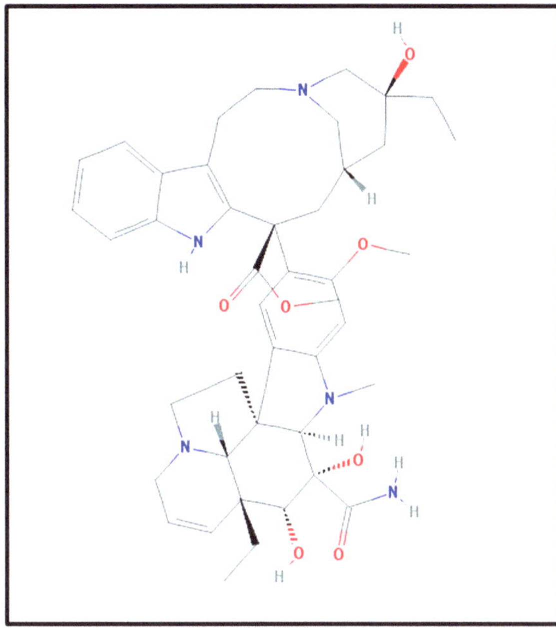

Fig. 5.8 Chemical structure of vindesine

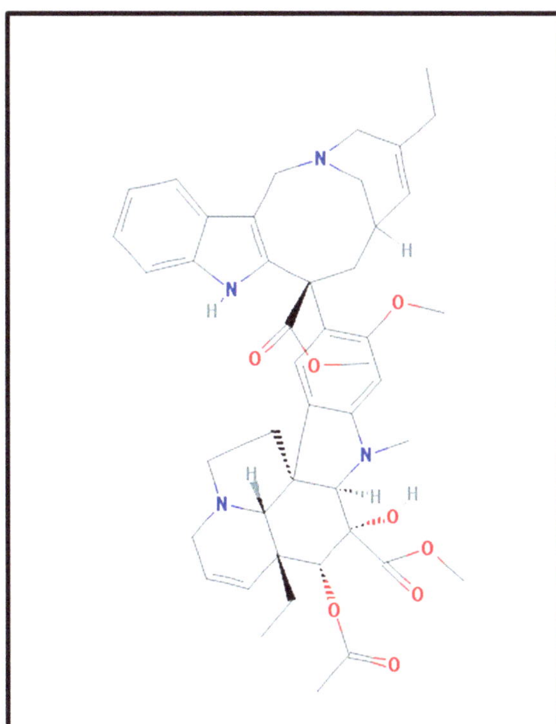

Fig. 5.9 Chemical structure of vinorelbine

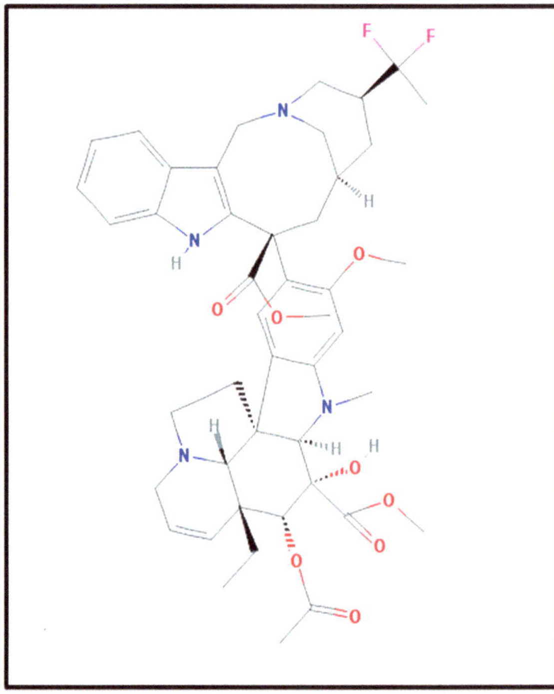

Fig. 5.10 Chemical structure of vinflunine

Vinca alkaloids, and it has been used in Europe for the treatment of second-line transitional cell carcinoma of the urothelium (TCCU). Furthermore, it has been applied for clinical development in a wide spectrum of solid tumors, and important activity has been observed in the treatment of transitional cell carcinoma of the urothelial tract, non-small cell lung cancer, and breast carcinoma. Vinflunine has been also tested in patients with TCCU and first-line advanced breast cancer (Moudi et al. 2013).

Anhydrovinblastine (Hydravin™, KRX-0403, 6) is an analogue of vinblastine that differs from its parent by one molecule of water. It can also be considered a homologue of vinorelbine with an additional carbon in the indole-piperidine bridge (Fig. 5.11). It entered phase I trial for the treatment of advanced solid tumors, including NSCLC, soft tissue sarcoma, and colorectal cancer (Lu et al. 2012; Moudi et al. 2013).

Other alkaloids of plant origin were investigated for anticancer activities. For instance, homoharringtonine (HHT) is an alkaloid with a cephalotaxine nucleus. It was first isolated from *Cephalotaxus harringtonii* and *Cephalotaxus fortunei* trees, whose bark extracts were used in Chinese traditional medicine to treat cancer. Homoharringtonine and other cephalotaxine derivatives can also be found in leaves, bark, and seeds of other *Cephalotaxus* species (Kantarjian et al. 2013). The cephalotaxine itself is very abundant in *Cephalotaxus* species leaves, which can be isolated and transformed by simple esterification into homoharringtonine. Since the

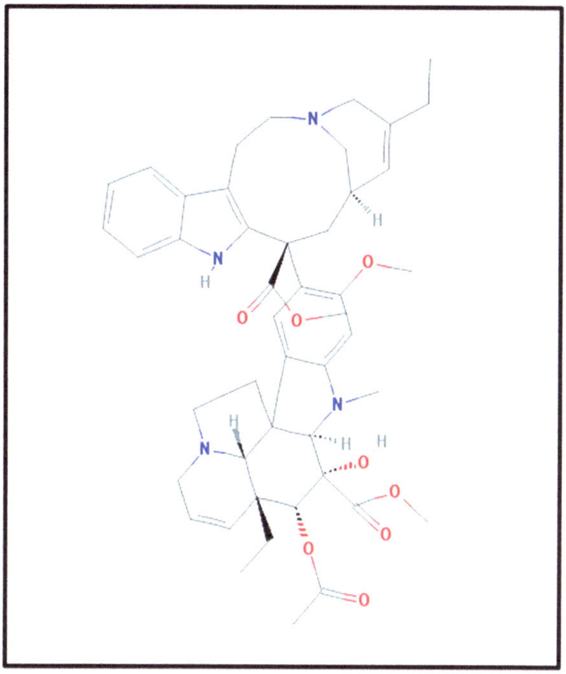

Fig. 5.11 Chemical structure of anhydrovinblastine

1970s, homoharringtonine or a mixture of cephalotaxus esters have been used in China to treat hematological malignancies (Lü and Wang 2014).

Homoharringtonine is a protein translation inhibitor, and its mechanism of action consists in the inhibition of the elongation step of protein synthesis. It binds to the A-site of the large ribosomal subunit, and this action blocks the access of the charged tRNA and consequently the peptide bond formation. Its success is mainly due to the fact that it can perturb proteins with rapid turnover such as the leukemic cell upregulated short-lived oncoproteins BCR-ABL1 and antiapoptotic proteins (Mcl-1, Myc) leading to cells apoptosis (Gandhi et al. 2014). Homoharringtonine, also called omacetaxine mepesuccinate (Fig. 5.12), was approved by FDA in 2012 (sold under the trade name Synribo®) and used in the treatment of chronic myeloid leukemia in patients with resistance and/or intolerance to two or more tyrosine kinase inhibitors; it is the only natural therapeutic agent approved as a commercial drug to treat chronic myeloid leukemia (Seca and Pinto 2018).

Berberine (Fig. 5.13) is an isoquinoline alkaloid extensively distributed in natural herbs (Chen et al. 2008) with a variety of biological activities such as anti-inflammatory, antibacterial, antidiabetes, antiulcer, sedation, protection of myocardial ischemia-reperfusion injury, expansion of blood vessels, inhibition of platelet aggregation, and hepatoprotective and neuroprotective effects (Lau et al. 2001; Yu et al. 2005; Han et al. 2010; Ji 2011). Studies have demonstrated that berberine possesses anticancer potentials by interfering with tumorigenesis and tumor progression in both in vitro and in vivo experiments (Sun et al. 2009; Diogo

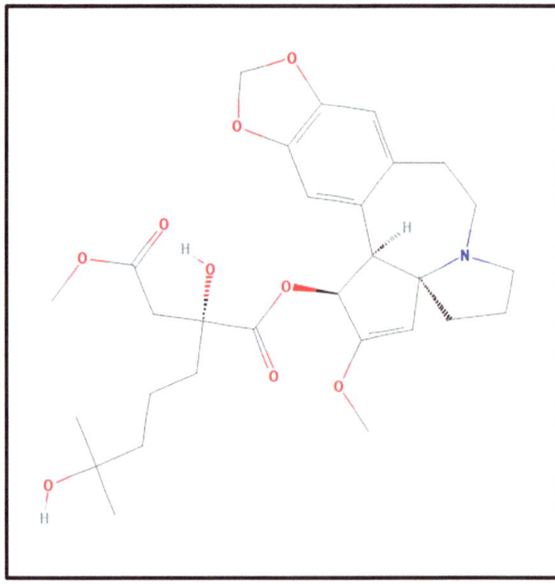

Fig. 5.12 Chemical structure of homoharringtonine also named omacetaxine mepesuccinate

et al. 2011; Tan et al. 2011). Berberine inhibits the proliferation of multiple cancer cell lines by inducing cell cycle arrest at G1 or G2/M phases and by apoptosis (Sun et al. 2009; Eom et al. 2010; Burgeiro et al. 2011) and endoplasmic reticulum stress (Eom et al. 2010) and autophagy (Wang et al. 2010) in cancer cells. Moreover, its combination with chemotherapeutic drugs or irradiation could enhance the therapeutic effects (Youn et al. 2008; Hur et al. 2009). Other mechanisms of berberine are mainly related to its effect on cell cycle arrest and apoptosis, including regulation of cyclin-dependent kinase (CDK) family of proteins (Mantena et al. 2006; Sun et al. 2009) and expression regulation of B-cell lymphoma 2 (Bcl-2) family of proteins (such as Bax, Bcl-2, and Bcl-xL) (Mantena et al. 2006; Sun et al. 2009; Eom et al. 2010) and caspases (Mantena et al. 2006; Eom et al. 2010). Furthermore, berberine inhibits the activation of the nuclear factor κ-light-chain enhancer of activated B cells (NF-κB) and promotes the formation of intracellular reactive oxygen species (ROS) in cancer cells (Sun et al. 2009; Eom et al. 2010).

Evodiamine (Fig. 5.14), a quinolone alkaloid, is one of the major bioactive compounds extracted from the Chinese herb *Evodia rutaecarpa*; this molecule exhibits anti-inflammatory, antiallergic, anti-obese, and anticancer effects. Anticancer activities were reported both in vitro and in vivo. Its mechanism of action is inducing cell cycle arrest or apoptosis, and it can inhibit angiogenesis, invasion, and metastasis in a variety of cancer cell lines (Ogasawara et al. 2001, 2002; Fei et al. 2003; Zhang et al. 2003; Shyu et al. 2006). Evodiamine also stimulates autophagy, a physiological process involved in the maintenance of cell homeostasis (Yang et al. 2008). Compared with other compounds, evodiamine is less toxic toward normal human cells, such as human peripheral blood mononuclear cells.

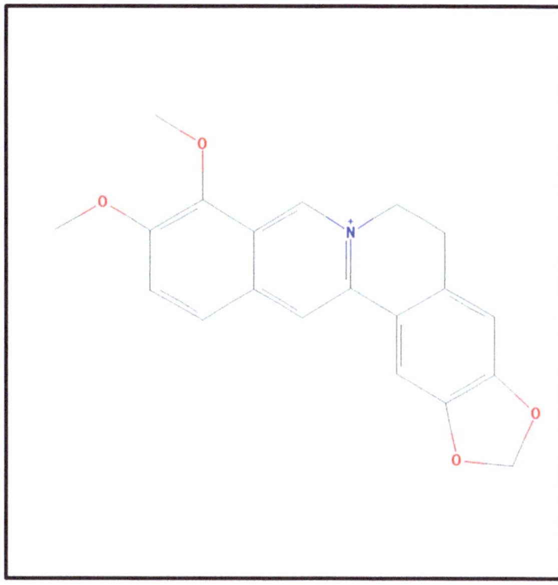

Fig. 5.13 Chemical structure of berberine

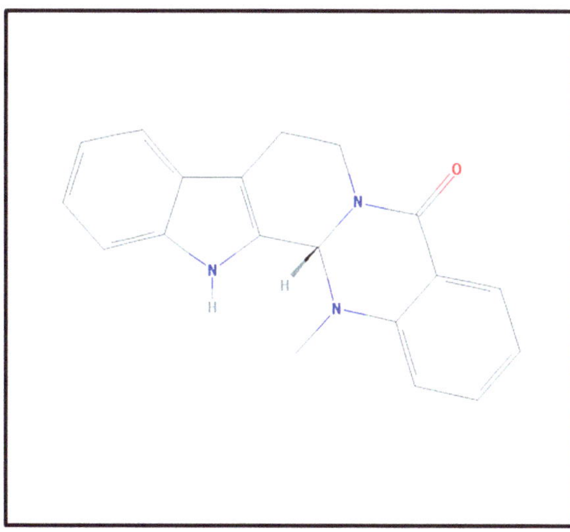

Fig. 5.14 Chemical structure of evodiamine

Matrine (Fig. 5.15) is a major alkaloid found in many *Sophora* plants, including *Sophora flavescens* Ait. (Lai et al. 2003). It exhibits a wide range of pharmacological properties such as antibacterial, antiviral, anti-inflammatory, antiasthmatic, antiarrhythmic, anti-obesity, cardioprotective effects, diuretic, choleretic, hepatoprotective, nephroprotective, and anticancer (Long et al. 2004; Zhang et al. 2007; Zheng et al. 2009; Han et al. 2010; Li et al. 2010; Xing et al. 2010; Zhang et al. 2011a, b). It inhibits the proliferation of various types of cancer cells mainly through

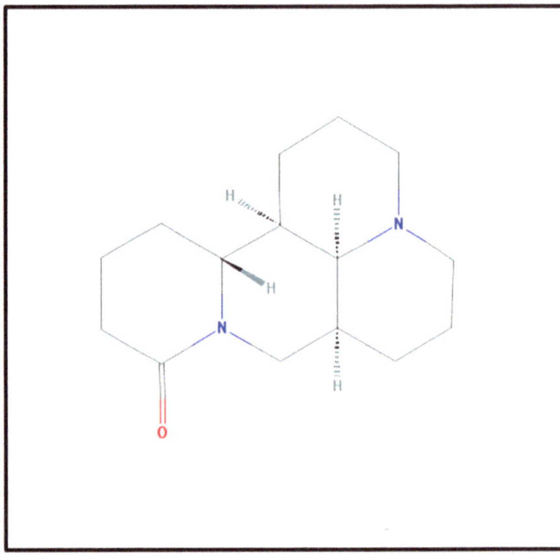

Fig. 5.15 Chemical structure of matrine

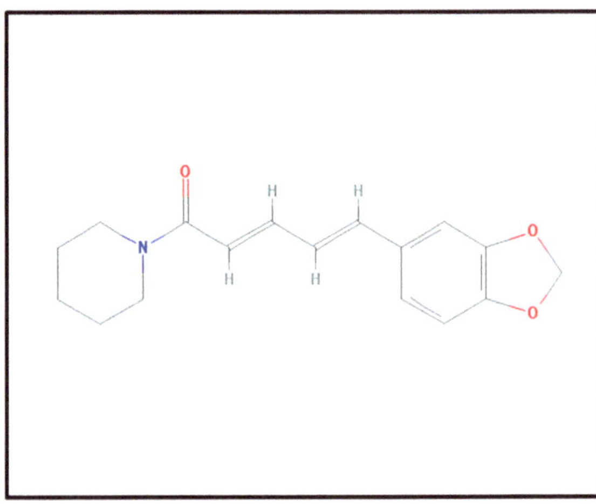

Fig. 5.16 Chemical structure of piperine

mediation of G1 cell cycle arrest or apoptosis (Jiang et al. 2007; Dai et al. 2009; Liu et al. 2010; Zhang et al. 2011a, b, 2012). It inhibits cancer cell invasion partially throughout the inhibition of MMP-2 and MMP-9 expression and modulation of the NF-κB signalling pathway (Yu et al. 2009; Luo et al. 2011; Yu et al. 2011).

Piperine (Fig. 5.16), a piperidine alkaloid isolated from *Piper nigrum* (also known as black pepper) and *Piper longum*, two famous spices that have been used for centuries (Szallasi 2005), exhibits antioxidant, anti-inflammatory, antidiarrheal, anticonvulsant, antimutagenic, and hypolipidemic effects, promoting bile secretion and tumor-inhibitory activities (Srinivasan 2007; Ji 2011; Bae et al. 2010). The che-

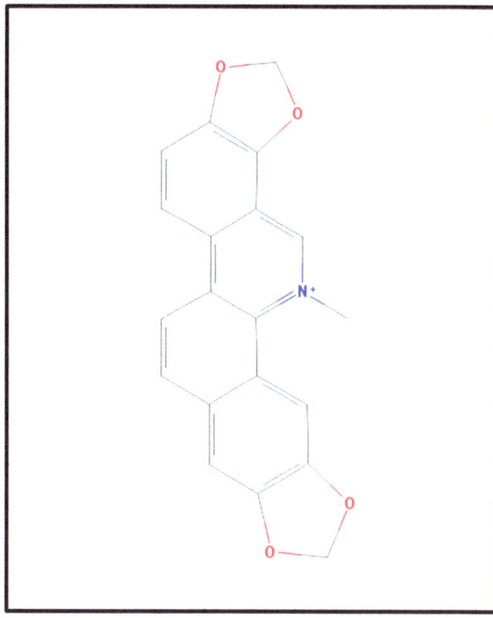

Fig. 5.17 Chemical structure of sanguinarine

mopreventive effects of piperine against several kinds of carcinogen, such as benzo(a)pyrene and 7,12-dimethyl benz(a)anthracene, show its potential as a cancer preventive agent (Khajuria et al. 1998; Selvendiran et al. 2003, 2005a, b; Selvendiran and Sakthisekaran 2004; Krishnakumar et al. 2009; Manoharan et al. 2009). A recent study has shown that piperine inhibits breast stem cell self-renewal and does not cause toxicity to differentiated cells (Kakarala et al. 2010).

Sanguinarine (Fig. 5.17) is a benzophenanthridine alkaloid isolated from the Papaveraceae family, which includes *Sanguinaria canadensis* L. (also known as bloodroot) and *Chelidonium majus* L. (Mahady and Beecher 1994; Vavrečková et al. 1996a, b). It has antibacterial, antifungal, antischistosomal, antiplatelet, anti-inflammatory (Lenfeld et al. 1981; Beuria et al. 2005; Jeng et al. 2007; Ji 2011), and also anticancer potentials (Debiton et al. 2003; Ahsan et al. 2007; Chang et al. 2007; Hussain et al. 2007). Sanguinarine induces cell cycle arrest at different phases or apoptosis in a variety of cancer cells lines (Adhami et al. 2003, 2004; Ahsan et al. 2007; Chang et al. 2007; Hussain et al. 2007; Kim et al. 2008). It remarkably sensitizes breast cancer cells to tumor necrosis factor-related apoptosis inducing ligand-mediated apoptosis (Kim et al. 2008). Sanguinarine also shows antiangiogenic effects in mice (5 mg/kg), presents anti-invasive effects, and overcomes P-gp-mediated MDR phenotype (Weerasinghe et al. 2006; De Stefano et al. 2009; Choi et al. 2009). Sanguinarine is a selective inhibitor of mitogen-activated protein kinase phosphatase 1 (MKP-1), which is overexpressed in many tumor cells (Vogt et al. 2005). Commercial uses of sanguinarine and bloodroot extract consist of dental hygiene products. FDA has approved the inclusion of sanguinarine in toothpastes as an antibacterial or anti-plaque agent (Kuftinec et al. 1990).

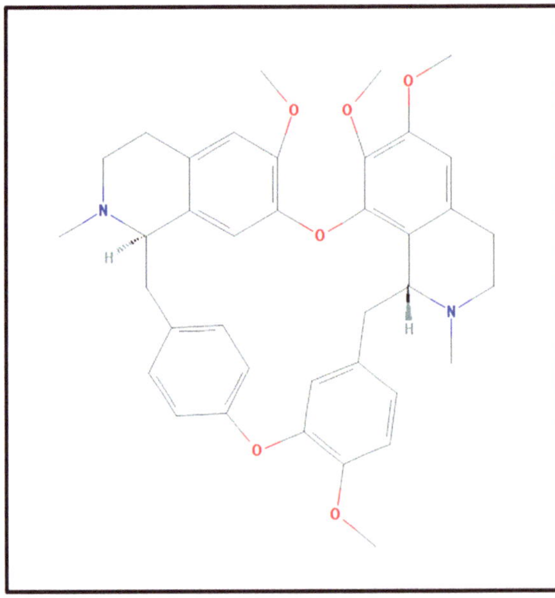

Fig. 5.18 Chemical structure of tetrandrine

Tetrandrine (Fig. 5.18), a bisbenzylisoquinoline alkaloid from the root of *Stephania tetrandra*, exhibits a broad range of pharmacological activities including immunomodulating, antihepatofibrogenetic, anti-inflammatory, antiarrhythmic, antiportal hypertension, anticancer, and neuroprotective activities (Li et al. 2001; Ji 2011). Tetrandrine induces cell cycle arrest (Kuo and Lin 2003; Meng et al. 2004; Ng et al. 2006) and apoptosis in many human cancer cells, including leukemia, bladder, colon, hepatoma, and lung (Lai et al.1998; Lee et al. 2002; Yoo et al. 2002; Kuo and Lin, 2003; Meng et al. 2004; Ng et al. 2006; Wu et al. 2010; Li et al. 2011; He et al. 2011). Co-administration of tetrandrine restores the sensitivity of MDR cancer cells to doxorubicin, paclitaxel, docetaxel, and vincristine (Fu et al. 2002, 2004; Zhu et al. 2005) through the inhibition of P-gp.

5.4 Conclusion

Since their discovery, alkaloids have had many applications as helpful pharmaceutical tools against pathologies, and investigations are running in labs to discover new alkaloid molecules and characterize their potential biological properties. In recent time the increasing of chemical investigations on natural compounds allowed to better understand on pharmaceutical potentialities of alkaloids and their derivates and by such molecules perspectives are now open for a more detailed characterization of cellular pathways and new useful applications in medicine.

References

Abou-Donia AH, Amer ME, Darwish FA, Kassem FF, Hammoda HM, Abdel-Kader MS, Zhou BN, Kingston DGI (2002) Two new alkaloids of the crinane series from *Pancratium sickenbergi*. Planta Med 68:379–381

Adhami VM, Aziz MH, Mukhtar H, Ahmad N (2003) Activation of prodeath Bcl-2 family proteins and mitochondrial apoptosis pathway by sanguinarine in immortalized human HaCaT keratinocytes. Clin Cancer Res 9(8):3176–3182

Adhami VM, Aziz MH, Reagan-Shaw SR, Nihal M, Mukhtar H, Ahmad N (2004) Sanguinarine causes cell cycle blockade and apoptosis of human prostate carcinoma cells via modulation of cyclin kinase inhibitor-cyclin-cyclin-dependent kinase machinery. Moll Cancer Thér 3(8):933–940

Ahsan H, Reagan-Shaw S, Breer J, Ahmad N (2007) Sanguinarine induces apoptosis of human pancreatic carcinoma AsPC-1 and BxPC-3 cells via modulations in Bcl-2 family proteins. Cancer Lett 249(2):198–208

Al-Douri NA (2000) A survey of medicinal plants and their traditional uses in Iraq. Pharm Biol 38:74–79

Aniszewski T (1994a) From iodine to enzyme: a critical review of chemical and biological methods of lupine alkaloids analysis. Sci Legum 1:25–36

Aniszewski T (1994b) The biological basis of quinolizidine alkaloids. Sci Legum 1:1–24

Aniszewski T (2007) Alkaloids-secrets of life alkaloid chemistry, biological significance, applications and ecological role. Elsevier, Amsterdam, pp 1–316

Antoun MD, Mendoza NT, Rios YR (1993) Cytotoxicity of *Hymenocallis expansa* alkaloids. J Nat Prod 56:1423–1425

Bae GS, Kim MS, Jung WS, Seo SW, Yun SW, Kim SG, Park RK, Kim EC, Song HJ, Park SJ (2010) Inhibition of lipopolysaccharide-induced inflammatory responses by piperine. Eur J Pharmacol 642(1–3):154–162

Bastida J, Codina C, Porras CL, Paiz L (1996b) Alkaloids from *Hippeastrum solandriflorum*. Planta Med 62:74–75

Bastida J, Sallés M, Codina C, Viladomat F, de la Luz JL (1996a) Alkaloids from *Behria tenuiflora* Greene. Planta Med 62:575–577

Bennouna J, Delord JP, Campone M, Nguyen L (2008) Vinflunine: a new microtubule inhibitor agent. Clin Cancer Res 14:1625–1632

Benyhe S (1994) Morphine: new aspects in the study of an ancient compound. Life Sci 55(13):969–979

Bernardini S, Tiezzi A, Laghezza Masci V, Ovidi E (2017) Natural products for human health: an historical overview of the drug discovery approaches. Nat Prod Res 6419:1–25

Beuria TK, Santra MK, Panda D (2005) Sanguinarine blocks cytokinesis in bacteria by inhibiting FtsZ assembly a bundling. Biochemist 44(50):16584–16593

Boronova ZS, Sultankhodzhaev MN (2000) Alkaloids of *Delphinium poltoratskii*. Chem Nat Compd 36(4):390–302

Bracca A, Bader A, Siciliano T, Morelli I, de Tommasi N (2003) New pyrrolizidine alkaloids and glycosides from *Anchusa strigosa*. Planta Med 69:835–841

Burgeiro A, Gajate C, Dakir EH, Villa-Pulgarín JA, Oliveira PJ, Mollinedo F (2011) Involvement of mitochondrial and B-RAF/ERK signaling pathways in berberine-induced apoptosis in human melanoma cells. Anti-Cancer Drugs 22(6):507–518

Butler MS (2005) Natural products to drugs: natural product derived compounds in clinical trials. Nat Prod Rep 22:162–195

Bynum WF, Porter R (1994a) Companion encyclopedia of the history of medicine, A chronology of medicine and related sciences, vol 1. Routledge, London/New York

Bynum WF, Porter R (1994b) In: Bynum (ed) Companion encyclopedia of the history of medicine, Doctors and discoveries: lives that created today's medicine, vol 2. Routledge, London/New York. Nature. 419(6903): 115–116

Camacho MR, Kirby GC, Warhurst DC, Croft SL, Phillipson JD (2000) Oxoaporphine alkaloids and quinones from *Stephania dinklagei* and evaluation of their antiprotozoal activities. Planta Med 66:478–480

Castellino F, Huang AY, Altan-Bonnet G, Stoll S, Scheinecker C, Germain RN (2006) Chemokines enhance immunity by guiding naive CD8+ T cells to sites of CD4 T cell-dendritic cell interaction. Nature 440(7086):890–895

Chang MC, Chan CP, Wang YJ, Lee PH, Chen LI, Tsai YL, Lin BR, Wang YL, Jeng JH (2007) Induction of necrosis and apoptosis to KB cancer cells by sanguinarine is associated with reactive oxygen species production and mitochondrial membrane depolarization. Toxicol Appl Pharmacol 218(2):143–151

Chen J, Hao HZ, Wang X, Lee FSC, Yang H, Zheng L (2008) Analysis of major alkaloids in Rhizoma coptidis by capillary electrophoresis-electrospray-time of flight mass spectrometry with different background electrolytes. Electrophoresis 29(10):2135–2147

Chen JJ, Duh CY, Huang HY, Chen IS (2003) Furoquinoline alkaloids and cytotoxic constituents from the leaves of *Melicope semecarpifolia*. Planta Med 69:542–546

Chen JJ, Fang HY, Duh CY, Chen IS (2005a) New indolopyridoquinazoline, benzophenanthridines and cytotoxic constituents from *Zanthoxylum integrifolium*. Planta Med 71:470–475

Chen YW, Li DG, Wu JX, Chen YW, Lu HM (2005b) Tetrandrine inhibits activation of rat hepatic stellate cells stimulated by transforming growth factor-beta in vitro via up-regulation of Smad 7. J Ethnopharmacol 100:299–305

Chenchen W, Wenlong W, Xiaoxue L, Feng M, Dandan C, Xiaowen Y, Shanshan W, Pengshuai G, Hao L, Baoyu Z (2014) Pathogenesis and preventive treatment for animal disease due to locoweed poisoning. Environ Toxicol Pharmacol 37:336–347

Choi YH, Choi WY, Hong SH, Kim SO, Kim GY, Lee WH, Yoo YH (2009) Anti-invasive activity of sanguinarine through modulation of tight junctions and matrix metalloproteinase activities in MDA-MB-231 human breast carcinoma cells. Chem Biol Interact 179(2–3):185–191

Clardy J, Walsh C (2004) Lessons from natural molecules. Nature 432(7019):829–837

Clayden J, Greeves N, Warren S, Wothers P (2001) Organic chemistry. Oxford University Press, Oxford/New York

Collins SL, Kunko PM, Ladenheim B, Cadet J-L, Carroll FI, Izenwasser S (2002) Chronic cocaine increases kappa-opioid receptor density: lack of effect by selective dopamine uptake inhibitors. Synapse 45:153–158

Courtwright DT (2009) Forces of habit drugs and the making of the modern world, 1st edn. Harvard University Press, Cambridge, MA, pp 36–37

Cushnie TPT, Cushnie B, Lamb AJ (2014) Alkaloids: an overview of their antibacterial, antibiotic-enhancing and antivirulence activities. Int J Antimicrob Agents 44:377–386

Dai ZJ, Gao J, Ji ZZ, Wang XJ, Ren HT, Liu XX, Wu WY, Kang HF, Guan HT (2009) Matrine induces apoptosis in gastric carcinoma cells via alteration of Fas/FasL and activation of caspase-3. J Ethnopharmacol 123(1):91–96

Dalbeth N, Lauterio TJ, Wolfe HR (2014) Mechanism of action of colchicine in the treatment of gout. Clin Ther 36(10):1465–1479

Dancey J, Steward WP (1995) The role of vindesine in oncology – recommendations after 10 years' experience. Anti-Cancer Drugs 6(5):625–636

Das NP, Ratty AK (1987) Studies on the effects of the narcotic alkaloids, cocaine, morphine, and codeine on nonenzymatic lipid peroxidation in rat brain mitochondria. Biochem Med Metab Biol 37:258–264

Dasgeb B, Kornreich D, McGuinn K, Okon L, Brownell I, Sackett DL (2018) Colchicine: an ancient drug with novel applications. Br J Dermatol 178(2):350–356

de Moura NF, Simionatto E, Porto C, Hoelzel SCS, Dessoy ECS, Zanatta N, Morel AF (2002) Quinoline alkaloids, cumarins and volatile constituents of *Helietta longifoliata*. Planta Med 68:631–634

De Stefano I, Raspaglio G, Zannoni GF, Travaglia D, Prisco MG, Mosca M, Ferlini C, Scambia G, Gallo D (2009) Antiproliferative and antiangiogenic effects of the benzophenanthridine alkaloid sanguinarine in melanoma. Biochem Pharmacol 78(11):1374–1381

Debiton E, Madelmont JC, Legault J, Barthomeuf C (2003) Sanguinarine-induced apoptosis is associated with an early and severe cellular glutathione depletion. Cancer Chemother Pharmacol 51(6):474–482

Dewick PM (2002) Medicinal natural products. A biosynthetic approach, Second edn. Wiley, Chichester/New York

Dias DA, Urban S, Roessner U (2012) A historical overview of natural products in drug discovery. Meta 2(2):303–336

Diogo CV, Machado NG, Barbosa IA, Serafim TL, Burgeiro A, Oliveira PJ (2011) Berberine as a promising safe anti-cancer agent—is there a role for mitochondria? Curr Drug Targets 12(6):850–859

Dupont C, Couillerot E, Gillet R, Caron C, Zeches-Hanrot M, Riou JF, Trentesaux C (2005) The benzophenanthridine alkaloid fagaronine induces erythroleukemic cell differentiation by gene activation. Planta Med 71:489–494

Dzhakhangirov FN, Bessonova IA (2002) Alkaloids of *Aconitum coreanum*. X. Curare-like activity-structure relationship. Chem Nat Compd 38(1):74–77

Eom KS, Kim HJ, So HS, Park R, Kim TY (2010) Berberine-induced apoptosis in human glioblastoma T98G cells is mediated by endoplasmic reticulum stress accompanying reactive oxygen species and mitochondrial dysfunction. Biol Pharm Bull 33(10):1644–1649

Erdemgil FZ, Telezhenetskaya MV, Baser KHC, Kirimer N (2000) Alkaloids of *Thalictrum orientale* growing in Turkey. Chem Nat Compd 36(2):223–224

Farsam H, Yassa N, Sarkhail P, Shafie A (2000) New pyrrolizidine alkaloids from Heliotropium crassifolium. Planta Med 66:389–391

Fei XF, Wang BX, Li TJ, Tashiro S, Minami M, Xing DJ, Ikejima T (2003) Evodiamine, a constituent of Evodiae Fructus, induces anti-proliferating effects in tumor cells. Cancer Sci 94(1):92–98

Fernandez LS, Sykes ML, Andrews KT, Avery VM (2010) Antiparasitic activity of alkaloids from plant species of Papua New Guinea and Australia. Int J Antimicrob Agents 36:275–279

Forgo P, Hohmann J (2005) Leucoverine and acetyllycovernine, alkaloids from *Leucojum vernum*. J Nat Prod 68:1588–1591

Frédérich M, Tits M, Golffin E, Philippe G, Grellier P, Mol PD, Hayette M-P, Angenot L (2004) In vitro and in vivo antimalarial properties of isostrychnopentamine, an indolomonoterpenic alkaloid from *Strychnos usambarensis*. Planta Med 70:520–525

Fu L, Liang Y, Deng L, Ding Y, Chen L, Ye Y, Yang X, Pan Q (2004) Characterization of tetrandrine, a potent inhibitor of P-glycoprotein-mediated multidrug resistance. Cancer Chemother Pharmacol 53(4):349–356

Fu LW, Zhang YM, Liang YJ, Yang XP, Pan QC (2002) The multidrug resistance of tumour cells was reversed by tetrandrine in vitro and in xenografts derived from human breast adenocarcinoma MCF-7/adr cells. Eur J Cancer 38(3):418–426

Gandhi V, Plunkett W, Cortes JE (2014) Omacetaxine: a protein translation inhibitor for treatment of chronic myelogenous leukemia. Clin Cancer Res 20:1735–1740

Ghawanmeh AA, Chong KF, Sarkar SM, Bakar MA, Othaman R, Khalid RM (2018) Colchicine prodrugs and codrugs: chemistry and bioactivities. Eur J Med Chem 144:229–242

Gören AC, Zhou B-N, Kingston DGI (2003) Cytotoxic and DNA damaging activity of some aporphine alkaloids from *Stephania dinklagei*. Planta Med 69:867–868

Gralla RJ, Gatzemeier U, Gebbia V, Huber R, O'Brien M, Puozzo C (2007) Oral vinorelbine in the treatment of non-small cell lung cancer: rationale and implications for patient management. Drugs 67(10):1403–1410

Haberer W, Witte L, Hartman T, Dobler S (2002) Pyrrolizidine alkaloids in *Pulmonaria obscura*. Planta Med 68:480–482

Han Y, Zhang S, Wu J, Yu K, Zhang Y, Yin L, Bi L (2010) Matrine induces apoptosis of human multiple myeloma cells via activation of the mitochondrial pathway. Leuk Lymphoma 51(7):1337–1346

He BC, Gao JL, Zhang BQ, Luo Q, Shi Q, Kim SH, Huang E, Gao Y, Yang K, Wagner ER, Wang L, Tang N, Luo J, Liu X, Li M, Bi Y, Shen J, Luther G, Hu N, Zhou Q, Luu HH, Haydon RC, Zhao Y, He TC (2011) Tetrandrine inhibits Wnt/beta-catenin signaling and suppresses tumor growth of human colorectal cancer. Mol Pharmacol 79(2):211–219

Hegnauer R (1966) Comparative phytochemistry of alkaloids. In: Swain T (ed) Comparative phytochemistry. Academic, New York, pp 211–212
Hegnauer R (1967) Chemical characters in plant taxonomy: some possibilities and limitations. Pure Appl Chem 14:173–187
Herrera MR, Machocho AK, Brun R, Viladomat F, Codina C, Bastida J (2001) Crinane and lycorane type alkaloids from *Zephyranthes citrina*. Planta Med 67:191–193
Himes RH (1991) Interactions of the catharanthus (Vinca) alkaloids with tubulin and microtubules. Pharmacol Ther 51:257–267
Hoelzel SCSM, Vieira ER, Giacomelli SR, Dalcol II, Zanatta N, Morel AF (2005) An unusual quinolinone alkaloid from *Waltheria douradinha*. Phytochemistry 66:1163–1167
Horie S, Koyama F, Takayama H, Ishikawa H, Aimi N, Ponglux D, Matsumoto K, Murayama T (2005) Indole alkaloids of a Thai medicinal herb, Mitragyna speciosa, that has opioid agonistic effect in guinea-pig ileum. Planta Med 71:231–236
Huang M, Gao H, Chen Y, Zhu H, Cai Y, Zhang X, Miao Z, Jiang H, Zhang J, Shen H, Lin L, Lu W, Ding J (2007) Chimmitecan, a novel 9-substituted Camptothecin, with improved anticancer pharmacologic profiles in vitro and in vivo. Clin Cancer Res 13(4):1298–1307
Hur JM, Hyun MS, Lim SY, Lee WY, Kim D (2009) The combination of berberine and irradiation enhances anti-cancer effects via activation of p38 MAPK pathway and ROS generation in human hepatoma cells. J Cell Biochem 107(5):955–964
Hussain AR, Al-Jomah NA, Siraj AK, Manogaran P, Al-Hussein K, Abubaker J, Platanias LC, Al-Kuraya KS, Uddin S (2007) Sanguinarine- dependent induction of apoptosis in primary effusion lymphoma cells. Cancer Res 67(8):3888–3897
Jakubke H-D, Jeschkeit H, Eagleson M (1994) Concise Encyklopedia chemistry. Walter de Gruyter, Berlin/New York
Jeng JH, Wu HL, Lin BR, Lan WH, Chang HH, Ho YS, Lee PH, Wang YJ, Wang JS, Chen YJ, Chang MC (2007) Antiplatelet effect of sanguinarine is correlated to calcium mobilization, thromboxane and cAMP production. Atherosclerosis 191(2):250–258
Ji YB (2011) Active ingredients of traditional Chinese medicine: pharmacology and application. People's Medical Publishing Hourse Cp., LTD.
Jiang H, Hou C, Zhang S, Xie H, Zhou W, Jin Q, Cheng X, Qian R, Zhang X (2007) Matrine upregulates the cell cycle protein E2F-1 and triggers apoptosis via the mitochondrial pathway in K562 cells. Eur J Pharmacol 559(2–3):98–108
Jiang Y, Li H, Cai P, Ye W (2005) Steroidal alkaloids from the bulbs of *Fritollaria puqiensis*. J Nat Prod 68:264–267
Joel S (1996) The comparative clinical pharmacology of vincristine and vindesine: does vindesine offer any advantage in clinical use? Cancer Treat Rev 21(6):513–525
Kakarala M, Brenner DE, Korkaya H, Cheng C, Tazi K, Ginestier C, Liu S, Dontu G, Wicha MS (2010) Targeting breast stem cells with the cancer preventive compounds curcumin and piperine. Breast Cancer Res Treat 122(3):777–785
Kantarjian HM, O'Brien S, Cortes J (2013) Homoharringtonine/omacetaxine mepesuccinate: the long and winding road to food and drug administration approval. Clin Lymphoma Myeloma Leuk 13:530–533
Khajuria A, Thusu N, Zutshi U, Bedi KL (1998) Piperine modulation of carcinogen induced oxidative stress in intestinal mucosa. Mol Cell Biochem 189(1–2):113–118
Khamidov II, Aripova SF, Karimov AK (2003) Berberis Alkaloids. XLI. Alkaloids from leaves of cultivated *Berberis oblonga*. Chem Nat Compd 39(4):407
Kim J, Buell CR (2015) A revolution in plant metabolism: genome-enabled pathway discovery. Plant Physiol 169(3):1532–1539
Kim S, Lee TJ, Leem J, Kyeong SC, Park JW, Taeg KK (2008) Sanguinarine-induced apoptosis: generation of ROS, down-regulation of Bcl-2, c-FLIP, and synergy with TRAIL. J Cell Biochem 104(3):895–907
Kingston DGI (2009) Tubulin-interactive natural products as anticancer agents. J Nat Prod 72(3):507–515

Koskine A (1993) Asymmetric synthesis of natural products. Wiley, Chichester/New York
Kramell R, Schmidt J, Hermann G, Schliemann W (2005) N-(jasmonoyl)tyrosine-derived compounds from flowers of broad beans (*Vicia faba*). J Nat Prod 68:1345–1349
Krishnakumar N, Manoharan S, Palaniappan PR, Venkatachalam P, Manohar MGA (2009) Chemopreventive efficacy of piperine in 7,12-dimethyl benz [a] anthracene (DMBA)-induced hamster buccal pouch carcinogenesis: an FT-IR study. Food Chem Toxicol 47(11):2813–2820
Krishnamurti C, Rao SC (2016) The isolation of morphine by Sertüner. Indian J Anaesth 60(11):861–862
Kuftinec MM, Mueller-Joseph LJ, Kopczyk RA (1990) Sanguinaria toothpaste and oral rinse regimen clinical efficacy in short- and long-term trials. J Can Dent Assoc 56(7):31–33
Kuo PL, Lin CC (2003) Tetrandrine-induced cell cycle arrest and apoptosis in Hep G2 cells. Life Sci 73(2):243–252
Lai JP, He XW, Jiang Y, Chen F (2003) Preparative separation and determination of matrine from the Chinese medicinal plant *Sophora flavescens* Ait by molecularly imprinted solid-phase extraction. Anal Bioanal Chem 375(2):264–269
Lai YL, Chen YJ, Wu TY, Wang SY, Chang KH, Chung CH, Chen ML (1998) Induction of apoptosisin human leukemic U937 cells by tetrandrine. Anti-Cancer Drugs 9(1):77–81
Lansiaux A, Bailly C, Houssier C, Colson P, de Pauw-Gillet M-C, Frédérich M, Tits M, Angenot L (2002) Apoptosis of H-60 leukemia cells induced by the bisindole alkaloids sungucine and isosungucine from *Strychos icaja*. Planta Med 68:591–595
Lau CW, Yao XQ, Chen ZY, Ko WH, Huang Y (2001) cardiovascular actions of berberine. Cardiovasc Drug Rev. 2001 19(3):234–244
Lee JH, Kang GH, Kim KC, Kim KM, Park DI, Choi BT, Kang HS, Lee YT, Choi YH (2002) Tetrandrine induced cell cycle arrest and apoptosis in A549 human lung carcinoma cells. Int J Oncol 21(6):1239–1244
Lee MR (2011) The history of Ephedra (ma-huang). J R Coll Physicians Edinb 41(1):78–84
Lenfeld J, Kroutil M, Marsalek E (1981) Antiinflammatory activity of quaternary enzophenanthridine alkaloids from *Chelidonium majus*. Planta Med 43(2):161–165
Leung YY, Hui LLY, Kraus VB (2015) Colchicine – update on mechanisms of action and therapeutic uses. Semin Arthritis Rheum 45(3):341–350
Lewis JR (1999) Miscellaneous alkaloids: amaryllidaceae, Sceletium, muscarine, imidazole, oxazole, peptide and other miscellaneous alkaloids. Nat Prod Rep 16:389–416
Lewis JR (2000) Amaryllidaceae, muscarine, imidazole, oxazole, thiazole, and peptide alkaloids, and other miscellaneous alkaloids. Nat Prod Rep 17:57–84
Li DG, Wang ZR, Lu HM (2001) Pharmacology of tetrandrine and its therapeutic use in digestive diseases. World J Gastroenterol 7(5):627–629
Li W, Shao Y, Hu L, Zhang X, Chen Y, Tong L, Li C, Shen X, Ding J (2007) BM6, a new semisynthetic Vinca alkaloid, exhibits its potent in vivo anti-tumor activities via its high binding affinity for tubulin and improved pharmacokinetic profiles. Cancer Biol Ther 6(5):787–794
Li X, Su B, Liu R, Wu D, He D (2011) Tetrandrine induces apoptosis and triggers caspase cascade in human bladder cancer cells. J Surg Res 166(1):e45–e51
Li X, Zhou R, Zheng P, Yan L, Wu Y, Xiao X, Dai G (2010) Cardioprotective effect of matrine on isoproterenol-induced cardiotoxicity in rats. J Pharm Pharmacol 62(4):514–520
Liu T, Song Y, Chen H, Pan S, Sun X (2010) Matrine inhibits proliferation and induces apoptosis of pancreatic cancer cells in vitro and in vivo. Biol Pharma Bull 33(10):1740–1745
Long Y, Lin XT, Zeng KL, Zhang L (2004) Efficacy of intramuscular matrine in the treatment of chronic hepatitis B. Hepatobiliary Pancreat Dis Int 3(1):69–72
Lou H, Yuan H, Yamazaki Y, Sasaki T, Oka S (2001) Alkaloids and flavonoids from peanut skins. Planta Med 67:345–349
Lu JJ, Bao JL, Chen XP, Huang M and Wang YT (2012) Alkaloids isolated from natural herbs as the anticancer agents. Evid Based Complement Alternat Med :12
Lü S, Wang J (2014) Homoharringtonine and omacetaxine for myeloid haematological malignancies. J Hematol Oncol 7:1–10

Lukesh JC, Carney DW, Dong H, Cross RM, Shukla V, Duncan KK, Yang S, Brody DM, Brutsch MM, Radakovic A, Boger DL (2017) Vinblastine 20′amides: synthetic analogues that maintain or improve potency and simultaneously overcome Pgp-derived efflux and resistance. J Med Chem 60:7591–7604

Luo C, Zhong HJ, Zhu LM, Wu XG, Ying JE, Wang XH, Lü WX, Xu Q, Zhu YL, Huang J (2011) Inhibition of matrine against gastric cancer cell line MNK45 growth and its antitumor mechanism. Mol Bio Rep 39(5):5459–5464

Macabeo AP, Krohn K, Gehle D, Read RW, Brophy JJ, Cordell GA, Franzblau SG, Aguinaldo AM (2005) Indole alkaloids from the leaves of Philippine *Alstonia scholaris*. Phytochemistry 66:1158–1162

Machocho A, Chabra SC, Viladomat F, Codina C, Bastida J (1998) Alkaloids from *Crinum stuhlmannii*. Planta Med 64:679–680

Mahady GB, Beecher CWW (1994) Quercetin-induced benzophenanthridine alkaloid production in suspension cell cultures of Sanguinaria canadensis. Planta Med 60(6):553–557

Mano M (2006) Vinorelbine in the management of breast cancer: new perspectives, revived role in the era of targeted therapy. Cancer Treat Rev 32(2):106–118

Manoharan S, Balakrishnan S, Menon VP, Alias LM, Reena AR (2009) Chemopreventive & efficacy of curcumin and piperine during 7,12-dimethylbenz [a]anthracene-induced hamster buccal pouch carcinogenesis. Singap Med J 50(2):139–146

Mantena SK, Sharma SD, Katiyar SK (2006) Berberine, a natural product, induces G1-phase cell cycle arrest and caspase-3-dependent apoptosis in human prostate carcinoma cells. Mol Cancer Ther 5(2):296–308

Matsumoto K, Yamamoto LT, Watanabe K, Yano S, Shan J, Pang J, Pang PK, Ponglux D, Takayama H, Horie S (2005) Inhibitory effect of mitragynine, an analgestic alkaloid from Thai herbal medicine, on neurogenic contraction of the vas deferens. Life Sci 78:187–194

Meng LH, Zhang H, Hayward L, Takemura H, Shao RG, Pommier Y (2004) Tetrandrine induces early G1 arrest in human colon carcinoma cells by down-regulating the activity and inducing the degradation of G 1-S-specific cyclin-dependent kinases and by inducing p53 and p21Cip1. Cancer Res 64(24):9086–9092

Moudi M, Go R, Yien CYS, Nazre M (2013) Vinca Alkaloids. Int J Prev Med 4(11):1231–1235

Nakaoji K, Nayeshiro H, Tanahashi T, Su Y, Nagakura N (1997) Bisbenzylisoquinoline alkaloids from *Stephania cepharantha* and their effects on proliferation of cultured cells from the murine hair apparatus. Planta Med 63:425–428

Nazrullaev SS, Bessonova IA, Akhmedkhodzhaeva KS (2001) Estrogenic activity as a function of chemical structure in Haplophyllum quinoline alkaloids. Chem Nat Compd 37(6):551–555

NCBI (2005) Alkaloids. NCBI. All Databases. WEB: National Library of Medicine

Ng LT, Chiang LC, Lin YT, Lin CC (2006) Antiproliferative and apoptotic effects of tetrandrine on different human hepatoma cell lines. Am J Chin Med 34(1):125–135

Ogasawara M, Matsubara T, Suzuki H (2001) Inhibitory effects of evodiamine on in vitro invasion and experimental lung metastasis of murine colon cancer cells. Biol Pharm Bull 24(8):917–920

Ogasawara M, Matsunaga T, Takahashi S, Saiki I, Suzuki H (2002) Anti-invasive and metastatic activities of evodiamine. Biol Pharm Bull 25(11):1491–1493

Orallo F (2004) Acute cardiovascular effects of (+)-nantenine, an alkaloid isolated from *Platycapnos spicata*, in anaesthetised normotensive rats. Planta Med 70:117–126

Pelser PB, de Vos H, Theuring C, Beuerle T, Vrieling K, Hartmann T (2005) Frequent gain and loss of pyrrolizidine alkaloids in the evolution of *Senecio* section *Jacobaea* (Asteraceae). Phytochemistry 66:1285–1295

Perez EA (2009) Microtubule inhibitors: differentiating tubulin-inhibiting agents based on mechanisms of action, clinical activity, and resistance. Mol Cancer Ther 8(8):2086–2095

Przybylak JK, Ciesiolka D, Wysocka W, Garcia-Lopez PM, Ruiz-Lopez MA, Wysocki W, Gulewicz K (2005) Alkaloid profiles of Mexican wild lupin and an effect of alkaloid preparation from *Lupinus exaltatus* seeds on growth and yield of paprika (*Capsicum annuum* L.). Ind Crop Prod 21(1):1–7

Rakotoson JH, Fabre N, Jacquemond-Collet I, Hannedouche S, Fourast I, Moulis C (1998) Alkaloids from *Galipea officinalis*. Planta Med 64:762–763

Ravishankara MN, Shrivastava N, Padh H, Rajani M (2001) HPTLC method for the estimation of alkaloids of *Cinchona officinalis* stem bark and its marketed formulations. Planta Med 67:294–296

Roeder E (1995) Medicinal plants in Europe containing pyrrolizidine alkaloids. Pharmazie 50:83–98

Roeder E, Eckert A, Wiedenfeld H (1996) Pyrrolizidine alkaloids from *Gynura divaricata*. Planta Med 62:386

Rowinsky E (2003) The Vinca alkaloids. In: Kufe DW, Pollock RE, Weichselbaum RR et al (eds) Holland-Frei cancer medicine, 6th edn. BC Decker, Hamilton

Sadre R, Magallanes-Lundback M, Pradhan S, Salim V, Mesberg A, Jones AD, Della Penna D (2016) Metabolite diversity in alkaloid biosynthesis: a multi-lane (diastereomer) highway for camptothecin synthesis in *Camptotheca acuminata*. Plant Cell 28:1926–1944

Said O, Khalil K, Fulder S, Azaizeh H (2002) Ethnopharmacological survey of medicinal herbs in Israel, the Golan Heights and the West Bank region. J Ethnopharmacol 83:251–256

Salimov BT (2001) Delcorinine, a new alkaloid from *Delphinium corymbosum*. Chem Nat Compd 37(3):272–273

Samuelsson G (2004) Drugs of natural origin: a textbook of pharmacognosy, 5th edn. Swedish Pharmaceutical Press, Stokkholm

Schutz FA, Bellmunt J, Rosenberg JE, Choueiri TK (2011) Vinflunine: drug safety evaluation of this novel synthetic Vinca alkaloid. Expert Opin Drug Saf 10:645–653

Schwarz A, Felippe EC, Bernardi MM, Spinosa HS (2005) Impaired female sexual behavior of rat offspring exposed to *Solanum lycocarpum* unripe fruits during gestation and lactation: lack of hormonal and fertility alterations. Pharmacol Biochem Behav 81:928–934

Seca AML, Pinto DCGA (2018) Plant secondary metabolites as anticancer agents: successes in clinical trials and therapeutic application. Int J Mol Sci 19(1)

Selvendiran K, Banu SM, Sakthisekaran D (2005a) Oral supplementation of piperine leads to altered phase II enzymes and reduced DNA damage and DNA-protein cross links in benzo(a) pyrene induced experimental lung carcinogenesis. Mol Cell Biochem 268(1–2):141–147

Selvendiran K, Sakthisekaran D (2004) Chemopreventive effect of piperine on modulating lipid peroxidation and membrane bound enzymes in benzo(a)pyrene induced lung carcinogenesis. Biomed Pharmacother 58(4):264–267

Selvendiran K, Singh JPV, Krishnan KB, Sakthisekaran D (2003) Cytoprotective effect of piperine against benzo[a]pyrene induced lung cancer with reference to lipid peroxidation and antioxidant system in Swiss albino mice. Fitoterapia 74(1–2):109–115

Selvendiran K, Thirunavukkarasu C, Singh JPV, Padmavathi R, Sakthisekaran D (2005b) Chemopreventive effect of piperine on mitochondrial TCA cycle and phase-I and glutathione-metabolizing enzymes in benzo(a)pyrene induced lung carcinogenesis in Swiss albino mice. Mol Cell Biochem 271(1–2):101–106

Sengbush V P (2003) Alkaloids. Botany online. WEB: University of Hamburg

Sheen WS, Tsai IL, Teng C-M, Ko FN, Chen IS (1996) Indolopyridoquinazoline alkaloids with antiplatelet aggregation activity from *Zanthoxylum integrifolium*. Planta Med 62:175–176

Shin HK, Do JC, Son JK, Lee CS, Lee CH, Cheong CJ (1998) Quinoline alkaloids from the fruits of *Evodia officinalis*. Planta Med 64:764–765

Shyu KG, Lin S, Lee CC, Chen E, Lin LC, Wang BW, Tsai SC (2006) Evodiamine inhibits in vitro angiogenesis: implication for antitumorgenicity. Life Sci 78(19):2234–2243

Siciliano T, De Leo M, Bader A, De Tommasi N, Vrieling K, Braca A, Morelli I (2005) Pyrrolizidine alkaloids from *Anchusa strigosa* and their antifeedant activity. Phytochemistry 66:1593–1600

Slobodnick A, Shah B, Krasnokutsky S, Pillinger MH (2018) Update on colchicine, 2017. Rheumatology 57(1):4–11

Sneader W (1990) Chronology of drug introductions. In: Hansch C et al (eds) Comprehensive medicinal chemistry. The rational design, mechanistic study and terapeutic applications of chemical compounds, General principles, vol I. Pergamon Press, Oxford, pp 11–80

Soto-Hernandez M, Jackson AH (1993) Studies of alkaloids in foliage of *Erythrinia berteroana* and *E. poeppigiana*: detection of β-erythroidine in goats milk. Phytochem Anal 4:97–99
Srinivasan K (2007) Black pepper and its pungent principle-piperine: a review of diverse physiological effects. Crit Rev Food Sci Nutr 47(8):735–748
Srivastava S, Singh MM, Kulshreshtha DK (2001) A new alkaloid and other anti-implantation principles from *Tabernaemontana heyneana*. Planta Med 67:577–579
Staerk D, Lemmich E, Christensen J, Kharazmi A, Olsen CE, Jaroszewski JW (2000) Leishmanicidal, antiplasmodial and cytotoxic activity of indole alkaloids from *Corynanthe pachyceras*. Planta Med 66:531–536
Suladze TS, Vachnadze VY (2002) Alkaloids of *Veratrum lobelianum* growing in Georgia. Chem Nat Compd 38(5):470
Sultankhodzaev MN, Tashkhodzhaev B, Averkiev BB, Antipin MY (2002) Secokaraconitine, a new diterpenoid alkaloid from *Aconitum karacolicum*. Chem Nat Compd 38(1):78–82
Sun Y, Xun K, Wang Y, Chen X (2009) A systematic review of the anticancer properties of berberine, a natural product from Chinese herbs. Anti-Cancer Drugs 20(9):757–769
Szallasi A (2005) Piperine: researchers discover new flavor in an ancient spice. Trends Pharmacol Sci 26(9):437–439
Szlávik L, Gyuris A, Minárovits J, Forgo P, Molnár J, Hohmann J (2004) Alkaloids from *Leucojum vernum* and antiretroviral activity of Ameryllidaceae alkaloids. Planta Med 70:871–873
Takanashi H, Umeda M, Hirono I (1980) Chromosomal aberrations and mutation in cultured mammalian cells induced by pyrrolizidine alkaloids. Mutat Res Toxicol 78:67–77
Takimoto CH, Calvo E (2008) Principles of oncologic pharmacotherapy. In: Pazdur R, Wagman LD, Camphausen KA, Hoskins WJ (eds) Cancer management: a multidisciplinary approach, 11 ed.
Tan W, Lu J, Huang M, Li Y, Chen M, Wu G, Gong J, Zhong Z, Xu Z, Dang Y, GuoJ CX, Wang Y (2011) Anti-cancer natural products isolated from Chinese medicinal herbs. Chin Med 6(1):27
Tanaka H, Etoh H, Shimizu H, Oh-Uchi T, Terada Y, Tateishi Y (2001) Erythrinan alkaloids and isoflavonoids from *Erythrina poeppigiana*. Planta Med 67:871–873
Tashkhodzhaev B, Saidkhodzhaeva Sh A, Bessonova IA, Antipin MY (2000) Arcutin – a new type of diterpene alkaloids. Chem Nat Compd 36(1):79–83
Tulyaganov TS, Allaberdiev FK (2001) Alkaloids of *Nitraria sibirica*. Dihydroschoberine and nitrabirine N-oxide. Chem Nat Compd 37(6):556–558
Tulyaganov TS, Nazarov OM, Levkovich MG, Abdullaev ND (2001) Alkaloids of the Nitraria genus. Komavine and acetylkomavine. Chem Nat Compd 37(1):61–62
Unver N, Gözler T, Walch N, Gözler B, Hesse M (1999) Two novel dinitrogenous alkaloids from *Galanthus plicatus* subsp. *Byzantinus* (Amaryllidaceae). Phytochemistry 50:1255–1261
Unver N, Kaya GI, Werner C, Verpoorte R, Gözler B (2003) Galanthindole: a new indole alkaloid from *Galanthus plicatus* ssp. *Byzantinus*. Planta Med 69:869–871
Varchi G, Battaglia A, Samori C, Baldelli E, Danieli B, Fontana G, Guerrini A, Bombardelli E (2005) Synthesis of deserpine from reserpine. J Nat Prod 68:1629–1631
Vavrečková C, Gawlik I, Müller K (1996a) Benzophenanthridine alkaloids of *Chelidonium majus*; I. inhibition of 5- and 12-lipoxygenase by a non-redox mechanism. Planta Med 62(5):397–401
Vavrečková C, Gawlik I, Müller K (1996b) Benzophenanthridine alkaloids of *Chelidonium majus*. II. Potent inhibitory action against the growth of human keratinocytes. Planta Med 62:491–494
Veeresham C (2012) Natural products derived from plants as a source of drugs. J Adv Pharm Technol Res 3(4):200–201
Vogt A, Tamewitz A, Skoko J, Sikorski RP, Giuliano KA, Lazo JS (2005) The benzo[c]phenanthridine alkaloid, sanguinarine, is a selective, cell-active inhibitor of mitogen-activated protein kinase phosphatase-1. J Biol Chem 280(19):19078–19086
Wall ME, Wani MC, Cook CE, Palmer KH, McPhail AI, Sim GA (1966) Plant antitumor agents. The isolation and structure of camptothecin, a novel alkaloidal leukemia and tumor inhibitor from *Camptotheca acuminate*. J Am Chem Soc 88(16):3888–3890
Waller GR, Nowacki EK (1978) The role of alkaloids in plants. In: Alkaloid biology and metabolism in plants. Springer, Boston, pp 143–181

Wang N, Feng Y, Zhu M, Tsang CM, Man K, Tong Y, Tsao SW (2010) Berberine induces autophagic cell death and mitochondrial apoptosis in liver cancer cells: the cellular mechanism. J Cell Biochem 111(6):1426–1436

Wanjala CCW, Juma BF, Bojase G, Gashe BA, Majinda RRT (2002) Erythrinaline. Alkaloids and antimicrobial flavonoids from *Erythrina latissima*. Planta Med 68:640–642

Weerasinghe P, Hallock S, Tang SC, Trump B, Liepins A (2006) Sanguinarine overcomes P-glycoprotein-mediated multidrug-resistance via induction of apoptosis and oncosis in CEM-VLB 1000 cells. Exp Toxicol Pathol 58(1):21–30

Wink M (1998) A short history of alkaloids. In: Roberts M, Wink M (eds) Alkaloids: biochemistry, ecology, and medicinal application. Plenum Press, New York, pp 11–44

Winterstein E, Tier G (1910) Die Alkaloide. Eine Monographie dernatürlichen Base. Borntr.ger, Berlin

Wright CM, van der Merwe M, DeBrot AH, Bjornsti MA (2015) DNA topoisomerase I domain interactions impact enzyme activity and sensitivity to camptothecin. J Biol Chem 290:12068–12078

Wu JM, Chen Y, Chen JC, Lin TY, Tseng SH (2010) Tetrandrine induces apoptosis and growth suppression of colon cancer cells in mice. Cancer Lett 287(2):187–195

Xing Y, Yan F, Liu Y, Liu Y, Zhao Y (2010) Matrine inhibits 3T3-L1 preadipocyte differentiation associated with suppression of ERK1/2 phosphorylation. Biochem Biophys Res Commun 396(3):691–695

Yang J, Wu LJ, Tashino SI, Onodera S, Ikejima T (2008) Reactive oxygen species and nitric oxide regulate mitochondria-dependent apoptosis and autophagy in evodiamine-treated human cervix carcinoma HeLa cells. Free Radic Res 42(5):492–504

Yoo SM, Oh SH, Lee SJ, Lee BW, Ko WG, Moon CK, Lee BH (2002) Inhibition of proliferation and induction of apoptosis by tetrandrine in HepG2 cells. J Ethnopharmacol 81(2):225–229

Youn MJ, So HS, Cho HJ, Kim HJ, Kim Y, Lee JH, Sohn JS, Kim YK, Chung SY, Park R (2008) Berberine, a natural product, combined with cisplatin enhanced apoptosis through a mitochondria/caspase-mediated pathway in HeLa cells. Biol Pharm Bull 31(5):789–795

Yu HB, Zhang HF, Li DY, Zhang X, Xue HZ, Zhao SH (2011) Matrine inhibits matrix metalloproteinase-9 expression and invasion of human hepatocellular carcinoma cells. J Asian Nat Prod Res 13(3):242–250

Yu HH, Kim KJ, Cha JD, Kim HK, Lee YE, Choi NY, You YO (2005) Antimicrobial activity of berberine alone and in combination with ampicillin or oxacillin against methicillin-resistant *Staphylococcus aureus*. J Med Food 8(4):454–461

Yu P, Liu Q, Liu K, Yagasaki K, Wu E, Zhang G (2009) Matrine suppresses breast cancer cell proliferation and invasion via VEGF-Akt-NF-kappaB signaling. Cytotechnology 59(3):219–229

Zhang B, Liu ZY, Li YY, Luo Y, Liu ML, Dong HY, Wang YX, Liu Y, Zhao PT, Jin FG, Li ZC (2011a) Anti-inflammatory effects of matrine in LPS-induced acute lung injury in mice. Eur J Pharm Sci 44(5):573–579

Zhang L, Wang T, Wen X, Wei Y, Peng X, Li H, Wei L (2007) Effect of matrine on HeLa cell adhesion and migration. Eur J Pharmacol 563(1–3):69–76

Zhang P, Wang Z, Chong T, Ji Z (2012) Matrine inhibits proliferation and induces apoptosis of the androgen independent prostate cancer cell line PC-3. Mol Med Rep 5(3):783–787

Zhang Y, Wu LJ, Tashiro SI, Onodera S, Ikejima T (2003) Intracellular regulation of evodiamine-induced A375-S2 cell death. Biol Pharm Bull 26(11):1543–1547

Zhang Z, Wang X, Wu W, Wang J, Wang Y, Wu X, Fei X, Li S, Zhang J, Dong P, Gu J, Liu Y (2011b) Effects of matrine on proliferation and apoptosis in gallbladder carcinoma cells (GBC-SD). Phytother Res 26(6):932–937

Zheng J, Zheng P, Zhou X, Yan L, Zhou R, Fu XY, Dai GD (2009) Relaxant effects of matrine on aortic smooth muscles of Guinea pigs. Biomed Environ Sci 22(4):327–332

Zhu X, Sui M, Fan W (2005) In vitro and in vivo characterizations of tetrandrine on the reversal of P-glycoprotein-mediated drug resistance to paclitaxel. Anticancer Res 25(3):1953–1962

Chapter 6
Nutraceutical and Bioactive Significance of Ferns with Emphasis on the Medicinal Fern *Diplazium*

Ammatanda A. Greeshma and Kandikere R. Sridhar

Abstract Being primitive vascular plants, pteridophytes (ferns) have wide geographic distribution in different climatic regimes by bridging the gap between lower and higher plants. Compared to the angiosperms, nutraceutical and bioactive potential of ferns attracted less attention. Many ferns are nutraceutically viable owing to their rich source of protein, fiber, minerals, vitamins, essential amino acids, and fatty acids. Besides bioherbicide potential, ferns are endowed with chemopreventive, hepatoprotective, cytotoxic, antihyperglycemic, leishmanicidal, trypanocidal, antimicrobial, antinociceptive, and immunomodulatory metabolites. Ferns are also valuable source of low-cost proteins, starch, and components of cosmeceutical significance. The genus *Diplazium* has pantropical distribution consisting of versatile nutraceutical and bioactive compounds. Future research should intensify toward exploitation of ferns as nutraceutical, healthcare, and industrial products to open up new avenues for food and pharmaceutical industries.

Keywords Bioactive compounds · Ethnic food · Leafy vegetable · Nutraceutical potential · Pteridophyte · Riparian fern

6.1 Introduction

Pteridophytes are the oldest primitive vascular plants constituting the second largest group of vascular plants contributing to the diversity of plant kingdom. Their global representation is more than 1200 species belonging to 204 genera. Pteridophytes are known to grow in varied climatic zones with different phytogeographical regimes depending on the microclimatic conditions. Being cosmopolitan, pteridophytes usually grow in the humid tropical, subtropical, temperate (Dryopteridaceae), alpine (Woodsiaceae), and arid (Cheilanthoideae and Pteridaceae) regions. Its sporophyte consists of roots, stem, leaves, and well-developed vascular strands. Although they prefer shady and moist habitats with moderate temperature regimes, they occupied

A. A. Greeshma · K. R. Sridhar (✉)
Department of Biosciences, Mangalore University, Mangalore, Karnataka, India

D. Egamberdieva, A. Tiezzi (eds.), *Medically Important Plant Biomes: Source of Secondary Metabolites*, Microorganisms for Sustainability 15,
https://doi.org/10.1007/978-981-13-9566-6_6

a wide range of habitats including high altitude biomes. Medicinal and ethnobotanical uses of pteridophytes are less versatile compared to the angiosperms (Chowdury 1973; Vyas and Sharma 1998). Liu et al. (2012) documented about 52 species of ferns, which are edible in China. Caius (1935) documented several medicinal applications of ferns occurring in the Indian subcontinent. The in vitro propagation of pteridophytes is of great interest for the preservation of economically valued, medicinal, edible, and ornamental ferns (Bharti et al. 2013).

In the Indian subcontinent, the pteridophyte diversity is notable especially in the Himalayas, Western Ghats, and Eastern Ghats (Dixit 2000). They are more specifically distributed in temperate Himalayas and tropical ecoregions of Deccan Plateau (Shankar and Khare 1992; Vasudev 1999; Shrivastava 2007). Surveys in Kigga forests of the central Western Ghats revealed occurrence of 31 species of pteridophytes belonging to 20 families (Deepa et al. 2013). Surveys in Mudigere region of the Western Ghats resulted in documentation of 26 fern species belonging to 17 families with highest frequency of *Selaginella monospora* followed by *Adiantum philippense*, *Pteris biaurita*, *Adiantum concinnum*, and *Tectaria paradoxa* (Parashurama et al. 2016). Maridass and Raju (2010) documented up to 272 species of ferns and fern allies (95 genera belong to 34 families) from southern Western Ghats. Among them, 89 species were common, 33 species were rare, 26 were endemic species, 12 were near threatened, and 9 were very rare. Benjamin and Manickam (2007) described 61 medicinal pteridophytes (belong to 31 families) in the Western Ghats utilized by the native tribes to treat ailments like cold, cough, fever, stomach disorders, rheumatics, diabetes, and poisonous bites. The aim of the present chapter is to consolidate the usefulness of ferns especially nutraceutical value, bioactive potential, and functional attributes with emphasis on the values and relevance of different species of *Diplazium*.

6.2 Nutraceutical Importance

Bracken (*Pteridium*) fiddle heads are nutritionally rich sources in Korea, and they contain significant amounts of protein, fiber, vitamins, and minerals (Copeland 1942; Thakur et al. 1998). Table 6.1 provides proximal composition of different parts of 20 edible ferns. The range of proximal composition of ferns includes moisture, 25.6–91.1%; crude protein, 0.6–24.1%; total lipids, 0.2–21.5%; crude fiber, 0.5–62.2%; ash, 0.6–19.5%; and carbohydrates, 3.1–57.2%. The crude protein content in *Pteridium aquilinum* (leaves) is comparable to those of legume seeds (21.9–24.1%). The total lipids of stem and leaves of *Dicksonia antarctica* were higher than other ferns (21.5–15.5 vs. 0.2–6%). The crude fiber content was high in *Azolla microphylla* (leaves, 13.4%), *A. pinnata* (whole plant, 14.7%), and *Acrostichum aureum* (rhizome, 10.3%). The carbohydrate content was highest in roots of *Diksonia antarctica* (57.2%) followed by *A. aureum* (rhizome, 53.3%), *A. microphylla* (leaves, 37.7%), *A. pinnata* (whole plant, 33.8%), and *D. antarctica* (stem, 25.3%). The high ash content in leaves of *A. microphylla* (24.3%), whole plant of *A.*

Table 6.1 Proximal composition of 20 edible ferns (%)

	Parts used	Moisture	Crude protein	Total lipids	Crude fiber	Ash	Carbohydrates
Acrostichum aureum[14]	Rhizome	66.8	4.9	0.7	10.3	4.3	53.3
Athyrium filix-femina[2]	Fiddle heads	91.1	3.2	0.2	–	0.6	4.9
Azolla microphylla[7]	Leaves	80.5	24.1	3.3	13.4	19.5	37.7
Azolla pinnata[9]	Whole plant	75.7	23.5	3.7	14.7	24.3	33.8
Ceratopteris cornuta[5]	Sterile fronds	87.0	4.2	–	–	–	3.1
Ceratopteris cornuta[5]	Fertile fronds	91.0	5.3	–	–	–	7.4
Crepidomanes intramarginale[10]	Leaves	–	0.6	–	–	3.2	13.7
Dicksonia antarctica[13]	Leaves	55.7	9.2	21.5	–	9.0	4.6
Dicksonia antarctica[13]	Stem	53.1	3.1	15.5	–	3.0	25.3
Dicksonia antarctica[13]	Root	25.6	5.3	6.0	–	6.0	57.2
Matteuccia struthiopteris [1]	Fiddle heads	87.0	4.2	0.5	1.1	4.0	3.1
Nephrolepis biserrata[4]	Leaves	80.0	6.1	0.3	0.9	1.9	10.9
Nephrolepis cordifolia[3]	Leaves	78.4	1.6	–	–	11.0	11.4
Nephrolepis cordifolia[8]	Leaflets	65.4	10.8	–	–	1.3	21.5
Nephrolepis cordifolia[3]	Rachis	68.5	3.8	–	–	6.3	8.1
Nephrolepis cordifolia[3]	Rhizome	66.0	11.9	–	–	8.0	9.3
Nephrolepis exaltata[11]	Unipinnate leaflets	78.5	12.5	0.5	0.5	0.9	7.1
Nephrolepis exaltata[11]	Bipinnate leaflets	82.7	9.5	0.5	0.6	0.9	6.0
Nephrolepis furcans[6]	Leaves	85.0	0.9	0.2	3.3	2.1	8.5
Pteridium aquilinum[12]	Leaves	–	21.9	2.7	4.1	7.1	–

[1]Bushway et al. 1982; [2]Drury 1985; [3]Gauchan et al. 2008; [4]Oloyede et al. 2008; [5]Oloyede et al. 2010; [6]Oloyede et al. 2012; [7]Chatterji et al. 2013; [8]Oloyede et al. 2013; [9]Cherryl et al. 2014; [10]Greeshma and Murugan 2014; [11]Oloyede et al. 2014; [12]Awe and Amobi 2015; [13]Ekwealor et al. 2015; [14]Lobo and Gulimane 2015; – not determined

pinnata (19.5%), and leaves of *Nephrolepis cordifolia* (11%) reveals their rich mineral composition (Table 6.1).

Several ferns are ideal source of major minerals (sodium, calcium, potassium, magnesium, and phosphorus) as well as trace elements (iron, copper, zinc, and manganese) (Oloyede et al. 2014). They possess remarkably high quantities of many essential amino acids (Deka et al. 2016; Zanariah et al. 1986; Sanginga and Hove 1989; Bhaskaran and Kannapan 2015). According to Delong et al. (2011), *Matteuccia struthiopteris* (ostrich fern) is known for higher quantity of rare fatty acids than green vegetable including ω-3 (arachidonic and eicosapentaenoic acids) and ω-6 fatty acids (γ-linolenic and dihomo-γ-linolenic acids). With ferns being nutritionally versatile, it is possible to preserve their fiddle heads (unfurled fronds) as source of food.

The process of preservation of fiddle heads is fairly simple, and cleaned fiddle heads should be steam boiled in hot water followed by sun drying or preserved in a salt layer, which could last long up to 2–3 years (Lee and Shin 2011). During requirement, the dried fiddle heads could be recooked (with mashed garlic, salt, sesame oil, and soybean sauces), while the salted fiddle heads should be rinsed with running water followed by cooking similar to dried ones. The fiddle heads are commonly boiled with butter followed by addition of cider or wine vinegar with a bit of pepper (Lee and Shin 2011).

Besides nutraceutical source to humans, some ferns such as *Dryopteris wallichiana*, *Nephrolepis biserrata*, and *Ophioglossum grande* are also useful as fodder in Nigeria (goat, sheep, and other ruminants) (Nwosu 2002; Oloyede et al. 2013). The aquatic fern *Azolla pinnata* is used as feed for broiler chicken, laying hens, goats, and calves of buffalo, while *Azolla filiculoides* serves as source of protein while rearing pigs (Becerra et al. 1990; Duran 1994; Samanta and Tamang 1995; Alalade et al. 2007; Balaji et al. 2009; Indira et al. 2009, Leterme et al. 2010; Cherryl et al. 2014).

6.3 Bioactive Potential

Although ferns bridge the gap between lower and higher plants, compared to angiosperms, ferns are underexplored as well as underutilized for their potency of phytochemicals (Cao et al. 2017). Ferns are also known to replace chemical herbicides owing to their inhibitory metabolites. Several studies documented therapeutic potential of ferns especially chemopreventive, hepatoprotective, cytotoxic, antihyperglycemic, leishmanicidal, trypanocidal, antimicrobial, antinociceptive, anti-inflammatory, and immunomodulatory properties (Wu et al. 2005; Wills and Asha 2006; Yonathan et al. 2006; Wills and Asha 2009; Radhika et al. 2010; Zheng et al. 2011a, b; Morais-Braga et al. 2013a, b; Socolsky et al. 2015; Cao et al. 2017). With expectation of a few (e.g., *Lycopodium*), ferns do not synthesize alkaloids. However, various phenolic compounds like acylphloroglucinols, nonprotein amino acids, cyanogenic glycosides, and flavonoids have been reported. Polyphenols are useful as antioxidants, and it is generally recognized to reduce the risks of many chronic diseases. Screening of 37 ferns and fern allies (*Polystichum lepidocaulon* and *P. polyblepharum*) showed presence of 13% total polyphenols in dried fronds and rhizomes (Shin 2010; Shin and Lee 2010). In addition, the dried fronds of *Davallia mariesii* and rhizomes of *Athyrium niponicum*, *Cyrtomium fortunei*, *Dicranopteris pedata*, and *Dryopteris nipponensis* possess >10% of total polyphenols.

6.3.1 Antioxidant Properties

In many countries, brackens are considered as poisoning plants owing to their carcinogenic and anti-thiamin properties (Somvanshi and Ravishankar 2004). However, the hot water extracts of dried bracken fiddle heads yielded acidic polysaccharides, which exhibits anticomplementary activities (Byeong et al. 1994). Antioxidant activities (DPPH radical and ABTS radical-scavenging) have been studied from the frond and rhizome extracts of several fern genera (*Adiantum*, *Athyrium*, *Coniogramme*, *Cytominum*, *Davallia*, *Dicranopteris*, *Dryopteris*, *Hypolepis*, *Lycopodium*, *Lygodium*, *Matteuccia*, *Onoclea*, *Osmunda*, *Pteridium*, *Polypodium*, *Polystichum*, *Pteris*, *Pyrrosia*, *Selaginella*, *Thelypteris*, and *Woodsia*) (Shin 2010). Many of them showed different antioxidant activities by scavenging DPPH and ABTS radicals. More specifically, fern families like Dryopteridaceae, Osmundaceae, and Woodsiaceae exhibited powerful antioxidant activities. The fiddle heads of *Athyrium acutipinulum* also showed strong antioxidant effects (Lee et al. 2005). Crude extracts of some ferns showed higher antioxidant activity than vitamin C (Soare et al. 2012). Thus, it is expected that analyzing antioxidant activities in ferns would result in the development of several healthcare products to combat aging as well as chronic diseases (Soare et al. 2012).

6.3.2 Antimicrobial Activities

Ferns are known for their efficient antimicrobial agents against several harmful microbes. They are also useful in developing antibiotic sprays, packing material, toothpaste, and handwash products and to protect human body from undesired microbes. The extracts obtained from ferns serve as effective antimicrobial agents against Gram-positive bacteria (e.g., *Bacillus subtilis* and *Staphylococcus aureus*), Gram-negative bacteria (e.g., *Escherichia coli*, *Pseudomonas aeruginosa*, and *Salmonella typhi*), and fungi (Banerjee and Sen 1980; Vincent and Kanna 2007; Lee et al. 2009). The genus *Dryopteris* exhibited vigorous antibiotic activities. *Dryopteris crassirhizoma* and *D. filix-mas* could be used to control methicillin-resistant *Staphylococcus aureus* (MRSA) (Lee et al. 2009), while *D. cochleata* serve against Gram-positive bacteria, Gram-negative bacteria, and fungi (Banerjee and Sen 1980). Because of high activity of *D. crassirhizoma* against *Streptococcus*, its anti-tooth decay substance has been patented in Korea. However, according to Shin (2010), fronds of *Athyrium niponicum* and *Hypolepis punctata* are more efficient than *D. crassirhizoma* against *Streptococcus mutans* and *S. sobrinus*.

6.3.3 Cosmeceutical Qualities

Ferns are also useful as cosmetic ingredients. For example, *Dryopteris* spp. possess strong inhibitory activities against acne causing *Propionibacterium acnes* on the skin (Yoon et al. 2006). Many fern extracts containing high phenolic compounds are currently used as ingredients of body and facial cosmetics (e.g., cleanser, toner, moisturizer, shampoo, and conditioner). Extracts of some ferns are useful and effective natural cosmetic ingredients. Furthermore, the phenolic compounds of ferns are beneficial in skin care owing to prevention of UV-induced skin damages, anti-wrinkle, and skin-whitening properties (Svobodová et al. 2003; Tanaka et al. 2004; An et al. 2005; Parvez et al. 2006). A number of ferns possess phytoecdysteroids (e.g., ecdysone) (which is not present in most of the angiosperms) helpful in promoting cell regeneration, refining skin texture, and strengthening skin barrier (Lin and Lin 1989; Meybeck et al. 1997). Since fern spores are not causing hay fever, cosmetics consisting of fern spores are patented in Korea, and such products are used as facial scrub (Moran 2004).

6.4 Functional Attributes

Physicochemical characteristics of proteins play a prominent role in quality control of food stuffs. The low-cost protein derived from green vegetables will meet the demand of food proteins in many industries. The food processing industries emphasize on leaf protein isolates owing to low lipid and fiber content (Rana et al. 2015). Green leaves possess protein up to 40–70% along with carotenoids, vitamin E, and minerals (Badar and Kulkarni 2011). Isolation of high quality of green leaf protein is simple, affordable, and favorable for commercial applications (Moure et al. 2002). However, the functional properties of fern proteins are less studied, and only countable research articles are available.

The rhizome starch fern is commonly employed in the food industry for preparation of vermicelli, cakes, and concentrated soups marketed in East Asian countries (Zhang et al. 2011). The fern starch is also useful in preparation of liquor and soft drinks in China (Liu et al. 2012). The starchy paste obtained from *Marsilea drummondii* is edible in Australia (Mannan et al. 2008). The starch purified from fern is significantly tough with suitable elastic properties (Cao et al. 2007). The starch granules from the rhizome of *Acrostichum aureum* possess high solubility (35%) as well as swelling power (12.3%) (Lobo and Gulimane 2015). The protein solubility is also remarkable in ferns, for example, high solubility is known in leaf protein concentrate of *Nephrolepis biserrata* (55.9%, pH 8), *Arthropteris orientalis* (45%, pH 10), and *Diplazium esculentum* (28.52%, pH 12) (Essuman et al. 2014; Rana et al. 2015). Ferns such as *Nephrolepis biserrata* and *Arthropteris orientalis* possess pH-dependent foam capacity and attain maximum foam capacity at pH 14. An epiphytic fern *Metapolypodium memeiense* is also used as a taste enhancer during cooking vegetables, and many ferns are available in local markets of China as dried fronds, salted fronds, packed fronds, fern starch, fern starch noodles, fern starch cakes, and fern leaf tea (Liu et al. 2012). Different parts of about 144 species of ferns are edible and utilized to replenish nutrition in China.

6.5 *Diplazium*: A Versatile Fern

Diplazium is a diverse genus belonging to the lady fern family (Athyriaceae). This fern has pantropical distribution and is separated from *Allantodia* based on the characteristic non-imparipinnate fronds and grooves from rachis to costa (Ching 1964). It is a terrestrial fern with erect to suberect rhizome forming caudex and strong roots. The fronds are large and pinnately compound with herbaceous lamina.

6.5.1 Diversity

The genus *Diplazium* consists of 400 species distributed in different pantropical regimes (e.g., Malesia, Neotropics, Afro-Madagascar region, and cold temperate regions of Eurasia) (Tryon and Tryon 1982; Wu and Ching 1991; Chu and He 1999; Pacheco and Moran 1999; Mickel and Smith 2004; Roux 2009). The rare and endangered endemic fern *Diplazium molokaiense* was reported in East Maui, Hawaii (Wood 2006). *Diplazium fimbriatum* has been reported as a new species from Brazil (Mynssen and Matos 2012). Several *Diplazium* spp. are known from Thailand (*D. bantamense*, *D.cordifolium*, *D. crenatoseratum*, *D. esculentum*, *D. polypodioides*, *D. riparium*, *D. silvaticum*, *D. simplicivenium*, *D. sorzogonense*, and *D. tomentosum*) (Boonkerd et al. 2008). Indonesia also possesses many *Diplazium* spp. (*D. asperum*, *D. cordifolium*, *D. esculentum*, *D. lomariaceum*, *D. pallidum*, *D. simplicivenium*, *D. sorzogonense*, and *D. tomentosum*) (Slik et al. 2006). Kholia (2011) has documented nearly 40 species of *Diplazium* in India, and up to 17 species have been reported in Sikkim alone. The common *Diplazium* in India include *D. bellum*, *D. dilatatum*, *D. doederleinii*, *D. esculentum*, *D. forrestii*, *D. kawakamii*, *D. latifolium*, *D. laxifrons*, *D. longifolium*, *D. maximum*, *D. medogense*, *D. polypodioides*, *D. sikkimense*, *D. spectabile*, *D. squamigerum*, and *D. stoliczkae*. The *Diplazium* species are also reported from the Western Ghats regions of Maharashtra, Karnataka, Kerala, and Tamil Nadu (Dudani et al. 2012, 2014; Das et al. 2013; Kavitha et al. 2015; Patil et al. 2016; Sathish and Vijayakanth 2016). It is possible to conserve *Diplazium* spp. by spore explant (initiation, multiplication, and differentiation), which is economical as well as rapid method of propagation (Nair et al. 2014).

6.5.2 Nutraceutical Value

Fiddle heads and tender pinna of *Diplazium esculentum* are commonly edible in the Western Ghats region, which is common along the rivers and swampy areas in the Western Ghats of India (Copeland 1942; Akter et al. 2014; Greeshma and Sridhar 2016; Sridhar and Karun 2017; Greeshma et al. 2018). It is one of the most commonly consumed fern in hilly tribes in Northeastern India and the Philippines (Copeland and Collado 1936). Fiddle heads of three *Diplazium* species reported from different regions are known to possess a wide range of crude protein (0.2–31.2%), total lipid (0.1–8.3%), crude fiber (0.4–12.7%), ash (1.4–17.6%), and carbohydrates (0.02–68.6%) (Table 6.2). The qualitative test showed presence of reducing sugars especially anthraquinones and anthranol glycosides (Tongco et al. 2014).

Fern shoots are excellent source of minerals and electrolytes (e.g., potassium, iron, manganese, and copper) (Seal 2012). Fresh shoots contain up to 7% of daily required quantities of potassium, which combats blood pressure as well as regulate heart contractions. *Diplazium* spp. are also composed of major minerals (calcium, sodium, potassium, magnesium, and phosphorus) and trace elements (iron, copper, zinc, and manganese) (Table 6.3). The mineral composition of *Diplazium* is dependent on the species as well as different segments used for analysis. Potassium and calcium are the most abundant minerals present in *Diplazium* spp. Different parts of *Diplazium esculentum* and *D. sammatti* fulfil the range of NRC-NAS (1989) standards for major minerals and trace elements required for infants, children, and adults. Besides, these ferns also meet the desired Na/K (<1) and Ca/P (>1) ratios; such ratios are known to combat the blood pressure and prevent the loss of calcium in urine to restore calcium in bones, respectively (Shills and Young 1988; Yusuf et al. 2007).

Leaves and fiddle heads of *Diplazium esculentum* are also known for several indispensable and dispensable amino acids (Table 6.4). The fiddle heads possess higher amount of amino acids compared to leaves with exception of sulfur amino acids. Interestingly, sulfur amino acids were substantially higher in leaves than fiddle heads (methionine, 2.1 vs. 1.4 g/110 g protein; cystine, 4.2 vs. 0.5 g/100 g protein). The leaves fulfil the FAO-WHO (1991) stipulated standard for sulfur amino acids (6.3 vs. 2.5 g/100 protein). This offers nutraceutical advantage of use of fiddle heads along with desired quantity of leaves to gain sulfur amino acids. The fiddle heads also fulfil FAO-WHO (1991) standard for indispensable amino acids like histidine, isoleucine, leucine, lysine, threonine, and valine. Among the dispensable amino acids, glutamic acid was highest in leaves, while the glycine was highest fiddle heads followed by glutamic acid, alanine, proline, aspartic acid, serine, and arginine.

Table 6.2 The proximal composition of three *Diplazium* species (%)

	Parts used	Habitat	Moisture	Crude protein	Total lipids	Crude fiber	Ash	Carbohydrates
Diplazium esculentum[5]	Fiddle heads	India (Sikkim)	92.4	31.2	8.3	4.6	16.2	44.3
Diplazium esculentum[8]	Fiddle heads	India (Karnataka)	–	16.1	7.5	2.3	12.1	19.3
Diplazium esculentum[2]	Leaves	India (Meghalaya)	87.6	14.4	0.1	3.9	12.2	8.4
Diplazium esculentum[6]	Leaves	India (Assam)	71.7	18.3	0.3	4.45	14.4	5.5
Diplazium esculentum[3]	Leaves	India (Arunachal Pradesh)	–	17.4	5.6	12.7	17.6	37.7
Diplazium esculentum[7]	Leaves	India (Himachal Pradesh)	–	0.2	–	–	–	0.02
Diplazium esculentum[4]	Leaves	Philippines	91.8	0.9	0.3	0.7	1.4	–
Diplazium esculentum[4]	Leaves	Philippines	–	10.7	3.4	9.1	17.4	–
Diplazium maximum[7]	Leaves	India (Himachal Pradesh)	–	0.2	–	–	–	0.02
Diplazium sammatii[1]	Young pinna and crozier	Nigeria	85.3	10.2	14.1	0.4	10	62.3
Diplazium sammatii[1]	Mature pinna	Nigeria	80.0	10.3	9.5	0.4	11.2	68.6

[1]Bassey et al. 2001; [2]Seal 2012; [3]Tag et al. 2014; [4]Tongco et al. 2014; [5]Pradhan et al. 2015; [6]Deka et al. 2016; [7]Wali et al. 2016; [8]Greeshma et al. 2018; – not determined

6.5.3 *Bioactive Potential*

It is believed by the native tribes in India that *D. esculentum* is useful to counteract constipation and serve as an appetizer (Kala 2005; Das et al. 2008). Concoction of this fern is used to cure hemoptysis and cough, while the rhizomes serve as insecticides (Kaushik et al. 2011, 2012). Studies on *D. esculentum* also showed its usefulness as anaphylactic shock and mast cell stabilizer (Das et al. 2012). The fiddle heads contain only 34 calories of energy per 100 g, and nonetheless their high-quality nutraceutical profile is composed of health promoting antioxidants, vitamins, and essential fatty acids (ω-3 and ω-6).

Fresh fronds possess very high quantities of antioxidant vitamin A and carotenes. Fiddle heads weighing 100 g contain 3617 IU (or 120%) of recommended daily requirements of vitamin A. Further, they also endowed with small to moderate quantities of many B vitamins (e.g., niacin, riboflavin, and thiamine). Besides vitamins

Table 6.3 Mineral composition of different parts of two *Diplazium* species (mg/100 g)

	Diplazium species and parts used for analysis								
	Diplazium esculentum[7] (fiddle heads)	*Diplaziumesculentum*[6] (fiddle heads)	*Diplazium esculentum*[4] (fiddle heads)	*Diplazium esculentum*[3] (leaves)	*Diplazium esculentum*[5] (leaves)	*Diplazium esculentum*[2] (fronds)	*Diplazium sammatti*[1] (young pinna)	*Diplazium sammatti*[1] (mature pinna)	NRC-NAS standards[8]
Na	145	29.0	360.0	118.0	9.5	79.0	520.0	560.0	120–500
K	3351	74.5	1120.0	4373.0	914.4	2370.0	1600.0	1600.0	500–2000
Ca	436	52.7	1290.0	873.0	192.7	1020.0	190.0	190.0	600–800
Mg	481	15.3	–	–	0.4	505.0	–	–	60–350
P	1050	–	80.0	–	–	500.0	6.8	7.2	500–800
Mn	123	21.1	–	5.1	–	–	–	–	
Fe	52	38.2	–	25.7	11.2	560.0	4.3	6.7	10–15
Zn	194	4.3	–	16.7	2.7	58.0	3.6	4.5	12–15
Cu	509	1.7	–	2.6	0.3	4.0	3.5	2.5	0.6–3
Na/K ratio	0.04	0.39	0.32	0.03	0.01	0.03	0.33	0.35	
Ca/P ratio	0.41	–	16.1	–	–	2.04	27.9	26.4	

[1]Bassey et al. 2001; [2]Sundriyal and Sundriyal 2001; [3]Seal 2012; [4]Tag et al. 2014; [5]Pradhan et al. 2015; [6]Deka et al. 2016; [7]Greeshma et al. 2018; [8]NRC-NAS 1989 range for infants, children and adults; – not determined

Table 6.4 Amino acid composition of *Diplazium esculentum* in comparison with FAO-WHO standard (g/100 g protein)

	D. esculentum[1] (leaves)	*D. esculentum*[2] (fiddle heads)	FAO-WHO standard[3]
Indispensable amino acids			
Histidine	0.2	2.3	1.9
Isoleucine	0.6	5.3	2.8
Leucine	0.7	8.1	6.6
Lysine	0.3	8.4	5.8
Methionine	2.1	1.4	[a]2.5
Cystine	4.2	0.5	
Phenylalanine	0.8	6.2	[b]6.3
Tyrosine	0.6	3.3	
Threonine	0.6	4.3	3.4
Tryptophan	–	–	1.1
Valine	0.2	6.3	3.5
Dispensable amino acids			
Alanine	0.4	7.7	
Arginine	0.2	5.2	
Aspartic acid	0.3	6.3	
Glutamic acid	4.6	8.0	
Glycine	0.1	10.5	
Proline	[c]	6.8	
Serine	[c]	5.5	

[1]Deka et al. 2016; [2]Greeshma et al. 2018; [3]FAO-WHO 1991
[a]Methionine + Cystine
[b]Phenylalanine + Tyrosine
[c]Not detectable

(essential for vision), fiddle heads also serve as powerful natural antioxidants required by the human body for maintaining the integrity of the skin and mucus membranes. Studies also suggested that natural foods rich in vitamin A help in protecting the body against lung and oral cancers. They are also excellent source of many natural polyphenolic flavonoids (α-carotene and β-carotene), and they will be converted into vitamin A in vitro. The fresh fiddle heads of 100 g contain 26.6 mg (44%) of daily required vitamin C content. Vitamin C is a moderately potential water-soluble antioxidant with flavonoid compound like carotenes, which helps scavenging harmful free radicals, offering protection from cancers, inflammation, viral cough, and cold. Unique sweet taste of fiddle heads comes from their richness in vitamin C.

Ethanolic extract of shade-dried fronds possesses high quantities of phenolics, flavonoids, and saponins than aqueous extract (Das et al. 2013). Further, the chromatogram of HPTLC revealed presence of different types of polyphenols as well as steroidal saponins. Aqueous and ethanolic extracts of the *D. esculentum* showed inhibitory effect against growth of many human as well as plant pathogenic bacteria. The positive results are comparable with tetracycline as standard antibiotic. It was found that mixing fern extract with antibiotic in equal proportions will be more effective against pathogenic bacteria than antibiotic alone (Amit et al. 2011).

6.6 Concluding Remarks

Pteridophytes distributed in a wide range of habitats prefer shady moist conditions with moderate temperature regimes. They are also versatile like angiosperms in ethnobotanical and medicinal properties. Fiddle heads of several ferns are nutritionally rich especially proteins, fiber, minerals, and vitamins. They are also known for remarkably high quantities of many essential amino acids and endowed with essential fatty acids. Starch from fern rhizomes is suitable for preparation of several foodstuffs and industrial products. In addition to the leaf protein and rhizome starch, ferns possess many valuable functional attributes useful in production of food as well as medicinal products. Many ferns are known for antimicrobial, cytotoxic, hepatoprotective, antihyperglycemic, antiprotozoal, antinociceptive, immunomodulatory, and chemopreventive properties. Ferns have the capability to serve as bioherbicide to replace chemical herbicides.

Diplazium is one of the versatile ferns distributed in pantropical regions. In India, this fern is widely distributed in the Western Ghats and Sikkim. The fiddle heads of *Diplazium* possess good proximal components and minerals. *Diplazium* shows high antioxidant activities due to presence of vitamin A and carotenoids. They possess many therapeutic potential like protection of mucus membranes, prevention of cancers, and protection against inflammation.

Pteridophytes have potential nutraceutical and bioactive components useful in human diet and therapeutics. Many ferns could be cultivated in different locations besides their native habitats like kitchen gardens, botanical gardens, and arboretum similar to those of orchids. Although *Diplazium esculentum* occurs naturally as riparian fern in the Western Ghats, it could be cultivated in lateritic soils of Southwest India without major efforts. Morphologically distinct 3–4 varieties (or landraces) of *D. esculentum* are known from the Western Ghats based on ethnic knowledge of locals and tribes. Serious efforts are warranted toward utilization of ferns in the direction of nutraceuticals, and health protective attributes would open up new avenues for food and pharmaceutical industries.

Acknowledgments The first author greatly acknowledges the award of INSPIRE Fellowship by the Department of Science and Technology, New Delhi, Government of India. The corresponding author is grateful to the University Grants Commission, New Delhi, India, for the award of UGC-BSR Faculty Fellowship.

References

Akter S, Hossain MM, Aral I, Akhtar P (2014) Investigation of *in vitro* antioxidant, antimicrobial and cytotoxic activity of *Diplazium esculentum* (Retz.). Sw. Int J Adv Pharm Biol Chem 3:723–733

Alalade OA, Iyayi EA, Alalade TO (2007) The nutritive value of *Azolla* (*Azolla pinnata*) meal in diets for growing pullets and subsequent effect on laying performance. J Poult Sci 44:273–277

Amit S, Sunil K, Bhatt SP, Arvind N (2011) Antibacterial activity of *Diplazium esculentum* (Retz.) Sw. Pharmacogn J 3:77–79

An BJ, Kwak JH, Park JM et al (2005) Inhibition of enzyme activities and the antiwrinkle effect of polyphenol isolated from the persimmon leaf (*Diospyros kaki* Folium) on human skin. Dermatol Surg 31:248–285
Awe S, Amobi OO (2015) Antibacterial, phytochemical and proximate analysis of *Pteridium aquilinum*. Int J Res Pharm Biosci 2:1–7
Badar KV, Kulkarni AU (2011) LPC is novel source of protein for human health and nutrition: a review. Cur Bot 2:5–7
Balaji K, Jalaludeen A, Churchil RR et al (2009) Effect of dietary inclusion of *Azolla* (*Azolla pinnata*) on production performance of broiler chicken. Ind J Poult Sci 44:195–198
Banerjee RD, Sen SP (1980) Antibiotic activity of pteridophytes. Econ Bot 34:284–298
Bassey ME, Etuk EUI, Ibe MM, Ndon BA (2001) *Diplazium sammatii*, Athyraceae (Nyama Idim): age-related nutritional and antinutritional analysis. Pl Foods Hum Nutr 56:7–12
Becerra M, Murgueitio E, Reyes G, Preston TR (1990) *Azolla filiculoides* as partial replacement for traditional protein supplements in diets for growing- fattening pigs based on sugar cane juice. Livestock Res Rural Dev 2:15–22
Benjamin A, Manickam VS (2007) Medicinal pteridophytes from the Western Ghats. Ind J Trad Know 6:611–618
Bharti SK, Manabendra DC, Behari MP (2013) *In vitro* propagation in pteridophytes – a review. Int J Res Ayurv Pharm 4:297–303
Bhaskaran SK, Kannapan P (2015) Nutritional composition of four different species of *Azolla*. Eur J Exp Biol 5:6–12
Boonkerd T, Chantanaorrapint S, Khwaiphan W (2008) Pteridophyte diversity in the tropical lowland rainforest of Khao Nan National park, Nakhon Si Thammarat province, Thailand. Nat Hist J Chulalongkorn Univ 8:83–97
Bushway AA, Wilson AM, McGann DF, Bushway RJ (1982) The nutrient composition of fresh fiddlehead greens. J Food Sci 47:666–667
Byeong-Mi O, Mi-Hyang G, Gyeong-Su N (1994) Isolation and characterization of acidic polysaccharides activation complement system from the hot water extracts of *Pteridium aquilinum* var. *latiusculum*. Kor J Food Nutr 7:159–168
Caius JF (1935) Medicinal and poisonous ferns of India. J Bombay Nat Hist Soc 83:341–361
Cao QM, Zhong HY, Li ZH, Sun CB (2007) Research on *Pteridium squilinum* starch paste properties. Food Sci Technol 11:36–39
Cao H, Chai TT, Wang X et al (2017) Phytochemicals from fern species, potential for medicine applications. Phytochem Rev 16:379–440
Chatterjee A, Sharma P, Ghosh MK et al (2013) Utilization of *Azolla microphylla* as feed supplement for crossbred cattle. Int J Agric Food Sci Technol 4:207–214
Cherryl DM, Prasad RMV, Jagadeeswararao S et al (2014) A study on the nutritive value of *Azolla pinnata*. Livestock Res Int 2:13–15
Ching RC (1964) On some confused genera of the family Athyriaceae. Acta Phytotax Sinica 9:41–84
Chowdhury NP (1973) Pteridophyte flora of the upper Gangetic plain. Navyug Traders, New Delhi
Chu WM, He ZR (1999) Monomelangium Hayata, Allantodia R. Br., Callipteris Bory, *Diplazium* Sw. In: Chu WM (ed) Flora Reipublicae Popularis Sinicae, Volume # 3. Science Press, Beijing, pp 346–349
Copeland TG (1942) Edible ferns. Am Fern J 32:121–126
Copeland TG, Collado TG (1936) Crop ferns. Philip J Agric 7:367–377
Das AK, Dutta BK, Sharma GD (2008) Medicinal plants used by different tribes of Cachar District Assam. Ind J Trad Know 7:446–454
Das T, Paul T, Apte KG et al (2012) Antianaphylactic and mast cell stabilizing activity of *Diplazium esculentum* Retz. on sensitized wistar rats. Inventi Impact Ethnopharmacol 3:136–141
Das B, Paul T, Apte KG et al (2013) Evaluation of antioxidant potential & quantification of polyphenols of *Diplazium esculentum* Retz. with emphasis on its HPTLC chromatography. J Pharm Res 6:93–100

Deepa J, Parashurama TR, Krishnappa M, Nataraja S (2013) Distribution of pteridophytes in Kigga forest, Central Western Ghats, Karnataka, South India. Ind Fern J 30:18–24

Deka S, Saha J, Chakraborty S (2016) A comparative study between response surface methodology and genetic algorithm in optimization and extraction of leaf protein concentrate from *Diplazium esculentum* of Assam. Int J Biotechnol Wellness Ind 5:111–120

DeLong JM, Hodges DM, Prange RK et al (2011) The unique fatty acid and antioxidant composition of ostrich fern (*Matteuccia struthiopteris*) fiddleheads. Can J Plant Sci 91:919–930

Dixit RD (2000) Conspectus of pteridophytic diversity in India. Ind Fern J 17:77–91

Drury HM (1985) Nutrients in native foods of southeastern Alaska. J Ethnobiol 5:87–100

Dudani SN, Chandran MS, Ramachandra TV (2012) Pteridophytes of Western Ghats. In: Kumar AB (ed) Biodiversity documentation and taxonomy. Narendra Publishing House, Delhi, pp 343–351

Dudani SN, Mahesh MK, Chandran MS, Ramachandra TV (2014) Pteridophyte diversity in wet evergreen forests of Sakleshpur in Central Western Ghats. Ind J Plants Sci 3:28–39

Duran AO (1994) Raw palm oil as the energy source in pig fattening diets and *Azolla filiculoides* as a substitute for soya bean meal. Livestock Res Rural Dev 6:1–8

Ekwealor KU, Ugwoke CE, Okeke CU et al (2015) Preliminary studies on the evaluation of phytochemistry and proximate composition of *Dicksonia antarctica* (Labill). Scholars Acad J Biosci 3:607–610

Essuman EK, Ankar-Brewoo GM, Barimah J, Ofosu IW (2014) Functional properties of protein isolate from fern fronds. Int Food Res J 21:2085–2090

FAO-WHO (1991) Protein quality evaluation. Reports of a joint FAO-WHO expert consultation. Food and nutrition paper #. Food and Agriculture Organization of the United Nations, Rome, p 51

Gauchan DP, Manandhar D, Shrestha N, Suwal SK (2008) Nutrient analysis of *Nephrolepsis cordifolia* (L.) C. Presl. Kathmandu Univ J Sci Eng Technol 4:68–72

Greeshma GM, Murugan K (2014) Proximal composition of *Crepidomanes intramarginale* (Hooker & Greville) Copeland. and its functional properties. World J Pharm Pharmaceut Sci 4:397–408

Greeshma AA, Sridhar KR (2016) Ethnic plant-based nutraceutical values in Kodagu region of the Western Ghats. In: Pullaiah T, Rani SS (eds) Biodiversity in India, vol 8. Regency Publications, New Delhi, pp 299–317

Greeshma AA, Sridhar KR, Pavithra M (2018) Nutritional prospects of edible fern of the Western Ghats of India. In: Öztürk M, Hakeem KR, Ashraf M, Ahmed MSA (eds) Global perspectives on underutilized crops. Springer Nature, Cham, pp 151–164

Indira D, Rao SK, Suresh J et al (2009) Azolla (*Azolla pinnata*) as feed supplement in buffalo calves on growth performance. Ind J Anim Nutr 26:345–348

Kala CP (2005) Ethnomedicinal botany of the Apatani in the eastern Himalayan region of India. J Ethnobiol Ethnomed 1:1–8

Kaushik A, Kaushik JJ, Das A et al (2011) Preliminary studies on anti-inflammatory activities of *Diplazium esculentum* in experimental models. Int J Pharmaceu Sci Res 2:1251–1253

Kaushik A, Jijta C, Kaushik JJ et al (2012) FRAP (ferric reducing ability of plasma) assay and effect of *Diplazium esculentum* (Retz) Sw. (a green vegetable of North India) on central nervous system. Ind J Nat Prod Res 3:228–231

Kavitha C, Manoj G, Murugan K (2015) Diversity of pteridophytes of Ponmudi hills. Int J f Biol Pharm Allied Sci 4:6180–6190

Kholia BS (2011) Pteridophytic wealth of Sikkim Himalaya. Biodivers Sikkim 3:43–64

Lee CH, Shin SL (2011) Functional activities of ferns for human health. In: Fernández H, Kumar A, Revilla A (eds) Working with ferns: issues and applications. Springer, New York, pp 347–359

Lee SO, Lee HJ, Yu MH et al (2005) Total polyphenol contents and antioxidant activities of methanol extracts from vegetables produced in Ullung island. Korean J Food Sci Technol 37:233–240

Lee HB, Kim JC, Lee SM (2009) Antibacterial activity of two phloroglucinols, flavaspidic acids AB and PB, from *Dryopteris crassirhizoma*. Arch Pharmacol Res 32:655–659

Leterme P, Londono AM, Ordonez DC et al (2010) Nutritional value and intake of aquatic ferns (*Azolla fillicoides* Lam. and *Salvinia molesta* Mitchell) in sows. An Feed Sci Technol 155:55–64

Lin N, Lin W (1989) β-Ecdysone containing skin-protecting cosmetics. Chem Abs 111:239323e

Liu Y, Wujisguleng W, Long C (2012) Food uses of ferns in China: a review. Acta Soc Bot Polo 81:263–270

Lobo SM, Gulimane K (2015) Nutritional analysis of rhizome and physicochemical characteristics of starch extracted from the mangrove fern *Acrostichum aureum* L. Starch 67:716–719

Mannan MM, Maridass M, Victor B (2008) A review on the potential uses of ferns. Ethnobot Leafl 12:281–285

Maridass M, Raju G (2010) Conservation status of pteridophytes, Western Ghats, South India. Indian J Biotechnol 1:42–57

Meybeck A, Bonte EF, Redziniak EG (1997) Use of an ecdysteroid for the preparation of cosmetic or dermatological compositions intended in particular, for strengthening the water barrier function of the skin or for the preparation of a skin culture medium, as well as to the compositions. US Patent: 5,609,873

Mickel J, Smith AR (2004) *Diplazium* Sw. In: Buck WR (ed) Pteridophytes of Mexico, Part #1. New York Botanical Garden Press, New York, pp 251–268

Morais-Braga MFB, Souza T, Santos KK et al (2013a) Cytotoxicity and antiparasitic activity of *Lygodium venustum* SW. Acta Toxicol Arg 21:50–56

Morais-Braga MF, Souza TM, Santos KK et al (2013b) Phenol composition, cytotoxic and anti-kinetoplastidae activities of *Lygodium venustum* SW. (Lygodiaceae). Exp Parasitol 134:178–182

Moran RC (2004) A natural history of ferns. Timber Press, Cambridge

Moure A, Rúa M, Sineiro J, Domínguez H (2002) Aqueous extraction and membrane isolation of protein from defatted *Gevuina avellana*. J Food Sci 67:688–696

Mynssen CM, Matos FB (2012) *Diplazium fimbriatum* (Athyriaceae), a new species from Brazil. Am Fern J 102:167–173

Nair AG, Pradeesh S, Nikhila GS et al (2014) Conservation of vegetable fern (*Diplazium Esculentum*) of Western Ghats through *in vitro* spore culture. J Aq Biol Fish 2:43–50

NRC-NAS (1989) Recommended dietary allowances. National Academy Press, Washington, DC

Nwosu MO (2002) Ethnobotanical studies on some pteridophytes of southern Nigeria. Econ Bot 56:255–259

Oloyede FA, Alaf BO, Oloyede FM (2008) Nutrient evaluation of *Nephrolepis biserrata* (Nephrolepidaceae. Pteridophyta). Bot Lithuanica 14:207–210

Oloyede F, Oloyede F, Saliu T, Ogunwole A (2010) Chemical composition of a tropical fern *Ceratopteris cornuta* (Parkeriaceae, Pteridophyta) in Nigeria. Acta Bot Hung 52:145–150

Oloyede F, Makinde A, Ajayi O (2012) Proximate analysis, nutritional and anti-nutritional compositions of a tropical fern, *Nephrolepis furcans* in Nigeria. Acta Bot Hung 54:345–354

Oloyede FA, Ajayi OS, Bolaji IO, Famudehin TT (2013) An assessment of biochemical, phytochemical and antinutritional compositions of a tropical fern, *Nephrolepis cordifolia* L. Life J Sci 15:645–651

Oloyede FA, Airemionkhale M, Olatunji TY (2014) Chemical, nutritional and anti-nutritional compositions of *Nephrolepis exaltata* (L.) Schott. (Nephrolepidaceae). Futa J Res Sci 10:252–257

Pacheco L, Moran RC (1999) Monograph of the neotropical species of *Callipteris* with anastomosing veins (Woodsiaceae). Brittonia 51:343–388

Parashurama TR, Deepa JA, Prakash K (2016) Pteridophyte diversity in Mudigere Taluk, Central Western Ghats, Karnataka, South India. Int J Cur Res 8:40339–40342

Parvez S, Kang M, Chung HS et al (2006) Survey and mechanism of skin depigmenting and lightening agent. Phytother Res 20:921–934

Patil S, Lavate R, Rawat V, Dongare M (2016) Diversity and distribution of pteridophytes from Satara District, Maharashtra (India). Plant Sci Today 3:149–156

Pradhan S, Manivannan S, Tamang JP (2015) Proximate, mineral composition and antioxidant properties of some wild leafy vegetables. J Sci Ind Res 74:155–159

Radhika NK, Sreejith P, Asha VV (2010) Cytotoxic and apoptotic activity of *Cheilanthes farinosa* (Forsk.) Kaulf. against human hepatoma, Hep3B cells. J Ethnopharmacol 128:166–171
Rana V, Bachheti RK, Gogoi S et al (2015) Physicochemical, functional and antioxidant properties of *Diplazium esculentum* leaf protein concentrate. Cur Trad Med 1:145–158
Roux JP (2009) Synopsis of the lycopodiophyta and pteridophyta of Africa, Madagascar and neighbouring islands. South African National Biodiversity Institute, Pretoria
Samanta G, Tamang Y (1995) Feeding value of *Azolla* (*Azolla pinnata*) in goats. In: Annales de Zootechnie, vol 44. Institut national de la recherche agronomique, Paris, pp 1960–2000
Sanginga N, Van Hove C (1989) Amino acid composition of *Azolla* as affected by strains and population density. Plant Soil 117:263–267
Sathish SS, Vijayakanth P (2016) New additions of fern flora to Kolli Hills, eastern Ghats, Tamilnadu, India. Ind J Plant Sci 5:56–64
Seal T (2012) Evaluation of nutritional potential of wild edible plants, traditionally used by the tribal people of Meghalaya state in India. Am J Plant Nutr Fert Technol 2:19–26
Shankar R, Khare PK (1992) Ethnobotanical observation on some ferns of Pachmarhi Hills. J Econ Taxon Bot Additional Ser 10:97–100
Shills MEG, Young VR (1988) Modern nutrition in health and disease. In: Neiman DC, Buthepodorth DE, Nieman CN (eds) Nutrition. WmC Brown, Dubugue, pp 276–282
Shin SL (2010) Functional components and biological activities of pteridophytes as healthy biomaterials. Chungbuk National University, Korea
Shin SL, Lee CH (2010) Antioxidant effects of the methanol extracts obtained from aerial part and rhizomes of ferns native to Korea. Korean J Plant Resour 23:38–46
Shrivastava K (2007) Importance of ferns in human medicine. Ethnobot Leafl 11:231–234
Slik JWF, Hovenkamp P, Iqbal M, Raes N (2006) Structure, plant species diversity and plant species composition of the Gunung Lumut protection forest. In: Delongh HH, Persoon GA, Kustiawan (eds) Options for biodiversity conservation and sustainable use in lowland forests of southeast Borneo, vol 19. Institute of Environmental Sciences, Leiden, pp 71–107
Soare LC, Ferdes M, Stefanov S et al (2012) Antioxidant activity, polyphenols content and antimicrobial activity of several native pteridophytes of Romania. Not Bot Hortic Agrobot Cluj-Napoca 40:53–57
Socolsky C, Salamanca E, Giménez A et al (2015) Prenylated acylphloroglucinols with leishmanicidal activity from the fern *Elaphoglossum lindbergii*. J Nat Prod 79:98–105
Somvanshi R, Ravisankar R (2004) Recent advances in bracken fern toxin research. Nat Prod Radiance 3:304–308
Sridhar KR, Karun NC (2017) Plant based ethnic knowledge on food and nutrition in the Western Ghats. In: Pullaiah T, Krishnamurthy KV, Bhadur B (eds) Ethnobotany of India, Western Ghats and west coast of Peninsular India, vol 2. Apple Academic Press, Cambridge, MA, pp 255–276
Sundriyal M, Sundriyal RC (2001) Wild edible plants of the Sikkim Himalaya: nutritive values of selected species. Econ Bot 55:377–390
Svobodová A, Psotová J, Walterová D (2003) Natural phenolics in the prevention of UV-induced skin damage – a review. Biomed Pap 147:137–145
Tag H, Tsering J, Hui PK et al (2014) Nutritional potential and traditional uses of high altitude wild edible plants in eastern Himalayas, India. Int J Biol Biomol Agric Food Biotechnol Eng 8:238–243
Tanaka S, Sato T, Akimoto N et al (2004) Preventation of UVB – induced photoinflammation and photoaging by a polymethoxy flavonoid, nobiletin, in human keratinocytes *in vivo* and *in vitro*. Biochem Pharmacol 68:433–439
Thakur RC, Hosoi Y, Ishii K (1998) Rapid *in vitro* propagation of *Matteuccia struthiopteris* (L.) Todaro – an edible fern. Plant Cell Rep 18:203–208
Tongco JV, Villaber RA, Aguda R, Razal R (2014) Nutritional and phytochemical screening, and total phenolic and flavonoid content of *Diplazium esculentum* (Retz.) Sw. from Philippines. J Chem Pharmaceut Res 6:238–242

Tryon RM, Tryon AF (1982) Ferns and allied plants with special reference to tropical America. Springer, New York

Vasudev SM (1999) Economic importance of pteridophytes. Ind Fern J 16:130–152

Vincent P, Kanna R (2007) Antibacterial activity of ferns - *Christilla parasitica* and *Cyclosorus interuptus* against *Salmonella typhi*. Siddha Papers: http://openmed.nic.in/2009/

Vyas MS, Sharma BD (1998) Ethnobotanical importance of the ferns of Rajasthan. In: Kaushik P (ed) Indigenous medicinal plants including microbes and fungi. Today and Tomorrows Printers and Publishers, New Delhi, pp 61–66

Wali A, Sharma S, Walia M et al (2016) Two edible ferns of Western Himalaya: a comparative *in vitro* nutritional assessment, antioxidant capacity and quantification of lutein by UPLC-DAD. Int J Food Nutr Sci 5:9–18

Wills PJ, Asha VV (2006) Protective effect of *Lygodium flexuosum* (L.) Sw. extract against carbon tetrachloride-induced acute liver injury in rats. J Ethnopharmacol 108:320–326

Wills PJ, Asha VV (2009) Chemopreventive action of *Lygodium flexuosum* extract in human hepatoma PLC/PRF/5 and Hep 3B cells. J Ethnopharmacol 122:294–303

Wood KR (2006) Conservation status of the Hawaiian endemic fern *Diplazium molokaiense* (Athyriaceae) in Honomanu, East Maui, Hawai'i. Endanger Species Res 2:31–35

Wu SH, Ching RC (1991) Fern families and genera of China. Science Press, Beijing

Wu MJ, Weng CY, Wang L, Lian TW (2005) Immunomodulatory mechanism of the aqueous extract of sword brake fern (*Pteris ensiformis* Burm.). J Ethnopharmacol 98:73–81

Yonathan M, Asres K, Assefa A, Bucar F (2006) In vivo anti-inflammatory and anti-nociceptive activities of *Cheilanthes farinosa*. J Ethnopharmacol 108:462–470

Yoon CS, Kim HJ, Lim HW, Choi SW (2006) Antimicrobial effect of ethanol extract of *Dryopteris crassirhizoma* Nakai on *Propionibacterium acnes*. J Soc Cosme Sci Korea 32:201–208

Yusuf AA, Mofio BM, Ahmed AB (2007) Proximate and mineral composition of *Tamarindus indica* Linn 1753 seeds. Sci World J 2:1–4

Zanariah J, Rehan AN, Rosnah O (1986) Protein and amino acid compositions of Malaysian vegetables. MARDI Res Bull 14:140–147

Zhang S, Zhong G, Liu B, Wang B (2011) Physicochemical and functional properties of fern rhizome (*Pteridium aquilinum*) starch. Starch 63:468–474

Zheng XK, Li YJ, Zhang L et al (2011a) Antihyperglycemic activity of *Selaginella tamariscina* (Beauv.) spring. J Ethnopharmacol 133:531–537

Zheng XK, Zhang L, Wang WW et al (2011b) Anti-diabetic activity and potential mechanism of total flavonoids of *Selaginella tamariscina* (Beauv.) Spring in rats induced by high fat diet and low dose STZ. J Ethnopharmacol 137:662–668

Chapter 7
Secondary Metabolite Production in Medicinal Plants Using Tissue Cultures

Bilal Ahmad, Aamir Raina, and Samiullah Khan

Abstract Plants are an incredible treasure of lifesaving drugs and other products of diverse applications. Plant tissue cultures can be established routinely under sterile conditions from explants like plant leaves, stems, roots, meristems, etc. for both ways for multiplication and extraction of secondary metabolites. Strain improvement, methods for the selection of high-producing cell lines, and medium optimizations can lead to an enhancement in secondary metabolite production. Production of natural as well as recombinant bioactive products of commercial importance through the exploitation of plant cells has attracted substantial attention over the past few decades. Swift acceleration in the production of explicit secondary metabolism compounds at a rate similar or superior to the intact plants has been discovered through innovative plant cell cultures in the last few years. In view of obtaining optimum yields suitable for commercial exploitation, isolation of the biosynthetic activities of cultured cells has been focused upon, which is being achieved by the optimization of the cultural conditions, selection of high-yielding strains, and employment of transformation methods, precursor feeding, and immobilization techniques. Production of secondary metabolites through hairy root system is based on *Agrobacterium rhizogenes* inoculation and has grabbed substantial attention during the past few decades as an efficient method of secondary metabolite production in the plant roots. Due to certain reasons like very slow growth of root systems of higher plants and very difficult harvesting, alternative methods of bioactive compound production have been utilized and promising results have been obtained. Root cultures constitute a promising option for the production of medicinally

B. Ahmad
Plant Physiology Laboratory, Department of Botany, Aligarh Muslim University, Aligarh, India

A. Raina (✉)
Botany Section, Women's College, Aligarh Muslim University, Aligarh, India

Mutation Breeding Laboratory, Department of Botany, Aligarh Muslim University, Aligarh, India

S. Khan
Mutation Breeding Laboratory, Department of Botany, Aligarh Muslim University, Aligarh, India

D. Egamberdieva, A. Tiezzi (eds.), *Medically Important Plant Biomes: Source of Secondary Metabolites*, Microorganisms for Sustainability 15,
https://doi.org/10.1007/978-981-13-9566-6_7

important bioactive compounds. Organ cultures and in vitro biomass production often have sites of synthesis and storage of secondary metabolites in separate compartments. Elicitors, compounds triggering the formation of secondary metabolites, can be abiotic or biotic. Natural elicitors include polysaccharides such as pectin and chitosan, which are also used in the immobilization and permeabilization of plant cells. The present chapter reviews the secondary metabolite production through hairy root cultures, organ cultures, elicitation, and economically valuable secondary metabolites produced through tissue culture.

Keywords Plant cell cultures · Secondary metabolites · Elicitation

7.1 Introduction

Plants are an incredible treasure of lifesaving drugs and other products of diverse applications. Nowadays numerous distinct phytochemicals serve as imperative drugs, which are currently used across the globe to cure a variety of perilous diseases. Most of the medicinally important phytochemicals are the products of secondary metabolism, which, in addition to their pharmaceutical applications, find extensive applications in flavor and fragrances, food additives, pesticides, and dye and pigments. The chief role of these bioactive secondary products in plants is to help them in combating various types of biotic and abiotic stresses (Rao and Ravishankar 2002; Ahmad et al. 2019a, b; Naikoo et al. 2019). Production of natural as well as recombinant bioactive products of commercial importance through the exploitation of plant cells has attracted substantial attention over the past few decades (Canter et al. 2005). The mounting commercial importance of the secondary metabolism products has attracted significant interest in this subject in the recent past, particularly in the likelihood of alteration in the production of various bioactive plant metabolites with the help of tissue culture technology. Plant culture systems (cell and tissue cultures) signify a potential treasure of valued secondary metabolites and hold immense promise for the controlled production of such countless and valuable secondary metabolites on demand which find extensive applications in food additives, pharmaceuticals, and nutraceuticals (Zhong 2001). The synthesis of secondary metabolites with the help of the cell cultures is independent of environmental fluctuations as compared to their biosynthesis in plants. The chemical synthesis of various valuable metabolites is either not achievable or economically unfeasible. Furthermore, the natural bioactive phytoproducts used as food additives are better valued and accepted by consumers as compared to their synthetic counterparts. Swift acceleration in the production of explicit secondary metabolism compounds at a rate similar or superior to the intact plants has been discovered through innovative plant cell cultures in the last few years. In view of obtaining optimum yields suitable for commercial exploitation, isolation of the biosynthetic activities of cultured cells has been focused upon which is being achieved by the optimization of the cultural conditions, selection of high-yielding strains, and employment of transformation methods, precursor feeding, and immobilization techniques (DiCosmo and Misawa 1995). The role of plant tissue culture in the production of secondary metabolites has been completely reformed by transgenic hairy root cultures. These are exceptional in their biosynthetic and genetic stability, swift in growth, and very easily maintained. With the help of this methodology, an extensive variety of phytochemical compounds of commercial value has been synthesized (Giri and Narasu 2000).

In order to carry out the efficient extraction and increased production of bioactive secondary metabolites, plant cell and tissue cultures can be constantly established from different explants (plant leaves, stems, roots, meristems) under sterile conditions (Vijaya et al. 2010). Optimization of the media, strain improvement, and selection of high-yielding cell lines can enrich the secondary metabolite production. These advances have enhanced the phytochemical production beyond expectations (Vijaya et al. 2010). The competence of plant cell cultures to produce and accumulate countless of the identical precious compounds as the parent plant finds recognition nearly since the commencement of in vitro technology. The persistently increasing demand for the natural products has attracted substantial attention toward the plant culture systems as potential biosynthetic machines for secondary metabolism products and has opened new doors of anticipation for novel research exploring expression of secondary products in vitro (Karuppusamy 2009). The most promising approach of large-scale sustainable production of secondary metabolites is with the help of the plant cell factories which offers an incessant supply with the help of large-scale culture (Rao and Ravishankar 2002).

Secondary metabolite production through cell cultures is advantageous over the conventional production because of its independence of the environmental factors and seasonal variations as the economically valuable bioactive secondary metabolite production is carried out in controlled conditions through the elimination of negative biological influences (microorganisms and insects) (Hussain et al. 2012; Canter et al. 2005; Rao and Ravishankar 2002). Moreover, selection of high-yielding cell lines and a defined production system ensuring uniform quality and continuous supply and yield is met through culture systems. Furthermore, production of novel compounds normally absent in the parent plant can be ensured through the tissue cultures (Hussain et al. 2012; Rao and Ravishankar 2002).

Production of precious secondary plant products with the help of plant cell cultures as compared to whole plant or in vivo production is followed by a series of distinct advantages (Vijaya et al. 2010). Some of these advantages include;

- Production of useful and valuable compounds independent of soil conditions or climatic changes.
- Cells cultured through varied culture systems would be microbe and insect-free.
- Plant cells of different origin (alpine or tropical) could be multiplied effortlessly to yield important and specific metabolites.
- Reduced labor expenses and improved productivity would result from coherent regulation of metabolite processes and programmed control of cell growth.
- Extraction of organic substances from callus cultures.

7.1.1 Secondary Metabolite Production Through Hairy Root Cultures

Production of secondary metabolites through hairy root system is based on *Agrobacterium rhizogenes* inoculation and has grabbed substantial attention during the past few decades as an efficient method of secondary metabolite production in the plant roots (Hussain et al. 2012; Karuppusamy 2009; Palazon et al. 1997). After the inoculation, the hairy root phenotype produced exemplifies swift hormone-independent growth, lateral branching, lack of geotropism, and genetic stability. Such a secondary metabolite production is edged as these secondary products produced are with similar or higher yields and identical to those produced by the intact roots of parent plants (Sevón and Oksman-Caldentey 2002). This attribute along with genetic stability as well as speedy growth in media lacking phytohormones makes them specifically appropriate for biochemical studies which usually are difficult to carry out in the root cultures of intact plants. A part of the DNA (T-DNA), located on the Ri plasmid, is transferred to the plant cells during the process of infection, and fascinatingly the transferred genes find expression in the same way as those of the endogenous plant cell genes. During the infection process, *A. rhizogenes* transfers a part of the DNA (T-DNA) located in the root-inducing plasmid Ri to plant cells, and the genes contained in this segment are expressed in the same way as the endogenous genes of the plant cells. Certain strains of *A. rhizogenes* have two sections in T-DNA, each finding its incorporation individually into the plant genome. The root induction process involves two sets of plasmid genes, the aux genes and the rol genes (Hussain et al. 2012). Usually the hairy roots are induced on the wounded plant parts after inoculating these with *A. rhizogenes*. Transformation mediated through *A. rhizogenes* is advantageous as any gene of interest can be transferred to the hairy root clone. This can prove very fruitful for secondary metabolite production. For example, 6-hydroxylase gene from *Hyoscyamus muticus* was introduced into *Atropa belladonna* using *A. rhizogenes*-mediated transformation (Hashimoto et al. 1993). Enhanced enzyme activity and about fivefold increase in the concentration of scopolamine in the engineered roots were observed.

7.1.2 Secondary Metabolite Production Through Organ Cultures

Due to slow growth of root systems of higher plants and very difficult harvesting, alternative methods of bioactive compound production have been utilized, and promising results have been obtained (Pence 2011). Root cultures constitute a promising option for the production of medicinally important bioactive compounds (Pence 2011). Some of the noteworthy secondary metabolites that have been produced quite well in root cultures include the tropane alkaloids (hyoscyamine and scopolamine) (Fazilatun et al. 2004). Moreover, other aerial parts of the plants like shoots can also be utilized for the production of important secondary metabolites (Nogueira and Romano 2002; Smith et al. 2002). Shoot cultures have been utilized for the commercial production of secondary metabolites so as to reduce or overcome the exploitation of natural plants (Karuppusamy 2009; Khanam et al. 2000). In addition, shoot cultures are aimed at inducing somaclonal variations and provide the chance for selecting clones capable of high secondary production (Dhawan et al. 2003). However, the organ cultures encounter some major problems when cultured at large scale (Kaimoyo et al. 2008). Different types of bioreactors have been used for the culture of plant roots and/or shoots (Kašparová et al. 2009; Kim et al. 2002). Compared to the cell suspension cultures, organ cultures generally display a lower sensitivity to shear stress, but they show a high degree of spatial heterogeneity in biomass production. Another problem is the quite high cost of these bioreactor systems for commercial large-scale production of plant secondary metabolites. As they have to compete with the cultivation of the whole plant, such a process in most cases is not economically viable (Zhao et al. 2010).

7.1.3 Economically Valuable Secondary Metabolites Produced Through Tissue Culture

Tissue culture holds immense potential of controlled production of myriad of economically valuable and pharmaceutically useful secondary products, and the field is quite intriguing. Swift acceleration has been witnessed in the discovery of cultures competent enough to produce explicit medicinal compounds at a similar or speedy rate to that observed in intact plants (Vijaya et al. 2010). Biosynthetic activities of cultured cells are of imperative importance in order to achieve significant yields appropriate for commercial production of the valuable pharmaceutical secondary metabolites, and isolation of such cultured cells is necessary for the optimization of the cultural conditions, selection of high-yielding strains, utilization of precursor feeding, and transformation and immobilization techniques (Vijaya et al. 2010; DiCosmo and Misawa 1995). Transgenic hairy root cultures have brought a new life to secondary metabolite production through plant tissue culture as such cultures are exceptional in their biosynthetic and genetic stability, swift in growth, and comparatively easier in maintenance. Utilizing such methodology, synthesis of a diverse range of chemical compounds has been achieved (Vijaya et al. 2010; Giri and Narasu 2000). Recent advances in the field of cell cultures have achieved significant success in the production of a diverse range of pharmaceuticals belonging to different classes of secondary metabolites including terpenoids, flavanoids, alkaloids, phenolics, saponins, steroids, and amino acids (Abdin and Kamaluddin 2006; Jordon and Wilson 1995). Some pharmaceutically imperative secondary metabolites are briefly discussed below.

Taxol (paclitaxel), an efficient and promising anticancer substance for its exceptional mode of action on cell cycle arrest by checking the microtubular assembly, is a complex diterpene alkaloid obtained from the *Taxus* tree bark (Hussain et al. 2012; Cragg et al. 1993). Currently, taxol production by different *Taxus* species through cell cultures is an extensively explored area of tissue culture in the recent times owing to the colossal commercial importance of the alkaloid, the insufficiency of source tree, and the expensive chemical synthesis (Suffness 1995; Fett-Neto et al. 1994). Aiming at increased production of the alkaloid through the cultures owing to its pharmaceutical importance, cultures were supplemented with different amino acids, and the results revealed that phenylalanine supplementation had a profound effect on the production of taxol in *Taxus cuspidata* cultures (Ciddi et al. 1995). Moreover, the influence of different biotic and abiotic elicitors has also been studied to enhance the production of taxol through cultures (Hussain et al. 2012; Yukimune et al. 1996; Strobel et al. 1992; Tam et al. 1980).

Papaver somniferum, commonly known as opium poppy, is a rich treasure of commercial natural analgesics (morphine and codeine). These alkaloids are significantly valuable and of widespread use in different pharmaceutical preparations. Cell and suspension cultures of opium poppy are being envisaged as valuable and alternative means for the commercial production of these imperative alkaloids. Studies have revealed the production of codeine and morphine alkaloids through morphologically undifferentiated cultures (Yoshikawa and Furuya 1985). Application of growth regulators to the cultures results in the reduced biosynthesis of morphine and codeine as revealed from the study during which it evolved that highest morphine and codeine contents were 2.5 mg/g dry weight and 3.0 mg/g dry weight, respectively, which is about three times greater than the cultures supplemented with hormones (Hussain et al. 2012). Furthermore, Furuya et al. (1972) during the biotransformation studies of codeinone to codeine with the immobilized cells of *Papaver somniferum* reported the conversion yield of about 70.4%.

L-3,4-dihydroxyphenylalanine, a precursor of alkaloids, betalain, and melanin, is an imperative intermediate in the secondary metabolic pathway in higher plants and has been isolated from different plants (Brain and Lockwood 1976; Daxenbichler et al. 1971). Importantly, it is also a precursor of catecholamines which are involved in different signaling and metabolic phenomena in animals besides finding usage as an effective drug against a progressive immobilizing and disabling disorder resulting from the insufficiency of dopamine in the brain tissues called as Parkinson's disease. In view of this imperative pharmaceutical significance, a demand for hefty quantities was felt which led to the alternative ways of enhanced production of this alkaloid among which production through cell cultures has achieved significant success in this regard (Brain and Lockwood 1976). *Mucuna pruriens* has been reported to accumulate 25 mg/L DOPA in the medium under the influence of ample concentrations of 2,4-D. Among the induced callus cultures of three species of *Mucuna* (*M. hassjoo*, *M. pruriens*, *Mucuna deeringiana*), Teramoto and Komamine (1988) observed that the callus tissues of *M. hassjoo* accumulated the highest concentration of DOPA when the medium was supplemented with 10 mg/L kinetin and 0.025 mg/L 2.4-D.

Capsaicin, an alkaloid obtained from green pepper fruits, is used chiefly as a spicy food additive in various formulated foods (Ravishankar et al. 2003). In addition, it finds usage in various pharmaceutical preparations for treating rheumatic disorders besides acting as a digestive stimulant (Sharma et al. 2008). *Capsicum frutescens* suspension cultures are known to produce low capsaicin contents, but immobilization of the cells in reticulated polyurethane foam leads to100-fold increase in its production. Moreover, improvements in the yields can be obtained through the supplementation of isocaproic acid-like precursors. Lindsey (1985) reported that improved capsaicin synthesis can be obtained by the treatments which suppress primary metabolism and cell growth. Holden et al. (1988) have reported that spores of *Gliocladium deliquescens* can elicit capsaicin biosynthesis in the *C. frutescens* cell cultures. Detailed study on the influence of nutritional stress on capsaicin production of *Capsicum annuum* in immobilized cell cultures were carried out by Ravishankar and Ramachandra Rao (2000). Biotransformation of exogenously sourced caffeic acid and protocatechuic aldehyde to capsaicin in immobilized cells cultures and freely suspended cells of *Capsicum frutescens* has also been carried out (Sanatombi and Sharma 2007).

Diosgenin, a pharmaceutically valuable alkaloid which acts as a precursor for the synthesis of a variety of steroidal drugs, finds extensive appliance in the pharmaceutical industry because of which its demand is continuously mounting (Hussain et al. 2012; Tal et al. 1983). In 1983, culture experiments of Tal et al. (1983) revealed that carbon and nitrogen concentrations significantly influenced accumulation of diosgenin in one of the cell lines of *Dioscorea.* Furthermore, immobilized cell cultures were established by Ishida (1988), and it was observed that reticulated polyurethane foam stimulated diosgenin production, leading to an increase of 40% in the cellular concentration and an increase of 25% total yield. Moreover, 8% increase in the levels of diosgenin was reported in the batch-grown cell suspensions of *D. deltoidea* (Hussain et al. 2012; Tal et al. 1983).

7.1.4 Secondary Metabolite Production Through Elicitation

Secondary metabolite accumulation in plants is an important adaptive mechanism established by them during evolution as a part of defense system against pathogens, which is stimulated through elicitation by different elicitors, acting as signaling compounds during the defense responses in plants (Zhao et al. 2005). Elicitation has proven to be an efficient technique for enhancing the production of plant secondary metabolites through biotechnological approach. Elicitors are usually those compounds which stimulate plant defense response and promote production of secondary metabolism products in order to protect the plant (Klarzynski and Friting 2001; Baenas et al. 2014). Elicitors are of various types and varied nature and elicit secondary metabolite production through elicitation, a phenomenon during which induction or enhancement of secondary metabolites in plants is stimulated to ensure their survival, competitiveness, and persistence (Namdeo 2007). These elicitors can be categorized into different types on the basis of origin, viz., abiotic (elicitors of non-biological origin like physical factors and inorganic substances) and biotic elicitors (plant hormones like methyl jasmonate, salicylic acid, brassinosteroids, bacterial- and fungal-derived proteins, and peptides) (Gorelick and Bernstein 2014; Namdeo 2007). Inorganic elicitors like metal ions or salts have been utilized for increased bioactive compound production by eliciting or stimulating secondary metabolism. Zinc ions and salts like $AlCl_3$, $AgNO_3$, $CaCl_2$, and $MgSO_4$ have been used in cell suspensions, hairy roots, and adventitious roots for secondary metabolism elicitation (Verpoorte et al. 2002). Bulk of the biotic elicitors is recognized and bound by specific cell membrane receptors. After its stimulation, the cell surface receptor transfers the stimulus to the cell leading to a signal transduction cascade (Baenas et al. 2014). Once the signal transduction cascade is stimulated, different signaling molecules are produced which lead to the biosynthesis of products of secondary origin like phytoalexins; such a signaling response is determined by various factors, predominantly physiological state and genetic characteristics.

Studies have revealed that exposure of plant cell cultures to different elicitors can lead to increased production of secondary metabolites (Staniszewska et al. 2003). Elicitors of frequent usage in culture systems include fungal carbohydrates, yeast extract, methyl jasmonate (MJ), and chitosan. MJ, a phytohormone and an imperative signal compound, has proven as an effective elicitor for the production of taxol (Wu and Lin 2002) and ginsenoside (Yu et al. 2002; Kim et al. 2004; Thanh et al. 2005) in the cell/organ culture. MJ has also proven promising in the production of secondary metabolites in cell/adventitious root cultures of *Bupleurum falcatum* L. (Aoyagi et al. 2006) and *Taxus* spp. (Yukimune et al. 1996; Ketchum et al. 1999). Additionally, MJ-induced elicitation led to significant increase in the eleutheroside content after *Eleutherococcus senticosus* embryo culture was supplemented with this elicitor (Shohael et al. 2007).

The possible mechanism of elicitor signaling may involve stimulation of elicitor-induced DNA synthesis via G-protein-coupled receptor (GPCR) and phosphoinositide-specific phospholipase C (PI-PLC) pathways (Boland et al. 2003a, b, 2006), and the elicitor-induced methyl jasmonate biosynthesis (Doares et al. 1995) might give indication regarding the activation of methyl jasmonate response elements in DNA due to elicitors (Fig. 7.1). Enhanced methyl jasmonate biosynthesis may have been induced by elicitors, which led to increased secondary metabolite production, thereby increasing active constituents and the yield of the plant. Increase in the yield of the active constituents, such as citral, is in line with other studies (Adams 2007; Dar et al. 2015), which report enhancement in citronellal and trigonelline contents in eucalyptus and fenugreek, respectively, in response to elicitors.

7.2 Conclusion

The advances in modern technology, especially protocols for plant tissue cultures, paved a way for the commercial production of even rare plants and the chemicals they provide. The main advantage of plant tissue culture is that it can ultimately

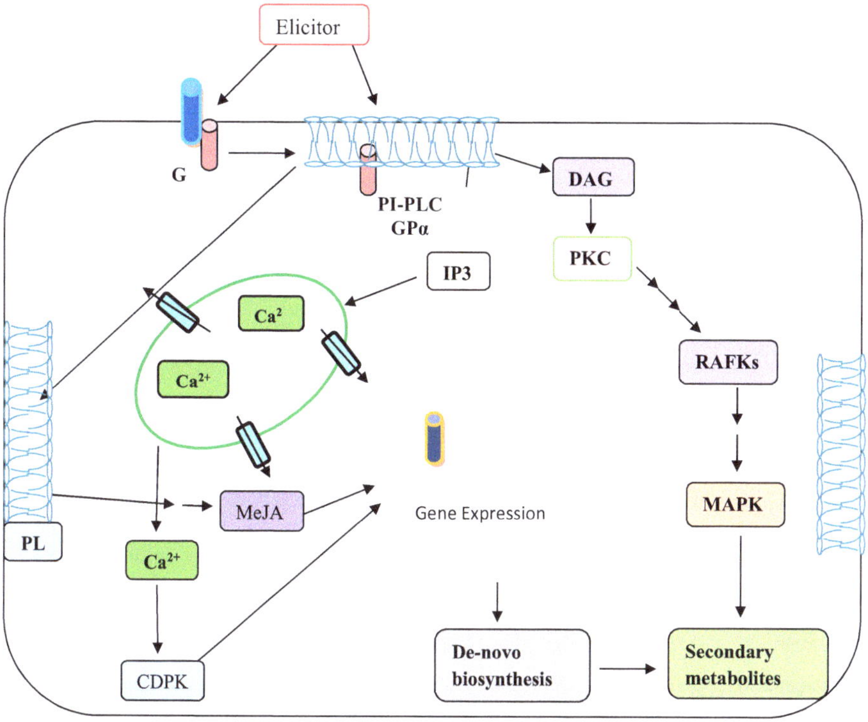

Fig. 7.1 A possible signaling mechanism of in vitro secondary metabolite production through elicitation. Elicitors perhaps activate G-protein-coupled receptor (GPCR) pathway. In the activated form, the Gα subunit of the GPCR binds the induced phosphoinositide-specific phospholipase C (PI-PLC) complex, which may lead to the activation of two different signaling molecules, inositol triphosphate (IP3) and diacylglycerol (DAG). Elicitors have earlier been known to generate PI-PLC signaling independent of the GPCR pathway as well as IP3, in turn, activating cyclin-dependent protein kinase (CDPK), which leads to the nucleus activating de novo DNA synthesis of enzymes involved in metabolic pathways. Protein kinase C (PKC), after activation, leads to nucleus activating gene expression through phosphorylation of transcription factors and DNA synthesis. It might also lead to activation of the MAP-kinase (MAPK) pathway, leading to increased biosynthesis of monoterpenes in lemongrass

provide a continuous, reliable source of natural products. The synthesis of bioactive secondary metabolites, running in controlled environment, independently from climate and soil conditions, is the main benefit of this method. This has a great strides building on advances in plant science has been accomplished by the use of in vitro plant cell culture for the production of secondary metabolites. Knowledge of biosynthetic pathways of desired phytochemicals in plants as well as in cultures is often still in its infancy, and consequently strategies are needed to develop an information based on a cellular and molecular level. The introduction of newer techniques of molecular biology, so as to produce transgenic cultures and to effect the expression and regulation of biosynthetic pathways, is also likely to be a significant step toward making cell cultures more applicable to the commercial production of secondary metabolites (Table 7.1).

Table 7.1 Secondary metabolite production through tissue culture

Active ingredient	Plant	Culture type	References
Flavonolignan	*Silybum marianum*	Root (LS + TDZ)	Alikaridis et al. (2000)
Saikosaponins	*Bupleurum falcatum*	Root (B5 + IBA)	Kusakari et al. (2000)
Anthraquinones	*Cassia acutifolia*	Suspension (MS + 2,4-D + kinetin)	Nazif et al. (2000)
Gallotannins	*Rhus javanica*	Root (LS + IAA +Kinetin)	Taniguchi et al. (2000)
Capsiacin	*Capsicum annuum*	Callus (MS + 2,4-D+ GA_3)	Varindra et al. (2000)
Reserpine	*Rauvolfia serpentina*	Callus (LS)	Gerasimenko et al. (2001)
Ramiflorin alkaloid	*Aspidosperma ramiflorum*	Callus (MS + 2-4,D + BAP + Sucrose)	Olivira et al. (2001)
Withaferin A	*Withania somnifera*	Shoot (MS + BA)	Ray and Jha (2001)
Indole alkaloids	*Catharanthus roseus*	Suspension (MS + NAA + Kinetin)	Zhao et al. (2001)
Diterpenoids	*Torreya nucifera*	Suspension (MS + 2,5-D)	Orihara and Furuya (1990)
Terpenoids	*Salvia officinalis*	Callus (MS + 2,4-D + BA)	Santos-Gomes et al. (2002)
Plumbagin	*Plumbago zeylanica*	Hairy root (MS + BAP + IBA)	Verma et al. (2002)
Plumbagin	*Plumbago rosea*	Callus (MS + $CaCl_2$)	Komaraiah et al. (2003)
Hypericin	*Hypericum perforatum*	Multiple shoot (MS + BA + TDZ)	Santarem and Astarita (2003)
Triterpenes	*Hyssopus officinalis*	Suspension (G5 + 2,4-D + IAA)	Skrzypek and Wysokinsku (2003)
Triterpenoid	*Ammi majus*	Suspension (MS + 2,4-D + BA)	Staniszewska et al. (2003)
Alkaloids	*Fritillaria unibracteata*	Multiple shoot (MS + 2,4-D + Kin)	Gao et al. (2004)
Corydaline	*Corydalis ambigua*	Embryo (MS + IAA + sucrose)	Hiraoka et al. (2004)
Asiaticoside	*Centella asiatica*	Shoot (MS + BAP + IAA)	Kim et al. (2004)
Berberine	*Coscinium fenestratum*	Suspension (MS + 2,4-D + GA3)	Narasimhan and Nair (2004)

(continued)

Table 7.1 (continued)

Active ingredient	Plant	Culture type	References
Vincamine	*Vinca major*	Hairy root (MS + BAP)	Tanaka et al. (2004)
Catechin	*Rheum ribes*	Callus (MS + IBA + BA)	Farzami and Ghorbant (2005)
Flavonoids	*Vaccinium myrtillus*	Callus culture (MS + BAP + NAA)	Hohtola et al. (2005)
Asiaticoside	*Centella asiatica*	Callus (MS + 2,4-D + Kin)	Kiong et al. (2005)
Hypericin	*Hypericum perforatum*	Suspension (Liquid MS + NAA + GA_3)	Hohtola et al. (2005)
7-Methyljuglone	*Drosera rotundifolia*	Shoot culture (MS + BAP + NAA)	Hohtola et al. (2005)
Lupeol, rutin	*Hemidesmus indicus*	Shoot culture (MS + BAP + NAA)	Misra et al. (2005)
Reserpine	*Rauvolfia tetraphylla*	Callus (MS + 2,4-D + Tryptophan)	Anitha and Kumari (2006)
Gymnemic acid	*Gymnema sylvestre*	Callus (MS + IAA + BA)	Devi et al. (2006)
Gymnemic acid	*Gymnema sylvestre*	Callus (MS + 2,4-D + IAA)	Gopi and Vatsala (2006)
Umbelliferone	*Ammi majus*	Shootlet (MS + BAP)	Krolicka et al. (2006)
Vincristine	*Catharanthus roseus*	Suspension (MS + 2,4-D + GA3)	Lee-Parsons and Rogce (2006)
Anthocyanin	*Vitis vinifera*	Suspension (MS + BAP + NAA)	Qu et al. (2006)
Essential oil	*Cymbopogon citratus*	Shoot (MS + IAA + GA3)	Quiala et al. (2006)
Sennosides	*Cassia senna*	Callus (MS + NAA + Kin)	Shrivastava et al. (2006)
Anthraquinone	*Saprosma fragrans*	Callus (MS + 2,4-D + NAA)	Singh et al. (2006)
Silymarin	*Silybum marianum*	Callus (MS + IAA + BA)	Tumova et al. (2006)
Flavonoid	*Momordica charantia*	Callus (MS + BAP + NAA)	Agarwal and Kamal (2007)
Rosmarinic acid	*Zataria multiflora*	Callus (MS + IAA + Kin)	Francoise et al. (2007)
Glycoside	*Panax ginseng*	Hairy root (MS + NAA + Kin)	Jeong and Park (2007)

(continued)

Table 7.1 (continued)

Active ingredient	Plant	Culture type	References
Hyperforin	*Hypericum perforatum*	Multiple shoot (MS + 2,4-D + Leusine)	Karppinen et al. (2007)
Asiaticoside	*Centella asiatica*	Hairy root (MS + 2,4-D)	Kim et al. (2007)
Hypericins	*Hypericum perforatum*	Multiple shoot (MS + BA + IAA)	Kornfeld et al. (2007)
Rosmarinic acid	*Agastache rugosa*	Hairy root (MS + 2,4-D + Kin + 3% sucrose)	Lee et al. (2007a)
Rutin	*Fagopyrum esculentum*	Hairy root (MS + NAA)	Lee et al. (2007b)
Saponins	*Primula veris*	Shoot (MS + BAP + GA_3)	Okrslar et al. (2007)
Eleutherosides	*Eleutherococcus senticosus*	Suspension (MS + 2,4-D)	Shohael et al. (2007)
Glucoside	*Gentiana macrophylla*	Hairy root (MS + IAA + Kin)	Tiwari et al. (2007)
Camptothecin	*Ophiorrhiza rugosa*	Shoot (MS + BA + Kin)	Vineesh et al. (2007)
Fixed oil	*Simmondsia chinensis*	Callus (MS + TDZ + GA_3)	Aftab et al. (2008)
Quercetin	*Pluchea lanceolata*	Callus (MS + NAA + BAP)	Arya et al. (2008)
Artemisinin	*Artemisia annua*	Callus (MS + NAA + Kinetin)	Baldi and Dixit (2008)
Flavonoid	*Salvia officinalis*	Multiple shoot (LMS + IAA + BAP)	Grzegorczyk and Wysokinska (2008)
Berberine	*Coscinium fenestratum*	Callus (MS + 2,4-D + BAP)	Khan et al. (2008)
Resveratrol	*Arachis hypogaea*	Hairy root (G5 + 2,4-D + Kin.)	Kim et al. (2008)
Tropane	*Brugmansia candida*	Hairy root (MS + 2,4-D + IAA)	Marconi et al. (2008)
Glycyrrhizin	*Glycyrrhiza glabra*	Hairy root (MS + 2,4-D + GA3)	Mehrotra et al. (2008)
Withanolide A	*Withania somnifera*	Hairy root (MS + IAA + Kin)	Murthy et al. (2008)
Reserpine	*Rauvolfia serpentina*	Callus (MS + IAA + Cu^{2+})	Nurchgani et al. (2008)
Azadirachtin	*Azadirachta indica*	Suspension (MS + 2,4-D + Cyanobacterial elicitor)	Poornasri Devi et al. (2008)

(continued)

Table 7.1 (continued)

Active ingredient	Plant	Culture type	References
Berberine	*Tinospora cordifolia*	Suspension (MS + IAA + GA_3)	Rama Rao et al. (2008)
Silymarin	*Silybum marianum*	Hairy root (MS + IAA + GA3)	Rahnama et al. (2008)
Catharanthine	*Catharanthus roseus*	Suspension (MS + 2,4-D + UV-B)	Ramani and Jayabaskaran (2008)
Serpentine	*Rauvolfia serpentina*	Callus (MS + BAP + IAA)	Salma et al. (2008)
Azadirachtin	*Azadirachta indica*	Suspension (MS + 2,4-D)	Sujanya et al. (2008)
Corydalin	*Cordyline terminalis*	Callus (MS + 2,4-D + BAP)	Taha et al. (2008)
Xanthone	*Gentianella austriaca*	Multiple shoot (MS + BAP)	Vinterhalter et al. (2008)
Cathin	*Brucea javanica*	Suspension (MS + IAA + GA3)	Wagiah et al. (2008)
Deoursin	*Angelica gigas*	Hairy root (MS (Liq.) + 2,4-D + GA_3)	Xu et al. (2008)
Piperine	*Piper solmsianum*	Suspension (MS + 2,4-D + BA)	Balbuena et al. (2009)
Myristin	*Myristica fragrans*	Shoot (MS + NAA + TDZ)	Indira Iyer et al. (2009)
Resveratrol	*Vitis vinifera*	Callus (MS + IAA + GA3 + UV)	Kin and Kunter (2009)
Podophyllotaxin	*Podophyllum hexandrum*	Shoot (MS + BAP + GA_3)	Li et al. (2009)
Flavonoid	*Crataegus sinaica*	Callus (MS + 2,4-D + NAA + BAP)	Maharik et al. (2009)
Steroidal lactone	*Withania somnifera*	Callus (MS + 2,4-D + BA)	Mirjalili et al. (2009)
Flavones	*Camellia chinensis*	Callus (MS + 2,4-D + NAA)	Nikolaeva et al. (2009)
Anthraquinone	*Rubia akane*	Hairy root (B5 + NAA + Kin)	Park and Lee (2009)
Stilbenes	*Cayratia trifoliata*	Suspension (MS + IAA + GA_3)	Roat and Ramawat (2009)
Isoflavones	*Psoralea cordifolia*	Multiple shoot (MS + TDZ + BAP)	Shinde et al. (2009)
Vasine	*Adhatoda vasica*	Shoot culture (MS + BAP + IAA)	Shalaka and Sandhya (2009)

References

Abdin MZ, Kamaluddin A (2006) Improving quality of medicinal herbs through physico-chemical and molecular approaches. In: Abdin MZ, Abrol YP, Narosa (eds) Traditional systems of medicine. Publishing House Pvt Ltd, India, pp 30–39

Adams RP (2007) Identification of essential oil components by gas chromatography/mass spectrometry. Allured, Carol Stream

Aftab F, Akram S, Iqbal J (2008) Estimation of fixed oil from various explants and *in vitro* callus cultures of Jojoba (*Simmondsia chinensis*). Pak J Bot 40:1467–1471

Agarwal M, Kamal R (2007) Studies on flavonoid production using *in vitro* cultures of *Momordica charantia* L. Indian J Biotechnol 6:277–279

Ahmad B, Raina A, Naikoo MI, Khan S (2019a) Role of methyl jasmonates in salt stress tolerance in crop plants. In: MIR K, Reddy PS, Ferrante A, Khan NA (eds) Plant signalling molecules. Woodhead Publishing, Elsevier, Duxford, United Kingdom, pp 371–384. https://doi.org/10.1016/B978-0-12-816451-8.00023-X

Ahmad B, Raina A, Samiullah K (2019b) Biotic and abiotic stresses, impact on plants and their response. In: Wani SH (ed) Disease resistance in crop plants. https://doi.org/10.1007/978-3-030-20728-1_1

Alikaridis F, Papadakis D, Pantelia K, Kephalas T (2000) Flavonolignan production from *Silybium marianum* transformed and untransformed root cultures. Fitoterapia 71:379–384

Anitha S, Kumari BDR (2006) Stimulation of reserpine biosynthesis in the callus of *Rauvolfia tetraphylla* L. by precursor feeding. Afr J Biotechnol 5:659–661

Aoyagi H, Akimoto-Tomiyama C, Tanaka H (2006) Preparation of mixed alginate elicitors with high activity for the efficient production of 5′-phosphodiesterase by *Catharanthus roseus* cells. Biotechnol Lett 28:1567–1571

Arya D, Patn V, Kant U (2008) *In vitro* propagation and quercetin quantification in callus cultures of Rasna (*Pluchea lanceolata* Oliver &Hiern.). Indian J Biotechnol 7:383–387

Baenas N, Garcia-Viguera C, Moreno DA (2014) Elicitation: a tool for enriching the bioactive composition of foods. Molecules 19:13541–13563

Balbuena TS, Santa-Catarina C, Silvera V et al (2009) *In vitro* morphogenesis and cell suspension culture establishment in *Piper solmsianum* DC. (Piperaceae). Acta Bot Bras 23:229–236

Baldi A, Dixit VK (2008) Enhanced artemisinin production by cell cultures of *Artemisia annua*. Curr Trends Biotechnol Pharmacol 2:341–348

Boland R, Vasconsuelo A, Picotto G, Giuletti AM (2003a) Involvement of the PLC/PKC pathway in chitosan-induced anthraquinone production by *Rubia tinctorum* cell cultures. Plant Sci 165:429–436

Boland R, Vasconsuelo A, Picotto G, Giuletti AM (2003b) Signal transductionevents mediating chitosan stimulation of anthraquinone synthesis in *Rubia tinctorum* L. Plant Sci 166:405–413

Boland R, Vasconsuelo A, Picotto G, Giuletti AM (2006) Involvement of G-proteins in chitosan-induced anthraquinone synthesis in *Rubia tinctorum* L. Plant Physiol 128:29–37

Brain KR, Lockwood GB (1976) Hormonal control of steroid levels in tissue cultures from *Trigonella foenum-graecum*. Phytochemistry 15:1651–1654

Canter PH, Thomas H, Ernst E (2005) Bringing medicinal plants into cultivation: opportunities and challenges for biotechnology. Trends Biotechnol 23:180–185

Ciddi V, Srinivasan V, Shuler VM (1995) Elicitation of Taxus cell cultures for production of taxol. Biotechnol Lett 17:1343–1346

Cragg GM, Schepartz SA, Suffness M, Grever MR (1993) The taxol supply crisis. New NCI policies for handling the large-scale production of novel natural product anticancer and anti-HIV agents. J Nat Prod 56:1657–1668

Dar TA, Uddin M, Khan MMA et al (2015) Effect of Co-60 gamma irradiated chitosan and phosphorus fertilizer on growth, yield and trigonelline content of *Trigonella foenum-graecum* L. J Radiat Res Appl Sci 8:446–458

Daxenbichler ME, VanEtten CH, Hallinan E et al (1971) Seeds as sources of L-DOPA. J Med Chem 14:463–465

Devi CS, Murugesh S, Srinivasan VM (2006) Gymnemic acid production in suspension calli culture of *Gymnema sylvestre*. J Appl Sci 6:2263–2268

Dhawan S, Shasany AK, Naqvi AA, Kumar S, Khanuja SPS (2003) Menthol tolerant clones of *Mentha arvensis*: approach for in vitro selection of menthol rich genotypes. Plant Cell Tissue Org Cult 75:87–94

Dicosmo F, Misawa M (1995) Plant cell and tissue culture: alternatives for metabolite production. Biotechnol Adv 13(3):425–453

Doares S, Syrovets HT, Weiler EW, Ryan CA (1995) Oligogalacturonides and chitosan activate plant defensive genes through the octadecanoid pathway. Proc Natl Acad Sci U S A 92:4095–4098

Farzami MS, Ghorbant M (2005) Formation of catechin in callus cultures and micropropagation of *Rheum ribes* L. Pak J Biol Sci 8:1346–1350

Fazilatun N, Nornisah M, Zhari I (2004) Superoxide radical scavenging properties of extracts and flavonoids isolated from the leaves of *Blumea balsamifera*. Pharm Biol 42(6):404–408

Fett-Neto AG, Stewart JM, Nicholson SA et al (1994) Improved taxol yield by aromatic carboxylic acid and amino acid feeding to cell cultures of *T. cuspidata*. Biotechnol Bioeng 44:967–971

Francoise B, Hossein S, Halimeh H, Zahra NF (2007) Growth optimization of *Zataria multiflora* Boiss. Tissue cultures and rosmarininc acid production improvement. Pak J Biol Sci 10:3395–3399

Furuya T, Ikuta A, Syono K (1972) Alkaloids from callus cultures of *Papaver somniferum*. Phytochemistry 11:3041–3044

Gao SL, Zhu DN, Cai ZH et al (2004) Organ culture of a precious Chinese medicinal plant – *Fritillaria unibracteata*. Plant Cell Tissue Organ Cult 59:197–201

Gerasimenko I, Sheludko Y, Unger M, Stöckigt J (2001) Development of an efficient system for the separation of indole alkaloids by high performance liquid chromatography and its applications. Phytochem Anal: An Int J Plant Chem Biochem Tech 12(2):96–103

Giri A, Narasu ML (2000) Transgenic hairy roots. Recent trends and applications. Biotechnol Adv 18:1–22

Gopi C, Vatsala TM (2006) *In vitro* studies on effects of plant growth regulators on callus and suspension culture biomass yield from *Gymnema sylvestre* R. Br Afr J Biotechnol 5:1215–1219

Gorelick J, Bernstein N (2014) Elicitation: an underutilized tool in the development of medicinal plants as a source of therapeutic secondary metabolites. In: Sparks DL (ed) Advances in agronomy. Elsevier, Amsterdam, pp 201–230

Grzegorczyk I, Wysokinska H (2008) Liquid shoot culture of *Salvia officinalis* L. for micropropagation and production of antioxidant compounds effect of triacontanol. Acta Soc Botanicorum Poloniae 73:99–104

Hashimoto T, Yun DJ, Yamada Y (1993) Production of tropane alkaloids in genetically engineered root cultures. Pyhtochemistry 32:713–718

Hiraoka N, Bhatt ID, Sakurai Y, Chang JI (2004) Alkaloid production by somatic embryo cultures of *Corydalis ambigua*. Plant Biotechnol 21:361–366

Hohtola A, Jalonen J, Tolnen A et al (2005) Natural product formation by plants, enhancement, analysis, processing and testing. In: Jalkanen A, Nygren P (eds) Sustainable use renewable natural resources from principles to practices. University of Helsinki Publication, Finland, pp 34–69

Holden RR, Holden MA, Yeoman MM (1988) The effects of fungal elicitation on secondary metabolism in cell cultures of *Capsicum frutescens*. In: Robins RJ, Rhodes MJC (eds) Manipulating secondary metabolism in culture. Cambridge University Press, Cambridge, UK, pp 67–72

Hussain MS, Fareed S, Ansari S et al (2012) Current approaches toward production of secondary plant metabolites. J Pharm Bioallied Sci 4:10–20

Ishida H (1988) Interfaces in polymer, ceramic, and metal matrix composites; proceedings of the second international conference on composite interfaces (ICCIII), Cleveland, OH, June 13–17, 1988

Iyer RI, Jayaraman G, Ramesh A (2009) *In vitro* responses and production of phytochemicals of potential medicinal value in nutmeg, *Myristica fragrans* Houtt. Indian J Sci Tech 2:65–70

Jeong GA, Park DH (2007) Enhanced secondary metabolite biosynthesis by elicitation in transformed plant root system. Appl Biochem Biotechnol 130:436–446

Jordon MA, Wilson L (1995) Microtuble polymerization dynamics, mitotic, and cell death by paclitaxel at low concentration. Am Chem Soc Symp Ser 583:138–153

Karppinen K, Hokkanen J, Tolonen A et al (2007) Biosynthesis of hyperforin and adhyperforin from amino acid precursors in shoot cultures of *Hypericum perforatum*. Phytochemistry 68:1038–1045

Karuppusamy S (2009) A review on trends in production of secondary metabolites from higher plants by in vitro tissue, organ and cell cultures. J Med Plants Res 3:1222–1239

Kašparová M, Siatka T, Dušek J (2009) Production of isoflavonoids in the *Trifolium pratense* L. suspension culture. Ceska Slov Farm 58(2):67–70

Ketchum RE, Gibson DM, Croteau RB, Shuler ML (1999) The kinetics of taxoid accumulation in cell suspension cultures of Taxus following elicitation with methyl jasmonate. Biotechnol Bioeng 62(1):97–105

Khan T, Krupadanam D, Anwar Y (2008) The role of phytohormone on the production of berberine in the calli culture of an endangered medicinal plant, turmeric (*Coscinium fenustratum* L.). Afr J Biotechnol 7:3244–3246

Khanam N, Khoo C, Khan AG (2000) Effects of cytokinin/auxin combinations on organogenesis, shoot regeneration and tropane alkaloid production in *Duboisia myoporoides*. Plant Cell Tissue Organ Cult 62:125–133

Kim Y, Wyslouzil BE, Weathers PJ (2002) Secondary metabolism of hairy roots in bioreactors. In Vitro Cell Dev Biol Plant 38:1–10

Kim OT, Kim MY, Hong MH et al (2004) Stimulation of asiticoside accumulation in the whole plant cultures of *Centella asiatica*(L.) urban by elicitors. Plant Cell Rep 23:339–344

Kim OT, Bang KH, Shin YS et al (2007) Enhanced production of asiaticoside from hairy root cultures of *Centella asitica* (L.) urban elicited by methyl jasmonate. Plant Cell Rep 26:1914–1949

Kim JS, Lee SY, Park SU (2008) Resveratol production in hairy root culture of peanut, *ArachyshypogaeaL.* transformed with different *Agrobacterium rhizogenes* strains. Afr J Biotechnol 7:3788–3790

Kin N, Kunter B (2009) The effect of callus age, VU radiation and incubation time on trans-resvertrol production in grapevine callus culture. Tarim Bilimleri Dergisi 15:9–13

Kiong AL, Mahmood M, Fodzillan NM, Daud SK (2005) Effects of precursor supplementation on the production of triterpenes by *Centella asiatica* callus culture. Pak J Biol Sci 8:1160–1169

Klarzynski O, Friting B (2001) Stimulation of plant natural defenses. C R Acad Sci III 324:953–963

Komaraiah P, Ramakrishna SV, Reddanna P, Kavikishore PB (2003) Enhanced production of plumbagin in immobilized cells of *Plumbago rosea* by elicitation and it situ adsorption. J Biotechnol 10:181–187

Kornfeld A, Kaufman PB, Lu CR et al (2007) The production of hypericins in two selected *Hypericum perforatum* shoot cultures is related to differences in black gland culture. Plant Physiol Biochem 45:24–32

Krolicka A, Kartanowicz R, Wosinskia S et al (2006) Induction of secondary metabolite production in transformed callus of *Ammi majus* L. grown after electromagnetic treatment of the culture medium. Enzyme Microbial Technol 39:1386–1389

Kusakari K, Yokoyama M, Inomata S (2000) Enhanced production of saikosaponins by root culture of *Bupleurum falcatum* L. using two step control of sugar concentration. Plant Cell Rep 19:1115–1120

Lee SY, Cho SJ, Park MH et al (2007a) Growth and rutin production in hairy root culture of buck weed (Fagopyruum esculentum). Prep Biochem Biotechnol 37:239–246

Lee SY, Xu H, Kim YK, Park SU (2007b) Rosmarinic acid production in hairy root cultures of *Agasta cherugosa* Kuntze. World J Microbiol Biotechnol 20:969–972
Lee-Parsons CWT, Rogce AJ (2006) Precursor limitations in methyl jasmonate-induced *Catharanthus roseus* cell cultures. Plant Cell Rep 25:607–612
Li W, Li M, Yang DL et al (2009) Production of podophyllotaxin by root culture of *Podophyllum hexandrum* Royle. Electron J Biol 5:34–39
Lindsey K (1985) Manipulation by nutrient limitation of the biosynthetic activity of immobilized cells of *Capsicum frutescens* Mill. ev. annum. Planta 165:126–133
Maharik N, Elgengaihi S, Taha H (2009) Anthocyanin production in callus cultures of *Crataegus sinaica* Bioss. Intrn J Acad Res 1:30–34
Marconi PL, Selten LM, Cslcena EN et al (2008) Changes in growth and tropane alkaloid production in long-term culture of hairy roots of *Brugmansia candida*. Elect J Integr Biosci 3:38–44
Mehrotra S, Kukreja AK, Khanuja SPS, Mishra BN (2008) Genetic transformation studies and scale up of hairy root culture of *Glycyrrhiza glabra* in bioreactor. 11: 717–728
Mirjalili MH, Moyano E, Bonfill M et al (2009) Steroidal lactones from *Withania somnifera*, an antioxidant plant for novel medicine. Molecules 14:2373–2393
Misra N, Misra P, Datta SK, Mehrotra S (2005) In vitro biosynthesis of antioxidants from *Hemidesmus indicus* R. Br. cultures. In vitro Dev Biol Plant 41:285–290
Murthy HN, Dijkstra C, Anthony P et al (2008) Establishment of *Withania somnifera* hairy root cultures for the production of Withanoloid. A J Integ Plant Biol 50:915–981
Namdeo AG (2007) Plant cell elicitation for production of secondary metabolites: a review. Pharmacogn Rev 1:69–79
Naikoo MI, Dar MI, Raghib F, Jaleel H, Ahmad B, Raina A, Khan FA, Naushin F (2019) Role and regulation of plants phenolics in abiotic stress tolerance: an overview. In: MIR K, Reddy PS, Ferrante A, Khan NA (eds) Plant signalling molecules. Woodhead Publishing, Elsevier, Duxford, United Kingdom, pp 157–168. https://doi.org/10.1016/B978-0-12-816451-8.00009-5
Narasimhan S, Nair M (2004) Release of berbeine and its crystallization in liquid medium of cell suspension cultures of *Coscinium fenustratum* (Gaertn.) Colebr. Curr Sci 86:1369–1371
Nazif NM, Rady MR, Seif MM (2000) Stimulation of anthraquinone production in suspension cultures of *Cassia acutifolia* by salt stress. Fitoterapia 71:34–40
Nikolaeva TN, Zagoskina NV, Zaprometov MN (2009) Production of phenolic compounds in callus cultures of tea plant under the effect of 2,4-D and NAA. Russ J Plant Physiol 56:45–49
Nogueira JMF, Romano A (2002) Essential oils from micropropagated plants of Lavandula viridis. Phytochem Anal 13:4–7
Nurchgani N, Solichatun S, Anggarwulan E (2008) The reserpine production and callus growth of Indian snake root (*Rauvolfia serpentina* (L.) Benth. exKurz.) cultured by addition of Cu2+. Biodiversitas 9:177–179
Okrslar V, Plaper I, Kovac M et al (2007) Saponins in tissue culture of *Primula veris*L. In Vitro Cell Dev Biol Plant 43:644–651
Olivira AJB, Koika L, Reis FAM, Shepherd SL (2001) Callus culture of *Aspidosperma ramiflorum* Muell.-Arg. Growth and alkaloid production. Acta Sci 23:609–612
Orihara Y, Furuya T (1990) Production of theanine and other – glutamyl derivatives by *Camellia sinensis* cultured cells. Plant Cell Rep 9:65–68
Palazon J, Pinol MT, Cusido RM et al (1997) Application of transformed root technology to the production of bioactive metabolites. Recent Res Dev Plant Phys 1:125–143
Park SU, Lee SY (2009) Anthraquinone production by hairy root culture of *Rubiaakane*Nakai: influence of media and auxin treatment. Sci Res Essays 4:690–693
Pence VC (2011) Evaluating costs for the in vitro propagation and preservation of endangered plants. In Vitro Cell Dev Biol Plant 47(1):176–187
Poornasri DB, Vimala A, Sai I, Chandra S (2008) Effect of cyanobacterial elicitor on neem cell suspension cultures. Ind J Sci Technol 1:1–5

Qu JG, Yu XJ, Zhang W, Jin MF (2006) Significant improved anthocyanins biosynthesis in suspension cultures of *Vitis vinifera* by process intensification. Sheng Wu Gong Cheng Xae Bae 22:299–305

Quiala E, Barbon R, Jimenez E, Feria MD, Chavez M, Capote A, Perez N (2006) Biomass production of *Cymbopogon citratus* (DC.) Stapf. A medicinal plant in temporary immerson systems. In Vitro Cell Dev Biol Plant 42:298–300

Rahnama H, Hasanloo T, Shams MR, Sepehrifar R (2008) Silymarin production by hairy root culture of *Silybium marianum* (L.) Gaertn. Iran J Biotechnol 6:113–118

Ramani S, Jayabaskaran C (2008) Enhanced catharathine and vindoline production in suspension cultures of *Catharanthus roseus* by ultraviolet-B light. J Mol Signal 3:9–14

Rao SR, Ravishankar GA (2002) Plant cell cultures: chemical factories of secondary metabolites. Biotechnol Adv 20:101–153

Rao R, Vijay BK, Amrutha D et al (2008) Effect of growth regulators, carbon source and cell aggregate size on berberine production from cell cultures of *Tinospora cordifolia* Miers. Curr Trends Biotechnol Pharm 2:269–276

Ravishankar GA, Rao R (2000) Biotechnological production of phytopharmaceuticals. J Biochem Mol Biol Biophys 4:73–102

Ravishankar GA, Suresh B, Giridhar P et al (2003) Biotechnological studies on capsicum for metabolite production and plant improvement. In: Amit DEK (ed) Capsicum: the genus Capsicum. Harwood Academic Publishers, Amsterdam, pp 96–128

Roat C, Ramawat KG (2009) Elicitor induced accumulation of stilbenes in cell suspension cultures of *Cayratia trifoliata* (L.) Domin. Plant Biotechnol Rep 3:135–138

Salma U, Rahman MSM, Islam S, Haque N, Jubair TA, Haque AKMF, Mukti IJ (2008) The influence of different hormone concentration and combination on callus induction and regeneration of *Rauvolfia serpentina* (L.)Benth. Pak J Biol Sci 11:1638–1641

Sanatombi K, Sharma GJ (2007) Micropropagation of *Capsicum frutescens* L. using axillary shoot explants. Sci Hortic 113:96–99

Santarem ER, Astarita LV (2003) Multiple shoot formation in *Hypericum perforatum* L and hypericin production. Braz J Plant Physiol 15:21–26

Santos-Gomes PC, Seabra RM, Andrade PB, Fernandes-Ferreira MM (2002) Phenolic antioxidant compounds produced by *in vitro* shoots of sage (*Salvia officinalis* L.). Plant Sci 162:981–987

Sevón N, Oksman-Caldentey KM (2002) *Agrobacterium rhizogenes* mediated transformation: root cultures as a source of alkaloids. Planta Med 68:859–868

Shalaka DK, Sandhya P (2009) Micropropagation and organogenesis in *Adhatoda vasica* for the estimation of vasine. Pharmacogn Mag 5:539–363

Sharma A, Kumar V, Giridhar P, Ravishankar GA (2008) Induction of in vitro flowering in *Capsicum frutescens* under the influence of silver nitrate and cobalt chloride and pollen transformation. Plant Biotechnol J 11:84–89

Shinde AN, Malpathak N, Fulzele DP (2009) Induced high frequency shoot regeneration and enhance a isoflavones production in *Psoralea corylifolia*. Rec Nat Prod 3:38–45

Shohael AM, Murthy HN, Hahn EJ, Paek KY (2007) Methyl jasmonate induced overproduction of eleuthrosides in somatic embryos of *Eleutherococcus senticosus* cultured in bioreactors. Electron J Biotechnol 10:633–637

Shrivastava N, Patel T, Srivastava A (2006) Biosynthetic potential of *in vitro* grown callus cells of *Cassia senna* L. var. *senna*. Curr Sci 90:1472–1473

Singh DN, Verma N, Raghnwanshi S et al (2006) Antifungal anthraquinone from *Saprosma fragrans*. Bioorg Med Chem Lett 16:4512–4514

Skrzypek Z, Wysokinsku H (2003) Sterols and titerpenes in cell cultures of *Hyssopus officinalis* L. Ver Lag der Zeitschrift fur Naturforschung D 58:308–312

Smith MAL, Kobayashi H, Gawienowski M, Briskin DP (2002) An in vitro approach to investigate medicinal chemical synthesis by three herbal plants. Plant Cell Tissue Organ Cult 70:105–111

Staniszewska I, Krolicka A, Mali E et al (2003) Elicitation of secondary metabolites in in vitro cultures of *Ammi majus* L. Enzyme Microbiol Technol 33:565–568

Strobel GA, Stierle A, van Kuijk JG (1992) Factors influencing the in vitro production of radiolabelledtaxol by Pacific yew, *Taxus brevifolia*. Plant Sci 84:65–74
Suffness M (1995) Taxol: science and applications. CRC Press, Boca Raton
Sujanya S, Poornasri DB, Sai I (2008) *In vitro* production of azadirachtin from cell suspension cultures of *Azadirachta indica*. J Biosci 33:113–120
Taha HS, El-Rahman A, Fathalla M et al (2008) Successful application for enhancement and production of anthocyanin pigment from calli cultures of some ornamental plants. Aust J Basic Appl Sci 2:1148–1156
Tal B, Rokem JS, Goldberg I (1983) Factors affecting growth and product formation in plant cells grown in continuous culture. Plant Cell Rep 2:219–222
Tam WH, Constabel F, Kurz WG (1980) Codeine from cell suspension cultures of *Papaver somniferum*. Phytochemistry 19:486–487
Tanaka N, Takao M, Matsumoto T (2004) Vincamine production in multiple shoot culture derived from hairy roots of *Vinca major*. Plant Cell Tissue Organ Cult 41:61–64
Taniguchi S, Yazaki K, Yabu-Uchi R, Kawakami KY, Ito H, Hatano T, Yoshida T (2000) Galloylglucoses and riccionidin a in Rhus javanica adventitious root cultures. Phytochemistry 53(3):357–363
Teramoto S, Komamine A (1988) L-DOPA production in plant cell cultures. In: Medicinal and aromatic plants I. Springer, Berlin, pp 209–224
Thanh NT, Murthy HN, Yu KW, Hahn EJ, Paek KY (2005) Methyl jasmonate elicitation enhanced synthesis of ginsenoside by cell suspension cultures of *Panax ginseng* in 5-l balloon type bubble bioreactors. Appl Microbiol Biotechnol 67(2):197–201
Tiwari KK, Trivedi M, Guang ZC et al (2007) Genetic transformation of *Gentiana macrophylla* with *Agrobacterium rhizogenes*: growth and production of secoiridoid glucoside gentiopicroside in transformed hairy root cultures. Plant Cell Rep 26:199–210
Tumova L, Rimakova J, Tuma J, Dusck J (2006) *Silybum marianum in vitro* – flavolignan production. Plant Cell Environ 52:454–458
Varindra S, Saikia R, Sandhu S, Gosal SS (2000) Effect of nutrient limitation on capsaicin production in callus culture derived from pericarp and seedling explants of *Capsicum annum* L. varieties. Plant Tissue Cult 10:9–16
Verma PC, Singh D, Rahman L, Gupta MM, Banerjee S (2002) *In vitro* studies in *Plumbago zeylanica*: rapid micropropagation and establishment of higher plumbag in yeilding hairy root cultures. J Plant Physiol 159:547–552
Verpoorte R, Contin A, Memelink J (2002) Biotechnology for the production of plant secondary metabolites. Phytochem Rev 1:13–25
Vijaya SN, Udayasri PV, Aswani KY et al (2010) Advancements in the production of secondary metabolites. J Nat Prod 3:112–123
Vineesh VR, Fijesh PV, Jelly LC et al (2007) In vitro production of camptothecin (an anticancer drug) through albino plants of *Ophiorrhiza rugosa* var. *decumbens*. Curr Sci 92:1216–1219
Vinterhalter B, Jankovic T, Sovikin L et al (2008) Propagation and xanthone content of *Gentianella austiaca* shoot cultures. Plant Cell Tissue Organ Cult 94:329–335
Wagiah ME, Alam G, Wiryowidagdo S, Attia K (2008) Improved production of the indole alkaloid cathin-6-one from cell suspension cultures of *Brucea javanica* (L.) Merr. Ind J Sci Technol 1:1–6
Wu J, Lin L (2002) Elicitor-like effects of low-energy ultrasound on plant (Panax ginseng) cells: induction of plant defense responses and secondary metabolite production. Appl Microbiol Biotechnol 59:51–57
Xu H, Kim YK, Suh SY et al (2008) Deoursin production from hariy root culture of *Angelica gigas*. J Korea Soc Appl Biol Chem 51:349–351
Yoshikawa T, Furuya T (1985) Morphinan alkaloid production by tissues differentiated from cultured cells of *Papaver somniferum*. Planta Med 2:110–113

Yu TW, Bai L, Clade D, Hoffmann D, Toelzer S, Trinh KQ, Xu J, Moss SJ, Leistner E, Floss HG (2002) The biosynthetic gene cluster of the maytansinoid antitumor agent ansamitocin from Actinosynnema pretiosum. Proc Natl Acad Sci U S A 99:7968–7973

Yukimune Y, Tabata H, Higashi Y, Hara Y (1996) Methyljasmonate-induced overproduction of paclitaxel and baccatin III in Taxus cell suspension cultures. Nat Biotechnol 14:1129–1132

Zhao J, Zhu W, Hu Q (2001) Enhanced catharanthine production in *Catharanthus roseus* cell cultures by combined elicitor treatment in shake flasks and bioreactors. Enzym Microb Technol 28:673–681

Zhao J, Davis LC, Verpoorte R (2005) Elicitor signal transduction leading to production of plant secondary metabolites. Biotechnol Adv 23(4):283–333

Zhao JL, Zhou LG, Wu JY (2010) Effects of biotic and abiotic elicitors on cell growth and tanshinone accumulation in Salvia miltiorrhiza cell cultures. Appl Microbiol Biotechnol 87(1):137–144

Zhong JJ (2001) Biochemical engineering of the production of plant-specific secondary metabolites by cell suspension cultures. In: Plant cells. Springer, Berlin, pp 1–26

Chapter 8
Endophytic Bacteria Associated with Medicinal Plants: The Treasure Trove of Antimicrobial Compounds

Dina Barman and Kaushik Bhattacharjee

Abstract Medicinal plants are recognized as prolific producer of bioactive compounds against an array of diseases. However, attention has been currently directed towards endophytic bacteria present inter- and/or intracellularly within host medicinal plants through symbiotic or parasitic interactions. They are the storehouse of wide variety of novel secondary metabolites that can serve as an excellent source of antimicrobial drugs. Hence, there is more opportunity to discover novel antimicrobial compounds from endophytic bacteria. In this scenario, it is of prime importance to focus research on the exploration of endophytic bacteria from medicinal plants and their utilization for the discovery of drug. Keeping on these importances, the intent of this chapter is to provide insights of the occurrence of medicinal plants with antimicrobial activities, exploration of medicinal plants for the isolation of endophytic bacteria and their potential to produce antimicrobial compounds against various pathogenic diseases.

Keywords Endophytic bacteria · Medicinal plants · Antimicrobial activity

8.1 Introduction

With increasing appearance of infectious pathogens, it is a big challenge to find new drugs where natural products have proved to be an attractive resource. Among the natural products, medicinal plants are of major importance. These plants are traditionally used worldwide as remedies for the treatment of various diseases which is due to the bioprospection of secondary metabolites produced by those plants (Egamberdieva et al. 2017). Especially, in developing countries, 80% of people rely on herbal drugs for primary healthcare (Chen et al. 2016). In comparison to modern

D. Barman
Microbial Ecology Laboratory, Department of Botany, Gauhati University, Guwahati, India

K. Bhattacharjee (✉)
Division of Life Sciences, Institute of Advanced Study in Science and Technology, Guwahati, Assam, India

D. Egamberdieva, A. Tiezzi (eds.), *Medically Important Plant Biomes: Source of Secondary Metabolites*, Microorganisms for Sustainability 15,
https://doi.org/10.1007/978-981-13-9566-6_8

synthetic drugs, these herbal medicines are economical and considerably safer. More than half of the pharmaceuticals being used today were derived from these natural products. However, due to major use of medicinal plants for drug discovery, it was reported that many of the medicinal plants are in endangered status or even in the verge of extinction; hence focus has turned towards endophytic bacteria which reside inter- and/or intracellularly within medicinal plants and proven to be the potential source of drug discovery (Venugopalan and Srivastava 2015). They form interactions with host plants ranging from mutualism to parasitism (Strobel 2002). The population structure of endophytic bacteria is strongly affected by genetic background of host plants, its fitness, ecological habitats where the plants live and soil nutrients (Jia et al. 2016). Consequently, it is also hypothesized that endophytic bacteria produce the same type of secondary metabolites as that of host plant species (Qin et al. 2011). Endophytic bacteria produce secondary metabolites of diverse pharmacological activities to protect host plant species from pathogens, to increase ability of plants to tolerate various types of abiotic and biotic stresses, improved nutrient acquisition, and plant growth promotion (Elsebai et al. 2014). Unexpectedly, it was observed that endophytic bacteria are more potential source of metabolites with high therapeutic potential than that of plants (Gouda et al. 2016). Additionally, microorganisms can be easily manipulated both physicochemically and genetically to increase yields of desired natural products (Elsebai et al. 2014).

However, among the different medicinal plants of the world, only a limited percentage were explored till now for endophytic bacterial population and their capacity to produce compounds with significant bioactivities. Consequently, the opportunity to discover new and fascinating endophytic bacteria among the myriad of medicinal plants is also exceptionally incredible. Hence, it is imperative to review the previous successes, ongoing research and latest developments in research associated with the presence of medicinal plants with antimicrobial activities, exploration of medicinal plants for the isolation of endophytic bacteria and their potential to produce antimicrobial compounds.

8.2 Medicinal Plants with Antimicrobial Activity

The plants containing useful concentration of medically active substances are known as medicinal plants. Such plants are traditionally utilized since ancient times for the treatment of different health problems (Nostro et al. 2000). As evidence from archaeological findings, clay tablets and ancient manuscripts, the peoples of Egypt, India, Greek, Roman, Summaria, Babylon and China developed their respective system of medicines from plants (Yaniv 2014). The medicinal properties of those plants are due to their capacity to synthesize a vast array of secondary metabolites such as alkaloids, resins, glycosides, triterpene alcohols, flavonoids, crotenoides and phenolic acids (Nascimento et al. 2000; Ramesh and Okigbo 2008). Mostly, medicinal plants are distributed in mega biodiversity countries of the world where India (which is considered as 'herbarium of world') and China containing utmost numbers of medicinal plants, followed by Colombia, South Africa, the United States and other 16 countries (Chen et al. 2016). Most of the medicinal plants are flowering plants comprising of 33% trees followed by herbs, shrubs, climbers and lower groups of plants (Nishteswar 2014) which are distributed into different families.

With the growing population of the world, the existence of multidrug-resistant antimicrobial compounds is being threatened to mankind. Hence, there is an urgent need to synthesize new drug with novel mechanism of action for new and remerging infections disease (Marasini et al. 2015). In search of novel drugs, scientists have found that medicinal plants are the suitable alternatives to pure pharmaceuticals and are a source of new antimicrobial agents with low toxicity and are also free from side effects caused by synthetic chemicals (Khan et al. 2013). In search of medicinal plants having antimicrobial properties, it may be highly imperative to accumulate knowledge of traditional medicines. It has been reported that about 60–80% of populations in the developing countries use traditional medicine which were derived from medicinal plants (Chen et al. 2016). In recent years, various investigations of traditional medicinal plants have been led in different countries which have provided the world with many of clinical drugs of today (Table 8.1).

Table 8.1 Traditional uses of medicinal plant species with their taxonomic classification

Plant species	Family	Parts used	Medicinal/traditional uses	References
Bidens pilosa	Asteraceae	Root, leaf, seed	Antibacterial	Rojas et al. (2006)
Jacaranda mimosifolia	Bignoniaceae	Root, bark	Treatment of syphilis	
Piper pulchrum	Piperaceae	Whole plant	Treatment of haemorrhagic venom effect from snakebite and antidote for snakebite	
Bixa orellana	Bixaceae	Leaf and seed	Treatment of various diseases	
Cecropia peltata	Urticaceae	Leaf	Treatment of asthma and rheumatism	
Cinchona officinalis	Rubiaceae	Bark	Treatment of bloating, fullness and other stomach problems	
Gliricidia sepium	Fabaceae	Whole plant	Treatment of colds, cough, fever, headache	
Justicia secunda	Acanthaceae	Leaf and stem	Treatment of anaemia, cough, cold, fever, amenorrhoea	
Spilanthes americana	Asteraceae	Whole plant	Antibacterial	
Hemidesmus indicus	Apocynaceae	Root	Antibacterial	Kumar et al. (2007)
Eclipta alba	Asteraceae	Whole plant	Treatment of cough, indigestion, toothache	

(continued)

Table 8.1 (continued)

Plant species	Family	Parts used	Medicinal/traditional uses	References
Coscinium fenestratum	Menispermaceae	Stem	Antimicrobial, antidiabetic, anti-inflammatory	
Cucurbita pepo	Cucurbitaceae	Fruit and seed	Anti-inflammatory, analgesic urinary disorders, antidiabetic, antioxidant	
Tephrosia purpurea	Fabaceae	Root	Treatment of diarrhoea, rheumatism, asthma	
Mentha piperita	Lamiaceae	Leaf	Treatment of colds, cough, nausea	
Pongamia pinnata	Fabaceae	Whole plant	Treatment of piles, skin diseases and wounds	
Symplocos racemosa	Symplocaceae	Bark, flower	Treatment of ulcer, skin disorder, bleeding disorder	
Euphorbia hirta	Euphorbiaceae	Whole plant	Antibacterial, antimalarial, antioxidant	
Tinospora cordifolia	Menispermaceae	Whole plant	Antiperiodic, antimicrobial, anti-inflammatory, antiallergic, antidiabetic	
Thespesia populnea	Malvaceae	Bark, fruit	Treatment of dysentery, diabetes, gonorrhoea	
Jasminum officinale	Oleaceae	Flower	Aphrodisiac, antiseptic, antidepressant, antispasmodic, analgesic	
Allium sativum	Amaryllidaceae	Bulb	Antibacterial	Ushimaru et al. (2007)
Zingiber officinale	Zingiberaceae	Rhizome	Antibacterial	
Caryophyllus aromaticus	Myrtaceae	Flower bud	Antibacterial	
Cymbopogon citratus	Poaceae	Leaf	Antibacterial	
Mikania glomerata	Asteraceae	Leaf	Antibacterial	
Psidium guajava	Myrtaceae	Leaf	Antibacterial	
Acacia pennivenia	Mimosaceae	Leaf	Used for women with mastitis	Mothana et al. (2009)
Acanthospermum hispidum	Astraceae	Leaf	Antibacterial	
Acridocarpus socotranus	Malpighiaceae	Stem and leaf	Treatment of headaches, paralysis and muscle or tendon pain	
Aloe perryi	Aloaceae	Root	To treat stomach problems, constipation, malaria, wounds, burns	
Ballochia atro-virgata	Acanthaceae	Stem and leaf	Antibacterial	

(continued)

Table 8.1 (continued)

Plant species	Family	Parts used	Medicinal/traditional uses	References
Blepharis spiculifolia	Acanthaceae	Leaf and stem	Antibacterial	
Boswellia dioscorides	Burseraceae	Bark	To treat common cold, bronchitis, asthma and rheumatism	
Boswellia socotrana	Burseraceae	Bark	To treat common cold, bronchitis, asthma and rheumatism	
Capparis cartilaginea	Capparaceae	Leaf	To treat itching, shortness of breath, head cold, tumour	
Commiphora ornifolia	Burseraceae	Bark	Antiseptic, to treat diarrhoea, dysentery	
Corchorus erodioides	Tiliaceae	Flower and leaf	Diuretic and urinary tract infections	
Croton socotranus	Euphorbiaceae	Fruit and leaf	For wounds	
Euclea divinorum	Ebenaceae	Root	For oral care, toothache	
Euphorbia socotrana	Euphorbiaceae	Leaf	For skin diseases and wounds	
Eureiandra balfourii	Cucurbitaceae	Leaf	Antibacterial	
Ficus cordata	Moraceae	Leaf	Antiseptic and for ulcers and wounds	
Glossonema revoili	Asclepiadaceae	Flower and leaf	Increase milk production in breastfeeding women	
Hibiscus noli-tangere	Malvaceae	Leaf and root	For snakebite and fever in children	
Hypoestes pubescens	Acanthaceae	Leaf	Fungal skin diseases and scabies	
Lannea transulta	Anacardiaceae	Leaf	Haemostatic for wounds and sores	
Leucas samhaensis	Labiatae	Leaf	For cough and cold	
Leucas virgata	Labiatae	Leaf	For persons with heartburn and indigestion	
Lycium sokotranum	Solanaceae	Leaf and stem	For stomach ailments	
Maerua angolensis	Capparaceae	Leaf	To treat fever, aches and general malaise	
Rhus thyrsiflora	Anacardiaceae	Fruit and leaf	To treat anorexia, general tonic, and for painful joints	
Teucrium sokotranum	Labiatae	Flower and leaf	As flavouring agent and for indigestion	
Zingiber officinale	Zingiberaceae	Rhizome	Analgesic, sedative, antipyretic and antibacterial	Sharma et al. (2009)

(continued)

Table 8.1 (continued)

Plant species	Family	Parts used	Medicinal/traditional uses	References
Cinnamomum cassia	Lauraceae	Bark	Antibacterial, circulatory, respiratory, uterotonic and stomachic	
Terminalia chebula	Combretaceae	Fruit	Laxative, stomachic, tonic and alternative	
Plantago ovata	Plataginaceae	Husk	Constipation, colitis, irritable bowel, cystitis	
Vachellia nilotica	Fabaceae	Leaf	Treating premature ejaculation	
Pimpinella anisum	Apiaceae	Seed	Antiseptic, digestive, galactagogue, pectoral, stimulant	
Ocimum sanctum	Laminaceae	Leaf	Antibacterial, cures cough, cold, skin diseases	
Azadirachta indica	Meliaceae	Fruit	Skin disease, blood disorder, antibacterial	
Phyllanthus fraternus	Euphorbiaceae	Leaf	Jaundice, liver disease, fever, genitourinary disease, oedema	
Coriandrum sativum	Apiaceae	Seed	Flatulence, colic, joint pain, antiseptic	
Abutilon indicum	Malvaceae	Stem	Demulcent, aphrodisiac, laxative, astringent and diuretic, analgesic	
Punica granatum	Lythraceae	Seed	Anthelminthic (esp. tapeworm), diarrhoea, dyspepsia	
Syzygium cumini	Myrtaceae	Bark	Astringent, stomachic, carminative, antiscorbutic, diuretic	
Cyperus scariosus	Cyperaceae	Root	Astringent, diaphoretic, desiccant, cordial and stomachic	
Andrographis paniculata	Acanthaceae	Bark	Laxative, antipyretic, antiperiodic, anti-inflammatory, antibacterial	
Mangifera indica	Anacardiaceae	Leaf	Supplement of sexual potency, antiallergic, hypoglycaemic and antidiabetic	
Achillea millefolium	Asteraceae	Flower	Analgesic, antidiarrheal, antiemetic, Anthelmintic	Frey and Meyers (2010)
Ipomoea pandurata	Convolvulaceae	Flower, leaf	Analgesic, cough, gastrointestinal	
Hieracium pilosella	Asteraceae	Flower, leaf, stem	Antidiarrheal	
Solidago canadensis	Asteraceae	Leaf	Analgesic, gastrointestinal, sedative	

(continued)

Table 8.1 (continued)

Plant species	Family	Parts used	Medicinal/traditional uses	References
Hesperis matronalis	Brassicaceae	Stem	Antibacterial	
Rosa multiflora	Rosaceae	Flower, leaf	Antibacterial	
Asphodelus tenuifolius	Liliaceae	Fruit	Diuretic agent, healing wound	Panghal et al. (2011)
Asparagus racemosus	Liliaceae	Root	Demulcent, diuretic, aphrodisiac, antiseptic antiparasitic, antitumor	
Balanites aegyptiaca	Balanitaceae	Fruit	To cure mouth ulcer, whooping cough, sleeping sickness and skin diseases	
Cordia dichotoma	Boraginaceae	Fruit	Anthelminthic, diuretic, purgative, useful in dry cough, for cure of jaundice	
Eclipta alba	Asteraceae	Whole plant	Alopecia, ringworm, hepatitis, jaundice	
Murraya koenigii	Rutaceae	Leaf, bark, root	Treatment of stomachache, stimulant, piles, influenza, rheumatism, traumatic injury	
Pedalium murex	Pedaliaceae	Fruit	Aphrodisiac, antiseptic, demulcent, diuretic	
Ricinus communis	Euphorbiaceae	Seed, fruit	Antidote, bactericide, expectorant, insecticide, larvicidal, laxative, purgative	
Trigonella foenum	Fabaceae	Leaf	Remedy for fever and swelling	
Piptadeniastum africana	Fabaceae	Leaf	Antibacterial	Assob et al. (2011)
Cissus aralioides	Vitaceae	Leaf	Antibacterial	
Hileria latifolia	Phytolaccaceae	Leaf	Antibacterial	
Phyllanthus muellerianus	Phyllanthaceae	Stem bark	Antibacterial	
Gladiolus gregasius	Iridaceae	Bulb	Antibacterial	
Aloe vera	Asphodelaceae	Whole plant	Antibacterial	Selvamohan et al. (2012)
Phyllanthus emblica	Phyllanthaceae	Whole plant	Antibacterial	
Phyllanthus niruri	Phyllanthaceae	Whole plant	Antibacterial	
Cynodon dactylon	Poaceae	Whole plant	Antibacterial	
Murrya koenigii	Rutaceae	Whole plant	Antibacterial	

(continued)

Table 8.1 (continued)

Plant species	Family	Parts used	Medicinal/traditional uses	References
Lawsonia inermis	Lythraceae	Whole plant	Antibacterial	
Adhatoda vasica	Acanthaceae	Whole plant	Treat asthma, bronchitis, tuberculosis and antibacterial	
Coscinium fenestratum	Menispermaceae	Stem	Antimicrobial, antidiabetic, anti-inflammatory	Kaewpiboon et al. (2012)
Sonneratia alba	Lythraceae	Leaf	Treat swellings and sprains	
Anacardium occidentale	Anacardiaceae	Leaf	To treat fever, malaria, toothache and gum problems	
Acacia karoo	Mimosoideae	Leaf, stem, bark	Mouth ulcers, oral thrush, diarrhoea,	Nielsen et al. (2012)
Erythrophleum lasianthum	Caesalpinioideae	Leaf, stem	Headaches, fever	
Salvia africana	Lamiaceae	Leaf	Colds, flu, bronchitis, abdominal and uterine trouble	
Curtisia dentate	Cornaceae	Leaf, stem	Stomach ailments, diarrhoea, blood purifier	
Ptaeroxylon obligun	Ptaeroxylaceae	Leaf, stem, bark	Snuff for headache, rheumatism, arthritis	
Hymenelobium petraeum	Fabaceae	Whole plant	Antibacterial	Oliveira et al. (2013)
Vatairea guianensis	Fabaceae	Bark, seed and leaf	Treatment of scabies, skin diseases	
Symphonia globulifera	Clusiaceae	Bark and leaf	To treat river blindness, coughs in children	
Lagerstroemia indica	Lythraceae	Leaf	Antibacterial	Chandra (2013)
Annona reticulata	Annonaceae	Leaf	Antibacterial	
Achyranthes aspera	Amaranthaceae	Leaf, stem	Antibacterial	Pandey et al. (2013)
Bergenia ciliata	Saxifragaceae	Root	Antibacterial	Khan et al. (2013)
Jasminum officinale	Oleaceae	Leaf	Antibacterial	
Santalum album	Santalaceae	Wood	Antibacterial	
Artocarpus integer	Moraceae	Stem, root, bark	Antibacterial	Dej-adisai et al. (2014)
Averrhoa bilimbi	Oxalidaceae	Juice	Antibacterial	
Citrus ichangensis	Rutaceae	Peel	Antibacterial	
Cudrania javanensis	Moraceae	Wood	Antibacterial	
Ficus racemosa	Moraceae	Wood	Antibacterial	

(continued)

Table 8.1 (continued)

Plant species	Family	Parts used	Medicinal/traditional uses	References
Hydnophytum formicarum	Rubiaceae	Root	Antibacterial	
Sauropus changiana	Euphorbiaceae	Leaf	Antibacterial	
Solanum ferox	Solanaceae	Branch	Antibacterial	
Bergenia ciliata	Saxifragaceae	Rhizome	It is used in washing ulcer, to cure backbone and in wound healing	
Punica granatum	Lythraceae	Fruit	It is used in piles, diarrhoea, dysentery, whooping cough	
Azadirachta indica	Meliaceae	Leaf	Antibacterial	Farjana et al. (2014)
Camellia sinensis	Theaceae	Leaf	Antibacterial	
Psidium guajava	Myrtaceae	Leaf	Antibacterial	
Calendula officinalis	Asteraceae	Leaf	Antibacterial	
Acorus calamus	Araceae	Rhizomes	Cough, respiratory tract infections, skin disease, toothache, dysentery	Marasini et al. (2015)
Adhatoda vasica	Acanthaceae	Leaves	Bronchitis, asthma, diarrhoea, dysentery, as anthelmintic	
Artemisia vulgaris	Compositae	Aerial parts	Antiseptic, diarrhoea, dysmenorrhea, asthma, as anthelmintic	
Asparagus racemosus	Liliaceae	Rhizome, stem	Urinary troubles, diarrhoea	
Centella asiatica	Umbelliferae	Whole plant	Urinary tract infection, leprosy, ulcers, indigestion	
Cinnamomum camphora	Lauraceae	Leaf, seed, bark	As antiseptic, bronchitis, bronchopneumonia, epilepsy	
Curcuma longa	Zingiberaceae	Rhizome	Antiseptic, cuts, wounds, as anthelmintic, jaundice, liver disorders	
Cuscuta reflexa	Cuscutaceae	Whole plant	Fever, stomachache, rheumatism, anthelmintic	
Cynodon dactylon	Poaceae	Whole plant	Cuts, wounds, indigestion, genitourinary disorders	
Eupatorium adenophorum	Compositae	Leaf	Antiseptic	
Ginkgo biloba	Ginkgoaceae	Leaf	Alzheimer's disease, as anticoldness, as antinumbness	
Psidium guajava	Myrtaceae	Leaf, bark	Diarrhoea, dysentery, cuts, wounds, piles, cholera	

(continued)

Table 8.1 (continued)

Plant species	Family	Parts used	Medicinal/traditional uses	References
Rauwolfia serpentina	Apocynaceae	Root	As antidysenteric, as antidote to snakebite, cuts, wounds and boils	
Swertia chirayita	Gentianaceae	Aerial part	Skin disease, eczema, as anthelmintic, as antidiarrheal, dyspepsia	
Corymbia intermedia	Myrtaceae	Leaf	For the treatment of wounds	Packer et al. (2015)
Lophostemon suaveolens	Myrtaceae	Leaf	Antiseptic purposes	
Syncarpia glomulifera	Myrtaceae	Leaf	Antiseptic purposes	
Aframomum corrorima	Zingiberaceae	Fruit	Antibacterial	Bacha et al. (2016)
Albizia schimperiana	Fabaceae	Root	Antibacterial	
Curcuma longa	Zingiberaceae	Rhizome	Antibacterial	
Erythrina brucei	Fabaceae	Stem, bark	Antibacterial	
Justicia schimperiana	Acanthaceae	Seed	Antibacterial	
Nigella sativa	Ranunculaceae	Seed	Antibacterial	
Ocimum sauve	Lamiaceae	Leaf	Antibacterial	
Vernonia amygdalina	Asteraceae	Leaf	Antibacterial	
Ferula songorica	Apiaceae	Root	Antioxidant, antiviral, antifungal	Liu et al. (2016)
Hypericum perforatum	Hypericaceae	Leaf	Antibacterial	Egamberdieva et al. (2017)
Teucrium polium	Lamiaceae	Leaf	Antibacterial	Hassan (2017)
Echinacea purpurea	Asteraceae	Root, leaf	Antibacterial	Maggini et al. (2017)

8.3 Endophytic Bacteria Associated with Medicinal Plants

Medicinal plants are the source of various bioactive compounds against different ailments for centuries, and association of bacteria have been proven to offer advantages to these plants with high therapeutic potentials (Gouda et al. 2016). Endophytes present in medicinal plants perhaps contribute in their metabolic pathways and produce analogous or novel bioactive compounds (Qin et al. 2011).

Endophytic bacteria are those which are present inter- and/or intracellularly within a plant species without causing any obvious negative harm to the host (Barman and Dkhar 2015). The existence of endophytes has been known for more than 125 years ago (Bacon and White 2000). In 1886, De Bary first introduced the term endophytes for microorganisms harbouring internal plant tissues (Stepniewska and Kuzniar 2013). Since then, Galippe (1887), Henning and Villforth (1940), Carrol (1986), Petrini (1991), Hirsch and Braun (1992) and Hallmann et al. (1997) have defined endophytes in different ways (Stepniewska and Kuzniar 2013). It is generally accepted that medicinal plants with an ethnobotanical history may harbour greater number of endophytic microbiome. Virtually, all the medicinal plant species on earth are the hosts of one or more types of endophytic bacteria (Strobel and Daisy 2003). These inhere in the living tissues of the host plant in a variety of relationships ranging from symbiotic mutualism to parasitism (Strobel 2002). Hence, the presence of endophytes in a plant species is considered as a sign of a healthy plant system (Barman and Dkhar 2018). Endophytic bacteria are promising sources of various secondary metabolites including antibiotics, immunosuppressant, antiparasitics, antioxidants, anticancer agents, plant growth-stimulating metabolites and enzymes which have important roles in plant development and health. They can also protect the plants by providing the ability to defend against predators and help their hosts to adapt in different stress conditions for survival (Qin et al. 2011) (Table 8.2).

Table 8.2 Bacterial endobiome associated with medicinal plant species (period: 2007–2018)

Parts used	Host plant	Group	Identified endophytic microorganisms	Reference
Root	*Panax ginseng*	Bacilli	*Bacillus* sp.	Cho et al. (2007)
		Bacilli	*Bacillus sphaericus*	
		Actinobacteria	*Kocuria carniphila*	
		Proteobacteria	*Rahnella* sp.	
		Actinobacteria	*Microbacterium phyllosphaerae*	
		Gammaproteobacteria	*Pseudomonas* sp.	
		Actinobacteria	*Pseudoclavibacter helvolus*	
		Bacilli	*Bacillus megaterium*	
		Bacilli	*Bacillus* sp.	
		Bacilli	*Paenibacillus polymyxa*	
		Actinobacteria	*Microbacterium hydrocarbonoxydans*	
		Gammaproteobacteria	*Erwinia persicina*	
		Gammaproteobacteria	*Pseudomonas* sp.	
		Gammaproteobacteria	*Serratia plymuthica*	
		Actinobacteria	*Pseudoclavibacter helvolus*	
		Gammaproteobacteria	*Pseudomonas poae*	
		Gammaproteobacteria	*Pantoea ananatis*	
		Gammaproteobacteria	*Serratia plymuthica*	
		Actinobacteria	*Kocuria carniphila*	

(continued)

Table 8.2 (continued)

Parts used	Host plant	Group	Identified endophytic microorganisms	Reference
	Salvia miltiorrhiza	*Gammaproteobacteria*	*Pseudomonas brassicacearum* subsp. *neoaurantiaca*	Vendan et al. (2010)
		Alphaproteobacteria	*Agrobacterium tumefaciens*	
		Gammaproteobacteria	*Pseudomonas thivervalensis*	
		Gammaproteobacteria	*Pseudomonas frederiksbergensis*	
		Bacilli	*Bacillus aryabhattai*	
		Alphaproteobacteria	*Novosphingobium resinovorum*	
	Suaeda maritima	*Actinobacteria*	*Hoeflea suaedae* sp. nov	Chung et al. (2013)
	Origanum vulgare	*Gammaproteobacteria*	*Leclercia* sp.	Bafana (2013)
		Gammaproteobacteria	*Pseudomonas* sp.	
		Gammaproteobacteria	*Stenotrophomonas* sp.	
		Gammaproteobacteria	*Stenotrophomonas* sp.	
		Bacilli	*Bacillus* sp.	
		Bacilli	*Solibacillus* sp.	
		Bacilli	*Lysinibacillus* sp.	
	Cassia tora	Bacilli	*Bacillus subtilis*	Kumar et al. (2015)
		Alphaproteobacteria	*Agrobacterium tumefaciens*	
		Bacilli	*Bacillus* sp.	
		Gammaproteobacteria	*Pseudomonas putida*	
		Gammaproteobacteria	*Pseudomonas* sp.	
	Stachys lavandulifolia	*Actinobacteria*	*Amycolatopsis tolypophora*	Beiranvand et al. (2017)
	Physalis alkekengi	Bacilli	*Bacillus thuringiensis*	
	Allium schoenoprasum	Bacilli	*Bacillus aryabhattai*	
	Mentha pulegium	Bacilli	*Planomicrobium* sp.	
	Marrubium vulgare	*Actinobacteria*	*Actinoallomurus acacia*	
	Falcaria vulgaris	*Actinobacteria*	*Actinoallomurus oryzae*	
	Ocimum basilicum	Bacilli	*Bacillus polyfermenticus*	
	Chenopodium album	Bacilli	*Bacillus pumilus*	
	Gundelia tournefortii	Bacilli	*Bacillus*sp.	
	Achillea millefolium	Bacilli	*Staphylococcus* sp.	

(continued)

Table 8.2 (continued)

Parts used	Host plant	Group	Identified endophytic microorganisms	Reference
	Zataria multiflora	*Alphaproteobacteria*	*Azospirillum brasilense*	
	Chenopodium album	Bacilli	*Bacillus velezensis*	
	Lavandula angustifolia	Bacilli	*Planomicrobium chinense*	
	Cymbopogon olivieri	*Actinobacteria*	*Nocardia niigatensis*	
	Teucrium polium	*Gammaproteobacteria*	*Pseudomonas graminis*	
	Cucumis sativus	*Actinobacteria*	*Nocardia cyriacigeorgica*	
	Coriandrum sativum	*Actinobacteria*	*Microbacterium testaceum*	
Rhizome	*Zingiber officinale*	Bacilli	*Bacillus* sp.	Jasim et al. (2014)
		Gammaproteobacteria	*Pseudomonas* sp.	
		Gammaproteobacteria	*Stenotrophomonas* sp.	
		Bacilli	*Staphylococcus* sp.	
Stem	*Panax ginseng*	Bacilli	*Bacillus pseudomycoides*	Vendan et al. (2010)
		Actinobacteria	*Micrococcus luteus*	
		Bacilli	*Bacillus thuringiensis*	
		Bacilli	*Bacillus pumilus*	
		Bacilli	*Lysinibacillus sphaericus*	
		Bacilli	*Bacillus megaterium*	
		Bacilli	*Bacillus acidiceler*	
		Gammaproteobacteria	*Pseudomonas marginalis*	
		Gammaproteobacteria	*Stenotrophomonas maltophilia*	
		Alphaproteobacteria	*Agrobacterium tumefaciens*	
		Bacilli	*Paenibacillus glucanolyticus*	
		Bacilli	*Staphylococcus epidermidis*	
		Gammaproteobacteria	*Pectobacterium carotovorum*	
	Ipomoea batatas	*Gammaproteobacteria*	*Enterobacter* sp.	Khan and Doty (2009)
		Gammaproteobacteria	*Rahnella aquatilis*	
		Gammaproteobacteria	*Pseudomonas* sp.	
		Gammaproteobacteria	*Rhodanobacter terrae*	
		Gammaproteobacteria	*Stenotrophomonas maltophilia*	

(continued)

Table 8.2 (continued)

Parts used	Host plant	Group	Identified endophytic microorganisms	Reference
		Alphaproteobacteria	*Phyllobacterium myrsinacearum*	
		Gammaproteobacteria	*Xanthomonas* sp.	
	Piper nigrum	Bacilli	*Bacillus firmus*	Jasim et al. (2013)
		Bacilli	*Paenibacillus dendritiformis*	
		Gammaproteobacteria	*Pseudomonas* sp.	
		Betaproteobacteria	*Bordetella* sp.	
		Bacilli	*Bacillus* sp.	
		Gammaproteobacteria	*Stenotrophomonas* sp.	
	Alcea amcheri	*Actinobacteria*	*Dietzia cercidiphylli*	Beiranvand et al. (2017)
	Allium ursinum	*Alphaproteobacteria*	*Azorhizobium caulinodans*	
	Phasaeolous vulgaris	*Firmicutes*	*Bacillus* sp.	
	Rheum rhaponticum	*Actinobacteria*	*Streptomyces artemisiae*	
Leaf	*Aloe vera*	*Actinobacteria*	*Micrococcus aloeverae*	Prakash et al. (2014)
	Hylomecon japonica	*Alphaproteobacteria*	*Sphingobium endophyticus*	Zhu et al. (2015)
	Aloe vera	*Actinobacteria*	*Arthrobacter globiformis*	Beiranvand et al. (2017)
	Teucrium polium	*Firmicutes*	*Bacillus cereus*	Hassan(2017)
		Firmicutes	*Bacillus subtilis*	
Bulbil	*Dioscorea bulbifera* L.	*Actinobacteria*	*Streptomyces dioscori* sp. nov.	Wang et al. (2018)
Plant tissues	*Panax notoginseng*	Bacilli	*Bacillus amyloliquefaciens* subsp. *plantarum*	Ma et al. (2013)
		Bacilli	*Bacillus methylotrophicus*	
	Ferula songorica	*Alphaproteobacteria*	*Sphingomonas* sp.	Liu et al. (2016)
		Gammaproteobacteria	*Acinetobacter* sp.	
		Gammaproteobacteria	*Pseudomonas* sp.	
		Alphaproteobacteria	*Methylobacterium* sp.	
		Alphaproteobacteria	*Rhizobium* sp.	
		Alphaproteobacteria	*Paracoccus* sp.	
		Betaproteobacteria	*Ralstonia* sp.	
		Alphaproteobacteria	*Brevundimonas* sp.	
		Bacilli	*Paenibacillus* sp.	
		Bacilli	*Bacillus* sp.	

(continued)

Table 8.2 (continued)

Parts used	Host plant	Group	Identified endophytic microorganisms	Reference
		Actinobacteria	*Dietzia* sp.	
		Actinobacteria	*Nocardioides* sp.	
		Actinobacteria	*Saccharopolyspora* sp.	
		Actinobacteria	*Pseudonocardia* sp.	
		Actinobacteria	*Streptomyces* sp.	
		Actinobacteria	*Rhodococcus* sp.	
		Actinobacteria	*Promicromonospora* sp.	
		Actinobacteria	*Brevibacterium* sp.	
		Actinobacteria	*Micrococcus* sp.	
		Actinobacteria	*Arthrobacter* sp.	
		Actinobacteria	*Microbacterium* sp.	
	Glycyrrhiza uralensis	*Actinobacteria*	*Brevibacterium frigoritolerans*	Li et al. (2018)
		Bacilli	*Bacillus mojavensis*	
		Betaproteobacteria	*Achromobacter spanius*	
		Gammaproteobacteria	*Stenotrophomonas rhizophila*	
		Bacilli	*Bacillus aryabhattai*	

8.3.1 Origin and Localization of Endophytes

Endophytes are supposed to originate from the epiphytic bacterial communities of the rhizosphere, phylloplane, endophyte-infested seeds or planting materials as well as natural openings or wounds (Hallmann et al. 1997). They enter and colonize in plants mainly through emergence points of lateral roots, the zone of differentiation and elongation near the root tip, stomata, lenticels and broken trichome (Zinniel et al. 2002). Due to the lack of penetration structures, bacteria are unable to exert mechanical or physical forces to penetrate the epidermal cells. Bacteria normally enter intact plant tissue by invagination of the root hair cell wall, by penetration of the junction between root hair and adjacent epidermal cells or by secreting cell wall-degrading enzymes. Plants are autotrophic organisms which are capable for transforming light energy into chemical (carbonaceous) compounds. By releasing these photo-assimilated compounds from plant root into the rhizosphere, they can attract different microorganisms to become endophyte (Bais et al. 2004). On entering into plants, microorganisms spread inside the host plant species via intercellular spaces or conducting elements and ultimately reach the flowers or fruits. Endophytic microorganisms can also reach seed via vascular connections from the maternal plant, directly through gametes colonizing the resulting embryo and endosperm or through colonized shoot meristems which eventually rise to ovules and thus seeds (Truyens et al. 2015). Some bacteria can directly interact with seeds present in soil.

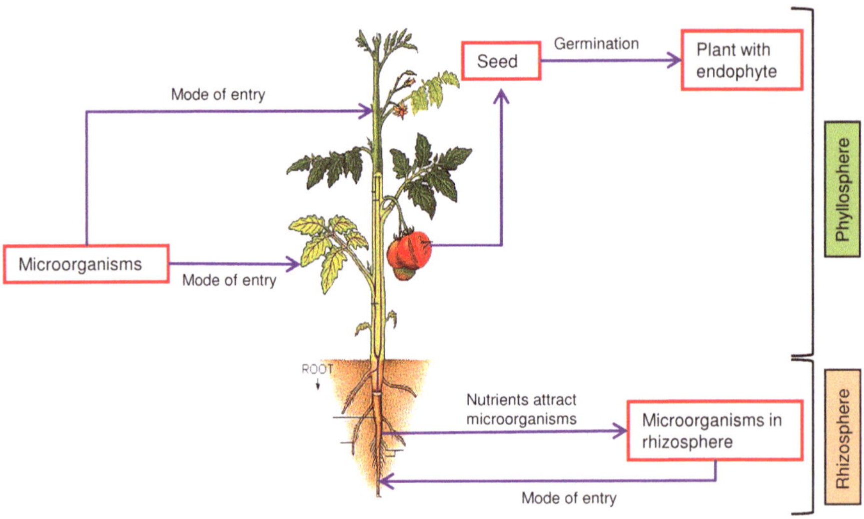

Fig. 8.1 Schematic representation of colonization routes of endophytic microorganisms

For that seed exudates released during imbibition and germination influence the bacterial population that can be supported in the spermosphere (Quadt-Hallman et al. 1997; Truyens et al. 2015) (Fig. 8.1).

The presence of endophytic bacterial colonization in tissues of plants can be documented based on microscopic study such as transmission electron microscopy (TEM), scanning electron microscopy (SEM) and confocal laser scanning microscopy by tagging them with autofluorescent protein (AFP) such as green fluorescent protein (GFP) and *Discosoma striata* red fluorescent protein (DsRed) (Gyaneshwar et al. 2001; Ryan et al. 2008; Thomas and Reddy 2013; Barman and Dkhar 2018). The colonization pattern and tracking of introduced endophytic microbes inside their niche can also be visualized by immunological methods on using monoclonal or polyclonal antibodies followed by ELISA, dot blot assay, tissue printing, immunogold labelling, fluorescent in situ hybridization (FISH), triphenyl tetrazolium chloride vital staining and fluorescence resonance energy transfer (FRET) (Hallmann and Kloepper 1996; Compant et al. 2011; Thomas 2011; Banik et al. 2016).

8.3.2 Culture-Dependent Analysis of Endophytic Bacteria

Isolation and characterization of endophytic bacteria have foremost importance to unearth antimicrobial compounds which mainly involve three steps - surface sterilization of collected plant parts, followed by fractionation of the plant material into small pieces or pestled and homogenization of plant material in a mortar and lastly plating on suitable bacteriological media. Surface sterilization of plant parts is the important step to get rid of epiphytic microorganisms and to ensure that isolated strains are endophytes (Martinez-Klimova et al. 2017). It can be done by washing the collected plant parts in running tap water to remove soil debris followed by treating them with suitable sterilizing reagents to completely remove the epiphytic population. Commonly used disinfecting agents were 70% ethanol (EtOH), sodium hypochlorite (0.9–5.25%), mercuric chloride (0.1%), hydrogen peroxide, Triton-X-100 and Tween 80. It is necessary to treat the plant parts with the disinfecting agents for a suitable period of time to reduce their detrimental effect on plant tissue which leads to hamper the isolation of endophytic bacteria. After each treatment with the disinfecting agents, it is also necessary to wash the plant parts in sterile water to remove the agents (Cao et al. 2004; Kukkurainen et al. 2005; Qin et al. 2011; Jasim et al. 2014). To ensure the effectiveness of surface sterilization procedure, it's essential to use an aliquot of final sterile water from the mixture of surface-sterilized samples and sterile water followed by plating on isolation media (Barman and Dkhar 2015).

After surface sterilization, the plant parts have to plate on suitable isolation media followed by incubation for suitable time at desired temperature. The composition of media mainly depends on energy and nutrients requirement for their growth. Some of the classical media used for isolation of endophytic bacteria includes nutrient agar, Luria-Bertani agar, R2A agar and tryptic soy agar (Gagne-Bourgue et al. 2012; Zhang et al. 2014; Barman and Dkhar 2018). It is also important to note that on using a portion of autoclaved plant extracts of the host plant species to the growth media, isolation of endophytic bacterial population can be enhanced (Murphy et al. 2015).

To identify endophytic bacteria, various micromorphological, biochemical, and molecular techniques with appropriate bioinformatics tools are useful. *Alphaproteobacteria*, *Betaproteobacteria*, *Gammaproteobacteria*, *Actinobacteria*, and *Firmicutes* were the most common groups of endophytic bacteria isolated (Mendes et al. 2007; Ulrich et al. 2008; Vendan et al. 2010). Preliminary identification of bacterial isolates to the genus or species level can be done using various cultural, morphological, biochemical, physiological and chemotypic analyses (Shirling and Gottlieb 1966; Holt et al. 1994; Zhang et al. 2014). Though these aspects can identify bacteria up to genera, sometimes it is not adequate in itself to differentiate between many genera. The advent of molecular criteria for the characterization of bacteria has provided taxonomists with a set of reliable and reproducible tools for studying the systematics. Molecular identification of bacteria can be done by 16S rRNA gene amplification of genomic DNA. Percentage of G+C content of DNA and DNA/DNA-hybridization techniques are also useful tools for the identification of microbes. To characterize taxa at and below the rank of species, the DNA-DNA relatedness, molecular fingerprinting and phenotypic techniques are methods of choice (Zhang et al. 2014).

8.3.3 Culture-Independent Analysis of Endophytic Bacteria

Since culture-based techniques of analysing the diversity of endophytic bacterial community is dependent on various factors including cultivation media, growth conditions and plant tissue manipulation, hence, culture-independent method of analysing the diversity of endophytic bacterial community is more specific and replicable. It provides greater insights of endophytic bacterial community (Yang et al. 2017). In this aspect, metagenomics study with next-generation sequencing (NGS) technology, metatranscriptomics, metaproteomics and metaproteogenomics are widely used (Kaul et al. 2016). During the process, the genetic material has to be isolated from the plant samples followed by amplification of V3-V4 hypervariable region of the bacterial 16S rRNA gene using universal primers followed by sequencing on 454/Roche or Illumina/Solexa (HiSeq, MiSeq) platforms, and finally, the reads (short fragments of genomes obtained in sequencing) are assembled and annotated (Redford et al. 2010; Yang et al. 2017). These technologies can also explore the genes associated with the production of secondary metabolites which may be for plant growth promotion, biocontrol, nutrition and niche adaptation. It helps us to understand their role and mechanism in host plant interaction and protection (Tian et al. 2015).

8.4 Endophytic Bacteria for Their Antimicrobial Potential

Usually, the selection of bacteria for their antimicrobial activity can be evaluated by measuring the minimum inhibitory concentration (MIC) mainly by diffusion methods, dilution methods and bioautography on using the cell-free culture supernatant of the isolates or using the organic extracts of the isolates (Choma and Grzelak 2011). Diffusion methods including disc method, cylinder method and hole plate assay method are mainly used for determination of antimicrobial susceptibility of the test compound preferably of polar ones (Choma and Grzelak 2011), whereas dilution methods (agar dilution and tube assay) are frequently used to estimate the concentration of the test compound (both polar and nonpolar samples) which in the form of complex extracts or pure substances in the agar medium or in the broth suspension (Choma and Grzelak 2011). Bioautography (contact bioautography, immersion bioautography and direct bioautography) is another screening method for detection of antimicrobial activity which is more or less similar to agar diffusion method. The main advantage of this method is that it can be combined with thin-layer chromatography (TLC), high-performance thin-layer chromatography (HPTLC), overpressured-layer chromatography (OPLC) and planar electrochromatography (PEC) (Choma and Grzelak 2011).

Functional gene-based screening of the isolates for antimicrobial potential can be performed by PCR amplification of nonribosomal peptide synthetases (NRPS) and polyketide synthases (PKS) biosynthetic systems within the genomic sequences of the isolates. Both of these systems are involved in the production of biologically active polyketide and peptide compounds including antibiotic having applications in medicine, agriculture and biochemical research (Amoutzias et al. 2008). Both of the systems are composed of multiple large peptides, each of them encoded by a variable number of modules. Each module can be further categorized into minimum three "domains" having special function (Amoutzias et al. 2008), out of which two are catalytical and one is carrier domain. As per current classification, PKSs have been grouped into type I, II and III (Hopwood 1997). Different PCR primers were used for the screening of NRPS and PKS systems including KS-BEF/KS-BER, A3F/A7R, K1F/M6R and K1F/K2R (Ayuso-Sacido and Genilloud 2005; Gonzalez et al. 2005).

8.5 Extraction and Characterization of Antimicrobial Compounds

After preliminary screening of the isolates for antimicrobial potentials as mentioned above, the next step is the fermentation and extraction of antimicrobial product. The culture medium selected for fermentation is mainly based on the species under investigation. After an optimum period of incubation, extraction can be performed. Preliminary low-polarity solvents extraction yields the more lipophilic components, while organic solvent extraction (methanol, ethanol and hexane) yields a larger spectrum of both nonpolar and polar materials. Traditionally extraction is mainly performed by Soxhlet extraction, maceration, percolation, turbo-extraction and sonication. However, due to some drawbacks, a number of new extraction methods have been developed including supercritical fluid extraction (SFE), pressurized liquid extraction (PLE), microwave-assisted extraction (MAE), solid-phase microextraction (SPME), ultrasound-assisted extraction (UAE), superheated liquid extraction and extraction with supercritical or subcritical water (Sticher 2008). Finally, active components can be isolated by an array of chromatographic methods depending on the solubility, volatility and stability of the compounds to be separated which are commonly considered as the bottleneck of the isolation process (Sticher 2008). Generally, precipitation-thin-layer chromatography, liquid preparation chromatography and column chromatography are used for the process of purification of the active compound (Hu et al. 2010). In the process of chromatography, selection of appropriate solvent system and packing material plays an important role. Silica, alumina, carbohydrates polyacrylamide and polystyrene are mainly used as stationary phases for purification.

Characterization of the purified active compound can be performed by a series of spectroscopic methods such as Fourier transform infrared spectrometer (FT-IR), UV-visible, nuclear magnetic resonance (NMR) and mass spectroscopy (MS) (Bhattacharjee et al. 2017). FT-IR spectrometer is used for characterization of functional groups of the drug molecule having diverse vibrational frequencies which help to identify the chemical constituents and reveal the structural compounds (Altemimi et al. 2017). UV-visible spectroscopy is commonly used to identify the certain classes of compounds. NMR is related to the magnetic properties of certain atomic nuclei which enabled the researchers to find the positions of these nuclei in the molecule (Altemimi et al. 2017). MS is used to find the relative molecular mass and molecular formula of the compound with high accuracy based on the knowledge of relative abundance of a fragmented ion against the ratio of mass/charge of these ions (Fig. 8.2).

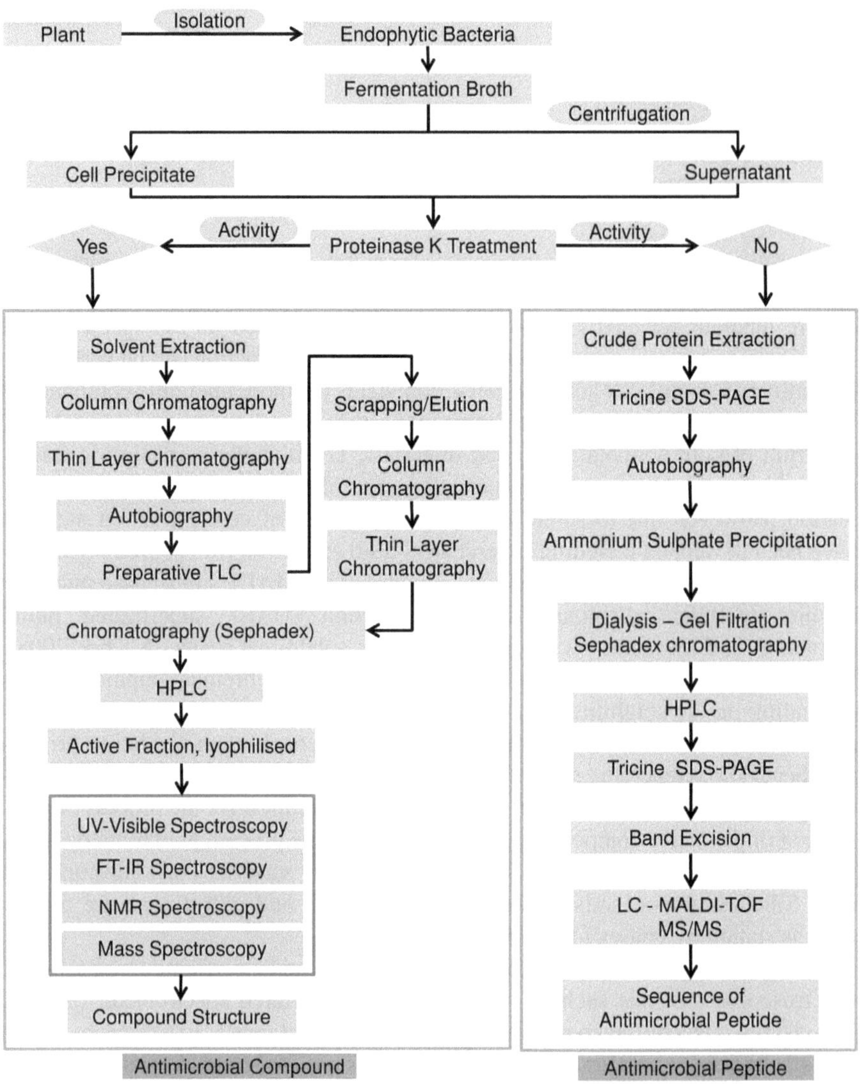

Fig. 8.2 Scheme representing the typical workflow for antimicrobial compounds from endophytic bacteria

8.6 Mechanism of Action of Antimicrobial Compounds

The antibiotics target bacterial cell death by inhibiting DNA synthesis, RNA synthesis, cell wall synthesis, protein synthesis and cell wall metabolism (Dzidic et al. 2008; Kohanski et al. 2010). To inhibit DNA synthesis, antimicrobial agents mainly quinolone class of antibiotics target on DNA gyrase (topoisomerase II) and topoisomerase IV. They bind to topoisomerase enzyme leading to DNA strand breakage. Similarly, to inhibit RNA synthesis, antimicrobial agents, for example, rifamycin can interfere with a DNA-directed RNA polymerase which is the main regulator of gene expression in prokaryotes. The inhibitor of protein synthesis is catagorized into the 50S inhibitors and 30S inhibitors. The 50S ribosome inhibitors (macrolide, lincosamide, streptogramin, amphenicol and oxazolidinone) can block initiation of protein translation or translocation of peptidyl-tRNAs which helps to inhibit the peptidyl transferase reaction that elongates the nascent peptide chain, whereas 30S ribosome inhibitors (tetracycline, aminoglycoside and aminocyclitol) can bind to 30S ribosome subunit and promoting tRNA mismatching which can result in protein mistranslation (Kohanski et al. 2010).

Bacterial cell wall mainly composed of peptidoglycan which is composed of peptide-linked β-(1–4)-N-acetyl hexosamine. Some of the antibiotics, for example, β-lactams, interfere with cell wall synthesis by inhibiting the peptide bond formation between the peptidoglycan units. Some of the antibiotics can inhibit peptidoglycan synthesis through binding with peptidoglycan units and by blocking transglycosylase and transpeptidase activity (Kohanski et al. 2010). Some of the antibiotics can also interfere with cell wall metabolism (Dzidic et al. 2008).

8.7 Antimicrobial Compounds Produced by Plant-Associated Bacteria

The human population is increasing with an alarming rate; ecosystems are deteriorating rapidly; and a variety of new types of health issues are popping up (Bhattacharjee et al. 2018). For instance, increase in number of drug-resistant bacteria is a cause of concern. In this perspective, bioprospecting for natural resources such as microorganisms, plants, algae and animals is the way for the discovery of new antibiotics (Martinez-Klimova et al. 2017). Among the natural sources of drug production, especially bacteria are the primary resource. However, for the discovery of drug, only a small percentage of bacteria have been explored (Bhattacharjee et al. 2018). Hence, it is extremely important to explore nature's hitherto untapped bacteria to achieve this objective. In this aspect, endophytic bacteria especially isolated from plants of ethnobotanical history are becoming the major trust area of research (Martinez-Klimova et al. 2017). The antimicrobial activity of endophytes was accounted for over 50 years when Smith (1957) isolated *Micromonospora* from the tomato plant which was reported to have antagonistic activity (Manikprabhu and Li

2015). Since then, endophytic bacteria were exploited to isolate various antimicrobial compounds.

Endophytic bacteria play an important role to produce a variety of antibiotics. They mainly produce those antibiotics to protect plants against stress, insects, pests and pathogenic microorganisms (Chandrakar and Gupta 2017). They have immense importance in various pharmaceutical industries, and they have agricultural applications also (Chandrakar and Gupta 2017). Among the antibiotic-producing endophytic bacteria, *Streptomyces* is the richest source of antibiotics, namely, Munumbicins A, Munumbicins B, Munumbicins C, Munumbicins D, Munumbicins E-4, Munumbicins E-5, Kakadumycin A, and Celastramycins A/B (Golinska et al. 2015). Some other bacterial species, including *Pseudomonas*, *Streptosporangium*, *Serratia*, *Bacillus*, *Azospirillum*, *Burkholderia* and *Azoarcus*, also subsidize a distinctive source of antibiotics. These antimicrobial compounds were found to be effective against a range of pathogenic bacteria, fungi and protozoa.

8.7.1 Munumbicins

Munumbicins A, B, C and D are some important antimicrobial peptides which showed activity against a wide spectrum of human as well as plant pathogenic fungi and bacteria and a *Plasmodium* sp. These antibiotics were obtained from *Streptomyces* sp. NRRL30562 which is an endophyte of *Kennedia nigricans*. All these antibiotics are peptides having common compositional features, and Munumbicins C and D represent a novel peptide where Munumbicins A and B are corresponding to actinomycin X2 and actinomycin D, respectively. All the Munumbicins A, B, C and D were found to be active against human-pathogenic bacterium and fungi *Pseudomonas syringae* and *Cryptococcus neoformans*, respectively, and some plant-pathogenic fungi *Pythium ultimum* and *Sclerotinia sclerotiorum* (Castillo et al. 2002). Also in the MIC test, Munumbicins A and C were effective against *Enterococcus faecalis* ATCC 51299, whereas Munumbicins C and D had bioactivity against a drug-sensitive strain of *Staphylococcus aureus* MH II and Munumbicin B against *Staphylococcus aureus* ATCC 33591. The Munumbicin B is of a special interest since it is active against multiple-drug-resistant *Mycobacterium tuberculosis* having IC_{50} value of 10 μgml^{-1}. Another outstanding activity of the Munumbicins was found against the malaria-causing pathogen *Plasmodium falciparum*. Though all the Munumbicins were active against *Plasmodium falciparum*, however, Munumbicins C and D were of special interest due to their low IC_{50} values.

Another two broad-spectrum antibiotics, namely, Munumbicins E-4 and Munumbicins E-5, were isolated from endophytic actinobacterium *Streptomyces* sp. NRRL3052 which was obtained from *Kennedia nigricans*, in the Northern Territory of Australia. Both the antibiotics were effective in the same range of biological activity against *Bacillus subtilis*, *Pythium ultimum* and *Staphylococcus aureus*. Munumbicin E-5 showed more effective than E-4 against *Burkholderia thailanden-*

sis; however, Munumbicin E-4 was more effective than E-5 against *Staphylococcus aureus* ATCC 29213 and *Staphylococcus aureus* 43000 (MRSA) (Castillo et al. 2006). The antimalarial activity of Munumbicins E-4 and E-5 is also reported to be double than that of chloroquine.

8.7.2 Kakadumycins

Streptomycete sp. (NRRL30566) an endophyte of *Grevillea pteridifolia* is the source of peptide antibiotics kakadumycins. The structure of the antibiotic is related to a quinoxaline antibiotic, echinomycin. Like echinomycin, kakadumycins also have same mode of action. It can preferentially inhibit DNA-directed enzymatic RNA synthesis along with it and also can inhibit protein synthesis and cell wall synthesis to some extent. The antibiotic is reported to have strong antimalarial and anti-*Bacillus anthracis* activities (Castillo et al. 2003).

8.7.3 Celastramycins

Celastramycins A and B were isolated from the *Streptomyces* sp. MaB-QuH-8 of the plant *Putterlickia retrospinosa*. Celastramycin A belongs to chloropyrrolo family of antibiotics, while Celastramycin B is an unusual chlorinated anthracyclinone metabolite (Fig. 8.3). On testing both the antibiotics against different pathogens, it was reported that Celastramycin A is more potent against multiresistant bacterial strain in comparison to Celastramycin B. Both Celastramycins A and B showed activity against *Mycobacterium vaccae* IMET 10670 and *Bacillus subtilis* ATCC 6633. In addition to that Celastramycin A was effective against some other pathogenic bacteria such as *Staphylococcus aureus* MRSA 134/93, *S. aureus* MR 994/93, *Enterococcus faecalis* V-r 1528, *Mycobacterium smegmatis* SG 98, *Mycobacterium aurum* SB 66 and *Mycobacterium fortuitum* (Pullen et al. 2002).

8.7.4 Coronamycins

Coronamycin is a novel group of peptide antibiotics which is active against various pythiaceous fungi, human fungal pathogen including *Pythium ultimum*, *Geotrichum candidum* and *Phytophthora cinnamomi*. The best bioactivity of Coronamycin was found against malarial parasite *Plasmodium falciparum* having IC_{50} values of 9 ± 7.3 ng ml^{-1}. It is produced by *Streptomyces* sp. which is an endophyte from an epiphytic vine, *Monstera* sp. Since this antibiotic is active against various pythiaceous fungi, it may be used for agricultural purposes (Ezra et al. 2004).

8.7.5 Xiamycins

Xiamycin A represents one of the novel pentacyclicindolosesquiterpene isolated from *Streptomyces* sp. strain HKI0595 from the stem segments of a mangrove tree, *Kandelia candel*. The research findings suggest that Xiamycin A has strong antimicrobial activities against *Pseudomonas aeruginosa*, *Staphylococcus aureus*, *Bacillus subtilis*, *Mycobacterium vaccae*, methicillin-resistant *Staphylococcus aureus* and vancomycin-resistant *Enterococcus faecalis* (Ding et al. 2011). Along with Xiamycin A, Ding et al. (2011) also isolated three new alkaloids, Xiamycin B, Indosespene and Sespenine, of which Indosespene and Sespenine have moderate or weak antimicrobial activities, respectively (Fig. 8.3).

8.7.6 Ecomycins

Ecomycin (Ecomycins A, Ecomycins B, Ecomycins C) is a novel family of lipopeptide antibiotics which was isolated from *Pseudomonas viridiflava*, an endophytic bacterium associated with grass species. Based on molecular weight and amino acid composition, Ecomycin A was found to be similar to Syringotoxin; however, Ecomycin B and C represented a unique set of related lipopeptides. The antibiotics showed bioactivity against human fungal pathogens including *Candida albicans* and *Cryptococcus neoformans* and some plant pathogenic fungi such as *Sclerotinia sclerotiorum*, *Fusarium oxysporum* and *Rhizoctonia solani*. The biological activities of the Ecomycin were found to be similar to Amphotericin B. However, since Amphotericin B is extremely toxic to human cells, hence, Ecomycin can be used as a suitable alternative to Amphotericin B (Miller et al. 1998).

8.7.7 Pseudomycin

Pseudomycin is an antibiotic which represent a family of lipopeptides and was reported from plant-associated *Pseudomonas syringae*. It was found to be active against an array of plant- and human-pathogenic fungi including *C. albicans*, *C. neoformans*, *Ceratocystis ulmi* and *Mycosphaerella fijiensis* (Harrison et al. 1991).

8.7.8 Efomycins

Streptomyces sp. BCC72023 isolated from *Oryza sativa* L. was found to be the source of three macrolides, Efomycin M, Efomycin G and Oxohygrolidin, and two polyethers, abierixin and 29-O-methylabierixin (Fig. 8.3). Efomycin M can inhibit

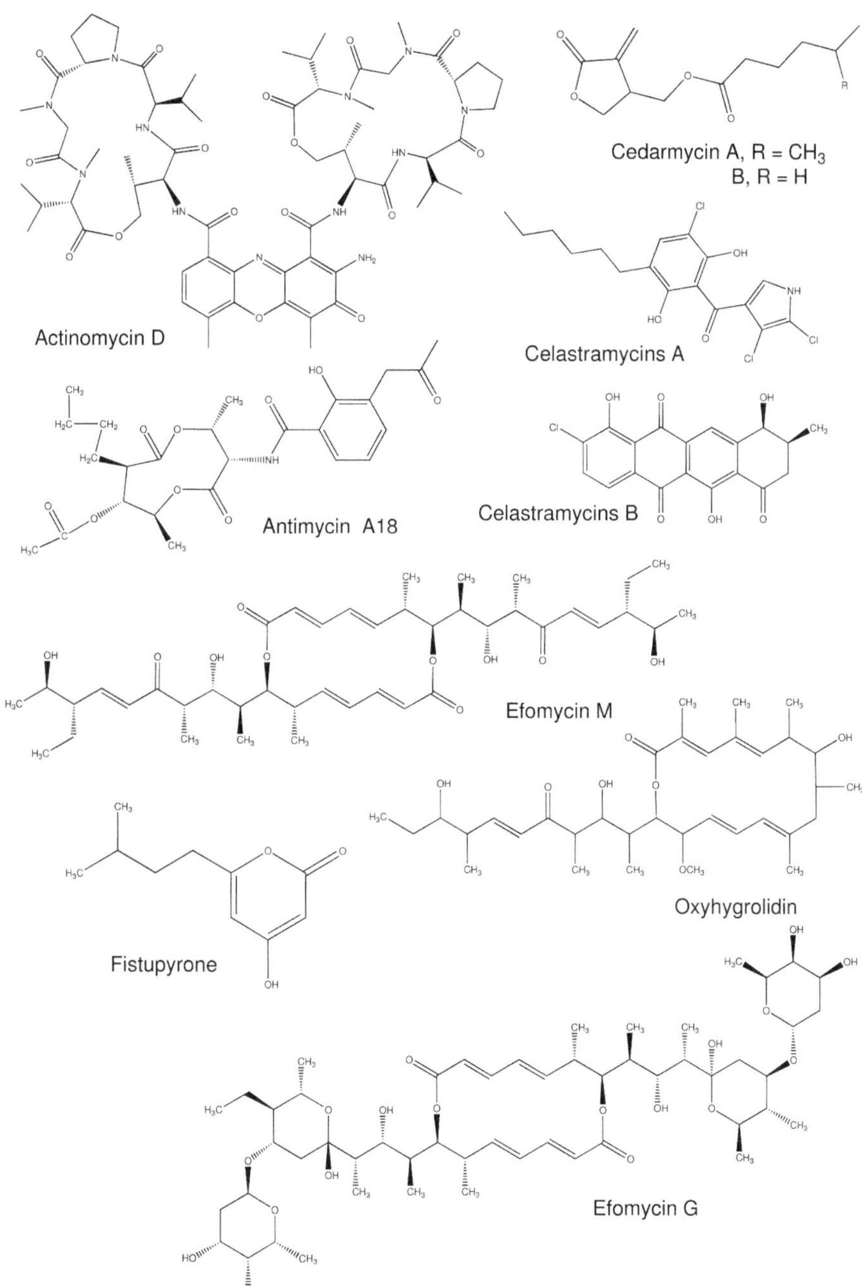

Fig. 8.3 Chemical structure of representative antimicrobial compounds obtained from endophytic bacteria

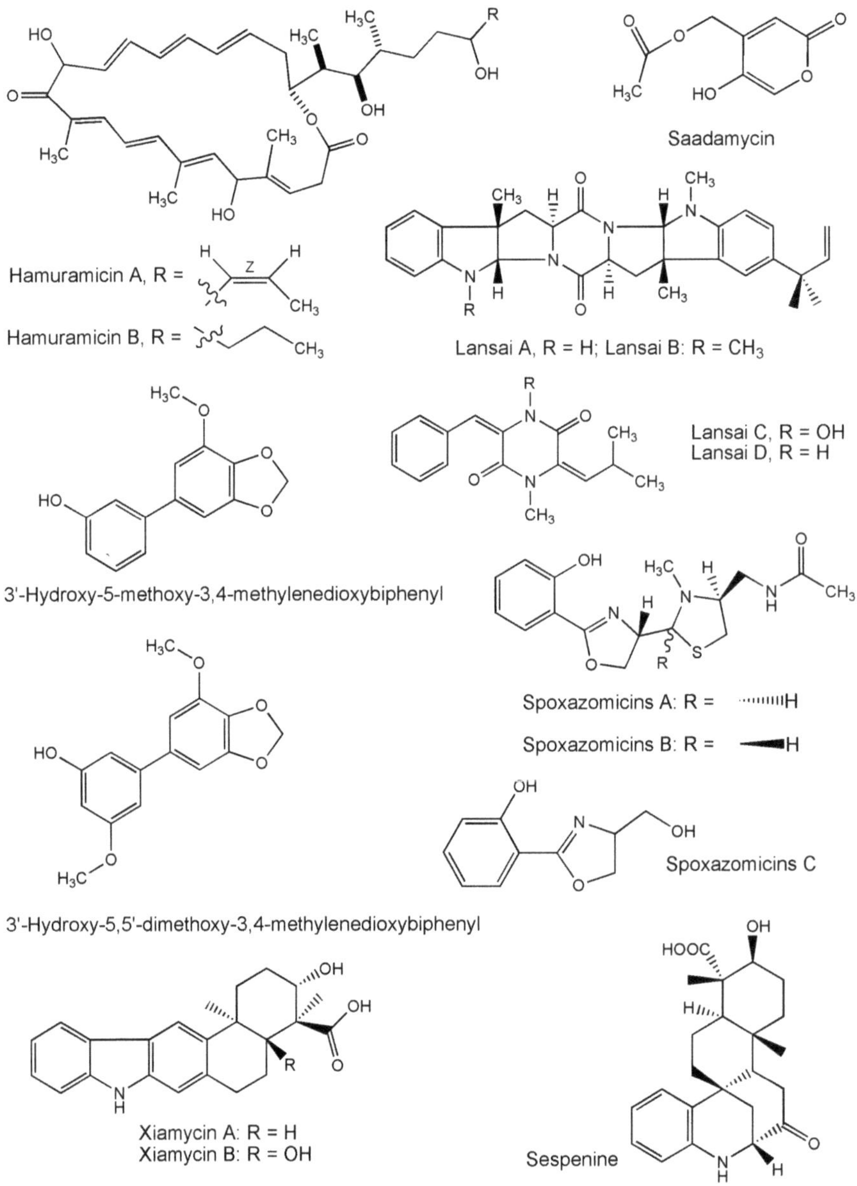

Fig. 8.3 (continued)

E- and P-selectin binding in vitro and thus prevented rolling of T cells. All the three macrolides Efomycin M, Efomycin G and Oxohygrolidin can inhibit the growth of both Gram-positive and Gram-negative bacteria and have antimalarial activity. Abierixin and 29-O-methylabierixin also displayed antimalarial activity (Supong et al. 2016).

8.7.9 Antimycin A18

Antimycin A18 was produced by an endophytic actinobacterium *Streptomyces albidoflavus* isolated from a leaf of a mangrove plant *Bruguiera gymnorrhiza* collected from Shankou, Guangxi Province, People's Republic of China. This compound was reported to have antifungal activities against plant pathogenic fungi (*Colletotrichum lindemuthianum*, *Botrytis cinerea*, *Alternaria solani* and *Magnaporthe grisea*) suggesting that it may use for protection of plants (Yan et al. 2010) (Fig. 8.3).

8.7.10 3′-Hydroxy-5-Methoxy-3,4-Methylenedioxybiphenyl

Streptomyces sp. BO-07 which was isolated from root tissue of the medicinal plant of Thailand *Boesenbergia rotunda* (L.) Mansf A was found to be the source of another two antimicrobial compounds, i.e., 3′-hydroxy-5-methoxy-3,4-methylenedioxybiphenyl and 3′-hydroxy-5,5′-dimethoxy-3,4-methylenedioxybiphenyl (Fig. 8.3). Both of these compounds have strong antibacterial activity against Gram-positive bacteria (*Staphylococcus aureus* ATCC25932, *Bacillus cereus* ATCC7064, *Bacillus subtilis* ATCC6633) and moderate inhibitors against Gram-negative bacteria (*Escherichia coli* ATCC10536, *Pseudomonas aeruginosa* ATCC27853, *Salmonella Typhi* ATCC19430, *Serratia marcescens* ATCC8100) (Taechowisan et al. 2017).

8.7.11 Spoxazomicins

Spoxazomicins were included in novel antitrypanosomal alkaloids of the pyochelin family which were isolated from *Streptosporangium oxazolinicum* K07-0460, an endophyte from the roots of orchid. In vitro antitrypanosomal activities revealed that both Spoxazomicins A and B have strong antitrypanosomal activity against GUTat 3.1 strain of *Trypanosoma brucei brucei* with an IC_{50} value of 0.11 μg ml^{-1} and 0.55 μg ml^{-1}, respectively. However, Spoxazomicins C showed weak antitrypanosomal activity, with an IC_{50} value of 3.0 μg ml^{-1} (Inahashi et al. 2011) (Fig. 8.3).

8.7.12 Cedarmycins

Cedarmycins are novel butyrolactone antibiotics isolated from *Streptomyces* sp. TP-A0456, a plant-associated actinobacteria from stem of *Cryptomeria japonica.* Both of these compounds showed weak to moderate antibiotic activity against both Gram-positive and Gram-negative bacteria, while more potent activity against *Candida glabrata* having IC_{50} of 0.40 and 1.60 μgml^{-1}, respectively (Sasaki et al. 2001) (Fig. 8.3).

8.7.13 Fistupyrone

It is a microbial metabolite isolated from plant-associated *Streptomyces* sp. TP-A0569. Fistupyrone can inhibit the *in vivo* infection of the seedlings of Chinese cabbage caused by *Alternaria brassicicola* TP-F0423 (Igarashi et al. 2000) (Fig. 8.3).

8.7.14 Saadamycin

Saadamycin is another antibiotic isolated from endophytic actinomycetes *Streptomyces* sp. Hedaya48 which is active against dermatophytes and other clinical fungi (El-Gendy and El-Bondkly 2010) (Fig. 8.3).

8.7.15 Lansai A–D

These antibiotics were isolated from *Streptomyces* sp. SUC1, an endophytic actinobacterium from the aerial roots of *Ficus benjamina.* All the antibiotics showed weak activity against *Colletotrichum musae* with MIC>100 μgml^{-1} (Tuntiwachwuttikul et al. 2008) (Fig. 8.3).

8.7.16 Actinomycin D

It is a potent antibiotic from *Streptomyces* sp. Tc022, an endophyte from roots of *Alpinia galangal.* It showed bioactivity against plant pathogenic fungi *Colletotrichum musae* (MIC = 10 $\mu g\ ml^{-1}$) and *Candida albicans* (MIC = 20 $\mu g\ ml^{-1}$) (Taechowisan et al. 2006) (Fig. 8.3).

8.7.17 *Clethramycin*

Clethramycin is structurally similar to Linearmycin and was isolated from *Streptomyces hygroscopicus* TP-A0623, endophyte from the root of *Clethra barbinervis* collected in Toyama, Japan. It was reported to have strong bioactivity against yeast (*Candida albicans*, *C. glabrata*) and fungus (*Aspergillus fumigatus*), however weak activity against Gram-positive and negative bacteria (Furumai et al. 2003).

8.7.18 *Hamuramicins*

Two new compounds containing 22-membered macrolide-containing triene and trienone with an alkyl side chain, designated as Hamuramicins A and B (Fig. 8.3), were isolated from the cultured broth of an endophytic actinomycete *Allostreptomyces* sp. K12-0794. Both of these compounds showed growth inhibition activity against *Kocuria rhizophila* and *Xanthomonas oryzae* pv. *oryzae* (Suga et al. 2018).

8.8 Conclusion

Over the centuries medicinal plant species were used by humans for traditional benefits. Perusal of literature also disclosed that medicinal plant species are the treasure of novel bioactive molecules, among which some led to the discovery of new drugs. Endophytic bacterial population of medicinal plants which are relatively poorly investigated serve as an important component of biodiversity and as a promising source of antimicrobial compounds. Combinations of different cultivation-dependent and cultivation-independent techniques increase our understandings of analysing the diversity of bacterial endobiome and consequently understanding the mechanisms underlying the plant-endophyte interaction. An extensive characterization and identification of the diverse population also help to discover new antimicrobial compounds from them which lead to solve the present-day problems like the appearance of various life-threatening diseases and resistance to existing drugs which ultimately prove to be safe and efficacious for human healthcare.

Acknowledgments KB gratefully acknowledges the financial assistance from the Science and Engineering Research Board, Department of Science and Technology (DST), Government of India, New Delhi. DB is grateful to the Department of Biotechnology, Government of India, New Delhi, for providing financial support.

References

Altemimi A, Lakhssassi N, Baharlouei A et al (2017) Phytochemicals: extraction, Isolation,and identification of bioactive compounds from plant extracts. Plants (Basel) 6(4):E42. https://doi.org/10.3390/plants6040042

Amoutzias GD, Van de Peer Y, Mossialos D (2008) Evolution and taxonomic distribution of non-ribosomal peptide and polyketide synthases. Future Microbiol 3(3):361–370. https://doi.org/10.2217/17460913.3.3.361

Assob JC, Kamga HL, Nsagha DS et al (2011) Antimicrobial and toxicological activities of five medicinal plant species from cameroon traditional medicine. BMC Complement Altern Med 11:70. https://doi.org/10.1186/1472-6882-11-70

Ayuso-Sacido A, Genilloud O (2005) New PCR primers for the screening of NRPS and PKS-I systems in actinomycetes: detection and distribution of these biosynthetic gene sequences in major taxonomic groups. Microb Ecol 49(1):10–24. https://doi.org/10.1007/s00248-004-0249-6

Bacha K, Tariku Y, Gebreyesus F (2016) Antimicrobial and anti-quorum sensing activities of selected medicinal plants of Ethiopia: implication for development of potent antimicrobial agents. BMC Microbiol 16(1):139. https://doi.org/10.1186/s12866-016-0765-9

Bacon CW, White JF (2000) Microbial endophytes. M. Dekker, New York

Bafana A (2013) Diversity and metabolic potential of culturable root-associated bacteria from *Origanum vulgare* in sub-Himalayan region. World J Microbiol Biotechnol 29(1):63–74. https://doi.org/10.1007/s11274-012-1158-3

Bais HP, Park SW, Weir TL et al (2004) How plants communicate using the underground information superhighway. Trends Plant Sci 9:26–32. https://doi.org/10.1016/j.tplants.2003.11.008

Banik A, Mukhopadhaya SK, Sahana A et al (2016) Fluorescence resonance energy transfer (FRET)-based technique for tracking of endophytic bacteria in rice roots. Biol Fertil Soils 52:277–282. https://doi.org/10.1007/s00374-015-1064-6

Barman D, Dkhar MS (2015) Amylolytic activity and its parametric optimization of an endophytic bacterium *Bacillus subtilis* with an ethno-medicinal origin. Biologia 70(3):283–293. https://doi.org/10.1515/biolog-2015-0047

Barman D, Dkhar MS (2018) Plant growth-promoting potential of endophytic bacteria isolated from *Costus speciosus* in tropical deciduous forest of eastern Himalaya. Proc Natl Acad Sci India Sect B Biol Sci. https://doi.org/10.1007/s40011-018-0998-5

Beiranvand M, Amin M, Hashemi-Shahraki A et al (2017) Antimicrobial activity of endophytic bacterial populations isolated from medical plants of Iran. Iran J Microbiol 9(1):11–18

Bhattacharjee K, Kumar S, Palepu NR et al (2017) Structure elucidation and in silico docking studies of a novel furopyrimidine antibiotics synthesized by endolithic bacterium *Actinomadura* sp. AL2. World J Microbiol Biotechnol 33(10):178. https://doi.org/10.1007/s11274-017-2343-1

Bhattacharjee K, Palepu NR, Rao KM et al (2018) Precursor-directed combinatorial biosynthesis of cephalosporin analogue by endolithic actinobacterium *Streptomyces* sp. AL51 by utilizing thiophene derivative. 3 Biotech 8(1):31. https://doi.org/10.1007/s13205-017-1051-8

Cao L, Qiu Z, Dai X et al (2004) Isolation of endophytic actinomycetes from roots and leaves of banana (Musa acuminata) plants and their activities against *Fusarium oxysporum* f. sp. *cubense*. World J Microbiol Biotechnol 20:501–504. https://doi.org/10.1023/B:WIBI.0000040406.30495.48

Castillo U, Strobel G, Ford E et al (2002) Munumbicins, wide- spectrum antibiotics produced by Streptomyces NRRL 30562, endophytic on *Kennedia nigriscans*. Microbiology 148:2675–2685. https://doi.org/10.1099/00221287-148-9-2675

Castillo U, Harper J, Strobel G et al (2003) Kakadumycins, novel anti biotics from *Streptomyces* sp. NRRL 30566, an endophyte of *Grevillea pteridifolia*. FEMS Microbiol Lett 224(2):183–190

Castillo UF, Strobel GA, Mullenberg K et al (2006) Munumbicins E-4 and E-5: novel broad-spectrum antibiotics from *Streptomyces* NRRL 3052. FEMS Microbiol Lett 255:296–300. https://doi.org/10.1111/j.1574-6968.2005.00080.x

Chandra M (2013) Antimicrobial activity of medicinal plants against human pathogenic bacteria. Int J Biotechnol Bioeng Res 4(7):653–658

Chandrakar S, Gupta AK (2017) Antibiotic potential of endophytic actinomycetes of medicinal herbs against human pathogenic bacteria. Proc Natl Acad Sci, India Sect B Biol Sci 87(3):905–915. https://doi.org/10.1007/s40011-015-0668-9

Chen SL, Yu H, Luo HM et al (2016) Conservation and sustainable use of medicinal plants: problems, progress, and prospects. Chin Med 11:37. https://doi.org/10.1186/s13020-016-0108-7

Cho KM, Hong SY, Lee SM et al (2007) Endophytic bacterial communities in ginseng and their antifungal activity against pathogens. Microb Ecol 54(2):341–351. https://doi.org/10.1007/s00248-007-9208-3

Choma IM, Grzelak EM (2011) Bioautography detection in thin-layer chromatography. J Chromatogr A 1218(19):2684–2691. https://doi.org/10.1016/j.chroma.2010.12.069

Chung EJ, Park JA, Pramanik P et al (2013) *Hoeflea suaedae* sp. nov., an endophytic bacterium isolated from the root of the halophyte *Suaeda maritima*. Int J Syst Evol Microbiol 63(Pt 6):2277–2281. https://doi.org/10.1099/ijs.0.045484-0

Compant S, Mitter B, Colli-Mull JG et al (2011) Endophytes of grapevine flowers, berries and seeds: identification of cultivable bacteria, comparison with other plant parts, and visualization of niches of colonization. Microb Ecol 62:188–197. https://doi.org/10.1007/s00248-011-9883-y

Dej-Adisai S, Meechai I, Puripattanavong J et al (2014) Antityrosinase and antimicrobial activities from Thai medicinal plants. Arch Pharm Res 37(4):473–483. https://doi.org/10.1007/s12272-013-0198-z

Ding L, Maier A, Fiebig HH et al (2011) A family of multi-cyclic indolosesquiterpenes from a bacterial endophyte. Org Biomol Chem 9(11):4029–4031. https://doi.org/10.1039/c1ob05283g

Dzidic S, Suskovic J, Kos B (2008) Antibiotic resistance mechanisms in bacteria: biochemical and genetic aspects. Food Technol Biotechnol 46(1):11–21

Egamberdieva D, Wirth S, Behrendt U et al (2017) Antimicrobial activity of medicinal plants correlates with the proportion of antagonistic endophytes. Front Microbiol 8:199. https://doi.org/10.3389/fmicb.2017.00199

El-Gendy MM, El-Bondkly AM (2010) Production and genetic improvement of a novel antimycotic agent, saadamycin, against dermatophytes and other clinical fungi from endophytic *Streptomyces* sp. Hedaya48. J Ind Microbiol Biotechnol 37(8):831–841. https://doi.org/10.1007/s10295-010-0729-2

Elsebai MF, Tejesvi MV, Pirttila AM (2014) Endophytes as a novel source of bioactive new structures. In: Verma VC, Gange AC (eds) Advances in endophytic research. Springer, India, pp 191–202

Ezra D, Castillo UF, Strobel GA et al (2004) Coronamycins, peptide antibiotics produced by a verticillate Streptomyces sp. (MSU-2110) endophytic on *Monstera* sp. Microbiology 150(Pt 4):785–793. https://doi.org/10.1099/mic.0.26645-0

Farjana A, Zerin N, Kabir S (2014) Antimicrobial activity of medicinal plant leaf extracts against pathogenic bacteria. Asian Pac J Trop Dis 4:S920–S923. https://doi.org/10.1016/S2222-1808(14)60758-1

Frey FM, Meyers R (2010) Antibacterial activity of traditional medicinal plants used by Haudenosaunee peoples of New York State. BMC Complement Altern Med 10:64. https://doi.org/10.1186/1472-6882-10-64

Furumai T, Yamakawa T, Yoshida R et al (2003) Clethramycin, a new inhibitor of pollen tube growth with antifungal activity from *Streptomyces hygroscopicus* TP-A0623. I. Screening, taxonomy, fermentation, isolation and biological properties. J Antibiot (Tokyo) 56(8):700–704

Gagne-Bourgue F, Aliferis KA, Seguin P et al (2012) Isolation and characterization of indigenous endophytic bacteria associated with leaves of switchgrass (*Panicum virgatum* L.) cultivars. J Appl Microbiol 114(3):836–853. https://doi.org/10.1111/jam.12088

Golinska P, Wypij M, Agarkar G et al (2015) Endophytic actinobacteria of medicinal plants: diversity and bioactivity. Antonie Van Leeuwenhoek 108(2):267–289. https://doi.org/10.1007/s10482-015-0502-7

Gonzalez I, Ayuso-Sacido A, Anderson A et al (2005) Actinomycetes isolated from lichens: evaluation of their diversity and detection of biosynthetic gene sequences. FEMS Microbiol Ecol 54(3):401–415. https://doi.org/10.1016/j.femsec.2005.05.004

Gouda S, Das G, Sen SK et al (2016) Endophytes: a treasure house of bioactive compounds of medicinal importance. Front Microbiol 7:1538. https://doi.org/10.3389/fmicb.2016.01538

Gyaneshwar P, James EK, Mathan N et al (2001) Endophytic colonization of rice by a diazotrophic strain of Serratia marcescens. J Bacteriol 183:2634–2645. https://doi.org/10.1128/JB.183.8.2634-2645.2001

Hallmann AQ, Kloepper JW (1996) Immunological detection and localization of the cotton endophyte Enterobacter asburiae JM22 in different plant species. Can J Microbiol 42:1144–1154. https://doi.org/10.1139/m96-146

Hallmann J, Qualt-Hallmann A, Mahaffee WF et al (1997) Bacterial endophytes in agricultural crops. Can J Microbiol 43:895–914. https://doi.org/10.1139/m97-131

Harrison L, Teplow DB, Rinaldi M et al (1991) Pseudomycins, a family of novel peptides from pseudomonas syringae, possessing broad spectrum antifungal activity. J Gen Microbiol 137(12):2857–2865. https://doi.org/10.1099/00221287-137-12-2857

Hassan SED (2017) Plant growth-promoting activities for bacterial and fungal endophytes isolated from medicinal plant of *Teucrium polium* L. J Adv Res 8(6):687–695. https://doi.org/10.1016/j.jare.2017.09.001

Holt JG, Krieg NR, Sneath PHA et al (1994) Bergey's manual of determinative bacteriology, 9th edn. Lippincott, Williams & Wilkins, Baltimore

Hopwood DA (1997) Genetic contributions to understanding polyketide synthases. Chem Rev 97:2465–2498

Hu HQ, Li XS, He H (2010) Characterization of an antimicrobial material from a newly isolated *Bacillus amyloliquefaciens* from mangrove for biocontrol of Capsicum bacterial wilt. Biol Control 54(3):359–365. https://doi.org/10.1016/j.biocontrol.2010.06.015

Igarashi Y, Ogawa M, Sato Y et al (2000) Fistupyrone, a novel inhibitor of the infection of Chinese cabbage by *Alternaria brassicicola*, from *Streptomyces* sp. TP-A0569. J Antibiot (Tokyo) 53(10):1117–1122

Inahashi Y, Iwatsuki M, Ishiyama A et al (2011) Spoxazomicins A-C, novel antitrypanosomal alkaloids produced by an endophytic actinomycete, *Streptosporangium oxazolinicum* K07–0460(T). J Antibiot (Tokyo) 64(4):303–307. https://doi.org/10.1038/ja.2011.16

Jasim B, Jimtha CJ, Jyothis M et al (2013) Plant growth promoting potential of endophytic bacteria isolated from *Piper nigrum*. Plant Growth Regul 71:1–11. https://doi.org/10.1007/s10725-013-9802-y

Jasim B, Joseph AA, John CJ et al (2014) Isolation and characterization of plant growth promoting endophytic bacteria from the rhizome of *Zingiber officinale*. 3. Biotech 4:197–204. https://doi.org/10.1007/s13205-013-0143-3

Jia M, Chen L, Xin HL et al (2016) A friendly relationship between endophytic fungi and medicinal plants: a systematic review. Front Microbiol 7:906. https://doi.org/10.3389/fmicb.2016.00906

Kaewpiboon C, Lirdprapamongkol K, Srisomsap C (2012) Studies of the in vitro cytotoxic, antioxidant, lipase inhibitory and antimicrobial activities of selected Thai medicinal plants. BMC Complement Altern Med 12:217. https://doi.org/10.1186/1472-6882-12-217

Kaul S, Sharma T, Dhar MK (2016) "Omics" tools for better understanding the plant–endophyte interactions. Front Plant Sci 7:955. https://doi.org/10.3389/fpls.2016.00955

Khan Z, Doty SL (2009) Characterization of bacterial endophytes of sweet potato plants. Plant Soil 322(1-2):197–207. https://doi.org/10.1007/s11104-009-9908-1

Khan UA, Rahman H, Niaz Z et al (2013) Antibacterial activity of some medicinal plants against selected human pathogenic bacteria. Eur J Microbiol Immunol 3(4):272–274. https://doi.org/10.1556/EuJMI.3.2013.4.6

Kohanski MA, Dwyer DJ, Collins JJ (2010) How antibiotics kill bacteria: from targets to networks. Nat Rev Microbiol 8(6):423–435. https://doi.org/10.1038/nrmicro2333

Kukkurainen S, Leino A, Vahamiko S et al (2005) Occurrence and location of endophytic bacteria in garden and wild strawberry. Hortic Sci 40(2):348–352

Kumar GS, Jayaveera KN, Kumar CKA et al (2007) Antimicrobial effects of Indian medicinal plants against acne-inducing bacteria. Trop J Pharm Res 6(2):717–723

Kumar V, Kumar A, Pandey KD et al (2015) Isolation and characterization of bacterial endophytes from the roots of *Cassia tora* L. Ann Microbiol 65(3):1391–1399. https://doi.org/10.1007/s13213-014-0977-x

Li L, Mohamad OAA, Ma J et al (2018) Synergistic plant-microbe interactions between endophytic bacterial communities and the medicinal plant *Glycyrrhiza uralensis* F. Antonie Van Leeuwenhoek. https://doi.org/10.1007/s10482-018-1062-4

Liu YH, Guo JW, Salam N et al (2016) Culturable endophytic bacteria associated with medicinal plant *Ferula songorica*: molecular phylogeny, distribution and screening for industrially important traits. 3 Biotechnol 6(2):209. https://doi.org/10.1007/s13205-016-0522-7

Ma L, CaoYH CMH et al (2013) Phylogenetic diversity of bacterial endophytes of *Panax notoginseng* with antagonistic characteristics towards pathogens of root-rot disease complex. Antonie Van Leeuwenhoek 103(2):299–312. https://doi.org/10.1007/s10482-012-9810-3

Maggini V, De Leo M, Mengoni A et al (2017) Plant-endophytes interaction influences the secondary metabolism in Echinacea purpurea (L.) Moench: an *in vitro* model. Sci Rep 7:16924. https://doi.org/10.1038/s41598-017-17110-w

Manikprabhu D, Li WJ (2015) Antimicrobial agents from actinomycetes chemistry and applications. In: Dhanasekaran D, Thajuddin N, Panneerselvam A (eds) Antimicrobials synthetic and natural compounds. CRC Press, Taylor and Francis, pp 99–116

Marasini BP, Baral P, Aryal P et al (2015) Evaluation of antibacterial activity of some traditionally used medicinal plants against human pathogenic bacteria. Biomed Res Int 2015(265425):1–6. https://doi.org/10.1155/2015/265425

Martinez-Klimova E, Rodríguez-Peña K, Sánchez S (2017) Endophytes as sources of antibiotics. Biochem Pharmacol 134:1–17. https://doi.org/10.1016/j.bcp.2016.10.010

Mendes R, Pizzirani-Kleiner AA, Araujo WL et al (2007) Diversity of cultivated endophytic bacteria from sugarcane: genetic and biochemical characterization of *Burkholderia cepacia* complex isolates. Appl Environ Microbiol 73(22):7259–7267. https://doi.org/10.1128/AEM.01222-07

Miller CM, Miller RV, Garton-Kenny D et al (1998) Ecomycins, unique antimycotics from *Pseudomonas viridiflava*. J Appl Microbiol 84(6):937–944

Mothana RA, Lindequist U, Gruenert R et al (2009) Studies of the in vitro anticancer, antimicrobial and antioxidant potentials of selected Yemeni medicinal plants from the island Soqotra. BMC Complement Altern Med 9:7. https://doi.org/10.1186/1472-6882-9-7

Murphy B, Batke SP, Doohan FM et al (2015) Media manipulation and the culture of beneficial fungal root endophytes. Int J Biol 7:94–102. https://doi.org/10.5539/ijb.v7n3p94

Nascimento GGF, Locatelli J, Freitas PC et al (2000) Antibacterial activity of plant extracts and phytochemicals on antibiotic-resistant bacteria. Braz J Microbiol 31(4):247–256. https://doi.org/10.1590/S1517-83822000000400003

Nielsen TRH, Kuete V, Jager AK et al (2012) Antimicrobial activity of selected south African medicinal plants. BMC Complement Altern Med 12:74. https://doi.org/10.1186/1472-6882-12-74

Nishteswar K (2014) Depleting medicinal plant resources: a threat for survival of Ayurveda. Ayu 35(4):349–350. https://doi.org/10.4103/0974-8520.158972

Nostro A, Germano MP, Angelo V et al (2000) Extraction methods and bioautography for evaluation of medicinal plant antimicrobial activity. Lett Appl Microbiol 30(5):379–348

Oliveira AA, Segovia JF, Sousa VY et al (2013) Antimicrobial activity of amazonian medicinal plants. Springerplus 2:371. https://doi.org/10.1186/2193-1801-2-371

Packer J, Naz T, Yaegl Community Elders, Harrington D et al (2015) Antimicrobial activity of customary medicinal plants of the Yaegl aboriginal community of northern New South Wales, Australia: a preliminary study. BMC Res Notes 8:276. https://doi.org/10.1186/s13104-015-1258-x

Pandey R, Sambasivarao Y, Gurumurthy (2013) Antibacterial activity of medicinal plants against pathogens from extracts of *Achyranthes Aspera*. Med Aromat Plants 2:5. https://doi.org/10.4172/2167-0412.1000135

Panghal M, Kaushal V, Yadav JP (2011) *In vitro* antimicrobial activity of ten medicinal plants against clinical isolates of oral cancer cases. Ann Clin Microbiol Antimicrob 10:21. https://doi.org/10.1186/1476-0711-10-21

Prakash O, Nimonkar Y, Munot H et al (2014) Description of *Micrococcus aloeverae* sp. nov., an endophytic actinobacterium isolated from *Aloe vera*. Int J Syst Evol Microbiol 64(Pt 10):3427–3433. https://doi.org/10.1099/ijs.0.063339-0

Pullen C, Schmitz P, Meurer K et al (2002) New and bioactive compounds from *Streptomyces* strains residing in the wood of Celastraceae. Planta 216:162–167. https://doi.org/10.1007/s00425-002-0874-6

Qin S, Xing K, Jiang JH et al (2011) Biodiversity, bioactive natural products and biotechnological potential of plant-associated endophytic actinobacteria. Appl Microbiol Biotechnol 89:457–473. https://doi.org/10.1007/s00253-010-2923-6

Quadt-Hallman A, Hallman J, Kloepper JW (1997) Bacterial endophytes in cotton: location and interaction with other plant associated bacteria. Can J Microbiol 43(3):254–259. https://doi.org/10.1139/m97-035

Ramesh P, Okigbo RN (2008) Effects of plants and medicinal plant combinations as anti-infectives. Afr J Pharm Pharmacol 2(7):130–135

Redford AJ, Bowers RM, Knight R et al (2010) The ecology of the phyllosphere: geographic and phylogenetic variability in the distribution of bacteria on tree leaves. Environ Microbiol 12:2885–2893. https://doi.org/10.1111/j.1462-2920.2010.02258.x

Rojas JJ, Ochoa VJ, Ocampo SA et al (2006) Screening for antimicrobial activity of ten medicinal plants used in Colombian folkloric medicine: a possible alternative in the treatment of non-nosocomial infections. BMC Complement Altern Med 6:2. https://doi.org/10.1186/1472-6882-6-2

Ryan RP, Germaine K, Franks A et al (2008) Bacterial endophytes: recent developments and applications. FEMS Microbiol Lett 278(1):1–9. https://doi.org/10.1111/j.1574-6968.2007.00918.x

Sasaki T, Igarashi Y, Saito N et al (2001) Cedarmycins A and B, new antimicrobial antibiotics from *Streptomyces* sp. TP-A0456. J Antibiot (Tokyo) 54(7):567–572

Selvamohan T, Ramadas V, Kishore SSS (2012) Antimicrobial activity of selected medicinal plants against some selected human pathogenic bacteria. Adv Appl Sci Res 3(5):3374–3381

Sharma A, Chandraker S, Patel VK et al (2009) Antibacterial activity of medicinal plants against pathogens causing complicated urinary tract infections. Indian J Pharm Sci 71(2):136–139. https://doi.org/10.4103/0250-474X.54279

Shirling EB, Gottlieb D (1966) Methods for characterization of *Streptomyces* species. Int J Syst Bacteriol 16:313–340. https://doi.org/10.1099/00207713-16-3-313

Stepniewska Z, Kuzniar A (2013) Endophytic microorganism-promising applications in bioremediation of greenhouse gases. Appl Microbiol Biotechnol 97(22):9589–9596. https://doi.org/10.1007/s00253-013-5235-9

Sticher O (2008) Natural product isolation. Nat Prod Rep 25:517–554. https://doi.org/10.1039/B700306B

Strobel G, Daisy B (2003) Bioprospecting for microbial endophytes and their natural products. Microbiol Mol Biol Rev 67(4):491–502. https://doi.org/10.1128/MMBR.67.4.491-502.2003

Strobel GA (2002) Rainforest endophytes and bioactive products. Crit Rev Biotechnol 22(4):315–333. https://doi.org/10.1080/07388550290789531

Suga T, Kimura T, Inahashi Y et al (2018) Hamuramicins A and B, 22-membered macrolides, produced by an endophytic actinomycete *Allostreptomyces* sp. K12-0794. J Antibiot (Tokyo). https://doi.org/10.1038/s41429-018-0055-x

Supong K, Thawai C, Choowong W et al (2016) Antimicrobial compounds from endophytic *Streptomyces* sp. BCC72023 isolated from rice (*Oryza sativa* L.). Res Microbiol 167(4):290–298. https://doi.org/10.1016/j.resmic.2016.01.004

Taechowisan T, Chaisaeng S, Phutdhawong WS (2017) Antibacterial, antioxidant and anticancer activities of biphenyls from *Streptomyces* sp. BO-07: an endophyte in *Boesenbergia rotunda*

(L.) Mansf A. Food Agric Immunol 28(6):1330–1346. https://doi.org/10.1080/09540105.2017.1339669

Taechowisan T, Wanbanjob A, Tuntiwachwuttikul P (2006) Identification of *Streptomyces* sp. Tc022, an endophyte in *Alpinia galanga*, and the isolation of actinomycin D. Ann Microbiol 56(2):113–117. https://doi.org/10.1007/BF03174991

Thomas P (2011) Intense association of non-culturable endophytic bacteria with antibiotic-cleansed in vitro watermelon and their activation in degenerating cultures. Plant Cell Rep 30:2313–2325. https://doi.org/10.1007/s00299-011-1158-z

Thomas P, Reddy KM (2013) Microscopic elucidation of abundant endophytic bacteria colonizing the cell wall–plasma membrane peri-space in the shoot-tip tissue of banana. AoB Plants. https://doi.org/10.1093/aobpla/plt011

Tian BY, Cao Y, Zhang KQ (2015) Metagenomic insights into communities, functions of endophytes, and their associates with infection by root-knot nematode, Meloidogyne incognita, in tomato roots. Sci Rep 5:17087. https://doi.org/10.1038/srep17087

Truyens S, Weyens N, Cuypers A et al (2015) Bacterial seed endophytes: genera, vertical transmission and interaction with plants. Environ Microbiol Rep 7(1):40–50. https://doi.org/10.1111/1758-2229.12181

Tuntiwachwuttikul P, Taechowisan T, Wanbanjob A et al (2008) Lansai A–D, secondary metabolites from *Streptomyces* sp. SUC1. Tetrahedron 64(32):7583–7586. https://doi.org/10.1016/j.tet.2008.05.104

Ulrich K, Ulrich A, Ewald D (2008) Diversity of endophytic bacterial communities in poplar grown under field conditions. FEMS Microbiol Ecol 63(2):169–180. https://doi.org/10.1111/j.1574-6941.2007.00419.x

Ushimaru PI, da Silva MTN, Di Stasi LC et al (2007) Antibacterial activity of medicinal plant extracts. Braz J Microbiol 38(4):717–719. https://doi.org/10.1590/S1517-83822007000400024

Vendan RT, Yu Y, Lee S et al (2010) Diversity of endophytic bacteria in ginseng and their potential for plant growth promotion. J Microbiol 48:559–565. https://doi.org/10.1007/s12275-010-0082-1

Venugopalan A, Srivastava S (2015) Endophytes as *in vitro* production platforms of high value plant secondary metabolites. Biotechnol Adv 33:873–887. https://doi.org/10.1016/j.biotechadv.2015.07.004

Wang Z, Tian J, Li X et al (2018) Streptomyces dioscori sp. nov., a novel endophytic actinobacterium isolated from bulbil of dioscorea bulbifera L. Curr Microbiol. https://doi.org/10.1007/s00284-018-1534-9

Yan LL, Han NN, Zhang YQ et al (2010) Antimycin A18 produced by an endophytic *Streptomyces albidoflavus* isolated from a mangrove plant. J Antibiot (Tokyo) 63(5):259–261. https://doi.org/10.1038/ja.2010.21

Yang R, Liu P, Ye W (2017) Illumina-based analysis of endophytic bacterial diversity of tree peony (*Paeonia* Sect. *Moutan*) roots and leaves. Braz J Microbiol 48(4):695–705. https://doi.org/10.1016/j.bjm.2017.02.009

Yaniv Z (2014) Wild edible plants in Israel tradition versus cultivation. In: Yaniv Z, Dudai N (eds) Medicinal and aromatic plants of the middle-east, vol 2. Springer, Dordrecht, pp 1–8

Zhang L, Wei L, Zhu L et al (2014) *Pseudoxanthomonas* gei sp. nov., a novel endophytic bacterium isolated from the stem of *Geum aleppicum*. Antonie Van Leeuwenhoek 105(4):653–661. https://doi.org/10.1007/s10482-014-0119-2

Zhu L, Xin K, Chen C et al (2015) *Sphingobium endophyticus* sp. nov., isolated from the root of *Hylomecon japonica*. Antonie Van Leeuwenhoek 107(4):1001–1008. https://doi.org/10.1007/s10482-015-0392-8

Zinniel DK, Lambrecht P, Harris NB et al (2002) Isolation and characterization of endophytic colonizing bacteria from agronomic crops and prairie plants. Appl Environ Microbiol 68(5):2198–2208. https://doi.org/10.1128/AEM.68.5.2198-2208.2002

Chapter 9
The Importance of Endophytic Fungi from the Medicinal Plant: Diversity, Natural Bioactive Compounds, and Control of Plant Pathogens

Laith Khalil Tawfeeq Al-Ani

Abstract Endophyte fungi are amazing organisms with the ability to grow inside the host plant, particularly the medicinal plant, without any side effects on the plant tissues. These endophyte fungi can colonize and proliferate without causing any damage to the tissues of the medicinal plant, although many of the fungi genera are isolated from the medicinal plants that can cause the diseases for the plants directly or for animals and human indirectly. But some endophytic fungi genera are using a biocontrol agent with other benefits. Here, the diversity of endophytic fungi was about 180 genera. Also, I suggested a simple equation for displaying the fungi genera available among isolates of endophyte fungi from the medicinal plants. 180 genera of endophytic fungi into three levels including low, middle, and high levels depending on the equation that determined the importance of some fungi genera of from other fungi. These fungi comprised *Trichoderma*, *Curvularia*, *Pestalotiopsis*, *Cladosporium*, *Chaetomium*, *Phomopsis*, *Diaporthe*, *Phoma*, *Penicillium*, *Alternaria*, *Colletotrichum*, *Fusarium*, and *Aspergillus*. The roles of endophyte fungi are shown in various fields such as producing (A) antibiotics against many microbes, (B) mycotoxins, (C) anticancer, (D) insecticides, (E) enzymes, and (F) many compounds used in different fields. Also, endophyte fungi are able to change the chemistry of medicinal plant and control of plant pathogens. Endophyte fungi of a group the dark septate endophytes (DSE) detected that are able to play an important role by producing very interesting compounds. The conclusion is 13 fungi genera more important from others. Many endophyte fungi have displayed the role in producing the natural bioactive products and using to biocontrol of plant pathogens. These fungi can be used to produce natural drugs, biopesticides, and biofertilizers that lead to decrease the dangers of synthetic chemicals. This can save the ecosystem and reduce the chemical residue in the environment.

L. K. T. Al-Ani (✉)
Department of plant protection, College of Agriculture engineering science, University of Baghdad, Baghdad, Iraq

School of Biology Science, Universiti Sains Malaysia, Minden, Pulau Pinang, Malaysia

D. Egamberdieva, A. Tiezzi (eds.), *Medically Important Plant Biomes: Source of Secondary Metabolites*, Microorganisms for Sustainability 15,
https://doi.org/10.1007/978-981-13-9566-6_9

9.1 Introduction

Medicinal plants are very interesting as a big source of natural compounds to use in the drug and pesticide industry; therefore, they are useful for treating the diseases of humans, animals, and plants as an alternative method with synthetic chemicals. But there are some of the prepared compounds friendly to the environment that can use it as antifungal (Al-Ani et al. 2012; Mohammed et al. 2012). Some microbes such as fungi attack medicinal plants without causing any diseases and damages for the host plant that is called an endophyte. Endophytic fungi are forming the important role for the medicinal plants to incur the biotic and abiotic stress with other benefits. Endophytic fungi are very diverse in the medicinal plant and very helpful for medicinal plants by enhancing (1) producing the different compounds and enzymes, as the source for natural compounds useful to manufacture biopesticides, (2) enhancing the plant growth, (3) controlling and protecting from infecting with plant pathogens. (4) Affecting on the plant chemistry (Al-Ani and Albaayit 2018a, b; Al-Ani 2019a, b, c, d, e, f, g, h, i; Egbuna et al. 2019). Endophytic fungi included many fungi genera such as *Penicillium*, *Curvularia*, *Colletotrichum*, *Piriformospora*, *Fusarium*, *Trichoderma*, and others (Teles et al. 2005; Vega et al. 2006; Yadav et al. 2010; Al-Ani and Salleh 2010; Al-Ani 2018a, b; Al-Ani et al. 2013a, b; Hiruma et al. 2016; Al-Ani 2017a; Voříšková et al. 2017). In additional of endophytic bacteria such as plant growth-promoting rhizobacteria (PGPR) or bacterial symbionts as that used in the agriculture to control of plant diseases and enhance the plant growth (Al-Ani 2006; Al-Ani and Al-Ani 2011; Mohammed et al. 2011, 2013, 2014; Al-Ani 2017a). But other fungi associated with plants may be affecting the animals and humans by secreting several dangerous mycotoxins (Attitalla et al. 2010a, b).

The medicinal plants are exposed to endophytic fungi normally. Endophytic fungi can grow through natural openings of plants like a stomata pore and natural wounds. Fungi can grow and do a part or all of effectiveness inside the plant as long the nutrients is available. Generally, the effectiveness is including production the secondary metabolites, enzymes, induce the defense and resistance in plants, and enhance the plant growth. Selim et al. (2012) could produce many compounds including the antimicrobial, antioxidant, antiviral, and anticancer that are useful in different fields such as environmental applications, agriculture, and pharmaceutical industry. Endophytic fungi of medicinal plants are producing the high fortune of a source of bioactive compounds comprising antitumor, antimicrobial, insecticide, antioxidant, growth-promoting, and antithrombotic that almost can apply in agriculture and pharmacy (Zhang et al. 2014). While, in the same status can change the useful for the organisms to be harmful depending on the conditions, such as (1) may be a plant pathogen, and (2) may produce mycotoxins or compounds, and enzymes causing disadvantages for other plants, particularly the economic plants such as crops and vegetables. (i) Endophytic fungi produce different secondary metabolites and enzymes that can be utilized in the different fields such as agriculture, industry, medicine, etc. Bhagat et al. (2012) found 63 endophytic fungi from 2 medicinal plants, *Sapindus detergens* and *Ocimum sanctum*, that produced several biomolecules including compounds that are antimicrobial and anticancer. (ii) Endophytic fungi may control plant pathogens directly and indirectly. (iii) Endophytic fungi have the ability to change the plant chemistry. (iv) Endophytic fungi are able to enhance plant growth through production of plant growth regulators such as gibberellins (GAs), indole-3-acetic acid (IAA), and abscisic acid (ABA) (Dai et al. 2008; Strobel et al. 2001). (v) Endophytic fungi have the ability to produce mycotoxins inside the parts for medicinal plants (Ashiq et al. 2014).

Therefore, these points need to the elucidation in this chapter that may cover some advantage of endophyte fungi for the medicinal plants. The important role of endophytic fungi is appearing everyday and attractive for many researchers to detect and isolate new natural compounds and new species of fungi. Endophytic fungi are including several characterizes such as the producing of natural compounds, and control of the plant pathogens with the increasing in secondary metabolites of the medicinal plant, and improvement the plant growth.

9.2 Diversity of Endophytic Fungi in the Medicinal Plants

Endophytic fungi are able to colonize different tissues of the plant. Diversity of endophytic fungi become clearer by spreading in different parts of the medicinal plants that can appear by isolating endophytic fungi from the different parts of the medicinal plants such as leaves, stems, roots, seeds, and fruit. Endophytic fungi were diverse in various medicinal plants. They registered many species of endophytic fungi in last papers. The medicinal plant such other plants are the repository of fungi by providing the nutrient or cytosol. The diversity of fungi in the medicinal plants comprises the fungi that are feeding as the form saprophytic, or saprophytic and parasitic facultative. This is being as a theory because they have been isolated too much species of fungi from the inner tissues. Some of these species are pathogenic for several economic plants. The diversity of endophytic fungi is summarized in Table 9.1.

Endophytic fungi could be isolated from different medicinal plants (Table 9.1). This table is showing very amazing results about 180 genera of endophytic fungi. Therefore, I suggested some levels showing the gap between the genera of endophytic fungi that have been mentioned in Table 9.1 due to the high number of genera that reached 180. Also, I suggested a simple equation that can show the distribution percentage of isolated genera compared to the number of the medicinal plants. This suggestion is more important for a decrease the big gap between high numbers of some fungi genera to other fungi very little and can give the idea about the most important of genera that tendency to live as endophytic or nonpathogenic fungi inside of the medicinal plants. This suggestion showed three levels of fungi genera that can be divided into low, middle, and high levels (Fig. 9.1) according to the per-

Table 9.1 The fungi genera isolated from the various parts of the medicinal plants

No.	Fungal isolates	Medicinal plant	Parts of plant	References
1.	*Curvularia clavata*, *C. lunata*, *C. pallescens*, and *Fusarium oxysporum*	*Adhatoda zeylanica*, *Bauhinia phoenicea*, *Callicarpa tomentosa*, *Clerodendrum serratum*, and *Lobelia nicotianifolia* in the Western Ghats of India	Bark, leaf, and stem segments	Raviraja (2005)
2.	*Xylaria*, *Daldinia*, *Hypoxylon*, *Colletotrichum*, *Phomopsis*, *Fusarium*, *Coprinus*, *Psathyrella*, *Nodulisporium*, and *Penicillium*	*Cassia fistula* (known as golden shower) of Nakhon Ratchasima and Bangkok Province (collected during two seasons such as winter and rainy) in Thailand	Healthy specimens of leaves	Ruchikachorn (2005)
3.	*Fusarium* sp., *Penicillium* sp., *Schizophyllum* sp., and *Diaporthe* sp.	*Camellia sinensis* (tea plant) was collected of Dr. Partomuan in Japan		Agusta et al. (2006)
4.	*Trametes hirsute*	*Podophyllum hexandrum* of the northwestern Himalayan region at Kashmir and Jammu in India	Dried rhizomes	Puri et al. (2006)
5.	*Phialocephala fortinii*	*Podophyllum peltatum*	Rhizomes	Eyberger et al. (2006)
6.	*Fusarium oxysporum*	*Juniperus recurva* of South Kashmir	Stem, leaves, and root	Kour et al. (2008)
7.	*Alternaria* sp., *Chaetomium* sp., *Fusarium* sp., *Colletotrichum* sp., *Cladosporium* sp., *Penicillium* sp., *Phyllosticta* sp., and *Xylaria* sp.	*Adathoda vasica*, *Calotropis gigantean*, *Carissa carandus*, *Cassia alata*, *Citrus medica*, *Datura metel*, *Ervatamia coronaria*, *Hibiscus rosa sinensis*, *Ixora coccinea*, *Jatropha curcus*, *Lantana camara*, *Nerium indicum*, *Punica granatum*, *Toddalia asiatica*, and *Vitex nigundo* in Malnad region of Western Ghats, Southern India	Healthy specimens of leaf segments	Shankar Naik et al. (2008)
8.	*Fusarium oxysporum*	*Juniperus recurva* (matured) of Gulmarg region at South Kashmir	Stem, leaves, and root	Kour et al. (2008)

(continued)

Table 9.1 (continued)

No.	Fungal isolates	Medicinal plant	Parts of plant	References
9.	*Colletotrichum gloeosporioides* (strain JGC-9)	*Justicia gendarussa* in Chennai City, India	Leaves	Gangadevi and Muthumary (2008)
10.	*Pestalotiopsis vismiae*, *Colletotrichum* sp., *Phomopsis* sp., *Pho. amygdale*, *Clonostachys rosea*, *Penicillium* sp., *Penicillium griseofulvum*, *Fusarium* sp., *F. solani*, *F. proliferatum*, *Trichoderma chlorosporum*, *Guignardia mangiferae*, *Botryosphaeria* sp., *Xylaria* sp., *Hypoxylon* sp., *Nemania* sp., and *Rhizoctonia* sp.	*Dendrobium nobile* (epiphytic orchid) of Nabanhe National Nature Reserve, Yunnan province, Xishuangbanna, at southwest in China	Healthy specimens of stems, leaves, and roots	Yuan et al. (2009)
11.	*Chaetomella raphigera* (strain TAC-15)	*Terminalia arjuna*	–	Gangadevi and Muthumary (2009)
12.	*Colletotrichum gloeosporioides*	*Plumeria acutifolia* Poiret (Apocynaceae) in Guindy Campus Chennai City, India	Leaves	Nithya and Muthumary (2009)
13.	*Fusarium* and *Acremonium*	*Dendrobium loddigesii* Rolfe of the medicinal orchid in Shillong, Meghalaya, India	Healthy specimens of roots	Chen et al. (2010)
14.	*Pestalotiopsis versicolor* and *Pestalotiopsis neglecta*	*Taxus cuspidata* of Japanese yew tree	Healthy specimens of leaves and bark	Kumaran et al. (2010)
15.	*Fusarium culmorum* SVJM072 strain	*Tinospora cordifolia*		Sonaimuthu et al. (2010)
16.	*Xylaria* sp.	*Piper aduncum* from at NuBBE's greenhouse, São Paulo, Brazil	Healthy specimens of adult leaves	Silva et al. (2010)
17.	*Penicillium verruculosum* RS7PF	*Potentilla fulgens* L.	Roots	Bhagobaty et al. (2010)
18.	*Chaetomium globosum*	*Ginkgo biloba*	Healthy specimens of leaves	Li et al. (2011c)

(continued)

Table 9.1 (continued)

No.	Fungal isolates	Medicinal plant	Parts of plant	References
19.	*Alternaria* sp., *Alternaria alternata*, *Aspergillus niger*, *As. fumigatus*, *Curvularia lunata*, *Fusarium oxysporum*, *F. roseum*, *Penicillium* sp., *Trichoderma* sp., *Stenella agalis*, *Phomopsis* sp., *Chaetomium* sp., *Chaetomium globosum*, and *Rhizoctonia* sp.	*Adenocalymma alliaceum* from campus of Banaras Hindu University, Varanasi, in India	Healthy specimens of stems, petioles, and leaves	Kharwar et al. (2011)
20.	*Fusarium solani*	*Taxus baccata* collected from Dibang Valley of Arunachal Pradesh, at a part of Eastern Himalaya, in India	Healthy specimens of inner bark	Tayung et al. (2011)
21.	*Epicoccum nigrum*, *Cladosporium tenuissimum*, *Rhizomucor variabilis*, *Paraconiothyrium variabile*, *Phaeoacremonium rubrigenum*, *Xylaria mali*, *Fusarium equiseti*, *F. solani*, *F. oxysporum*, *F. avenaceum*, *Lasiodiplodia theobromae*, *Phoma herbarum*, *Coniothyrium nitidae*, *Chaetomium globosum*, *Pichia guilliermondii*, *Leptosphaerulina chartarum*, and *Hypocrea lixii*	*Aquilaria sinensis* (agarwood) of the tropical rainforests at Hainan and Yunnan provinces in China	Stem	Cui et al. (2011)
22.	*Emericella* sp.	*Aegiceras corniculatum*	Inner bark	Zhang et al. (2011)
23.	*Colletotrichum gloeosporioides*, *Alternaria alternata*, *Guignardia bidwelli*, *Phomopsis archeri*, *Curvularia pallescens*, *Fusarium lateritium*, *Paecilomyces variotti*, *Periconia byssoides*, *Ulocladium oudemansii*, *Dreschlera dematioidea*, *Microascus desmosporum*, and *Phoma tracheiphila*	*Lippia sidoides* Cham. from Agropecuary Research Company – Pernambuco (tropical moist) at Carpina, PE, in Brazil	Healthy specimens of stem and leaf	de Siqueira et al. (2011)
24.	*Colletotrichum gloeosporioides* Penz.	*Salacia chinensis L.* (Celastraceae) of the herbal garden of Mangalore University campus in India	Healthy specimens of stem and leaf	Bhagya et al. (2011)
25.	*Cochliobolus intermedius*, *Alternaria* sp., *Alt. alternata*, *Curvularia* sp., *Cur. affinis*, *Phomopsis* sp., *Pho. chimonanthi*, *Pho. micheliae*, *Diaporthe helianthi*, *Ascomycota* sp., and *Phoma* sp.	*Sapindus saponaria* tree (15 years old) from Maringa, Parana, Brazil	Healthy specimens of leaves	García et al. (2012)

(continued)

Table 9.1 (continued)

No.	Fungal isolates	Medicinal plant	Parts of plant	References
26.	*Trichoderma citrinoviride*, *Colletotrichum panacicola*, *Phoma radicina*, *Fusarium acuminatum*, *F. oxysporum*, *F. solani*, *Leptodontidium orchidicola*, *Cylindrocarpon destructans*, and *Colletotrichum pisi*	*Panax ginseng* Meyer (ginseng) of the National Institute of Horticultural and Herbal Science at Eumseong, in Korea	Roots of 1-, 2-, 3-, and 4-year-old plants	Park et al. (2012a)
27.	*Rhizopus oryzae*, *Aspergillus niger*, and *A. flavus*	*Terminalia brownie* (local name: Weiba) of Ghindae sub-zone at Eritrea, Northeast Africa	Healthy specimens of leaves	Basha et al. (2012)
28.	*Talaromyces flavus*, *Mortierella hyaline*, *Paecilomyces variabilis*, and *Penicillium* sp.	*Potentilla fulgens*, *Camellia caduca*, *Osbeckia stellata*, *Os. chinensis*, and *Schima khasiana* collected from Sohra, Mawphlang, Lum Shyllong, and Umsaw Nongkhrai "sacred groves" at Meghalaya, in India	Healthy specimens of stems and roots	Bhagobaty and Joshi (2012)
29.	*Alternaria alternata*, *Aspergillus fumigatus*, *A. niger*, *Cladosporium cladosporioides*, *Colletotrichum dematium*, *Chaetomium globosum*, *Curvularia fallax*, *C. lunata*, *C. oryzae*, *Drechslera ellisii*, *Fusarium oxysporum*, *Humicola grisea*, *Acremonium* sp., *Nigrospora oryzae*, *Penicillium* sp., *Phomopsis* sp., and *Rhizoctonia* sp.	*Nyctanthes arbor-tristis* from botanical garden from Banaras Hindu University, Varanasi, in India	Healthy specimens of stems and leaves	Gond et al. (2012)
30.	*Fusarium solani*	*Podophyllum hexandrum* of ten different locations around the village Jhuni, at Bageshwar Distr., Uttarakhand, in India	Rhizomes and roots	Nadeem et al. (2012)
31.	*Fusarium oxysporum*, *Phoma radicina*, *Ascomycota* sp., and *Setophoma terrestris*	*Panax ginseng* Meyer (ginseng) three cultivars such as Chunpoong, Yunpoong, and Gumpoong of Gangwon Province in Korea	Roots	Park et al. (2012b)

(continued)

Table 9.1 (continued)

No.	Fungal isolates	Medicinal plant	Parts of plant	References
32.	*Cladosporium cladosporioides*, *Glomerella cingulata*, *Ceratobasidium* sp., *Colletotrichum* sp., *C. gloeosporioides*, *C. trifolii*, *Fusarium* spp., *F.* oxysporum, *F. solani*, and *Mycoleptodiscus indicus*	*Echinacea purpurea* (clones) seed bought from the North Central Regional Plant Introduction Station (NCRPIS) in Ames, Iowa	Healthy specimens of leaves, lateral shoots, and roots	Rosa et al. (2012)
33.	*Chaetomium* sp., *Pestalotiopsis* sp., *Phyllosticta nobilis*, *Phomopsis* sp., *Phacidium* sp., *Alternaria cinerariae*, *Arthrobotrys* sp., *Arthrinium* sp., *Aspergillus fumigatus*, *As. niger*, *Chaetophoma*, *Curvularia lunata*, *Cladosporium tennuissimum*, *Drechslera* sp., *Gliomastix* sp., *Humicola greseia*, *Nigrospora oryzae*, *Penicillium* spp., *Periconia* sp., *Stachybotrys* sp., *Trichoderma harzianum*	*Cinnamomum camphora* (L.) Presl from the campus of Banaras Hindu University, Varanasi, in India	Young and mature tissues of stem, leaves, and petiole	Kharwar et al. (2012)
34.	*Rhizoctonia* sp., *Xylaria* sp., *Fusarium* sp., *Trichoderma* sp., *Colletotrichum* sp., *Pestalotiopsis* sp., and *Phomopsis* sp.	*Dendrobium nobile* and *Den. chrysanthum* as orchid species in China	Roots	Chen et al. (2012)
35.	*Alternaria* sp., *Phoma* sp., *Colletotrichum* sp., *Fusarium* sp., *Xylaria* sp., and *Entrophospora* sp.	*Panax ginseng* Meyer (3–4-year-old ginseng plants) in Korea	Leaf, petiole, root, stem, and flower stalk	Park et al. (2012c)
36.	*F. oxysporum* BH-3	*Lilium lancifolium* of Lanzhou, Gansu Province, in China	Healthy bulbs	Liu et al. (2012)
37.	*Fusarium oxysporum*	*Ginkgo biloba* from the forest site that located in the Changbai Mountain, China	Healthy specimens of the root bark	Cui et al. (2012)
38.	*Alternaria* sp.	*Datura stramonium* L.	–	Sun et al. (2012)
39.	*Rhizophus* spp., *Aspergillus* spp. 1, *Aspergillus* spp. 2	*Terminalia brownie* and local name: Weiba (Eritrean medicinal plant) collected from Ghindae sub-zone	Healthy specimens of mature leaves	Basha et al. (2012)
40.	*Alternaria* sp., *Exserohilum* sp., and *Phoma sojicola*	*Ocimum sanctum* and *Sapindus detergens* in Amritsar, India	Healthy specimens of stems, and leaves	Bhagat et al. (2012)
41.	*Diaporthe helianthi*, *Cordyceps memorabilis*, *Phomopsis* sp., and *Pho. longicolla*	*Trichilia elegans* (Meliaceae) is a native tree in several regions of Brazil	Healthy specimens of leaves	Rhoden et al. (2012)

(continued)

Table 9.1 (continued)

No.	Fungal isolates	Medicinal plant	Parts of plant	References
42.	*Phomopsis* sp. and *Phyllosticta* sp.	*Ginkgo biloba* L. (three healthy ginkgo trees) of Mie University, Tsu, Mie, Japan	Healthy specimens of twigs, petioles, and leaves	Thongsandee et al. (2012)
	Diaporthe cf. *phaseolorum*, *Guignardia* cf. *camelliae*, *Preussia pseudominima*, *Phomopsis* sp., *Xylaria hypoxylon*, *Diaporthe* sp., *Pseudofusicoccum stromaticum*, *Cytospora* sp., *Nigrospora oryzae*, *Neofusicoccum* cf. *ribis*, *Colletotrichum* sp., *Guignardia mangiferae*, *Xylaria* sp., *Pestalotiopsis clavispora*, *Preussia africana*, *Sporormiella isomera*, *Alternaria alternata*, *Cladosporium cladosporioides*, *Colletotrichum* cf. *gloeosporioides*, *Penicillium glabrum*, *Preussia* sp., *Sordaria fimicola*, *Aspergillus* cf. *ustus*, *Colletotrichum* cf. *boninense*, *Nigrospora* sp., *Penicillium* sp., *Pestalotiopsis microspora*, *Aspergillus* cf. *flavipes*, *Arthrobotrys* sp., *Botryosphaeria* sp., *Coniochaeta* sp., *Fimetariella rabenhorstii*, *Massarina igniaria*, *Muscodor* sp., *Paraconiothyrium brasiliense*, *Pestalotiopsis* cf. *cocculi*, *Preussia* cf. *isomera*, *Sordaria* cf. *tomento-alba*, and *Trichoderma* sp.	*Stryphnodendron adstringens* of two protected reserves: (1) "Cerrado" (a typical Brazilian savanna), the National Park of Serra of the Cipó (summer), and (2) Serra of São José (winter), both in Minas Gerais State, Brazil	Leaves and barks	Carvalho et al. (2012)
43.	*Colletotrichum gloeosporioides*	*Barringtonia acutangula* of the river bed in Kovilanchery, Chennai	Healthy specimens of leaves	Lakshmi and Selvi (2013)
44.	*Colletotrichum* sp., *Phoma* sp., *Fusarium* sp., *Phyllosticta* sp., *Phomopsis* sp., *Alternaria* sp., *Xylaria* sp., *Nodulisporium* sp., *Penicillium* sp., *Cladosporium* sp., *Aspergillus* sp., and *Daldinia* sp.	*Mitragyna javanica* of Koord and Val, collected in the wet season from Pathumthani and Ayuthaya provinces in Thailand	Healthy specimens of leaves	Pharamat et al. (2013)
45.	*Colletotrichum truncatum*, *Chaetomium* sp., *Guignardia cammillae*, *Nigrospora oryzae*, *Fusarium proliferatum*, and *Alternaria destruens*	*Jatropha curcas* of TERI, New Delhi, in India	Healthy specimens of leaves and petiole	Kumar and Kaushik (2013)

(continued)

Table 9.1 (continued)

No.	Fungal isolates	Medicinal plant	Parts of plant	References
46.	*Claviceps purpurea* and *Chaetomium globosum* Kunze ex Fr.	*Achnatherum inebrians* (Hance) Keng of a mountain near Urumuqi City at Xinjiang Province, in China	Healthy specimens of stems, roots, seeds, and leaves	Shi et al. (2013)
47.	*Fusarium solani* T-7	*Ginkgo biloba* L. of the Chinese Department of Hefei University, plants living body	–	Shimo et al. (2013)
48.	*Chaetomium globosum*	*Ginkgo biloba*, from Linyi, Shandong Province, in China	Healthy specimens of barks	Zhang et al. (2013)
49.	*Colletotrichum gloeosporioides*, *Aspergillus awamori*, and *Penicillium* sp.	*Rauwolfia serpentina* Benth (Sarpagandha or Indian snakeroot) of Northeast India	Healthy specimens of stem and leaf	Nath et al. (2013)
50.	*Alternaria* sp., *Phomopsis* sp., Sphaeropsis sp., *Guignardia* sp., *Botrytis* sp., *Penicillium* sp., *Cladosporium* sp., *Fusarium* sp., *Mucor* sp., *Aspergillus* sp., *Gloeosporium* sp., *Colletotrichum* sp., and *Chaetomium* sp.	*Ginkgo biloba* L. from Nanjing and Taixing in Jiangsu Province and Chengdu in Sichuan Province, PR China	Healthy specimens of leaves and small branches	Xiao et al. (2013)
51.	*Colletotrichum gloeosporioides* MKL1 and *Aspergillus oryzae* CeR1	*Centella asiatica* and *Murraya koengii*	Healthy specimens of leaves, stems, roots, and branches	Nath et al. (2014)
	Fusarium sp.	*Fritillaria unibracteata* var. *wabensis* of Mao County, Sichuan Province, China	Fresh bulbs	Pan et al. (2014)
52.	*Aspergillus niger*, *A. flavus*, *A. clavatus*, *A. variecolor*, *Penicillium chrysogenum*, *Curvularia lunata*, *Haplosporidium* sp., *Alternaria alternata*, *Phoma* sp., *Nigrospora* sp., *Colletotrichum* sp., *Geotrichum* sp., *Phomopsis* sp., *Trichoderma* sp., *Rhizopus* sp., *Cladosporium* sp., *Stemphylium* sp., and *Fusarium* sp.	*Adhatoda vasica*, *Ocimum sanctum*, *Viola odorata*, *Cannabis sativa*, and *Withania somnifera* of Mandi district, at Himachal Pradesh, in India	Healthy specimens of leaves	Gautam (2014)

(continued)

Table 9.1 (continued)

No.	Fungal isolates	Medicinal plant	Parts of plant	References
53.	*Chaetomium globosum*	*Ginkgo biloba*	Healthy specimens of leaves	Li et al. (2014a, b)
54.	*Aspergillus* sp., *A. tamari*, *A. pulvinus*, *A. parasiticus*, *A. terreus*, *A. niger*, *A. flavus*, *Curvularia* sp., *Emericella sp.*, *E. bicolour*, *E. rugulosa*, *E. nidulans*, and *Chaetomium* sp.	*Datura stramonium*, *Prosopis chilensis*, and *Moringa oleifera* of Khartoum State, at Central Sudan	Healthy specimens of fresh leaves and stems	Mahdi et al. (2014)
55.	*Alternaria alternata*, *Alternaria* sp., *Aspergillus* sp., *A. flavus*, *A. fumigatus*, *A. niger*, *A. terreus*, *Cladosporium cladosporioides*, *Cladosporium* sp., *Corynespora* sp., *Fusarium graminearum*, *Fusarium* sp., *Monodictys* sp., *Penicillium* sp., *Gliocladium* sp., *Taeniolella* sp., *Trichoderma* sp., *Aspergillus tubingensis*, *Aspergillus lentulus*, *Fusarium* sp., *Aschersonia* sp., *Botryodiplodia* sp., *Colletotrichum coccodes*, *C. gloeosporioides*, *Macrophoma* sp., *Pestalotia* sp., *Phlyctaena* sp., *Phomopsis* sp., *Phomopsis liquidambari*, *Botryosphaeria rhodina*, *Chaetomium* sp., *Pleospora* sp., *Preussia* sp., *Fusicoccum* sp., *Emericella* sp., and *Xylaria* sp.	*Madhuca indica* Gmel. from (Location 1) the campus of Banaras Hindu University at Varanasi district (less polluted area and relatively moist); (Location 2) agriculture lands of Chandauli district, UP, India; and (Location 3) the Hathinala forest at Sonebhadra district (natural forest)	Healthy specimens of stem, bark, and leaf	Verma et al. (2014)
56.	*Chaetomium globosum*, *Aureobasidium pullulans*, *Aspergillus niger*, *Curvularia lunata*, *Penicillium* spp., *Pestalotiopsis* spp., *Trichoderma viride*, *Fusarium* spp., and *Cladosporium cladosporioides*	*Achyranthes aspera* L., *Eclipta alba* L. Hassk, *Elephantopus scaber* L., *Leucas aspera* Spreng., *Ocimum sanctum* L., *Phyllanthus niruri* L., and *Sida acuta* Burm. f. of the Malnad region in India	Healthy specimens of leaves	Shankar Naik et al. (2014)

(continued)

Table 9.1 (continued)

No.	Fungal isolates	Medicinal plant	Parts of plant	References
57.	*Lecythophora* sp., *Lecythophora hoffmannii*, *Coniochaeta ligniaria*, *Hypoxylon perforatum*, *Hypo. ticinense*, *Hypo. notatum*, *Coniochaeta* sp., *Nemania* sp., *Leptospira rubella*, *Lophiostoma* sp., *Cerraena consors*, *Colletotrichum* sp., *Coll. panacicola*, *Acremonium* sp., *Stemphylium solani*, *Cladosporium* sp., *Gibberella moniliformis*, *Aureobasidium* sp., *Hansfordia* sp., *Daldinia childiae*, *Aspergillus* sp., *Coll. gloeosporioides*, *Edenia gomezpompae*, *Alternaria alternata*	*Panax ginseng* (Korean ginseng) from field-cultivated ginseng and mountain-cultivated ginseng at four sites of Chungbuk Province in Korea	Healthy specimens of leaves	Eo et al. (2014)
	Alternaria sp., *Cochliobolus* sp., *Diaporthe* sp., *Diaporthe phaseolorum*, *Epicoccum* sp., *Guignardia* sp., *Nigrospora* sp., *Pestalotiopsis* sp., *Phoma* sp., *Phomopsis* sp., *Preussia Africana*, *Pre. pseudominima*, *Podospora* sp., *Preussia* sp., *Chaetomium* sp., *Sporormiella* sp., and *Xylaria* sp.	*Baccharis trimera* of Serra of Ouro Branco at protected area from Minas Gerais State, in Brazil	Healthy specimens of leaves	Vieira et al. (2014)
58.	*Acremonium* sp., *Fusarium* sp., *Colletotrichum* sp., *Pestalotiopsis* sp., *Chaetomium* sp., *Myrothecium* sp., and *Phomopsis* sp.	*Tylophora asthmatica* (W. and A.), *Rubia cordifolia* L., *Plumbago zeylanica* L., *Phyllanthus amarus* (Schum. andThonn.), *Eryngium foetidum* L., *Zingiber* sp., and *Centella asiatica* L. of the Western Ghats region in India	Healthy specimens of root, inflorescence, and rhizome	Nalini et al. (2014)
59.	*Alternaria alternata*, *A. brassicae*, *Coll. gloeosporioides*, *Corynespora cassiicola*, *Fusarium oxysporum*, *F. solani*, *Lasiodiplodia crassispora*, *Lasid. pseudotheobromae*, *Lasid. theobromae*, *Nectria mauritiicola*, *Periconia byssoides*, and *Phialemonium* sp.	*Vitex negundo* L. of the Western Ghats region in India	Bark, twig, and leaf	Sunayana et al. (2014)

(continued)

Table 9.1 (continued)

No.	Fungal isolates	Medicinal plant	Parts of plant	References
60.	*Fusarium* sp., *Coprinopsis cinerea*, *Penicillium spinulosum*, *Aspergillus* sp., *A. flavus*, *A. peyronelii*, *A. niger*, *A. tubingensis*, *A. japonicus*, *A. terreus*, *A. aff. fumigatus*, *Curvularia lunata*, *Alternaria alternata*, *Syncephalastrum racemosum*, *Choanephora* sp., *Chaetomium* sp., *Trichoderma Longibrachiatum*, and *Paecilomyces formosus*	*Eugenia jambolana*	Leaf, petiole, and stem	Yadav et al. (2014)
61.	*Acremonium curvulum*, *Aspergillus ochraceus*, *Gibberella fujikuroi*, *Myrothecium verrucaria*, *M. schulzeri*, *Phoma putaminum*, *Gibberella baccata*, *Penicillium commune*, *P. glabrum*, and *Trichoderma piluliferum*	*Bauhinia forficata* of Didactic Garden from the Center of Biological Sciences, Federal University of Pernambuco, Recife, Brazil	Healthy specimens of leaves, stems, sepals, and seeds	Bezerra et al. (2015)
62.	*Alternaria* sp., *Bipolaris* sp., *Curvularia* sp., *Chaetomium* sp., *Drechslera* sp., *Emericella* sp., *Aspergillus* sp., *Cladosporium* sp., *Paecilomyces* sp., and *Phoma* sp.	*Calotropis procera*, *Catharanthus roseus*, *Euphorbia prostrate*, *Vernonia amygdalina*, and *Trigonella foenum-graecum* of Khartoum local market	Stems, leaves (*V. amygdalina*, *C. procera*, *C. roseus*, *E. prostrata*), and seeds (*T. foenum-graecum*)	Khiralla et al. (2015)
63.	*Penicillium chrysogenum* Pc_25 and *Alternaria alternata* Aa_27	*Asclepias sinaica* of Ain Shakaya, Saint Catherine, South Sinai, in Egypt	Leaves	Fouda et al. (2015)
64.	*Alternaria* sp., *Curvularia* sp., *Colletotrichum* sp., *Acremonium* sp., *Phoma* sp., *Ulocladium* sp., *Helminthosporium* sp., *Chaetomium* sp., *Aspergillus* sp., *Penicillium* sp., *Verticillium* sp., and *Cladosporium* sp.	*Centella asiatica* Linn, *Phlogacanthus Thyrsiflorus nees* L, *Gynura angulosa* De(L), *Ageratum conyzoides* Linn, *Eupatorium birmanicum* De, *Houttuynia cordata* Thunb, *Allium hookeri*, *Allium odorum* Linn, *Plantago major* Linn, and *Mimosa pudica* Linn of Imphal West in India	Healthy specimens of stem, leaves, and roots	Devi and Singh (2015)

(continued)

Table 9.1 (continued)

No.	Fungal isolates	Medicinal plant	Parts of plant	References
65.	*Colletotrichum siamense*, *Coll. gloeosporioides*, *Penicillium simplicissimum*, *Fusarium proliferatum*, *F. verticillioides*, *F. oxysporum*, *Phoma* sp., and *Ascomycota* sp.	*Pereskia bleo*, *Oldenlandia diffusa*, *Cymbopogon citratus*, and *Murraya koenigii* of Bandar Tun Hussein Onn, Selangor, Malaysia	Stems and leaves	Chow and Ting (2015)
66.	*Alternaria alternata*, *A. tenuissima*, *Aspergillus niger*, *Bipolaris maydis*, *Chaetomium coarctatum*, *Colletotrichum* sp., *Curvularia lunata*, *Diaporthe phaseolorum*, *Fusarium proliferatum*, *F. solani*, *F. verticillioides*, *Hypocrea* sp., *Hypoxylon* sp., *Macrophomina phaseolina*, *Meyerozyma guilliermondii*, *Meyerozyma* sp., *Penicillium crustosum*, *Penicillium* sp., *Rhizoctonia bataticola*, *Rhizopus oryzae*, *Setosphaeria rostrata*, and *Sympodiomyces* sp.	*Ocimum sanctum* (India's Queen of herbs Tulsi) of Southern Plateau and Hills Region, Hyderabad; Western Himalayan region, Trans-Gangetic plains and Mukteshwar; Gwal Pahari and Delhi in two different sampling time: summer (April–June) and autumn (August–September) in India	Healthy specimens of stems and leaves	Chowdhary and Kaushik (2015)
67.	*Nigrospora sphaerica*, *Beauveria bassiana*, *Fusarium oxysporum*, and *Gibberella moniliformis*	*Crescentia cujete* L. of Bharathidasan University campus in India	Healthy leaf	Prabukumar et al. (2015)
68.	*Verticillium dahliae*, *Penicillium janthinellum*, *Pen. skrjabinii*, *Pen. aculeatum*, *Colletotrichum gloeosporioides*, *Phomopsis asparagi*, *Eupenicillium brefeldianum*, *Umbelopsis dimorpha*, *Fusarium proliferatum*, *F. oxysporum*, *Trichoderma asperellum*, *T. spirale*, *Hypoxylon fragiforme*, *Trametes versicolor*, and *Xylaria venosula*	*Kadsura angustifolia* of Xichou and Maguan of Yunnan province of China	Healthy specimens of stems and roots	Huang et al. (2015)
69.	*Fusarium* sp., *Nigrospora* sp., *Botrytis* sp., *Phoma* sp., *Pestalotiopsis* sp., *Cladosporium* sp., *Aspergillus* sp., *Curvularia* sp., *Colletotrichum* sp., *Alternaria* sp., *Penicillium* sp., *Phyllosticta* sp., *Trichoderma* sp., and *Phomopsis* sp.	(1) *Coryllus avelana* (Hazel)	(1) Twigs, shells, and leaves from *Corylus avellana*	Michalczyk et al. (2015)
		(2) *Ocimum basilicum* (Basil) of Botanical Garden, at Polish Academy of Science, in Powsin, Poland	(2) Young and mature leaves from *Ocimum basilicum*	

(continued)

Table 9.1 (continued)

No.	Fungal isolates	Medicinal plant	Parts of plant	References
70.	*Colletotrichum* sp., *Diaporthe* cf. *mayteni*, *Pestalotiopsis* sp., *Diaporthe melonis*, *Phomopsis* sp., *Endomelanconiopsis endophytica*, *Diaporthe citri*, *Beltrania* sp., *Colletotrichum* sp., *Colletotrichum karstii*, *Colletotrichum alienum*, *Guignardia mangiferae*, *Aspergillus* sp. *Pestalotiopsis* sp., *Trichoderma* sp., *Xylaria* sp., *Botryosphaeria mamane*, *Colletotrichum fructicola*, *Colletotrichum siamense*, *Diaporthe* sp., *Fusarium* sp., *Penicillium citrinum*, *Penicillium* sp., *Pilidiella wangiensis*, and *Xylaria cubensis*	*Carapa guianensis* of municipality of São João da Baliza, from a particular area in Roraima State at the Amazonian region in Brazil	Healthy specimens of leaves	Ferreira et al. (2015)
71.	*Cladosporium sphaerospermum*, *Colletotrichum gloeosporioides*, *Coll. lindemuthianum*, *Aspergillus flavus*, *Nigrospora* sp., *Nig. sphaerica*, *Fusarium* sp., *F. semitectum*, *Penicillium* sp., *Curvularia borreriae*, *Phoma glomerata*, *Phomopsis* sp., *Pho. archeri*, *Alternaria raphani*, *Monodictys paradoxa*, *Mucor hiemalis*, *Curvularia borreriae*, *Drechslera australiensis*, and *Trichoderma harzianum*	*Ocimum sanctum*, *Vitex negundo*, and *Barleria prionitis* of Yashwantrao Mohite College, Bharati Vidyapeeth Deemed University, in India	Healthy specimens of leaves, stems, and roots	Desale (2016)
72.	*Albonectria rigidiuscula*, *Alternaria porri*, *Colletotrichum karstii*, *Coll. tropicicola*, *Coll. thailandicum*, *Coll. magnisporum*, *Corynespora cassiicola*, *Clonostachys agraualii*, *Diaporthe actinidiae*, *Di. kyushuensis*, *Guignardia mangiferae*, *Leptostroma* sp., *Neonectria macroconidialis*, *Nemania diffusa*, *Phomopsis asparagi*, *Pho. vaccinii*, *Pho. fukushii*, *Periconia* sp., *Penicillium citrinum*, *Pen. copticola*, *Pen. soppii*, and *Trametes polyzona*	*Cephalotaxus hainanensis* Li. of (1) Hainan Tropical Botanical Garden (codes A and B), (2) Hainan Jianfengling Nature Reserve (codes C, D, and E), and (3) Bawangling Nature Reserve (code F) in China	Healthy specimens of bark	Liu et al. (2016)
73.	*Periconia* sp.	*Annona muricata*		Zhang et al. (2016)

(continued)

Table 9.1 (continued)

No.	Fungal isolates	Medicinal plant	Parts of plant	References
74.	*Acremonium* sp., *Diaporthe* sp., *Pilidiella* sp., *Chaetomium* sp., *Colletotrichum* sp., *Phomopsis* sp., *Fusarium* sp., *Gibberella* sp., *Nodulisporium* sp., *Penicillium* sp., *Leptosphaeria* sp., *Corynespora* sp., *Mortierella* sp., and *Gongronella* sp.	*Melastoma malabathricum* L. of Phawngpui National Park, Mizoram, at the highest mountain peak (rising about 2157 m high near the Myanmar border), Northeast India	Leaves, roots, and stems	Mishra et al. (2016)
75.	*Chaetomium globosum*	*Houttuynia cordata* Thunb of Yaan City, at southwest China	Healthy specimens of tubers, leaves, and blades	Pan et al. (2016)
76.	*Penicillium oxalicum*, *Pestalotiopsis neglecta*, *Alternaria alternata*, and *Daldinia* sp.	*Cupressus torulosa* of Pauri Garhwal region, in India	Mature and healthy specimens of leaves	Sharma et al. (2016)
77.	*Acremonium* sp., *Alternaria* sp., *Aspergillus* sp., *Bipolaris* sp., *Chaetomium* sp., *Cladosporium* sp., *Colletotrichum* sp., *Curvularia* sp., *Cylindrocephalum* sp., *Drechslera* sp., *Fusarium* sp., *Lasiodiplodia* sp., *Mucor* sp., *Myrothecium* sp., *Nigrospora* sp., *Paecilomyces* sp., *Penicillium* sp., *Pestalotiopsis* sp., *Phoma* sp., *Phomopsis* sp., *Pithomyces* sp. *Rhizopus* sp., *Sordaria* sp., *Torula* sp., and *Trichoderma* sp.	*Tinospora cordifolia*, *Piper nigrum* L., *Piper longum* L., *Zingiber officinale*, *Hedychium coronarium*, and *H. flavescens* of Bisle region, Western Ghats of Karnataka in India	Healthy specimens of leaf, petiole, stems, and roots	Uzma et al. (2016)
78.	*Trichoderma gamsii* YIM PH30019	*Panax notoginseng* (2 years old) from Wenshan, China	Healthy specimens of roots	Chen et al. (2016 a)
79.	*Diaporthe* sp., *Phoma* sp., *Bipolaris* sp., and *Saccharicola* sp.	*Luehea divaricata* (traditional Brazilian medicine), *Sapindus saponaria* (American Southwest), *Trichilia elegans* (traditional Brazilian medicine), and *Saccharum* spp. (Philippine medicinal plants) from Microbial Biotechnology of Universidade Estadual de Maringá, Maringá, PR, Brazil	–	Alberto et al. (2016)

(continued)

Table 9.1 (continued)

No.	Fungal isolates	Medicinal plant	Parts of plant	References
80.	*Trichoderma asperellum*, *Penicillium radicum*, *Fusarium culmorum*, *Aspergillus flavus*, and *A. fumigatus*	*Carpobrotus edulis* of fully matured plant in Atlantic coast region of Rabat City in Morocco	Whole healthy plant	Idrissi et al. (2016)
81.	*A. niger*	*Terminalia catappa* Linn. (tropical almond) of the University of Yaoundé 1 main campus, in Cameroon	Healthy specimens of roots	Toghueo et al. (2016 a)
82.	*Trichoderma atroviride*	*Terminalia catappa* of University of Yaoundé 1, in Cameroon	Bark	Toghueo et al. (2016 b)
	Penicillium sp., *Colletotrichum gloeosporioides*, *Cladorrhinum* sp., *Colletotrichum* sp., *Phomopsis* sp., *Fusarium oxysporum*, *Fusarium lateritium*, *Curvularia lunata*, *Fusarium* sp., *Aspergillus* sp., *Botryosphaeria* sp., *Diaporthe* sp., *Purpureocillium lilacinum*, and *Corynespora cassiicola*	*Oroxylum indicum* of Gauhati University (GU) campus, Assam, in India	Healthy specimens of leaf, stem, root bark, seed, and stem bark	Das and Narzary (2017)
83.	*Alternaria* sp., *Aspergillus flavus*, *Cladosporium* sp., *Colletotrichum truncatum*, *Fusarium* sp., and *Penicillium* sp.	*Azadirachta indica* of (Location A) Zalta corner, (Loc. B) the Shendra MIDC, (Loc. C) Osmanpura, (Loc. D) Bidkin area, and (Loc. E) Khultabad in India	Leaves, stem, and petiole	Taware et al. (2017)
84.	*Phoma* sp., *Bipolaris* sp., *Alternaria* sp., and *Cladosporium* sp.	*Caralluma acutangula*, *Moringa peregrine*, and *Rhazya stricta* of Jabal Al Akhdar at the Sultanate of Oman	Stems, roots, and leaves	Khan et al. (2017a)

(continued)

Table 9.1 (continued)

No.	Fungal isolates	Medicinal plant	Parts of plant	References
85.	*Cladosporium* sp., *Clad. perangustum*, *Phoma* sp., *Phoma herbarum*, *Epicoccum* sp., *Alternaria alternata*, *Rhytidhysteron* sp., *Trichosporon asahii*, *Fomitopsis* sp., *Schizophyllum commune*, *Mortierella alpina*, *Mucor circinelloides*, *Aureobasidium pullulans*, *Rhexocercosporidium* sp., *Phialophora mustea*, *Cryptosporiopsis radicicola*, *Talaromyces funiculosus*, *Tala. pinophilus*, *Tala. verruculosus*, *Penicillium spinulosum*, *Pen. toxicarium*, *Pen. sclerotiorum*, *Pen. coffeae*, *Sagenomella* sp., *Aspergillus flavus*, *As. versicolor*, *Lasiodiplodia theobromae*, *Chaetomium aureum*, *Phialocephala humicola*, *Chaetosphaeria* sp., *Colletotrichum simmondsii*, *Purpureocillium lilacinum*, *Trichoderma* spp., *T. asperellum*, *Hypocrea nigricans*, *Myrothecium verrucaria*, *Metarhizium anisopliae*, *Fusarium solani*, *F. oxysporum*, *Trichosporon asahii*, and *Nectria haematococca*	*Sophora tonkinensis* collected in three periods of three different localities of traditional geo-authentic-producing areas in Guangxi Province of south China: (1) Tiandeng county, (2) Jingxi county, (3) Guangxi university	Healthy specimens of roots in three periods	Yao et al. (2017)
86.	*T. koningiopsis* QA-3	*Artemisia argyi* from Qichun of the Hubei Province in the Central China	Inner tissues	Shi et al. (2017)
87.	*Aspergillus japonicus*, *A. terreus*, *A. sydowii*, *Curvularia lunata*, *Nigrospora* spp., *Colletotrichum gloeosporioides*, *Rhizoctonia* spp., *Xylaria* spp., *Trichoderma* spp., *Fusarium chlamydosporum*, *F. oxysporum*, *Penicillium citrinum*, *Pen. purpurogenum*, *Pen. chrysogenum*, *Helminthosporium* spp., *Curvularia* spp., *Cladosporium* sp., *Alternaria alternata*, *Colletotrichum truncatum*, *Bipolaris* spp., *Talaromyces rotundus*, and *Cylindrocephalum* spp.	*Cymbidium aloifolium* L. from Western Ghats of Karnataka comprising Kemmanagundi, Chikkamagaluru, Shivamoga, and Sringeri in India	Leaf, flower, and roots	Shubha and Srinivas (2017)

(continued)

Table 9.1 (continued)

No.	Fungal isolates	Medicinal plant	Parts of plant	References
88.	*Cladosporium cladosporioides*, *Clad. oxysporum*, *Phialophora mustea*, *Phomopsis columnaris*, *Aspergillus versicolor*, *Penicillium chrysogenum*, *Pen. crustosum*, *Pen. cordubense*, *Pen. lapidosum*, *Colletotrichum gloeosporioides*, *Botryotinia fuckeliana*, *Acremonium implicatum*, *Fusarium acuminatum*, *F. oxysporum*, *F. solani*, *Ilyonectria macrodidyma*, *Myrothecium* sp., *Plectosphaerella* sp., *Trichoderma Longibrachiatum*, *T. koningiopsis*, *T. spirale*, *Periconia byssoides*, *Alternaria* sp., *Alt. alternata*, *Alt. tenuissima*, *Dictyosporium digitatum*, *Phoma draconis*, *Pho. radicina*, *Chaetomium globosum*, *Humicola fuscoatra*, *Thielavia arenaria*, *Arthrinium arundinis*, *Pestalotiopsis vismiae*, *Pest. uvicola*, and *Mucor hiemalis*	*Panax notoginseng* (3 years old) of a plantation in Wenshan, Yunnan, Southwest China	Healthy specimens of seed, leaf, stem, and roots	Zheng et al. (2017)
89.	*Absidia* sp., *Aspergillus* sp., *Cladosporium* sp., *Cunninghamella* sp., *Fusarium* sp., *Nigrospora* sp., *Paecilomyces* sp., *Penicillium chrysogenum*, and *Rhizopus* sp.	*Meyna spinosa* from Jorhat district, Assam, India	Healthy specimens of leaves, stems, and roots	Bhattacharyya et al. (2017)
90.	*Trichoderma* sp., *Fusarium* sp., *Umbelopsis* sp., *Irpex* sp., *Peniophora* sp., *Penicillium* sp., *Phomopsis*, *Ascomycota*, *Phoma*, *Alternaria* sp., *Geomyces* sp., *Bjerkandera* sp., *Ceratobasidium* sp., *Ceriporia* sp., *Hydnochaete* sp., *Resinicium* sp., *Mortierella* sp., *Mucor* sp., *Umbelopsis* sp., and *Zygorhynchus* sp.	*Panax ginseng* Meyer (4-year-old mountain-cultivated ginseng plants) from mountain areas of 24 different sites in Republic of Korea during the growing season	Healthy specimens of leaves, stems, and roots	Park et al. (2017)
91.	*Rhizoctonia* sp., *Colletotrichum* sp., *Mucor* sp., *Phomopsis* sp., *Aspergillus* sp., *Penicillium* sp., *Pestalotiopsis* sp., *Fusarium* sp., and *Cunninghamella* sp.	*Cinnamomum mercadoi* of Barangay Patag, at Baybay, Leyte, Philippines	Healthy specimens of bark	Marcellano et al. (2017)

(continued)

Table 9.1 (continued)

No.	Fungal isolates	Medicinal plant	Parts of plant	References
92.	*Setosphaeria rostrata* (teleomorph of *Exserohilum rostratum*), *Botryosphaeria dothidea*, *Phomopsis phyllanthicola*, *Albonectria rigidiuscula*, *Xylaria feejeensis*, *Phomopsis* sp., and *Fusarium* sp.	*Setosphaeria rostrata* of urban home garden at Colombo, Sri Lanka	Healthy and fresh specimens of leaf and stem	Centko et al. (2017)
93.	*Aspergillus* sp. and *Penicillium* sp.	*Calotropis procera* from Udaipur Rajasthan, in India	Leaf, flower, and stem	Nagda et al. (2017)
94.	*Aplosporella prunicola*, *Botryosphaeria* sp., *Bot. dothidea*, *Diaporthe* sp., *Di. arctii*, *Di. ceratozamiae*, *Di. eres*, *Di. hordei*, *Di. longicolla*, *Di. novem*, *Neofusicoccum australe*, *Neo. parvum*, *Biscogniauxia mediterranea*, *Chaetomium strumarium*, *Daldinia loculata*, *Nigrospora* sp., *Nig. oryzae*, *Nig. sphaerica*, *Pestalotiopsis* sp., *Sordaria fimicola*, *Thielavia arenaria*, *Thie. microspora*, *Candida* sp., *Colletotrichum capsici*, *Coll. gloeosporioides*, *Microascus intricatus*, *Nectria mauritiicola*, *Simplicillium lamelicola*, *Stachybotrys longispora*, *Camarosporium brabeji*, *Coniothyrium* sp., *Curvularia geniculata*, *Cur. intermedia*, *Cur. spicifera*, *Epicoccum nigrum*, *Leptosphaerulina* sp., *Microdiplodia hawaiiensis*, *Neoplatysporoides aloicola*, *Paraphoma chrysantemicola*, *Phoma* sp., *Pho. tracheiphila*, *Stemphylium* sp., *Stemphylium solani*, and *Tremateia* sp.	*Artemisia absinthium*, *A. austriaca*, *A. vulgaris*, *A. subulata*, *A. lavandulifolia*, *A. tangutica*, *A. argyi*, *A. scoparia*, *A. gorgonum*, *A. brachyloba* and *A. thuscula* collected from Romania, China – Wuhan and Qichun (10), Canary Islands (Fuerteventura, La Palma and Tenerife), and Cabo Verde	Healthy specimens of stem and plant fragment	Cosoveanu et al. (2017)
95.	*Alternaria alternata*, *Penicillium citrinum*, *Aspergillus niger*, *Cladosporium* sp., *Rhizopus* sp., *Curvularia vermiformis*, and *Fusarium* sp.	*Helicteres isora* L. of different places from Udupi district at Karnataka State, in India	Healthy and mature specimens of leaf and stem	Pai and Chandra (2017)

(continued)

Table 9.1 (continued)

No.	Fungal isolates	Medicinal plant	Parts of plant	References
96.	*Trichoderma* sp. 307	*Clerodendrum inerme* of Zhanjiang Mangrove National Nature Reserve at Guangdong Province, in China	The stem bark	Zhang et al. (2017)
97.	*Lasiodiplodia* sp., *Colletotrichum* sp., *Fusarium* sp., *Trichoderma* spp., *Aspergillus* spp., *A. fumigatus*, *A. niger*, *A. terreus*, *Pestalotia* sp., *Curvularia* sp., and *Gliocladium* sp.	*Prunus africana* (Hook. F.) Kalkman of four sites in Cameroon: (A) Mount Cameroon Forest, (B) Limbe Botanic Garden in the South West Region, (C) Mount Kilum-Ijim forest, (D) Bamendankwe in the Bamenda highlands of North West region	Healthy specimens of leaf, stem, seeds bark, and roots	Ntuba-Jua et al. (2017)
98.	*Epicoccum nigrum*, *Colletotrichum gloeosporioides* f. sp. *camelliae*, *Peyronellaea glomerata*, *Botryosphaeria dothidea*, *Cladosporium asperulatum*, *Diaporthe eres*, *Alternaria mali*, *Guignardia mangiferae*, *Setophoma chromolaena*, *Diaporthe* sp., *Di. nobilis*, *Di. pustulata*, *Di. sackstonii*, *Phomopsis* sp., *Pho. amygdale*, *Pho. subordinaria*, *Acremonium strictum*, *Glomerella* sp., *Pestalotiopsis* sp., *Pest. camelliae*, *Paraphaeosphaeria neglecta*, *Nemania* sp., *Melanconiella* sp., *Microdiplodia hawaiiensis*, *Phoma herbarum*, *Phialemonium dimorphosporum*, *Plenodomus* sp., *Pseudocercospora* sp., *Stagonosporopsis cucurbitacearum*, *Trichoderma koningiopsis*, and *Peniophora incarnata*	Three Japanese tea cultivars *Camellia sinensis* (Hokumei, Sayamakaori, and Yabukita) of Japan	Healthy specimens of new leaf, old leaf, xylem, and bark	Win et al. (2018)

(continued)

Table 9.1 (continued)

No.	Fungal isolates	Medicinal plant	Parts of plant	References
99.	*Phoma* sp., *Diaporthe* sp., *Di. phaseolorum*, *Alternaria* sp., *Alt. alternata*, *Diaporthe novem*, *Nigrospora* sp., *Nig. oryzae*, *Camarosporium brabeji*, *Cam. bradgi*, *Coniothyrium* sp., *Aspergillus flavus*, *Tremateia* sp., *Neoplatysporoides aloicola*, *Paraphoma chrysanthemicola*, *Nectria mauritiicola*, *Stachybotrys longispora*, *Stemphylium solani*, *Aplosporella prunicola*, *Biscogniauxia mediterranea*, *Neofusicoccum* sp., *Neo. parvum*, *Neo. australe*, *Pestalotiopsis* sp., *Penicillium viridicatum*, *Aureobasidium pullulans*, *Cladosporium* sp., *Preussia* sp., *Pre. australis*, *Biscogniauxia mediterranea*, *Curvularia lunata*, *Thielavia* sp., and *Macrophomina phaseolina*	*Artemisia thuscula* of Canary Islands (La Palma and Tenerife)	Healthy specimens of stem	Cosoveanu et al. (2018)
100.	*Chaetomium globosum D38*	*Salvia miltiorrhiza* of Shangluo, Shanxi, in China	Roots	Zhai et al. (2018)
101.	*Neurospora* sp., *Trichoderma* sp., *Cladosporium* sp., *Penicillium* sp., *Phomopsis* sp., *Colletotrichum* sp., *Phoma* sp., *Mucor* sp., *Purpureocillium* sp., *Rhizoctonia* sp., *Cladosporium* sp., *Chaetomium* sp., *Talaromyces* sp., *Ceratobasidium* sp., *Rhizopus* sp., *Plectosphaerella* sp., *Fusarium* sp., *Alternaria* sp., and *Aspergillus* sp.	*Pelargonium sidoides* of commercial flower gardeners at different locations within four provinces of South Africa	Healthy specimens of roots	Manganyi et al. (2018)
102.	*Nigrospora sphaerica*, *Acremonium falciforme*, *Acrophialophora* sp., *Penicillium chrysogenum*, *Periconia hispidula*, *Allomyces arbuscular*, *Chaetomium* sp., and *Aureobasidium* sp.	*Litsea cubeba* of the Botanical Garden, at Department of Botany, in Gauhati University, at Guwahati, Assam, in India	Healthy specimens of leaves and barks	Deka and Jha (2018)
103.	*Aspergillus flavus* IBRL-C8	*Cassia siamea* Lamk. from Industrial Biotechnology Research Laboratory, School of Biological Sciences, Universiti Sains Malaysia, Penang, in Malaysia	Healthy specimens of leaves	Darah Ibrahim and Hong (2018)

(continued)

Table 9.1 (continued)

No.	Fungal isolates	Medicinal plant	Parts of plant	References
104.	*Aspergillus nidulans*, *A. flavus*, *A. niger*, *A. oryzae*, *A. focus*, *A. fumicates*, *A. rugulosus*, *A. terreus*, *Alternaria tenuis*, *Rhizopus* sp., *Penicillium notatum*, *P. citrinum*, *Helminthosporium* sp., *H. oryzae*, *Bipolaris* sp., and *Fusarium* sp.	*Ludwigia perennis*, *Ocimum basilium*, *Cissus quadrangulaircs*, *Alpinia galanga*, *Bryophyllum pinnatum*, *Caesalpinia pulcherrima*, *Piper betle*, *Sansevieria laxburghi* = Medicinal plant, *Cymbopogon citratus*, *Sauropus androgynus*, *Cicca acida*, *Rauvolfia tetraphyhlla*, *Ficus nervosa*, *Clitoria ternatea*, and *Tridax prochumbins* of STET Women's college, STET Herbal Garden, Mannargudi, Thiruvarur (DT), Tamil Nadu in India	Healthy and mature specimens of leaf	Uma Maheswari and Saranya (2018)

cent of the genus number isolated from the medicinal plants by using an equation, as follows:

$$\mathrm{EFG\%} = \frac{\mathrm{EFI}}{\mathrm{EFW}} \times 100$$

EFG % = The percentage of genus isolated from the medicinal plants depending on Table 9.1
EFI = Sum of one genus in Table 9.1
EFW = Sum of the medicinal plants in Table 9.1 that accounted about 205 plants

The percentage of genera of endophyte fungi isolated from the medicinal plants can be divided into three levels depending on Table 9.1 comprising the low (0.5–9.9% genera including between 1 and 19 isolates/genus), middle (10–19.9% genera including between 20 and 59 isolates/genus), and high (20–40% genera including between 60 and 100 isolates/genus) levels (Figs. 9.1, 9.2, and 9.3). This division is assuming to give an idea which genus is more colonizing to medicinal plants by the endophytic fungi. The low levels are showing many fungi genera comprising 166 genera including other fungi in Fig. 9.2 such as *Chaetomella*, *Nemania*, *Botryosphaeria*, *Guignardia*, *Phyllosticta*, *Rhizomucor*, *Xylaria*, *Rhizoctonia*, *Daldinia*, *Acremonium*, *Hypoxylon*, *Clonostachys*, *Coprinus*, *Drechslera*, *Ascomycota*, *Microascus*, *Mortierella*, *Cylindrocarpon*, *Phacidium*, *Beauveria*, *Arthrobotrys*, *Chaetophoma*, *Arthrinium*, *Nigrospora*, *Exserohilum*, *Talaromyces*, *Pestalotia*, *Phlyctaena* sp., *Psathyrella*, *Ulocladium*, *Periconia*, *Paecilomyces*,

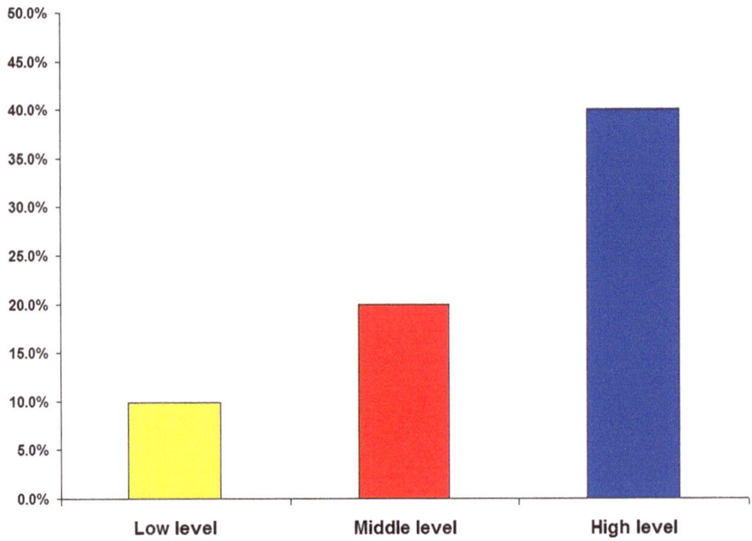

Fig. 9.1 Division of the fungi genera isolated from the medicinal plant into three levels: low (0.5–9.9%), middle (10–20%), and high levels (20.1–40%) depending on Table 9.1

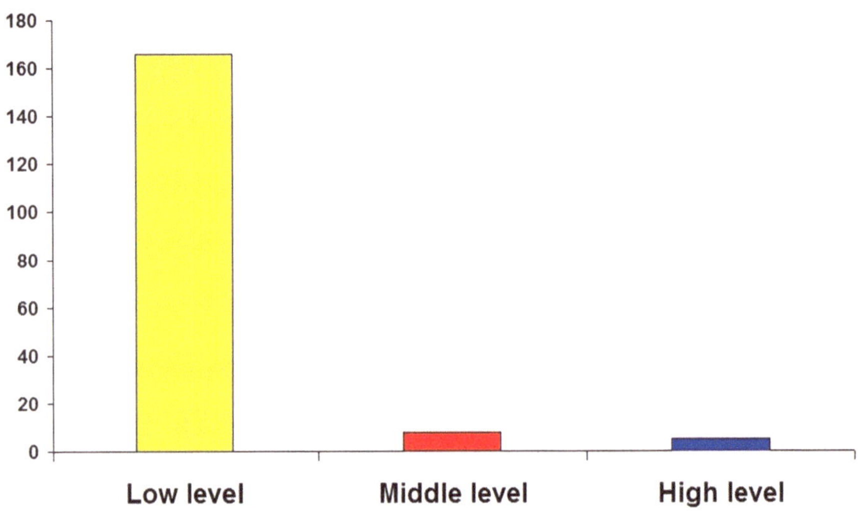

Fig. 9.2 Number the fungi genera isolated from the medicinal plant depending on Table 9.1 distributed into three levels: low, middle, and high levels. Low level = 166 genera, middle level = 8 genera, and high level = 5 genera

Nodulisporium, *Schizophyllum*, *Lasiodiplodia*, *Trametes*, *Phialemonium*, *Nectria*, *Leptodontidium*, *Botryodiplodia*, *Entrophospora*, *Phialocephala*, *Aschersonia*, *Rhizopus*, *Stachybotrys*, *Stenella*, *Paraconiothyrium*, *Phaeoacremonium*, *Coniochaeta*, *Coniothyrium*, *Pichia*, *Leptosphaerulina*, *Hypocrea*, *Emericella*, *Setophoma*, *Humicola*, *Taeniolella*, *Gliocladium*, *Monodictys*, *Cochliobolus*,

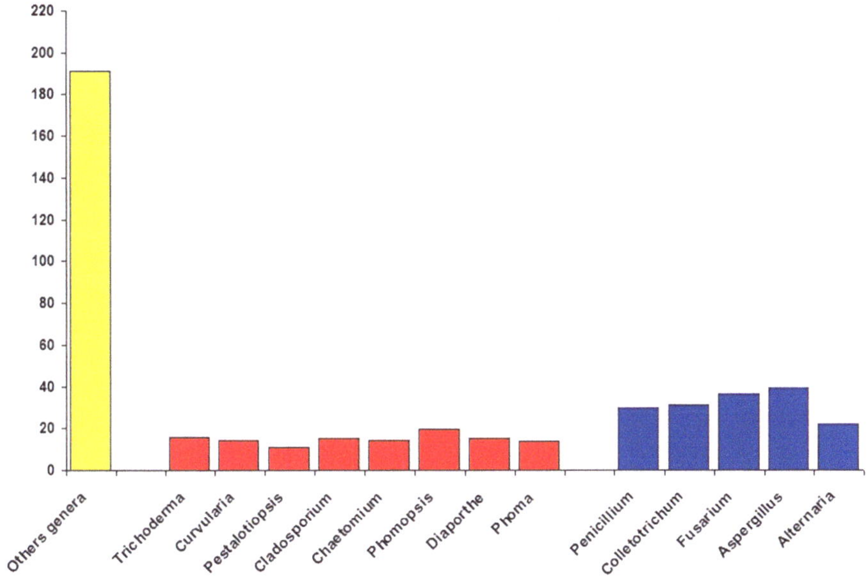

Fig. 9.3 The fungi genera distributed into three levels, low, middle, and high levels. Low level (yellow color) = 1–19 isolates, middle level (red color) = 20–59 isolates, and high level (blue color) = 60–100 isolates

Aureobasidium, *Glomerella*, *Ceratobasidium*, *Cordyceps*, *Nodulisporium*, *Mycoleptodiscus*, *Claviceps*, *Sphaeropsis*, *Preussia*, *Botrytis*, *Fusicoccum*, *Mucor*, *Gloeosporium*, *Haplosporidium*, *Stemphylium*, *Corynespora*, *Fomitopsis*, *Resinicium*, *Hydnochaete*, *Ceriporia*, *Trichosporon*, *Rhytidhysteron*, *Macrophoma*, *Lecythophora*, *Leptospira*, *Cerraena*, *Gibberella*, *Hansfordia*, *Cryptosporiopsis*, *Edenia*, *Phialophora*, *Myrothecium*, *Coprinopsis*, *Rhexocercosporidium*, *Syncephalastrum*, *Albonectria*, *Choanephora*, *Longibrachiatum*, *Pilidiella*, *Bipolaris*, *Helminthosporium*, *Verticillium*, *Macrophomina*, *Meyerozyma*, *Umbelopsis*, *Aplosporella*, *Eupenicillium*, *Setosphaeria*, *Zygorhynchus*, *Sympodiomyces*, *Endomelanconiopsis*, *Beltrania*, *Coniothyrium*, *Camarosporium*, *Peniophora*, *Leptostroma*, *Irpex*, *Neonectria*, *Leptosphaeria*, *Gongronella*, *Cylindrocephalum*, *Pithomyces*, *Plectosphaerella*, *Sordaria*, *Paraphaeosphaeria*, *Torula*, *Saccharicola*, *Epicoccum*, *Sagenomella*, *Chaetosphaeria*, *Purpureocillium*, *Metarhizium*, *Botryotinia*, *Ilyonectria*, *Dictyosporium*, *Thielavia*, *Absidia*, *Cunninghamella*, *Geomyces*, *Bjerkandera*, *Neofusicoccum*, *Biscogniauxia*, *Candida*, *Simplicillium*, *Microdiplodia*, *Paraphoma*, *Tremateia*, *Peyronellaea*, *Melanconiella*, *Phialemonium*, *Plenodomus*, *Pseudocercospora*, *Stagonosporopsis*, *Peniophora*, *Neoplatysporoides*, *Neurospora*, *Acrophialophora*, *Pseudofusicoccum*, *Podospora*, *Cytospora*, *Massarina*, *Muscodor*, *Sporormiella*, *Allomyces*, *Cladorrhinum*, and *Lophiostoma*, while 8 fungi genera were showed at the middle level such as *Trichoderma*, *Curvularia*, *Pestalotiopsis*, *Cladosporium*, *Chaetomium*, *Phomopsis*, *Diaporthe*, and *Phoma* (Fig. 9.2). The detected four fungi genera at the high level were comprised of *Penicillium*, *Alternaria*, *Colletotrichum*, *Fusarium*, and *Aspergillus* (Fig. 9.2).

9.3 The Role of Endophyte Fungi in Producing Different Compounds

Endophytic fungi of the medicinal plants are very interesting due to the several compounds produced such as secondary metabolites, enzymes, and others that can be utilized in various fields. The role of endophytic fungi isolated of the medicinal plants having the possibility to produce many novel compounds (Kaul et al. 2013). Endophytic fungi isolated from medicinal plants or other plants are utilized to produce many compounds including steroids, peptides, terpenoids, quinones, alkaloids, phenols, and flavonoids (Yu et al. 2010). The fields comprise agricultural, medical, and industrial. Endophyte fungus *Phomopsis archeri*a was able to produce various bioactive compounds comprising antifungal, antiprotozoal, anticancer, antiviral, antibacterial, and antioxidant activities (Desale 2016). *Penicillium copticola* was showing antimicrobial and anticancer activity (Liu et al. 2016).

9.3.1 Antimicrobial

The secondary metabolites of endophytic fungi isolated from *Cassia fistula* showed the antimicrobial activity against fungi and bacteria (Ruchikachorn 2005). *Fusarium proliferatum* of *Celastrus angulatus* was able to produce antimicrobial compounds (Ji et al. 2005). *Fusarium* sp. DL26 and *Pyrenochaeta* sp. DL351 produced antimicrobial against one bacterium (*Bacillus subtilis* As 1.308) and one fungus (*Aspergillus fumigatus* As 3.2910) (Chen et al. 2010). *Muscodor albus* strain GBA could produce the volatile compounds such as 3-methyl-, acetate, 1-butanol, and terpenoid that are affecting some fungi (*Candida albicans*, *Saccharomyces cerevisiae*, and *Aspergillus fumigatus*) and bacteria (*Bacillus subtilis* and *Escherichia coli*). Several endophytic fungi of agarwood could produce compounds having antimicrobial activity against *Escherichia coli*, *Bacillus subtilis*, *Aspergillus fumigatus*, and *Staphylococcus aureus* (Cui et al. 2011). Four species isolated from *Lippia sidoides* such as *Alternaria alternata*, *Phomopsis archeri*, *Drechslera dematioidea*, and *Colletotrichum gloeosporioides* confronted some bacteria and fungi by producing antimicrobial compounds (de Siqueira et al. 2011). The crude metabolite of *F. solani* detected several compounds such as 1-tetradecene, 8-octadecanone, 10-nonadecanone, 8-pentadecanone, and octylcyclohexane that were showing antimicrobial activity against *Staphylococcus aureus*, *St. epidermidis*, *Bacillus subtilis*, *Klebsiella pneumonia*, *Escherichia coli*, *Shigella flexneri*, *Candida albicans*, and *Ca. tropicalis* (Tayung et al. 2011). *Pestalotiopsis* sp. was able to inhibit the growth of some human microbes such as *Microsporum nanum* (51.4%), *M. gypseum* (48.5%), *Trichophyton rubrum* (49.7%), and *Pseudomonas fluorescence* (47.1%) (Kharwar et al. 2012).

However, the crude extract of *Nodulisporium* sp. PT11 was very efficient in inhibiting the bacteria and fungi (Pharamat et al. 2013). *Aspergillus awamori* showed the antimicrobial activity against several microbial pathogens such as bacteria (*Streptococcus pyogenes*, *Salmonella enterica* ser. *paratyphi*, *Escherichia coli*, and *Enterococcus faecalis*) and fungi (*Candida albicans*, *Fusarium oxysporum*, and *Emericella nidulans* var. *nidulans*) (Nath et al. 2013). The ethanolic extract of *Aspergillus oryzae* CeR1 inhibited the growth of many microbes including fungal strain *Candida albicans* and some bacteria such as *Staphylococcus aureus*, *Enterococcus faecalis*, and *Streptococcus pyogenes* (Nath et al. 2014). *Penicillium chrysogenum* Pc_25, *Alternaria alternata* Aa_27, and sterile hyphae Sh_26 showed the antimicrobial activity against human microbe pathogen both of bacteria (*Staphylococcus aureus*, *Salmonella typhimurium*, *Escherichia coli*, *Bacillus subtilis*, and *Pseudomonas aeruginosa*) and one fungal *Candida albicans* (Fouda et al. 2015). Some endophytic fungi isolated from *Edgeworthia chrysantha* showed antimicrobial activity against human pathogen *Candida albicans*, *Escherichia coli*, and *Staphylococcus aureus* (Zhang et al. 2015).

9.3.2 Antiviral

Antiviral compounds that are produced from endophytic fungi are very interesting to confront some viral diseases. The endophytic fungus *Cytonaema* sp. isolated from the medicinal plant *Quercus* sp. produced two novel tridepside inhibitors such as (1) cytonic acids A and (2) cytonic acids B for protease human cytomegalovirus (HCMV) (Guo et al. 2000). Endophytic fungus *Emericella* sp. (HK-ZJ) was isolated from *Aegiceras corniculatum* that could secrete some compounds as antiviral (anti-influenza A viral (H1N1)) including six isoindolone derivatives termed as (1) emeriphenolicins (A and D), (2) emerimidine (A and B) with (1) aspernidines A and B, (3) austin, (4) austinol, (5) dehydroaustin, and (6) acetoxydehydroaustin (Zhang et al. 2011). Endophyte fungi *Periconia* spp. produced two anti-HIV compounds (1) periconiasin H and (2) periconiasin G (Zhang et al. 2016). *Guignardia mangiferae*, *Pestalotiopsis* sp., *Diaporthe melonis*, and *Colletotrichum* sp. affected and inhibited the proliferation of yellow fever virus (Ferreira et al. 2015).

9.3.3 Antifungal

Endophytic fungi are able to produce antifungals against many fungi. *Xylaria* sp. produced two compounds of eremophilane sesquiterpenes (3) phaseolinone and (4) phomenone tested against two strains of *Cladosporium* such as *C. sphaerospermum* (Perzig) SPC 491 and *C. cladosporioides* (Fresen) de Vries SPC 140 (Silva et al. 2010). *Colletotrichum dematium* could affect the growth of some fungi including *Alternaria alternata*, *Curvularia lunata*, *Cladosporium cladosporioides*, *Fusarium udum*, and *Microsporum gypseum* (Gond et al. 2012). An endophytic fungus of *Nigrospora* cf. *oryzae* was secreting the antifungal that is able to inhibit the growth of two fungi *Cladosporium sphaerospermum* and *Candida albicans* (Carvalho et al. 2012). The crude extracts of endophytic fungi from *Baccharis trimera* exhibited antifungal activity against *Cryptococcus neoformans*, *C. gattii*, and *Paracoccidioides brasiliensis* (Vieira et al. 2014). *Phomopsis* sp. was showing antifungal activity against five fungi including *Aspergillus niger*, *Hormodendrum compactum*, *Pyricularia oryzae*, *Candida albicans*, and *Fusarium avenaceum* (Huang et al. 2018). *Aspergillus flavus* IBRL-C8 was having the ability to affect on the growth of *Candida albicans* (Darah Ibrahim and Hong 2018).

9.3.4 Antibacterial

Some of the compounds extracted from endophytic fungi show antibacterial activity. Four endophytic fungi of *Adenocalymma alliaceum* such as *Alternaria alternata*, *Chaetomium globosum*, *Curvularia lunata*, and *Penicillium* sp. were showing antibacterial activity against four human bacteria comprising (A) *Shigella flexneri*, (B) *Salmonella enteritidis*, (C) *S. paratyphi*, and (D) *Pseudomonas aeruginosa* but could not inhibit one human bacterium *Morganella morganii* (Kharwar et al. 2011). The media extraction of two endophytic fungi such as *Phomopsis* sp. and *Cochliobolus intermedius* of *Sapindus saponaria* L. (Tree) showed antibacterial activity against human bacterial pathogens (Garcia et al. 2012). The strain BH-3 secreted protein at molecular mass 55 KD as antibacterial (Liu et al. 2012). Fermentation aliquots of two *Aspergillus* species such as *A. niger* and *A. flavus* were affected by the human bacterial pathogen *Staphylococcus aureus* (Basha et al. 2012). The culture filter of *Aspergillus protuberus* isolate Pg30 inhibited for bacteria *Staphylococcus aureus*, also the fungus *Verticillium* sp. isolate Pg42-1 wasan able of high inhibitory for bacteria *Klebsiella pneumoniae* (Wu et al. 2013). Three bacteria such as *Pseudomonas aeruginosa*, *Escherichia coli*, and *Staphylococcus aureus* inhibited by treating with the activity compound of antibacterial from endophyte fungus *Diaporthe phaseolorum* var. *meridionalis* (Mishra et al. 2016).

In addition, the crude extract of endophytic fungi isolated from *Prosopis chilensis* showed antibacterial activity against *Pseudomonas aeruginosa*, *Escherichia coli*, *Staphylococcus aureus*, *Klebsiella pneumonia*, and *Salmonella typhi* (Mahdi et al. 2014). *Gibberella baccata*, *Aspergillus ochraceus*, *Penicillium commune*, and *P. glabrum* showed antibacterial activity (Bezerra et al. 2015). Many endophytic fungi such as *Aspergillus* sp. (MIL04), *Aschersonia* sp. (MIL13), *Botryosphaeria rhodina* (MIB01), *Pleospora* sp. (MIB04), and *Fusarium* sp. (MIB05) showed efficacy in reducing the growth of human bacterial pathogens both of Gram-negative and Gram-positive species at range 50–70% (Verma et al. 2014). *Cladosporium* sp. and *Alternaria alternata* showed the antibacterial activity by producing some compounds such as saponins, phenolics, steroids, anthraquinones, tannins, and cardiac glycosides (Selvi and Balagengatharathilagam 2014). Several activity compounds of antibacterial that affected *Escherichia coli*, *Salmonella typhimurium*, *Staphylococcus aureus*, and *Bacillus subtilis* extracted from endophytic fungi *Pestalotiopsis neglecta* (Sharma et al. 2016).

Trichoderma koningiopsis QA-3 secreted seven antibacterial compounds against eight human bacterial pathogens including *Escherichia coli* and seven bacteria of marine-derived aquatic (*Pseudomonas aeruginosa*, *Vibrio alginolyticus*, *V. anguillarum*, *V. parahaemolyticus*, *V. vulnificus*, *Edwardsiella tarda*, and *Micrococcus luteus*) (Shi et al. 2017). The endophyte fungi showed antibacterial activity against some pathogenic bacteria by using the EtOAc extract (Marcellano et al. 2017). The crude metabolites of *Acremonium falciforme* showed the antibacterial activity against *Staphylococcus epidermidis* (MTCC 435) at 12.3 ± 0.50 mm (inhibition zone) (Deka and Jha 2018).

9.3.5 Antioxidant

The endophytic fungi could produce the novel natural antioxidants and correlated with medicinal plant in producing several bioactive compounds including phenolic acids, flavonoids, quinones, volatile compounds, and aliphatic compounds (Huang et al. 2007). Also, Mahdi et al. (2014) found the phenolic content and antioxidant activity of endophytic fungi. *Chaetomium* sp., *Aspergillus* sp., *A. niger*, and *A. peyronelii* were the sources of antioxidant activity (Yadav et al. 2014). *Penicillium* sp. and *Aspergillus* sp. might a potential source for several the antioxidant compounds by producing the compounds of the phenolics and flavonoids (Nagda et al. 2017).

9.3.6 Anticancer

F. oxysporum from *Juniperus recurva* produced podophyllotoxin (Kour et al. 2008). A Toxal compound was produced by some endophytic fungi comprising *Colletotrichum gloeosporioides* (Gangadevi and Muthumary 2008; Nithya and Muthumary 2009), a strain TAC-15 of *Chaetomella raphigera* (Gangadevi and Muthumary 2008), *F. culmorum* SVJM072 (Sonaimuthu et al. 2010), *Pestalotiopsis versicolor* and *Pe. neglecta* (Kumaran et al. 2010), and *Fusarium redolens* (Garyali et al. 2014). Podophyllotoxin compound could be produced by several endophytic fungi such as *Trametes hirsute* (Puri et al. 2006), *Phialocephala fortinii* (Eyberger et al. 2006), *Fusarium oxysporum* (Kour et al. 2008), and *F. solani* (Nadeem et al. 2012). It is a very important compound in preparing anticancer drug.

Indeed, the extract of two endophyte fungi such as *Xylaria* sp. and *Diaporthe* cf. *phaseolorum* exhibited anticancer activity by showing the cytotoxic activity against tumor cells of the kidney (Carvalho et al. 2012). *Colletotrichum gloeosporioides* affected the cancer cells of human colon and caused 52% cytotoxicity for the cancer cells (Lakshmi and Selvi 2013). The culture broth of *Phoma* sp. PT01 was extracted and used against the cancer cells that showed cytotoxicity and apoptosis (Pharamat et al. 2013). *Chaetomium globosum* produced chaetoglobosins having cytotoxicity activity against a human colon cancer cell (Li et al. 2014a, b). *Alternaria* isolate C-9 secreted paclitaxel compound as antitumor (Michalczyk et al. 2015). Also, four depsidones such as botryorhodines H, C, D, and G were extracted from *Trichoderma* sp. 307, and a compound I showed high cytotoxicity against rat prolactinoma (Zhang et al. 2017). The endophyte strain of *Phomopsis* sp. YE3250 produced seven compounds which are called phomopoxides A–G (1–7) showing cytotoxicity against the human tumor cell (Huang et al. 2018).

9.3.7 Insecticidal

Mycoleptodiscus indicus could produce compounds that showed larvicidal activity against *A. aegypti* (Rosa et al. 2012). The different compounds that extraction from cultural filter of the endophyte fungus *Claviceps purpurea* impacted on *Aphis gossypii* Glover (Cotton aphid) and they were caused mortality rates more than 90% (Shi et al. 2013).

9.3.8 Mycotoxins

The synthetic of *Datura stramonium* L. could induce the dormant gene in *Alternaria* sp. to produce mycotoxin comprising alternariol, altenusin, 30-hydroxyalternariol-5-O-methyl ether, tenuazonic acid, alternariol-5-O-methyl ether, and altertoxin II (Sun et al. 2012). The strain MAFF744001 of *Fusarium oxysporum* f. sp. *conglutinans* of *Datura stramonium* L. could produce fusaric acid by promoting the dormant gene and using the chemical epigenetic modifiers (Chen et al. 2013). *Chaetomium globosum* produced five mycotoxins comprising chaetoglobosins A, G, V, Vb, and C (Li et al. 2014a, b).

9.3.9 Enzymes

Endophytic fungi can secrete many enzymes. Endophytic fungi of five medicinal plants secreted five enzymes amylase, cellulase, protease, lipase, and xylanase (Bhagobaty and Joshi 2012). *Penicillium* sp. nirjan22 of *Centella asiatica* secreted cellulase enzyme (Devi et al. 2012). The two enzymes of amylase and lipases were detected in culture of *Geotrichum* sp. (Mbouobda et al. 2014). Endophytic fungi were isolated from *Bauhinia forficata* showing proteolytic activity, lipolytic activity, and positivity with cellulase and xylanase (Bezerra et al. 2015). Three endophytic fungi of *Asclepias sinaica* such as *Penicillium chrysogenum* Pc_25, *Alternaria alternata* Aa_27, and sterile hyphae Sh_26 were secreted by some extracellular enzymes comprising amylase, tyrosinase, pectinase, gelatinase, cellulase, and xylanase (Fouda et al. 2015). Many endophytic fungi of six wild medicinal plants could secrete four extracellular enzymes comprising cellulase, amylase, asparaginase, and pectinase (Uzma et al. 2016). Many extracellular enzymes such as amylase, cellulase, pectinase, and protease were detected in crude extracted from various endophytic fungi (Alberto et al. 2016). *Aspergillus niger* secreted several secondary compounds including (1) oxalic acid, (2) isobutyl propyl ester, (3) cyclohexanecarboxaldehyde,3,3-dimethyl-5-oxo-, (4) 9-octadecenoic acid (Z)-, and (5) tetradecanoic acid, 12-methyl-, methylester (Toghueo et al. 2016a).

In addition, several of endophytic fungi isolation from the flower of *Cymbidium aloifolium* L. produced many extracellular enzymes including amylase, laccase, lipase, and phosphatase but more endophytic isolated of root same the host plant producing three extracellular enzymes comprising protease, cellulase, and pectinase (Shubha and Srinivas 2017). Endophytic fungi of three medicinal plants were able to secrete extracellular enzymes including amylase, laccase, cellulase, and lipase (Toghueo et al. 2017). Several species of Endophytic fungi that isolated of the medicinal plant *Azadirachta indica* secreting many enzymes but different among species included (1) detected in the crude extract of *Alternaria* sp. five enzymes as amylase, laccase, cellulase, tyrosinase, xylanase but two enzymes of *Alternaria* sp. such as asparaginase and cellulase. (2) *Aspergillus flavus* produced three enzymes such as amylase, xylanase, and cellulase. (3) *Cladosporium* sp. produced two enzymes such as amylase and cellulase. (4) *Colletotrichum truncatum* secreted two enzymes as cellulase and tyrosinase. (5) *Fusarium* sp. and *Penicillium* sp. could produce only one enzyme cellulase (Taware et al. 2017). Endophyte *Alternaria alternata* RSF-6L was able to protect *Solanum nigrum* from the damages of oxidative stress that happens by accumulation of cadmium (CD) through reduction in activities of polyphenol peroxidase (PPO) and peroxidase (POD) with increased activity of catalase (CAT) (Khan et al. 2017b).

9.3.10 Different Compounds Used in Various Fields

Some endophytic fungi produce compounds that can be used in different fields. Cui et al. (2012) found Ginkgolide B was extracted from *Fusarium oxysporum*. The phenolic and antioxidant were produced by *Aspergillus awamori* (Nath et al. 2013). Endophytic fungi can produce several phytochemicals that can be used in industry, pharmacy, agriculture, etc. Devi et al. (2012) detected several phytochemical compounds including cardiac glycosides, flavonoids, steroids, alkaloids, phenolic compounds, tannins, and phenolic compounds in crude extract of *Penicillium* sp. Also, *Fusarium* isolate Pg27 produced saponins (Wu et al. 2013). *Fusarium* sp. produced two alkaloid compounds like its host plant (*Fritillaria unibracteata* var. *wabensis*) such as peimisine and peiminine (Pan et al. 2014). Endophytic fungi of *Nothophoma multilocularis* of medicinal plant *Rhazya stricta* could produce many compounds including di-n-octyl phthalate, 2-allyl-3,4-dimethoxybenzaldehyde, maltol, cetene, 1-tetradecene, E-15-heptadecenal, 2,5-cyclohexadien-1-one, 1-octadecene, diethyldithiophosphinic acid, and phenol, 2,4-di-t-butyl-6-nitrophenol (Abdel-Wahab et al.

2017). *Geotrichum* sp. of plant *Lantana camara* produced the tannin and cardiac glycoside compounds (Mbouobda et al. 2014). Two endophytic fungi such as *Endomelanconiopsis endophytica* and *Diaporthe* cf. *mayteni* were able to antagonistic *Trypanosoma cruzi* (Protozoan parasite) by producing trypanocidal that is high effecting on amastigote forms (Ferreira et al. 2015). Many endophyte fungi isolated from *Helicteres isora* L. showed a positive reaction with some phytochemicals such as tannins, phenolics, carbohydrate compounds, and alkaloids (Pai and Chandra 2017). Endophytic fungi such as *Setosphaeria rostrata* secreted three new thiodiketopiperazine derivatives, (1) rostratazine A, (2) rostratazine C, and (3) rostratazine B, along with (4) exserohilone and (5) boydine A (Centko et al. 2017).

9.4 The Role of Endophyte Fungi in Control of the Plant Pathogens

Endophytic fungi have the ability to confront the plant pathogens by inhibiting the mycelium growth directly and indirectly. The ability of endophytic fungi in the biocontrol of plant pathogens is very interesting as they decrease the residue of pesticides and other synthetic chemicals (Al-Ani 2017b, 2018a, b). The contamination with many dangerous chemicals inside of medicinal herbs is mention by review of Tripathy et al. (2015) found several toxic heavy metals including Cd, As, Pb and Cr; persistent pesticides, particularly organochlorine, mycotoxins, organophosphates, fumigants, and polycyclic aromatic hydrocarbons. Also, the secondary compounds of endophytic fungi could enhance the resistance of medicinal plants against abiotic and biotic stresses such plant pathogens (Chowdhary et al. 2012). Directly, some endophytic fungi attack the plant pathogens through confronting in dual culture test and are able to inhibit the mycelium growth of plant pathogens by direct contact. The direct effect meaning a mycoparasitism that it fungus can contact and penetrate the hyphae for another fungus and kill it. Indirectly, the fungus can inhibit the growth of another fungus without direct contact including (1) producing the antibiotics such as enzymes and volatile and nonvolatile compounds and (2) inducing the defense of plants against plant pathogens.

9.4.1 The Direct Antagonistic

Paecilomyces lilacinus, *Penicillium* sp., and *Pen. copticola* were highly antagonistic in different media against *Trichothecium roseum* and *Botrytis cinerea* (Kusari et al. 2013). Three endophyte fungi *Alternaria*, *Fusarium*, and *Cladosporium* inhibited the growth of two plant pathogens such as *Phytophthora cactorum* and *Phy. capsici* and showed very efficacy in a confrontation by testing in the dual culture (Michalczyk et al. 2015). The isolate *Trichoderma atroviride* from the medicinal plant *Terminalia catappa* inhibited the spore germination and mycelium growth of *Fusarium solani* cause root rot disease for common bean (*Phaseolus vulgaris* L.) and the detection was execution by two methods such as 1. spore-spore confrontation and 2. dualculture (Toghueo et al. 2016b). The *Trichoderma gamsii* strain YIM PH30019 of the root of medicinal plant *Panax notoginseng* root was affected on four phytopathogenic of *P. notoginseng* including (A) *Epicocum nigrum*, (B) *Scytalidium lignicola*, (C) *Phoma herbarum*, and (D) *Fusarium flocciferum* by producing the mycoparasitism (Chen et al. 2016a). The two species of endophytic fungi *Rhexocercosporidium* sp. and *Fusarium solani* inhibited three plant fungal pathogens *F. solani*, *Colletotrichum gloeosporioides*, and *Alternaria panax* using the coculture method (Yao et al. 2017). Endophyte fungi isolated from the medicinal plant *Oroxylum indicum* showed antagonistic activity against four plant fungal pathogens including *Alternaria alternata*, *Pyricularia oryzae*, *Colletotrichum capsici*, and *Rhizoctonia solani* using the dual culture test (Das and Narzary 2017).

9.4.2 The Indirect Antagonistic

The indirect antagonistic is the very interesting method to inhibit the growth of plant pathogens by different tests. Endophytic fungi of the medicinal plants are showing the high inhibition against many plant fungal pathogens through using two methods: (1) producing the antibiotics and (2) inducing plant defenses, as follows.

9.4.2.1 Antibiotics

Many endophyte fungi can secrete a lot of bioactive compounds and enzymes having the inhibition activity against the plant pathogens. *Muscodor tigerii* produced volatile compounds such as 4-octadecylmorpholine, 1-tetradecanamine N, N-dimethyl, and 1,2-benzenedicarboxylic acid, mono(2-ethylhexyl) ester that affected *Cercospora beticola* and *Alternaria alternata* (Saxena et al. 2015). The GBA strain of *Muscodor albus* inhibited many plant fungal pathogens such as *Pythium ultimum*, *Phytophthora cinnamomi*, *Geotrichum candidum*, *Verticillium dahliae*, *Cercospora beticola*, *Sclerotinia sclerotiorum*, *Botrytis cinerea*, *Fusarium solani*, and *Rhizoctonia solani* by producing the volatile compounds (Banerjee et al. 2010). *Chaetomium globosum* produced a gliotoxin compound as anti-phytopathogenic fungi (Li et al. 2011c). The crude extract of some *Fusarium* isolates such as *Fusarium* spp., *F. solani*, and *F. oxysporum* was inhibited by three *Colletotrichum* species such as *C. acutatum*, *C. fragariae*, and *C. gloeosporioides* (Rosa et al. 2012). The growth inhibition of two important pathogens such *Pythium* (*Pythium oligandrum* FC011, *P. ultimum* FC036, *P. aphanidermatum* FC001) and *Phytophthora* (*Phytophthora cryptogea* FC509, *P. cactorum* FC514, and *P. infestans* DJF21) were significant after treatment with the endophytic fungi *Pestalotiopsis* sp. (Kharwar et al. 2012).

In addition, the chaetoglobosin A compounds of *Chaetomium globosum* showed high antifungal activity against two plant pathogens such as *Coniothyrium diplodiella* and *Rhizopus stolonifer* (Zhang et al. 2013). *Colletotrichum truncatum* EF13 showed high efficacy in growth inhibition of two plant pathogens comprising *Sclerotinia sclerotiorum* and *Fusarium oxysporum* (Kumar and Kaushik 2013). The culture filter of *Fusarium* isolate Pg27 and *Aspergillus protuberus* isolate Pg30 affected the growth of *Rhizoctonia solani* (Wu et al. 2013). The culture filter of *Chaetomium globosum* CDW7 was showing antifungal activity against three phytopathogenic fungi in vitro including *Fusarium graminearum*, *Phytophthora capsici*, and *Sclerotinia sclerotiorum* by producing 1,2-benzenedicerboxaldehyde-3,4,5-trihydroxy-6-methyl compounds (flavipin) (Xiao et al. 2013). Also, the endophytic fungi can serve the medicinal plants by protecting them against the plant pathogens. The CDW7 strain affected *Fusarium* head blight disease that led to a curative efficacy at 48.8% and a protective efficacy at 54.9% (Xiao et al. 2013). The media broth of CDW7 strain diluted threefold could inhibit the conidia germination and the growth of mycelium completely of *F. graminearum* in vitro (Xiao et al. 2013). The fermentation broth of *Fusarium solani* T-7 affected the growth of five phytopathogens including (A) *F. oxysporum* of tomato, (B) *F. graminearum* of wheat, (C) *Venturia inaequalis* of pear, and (D) *Cytospora mandshurica* and (F) *Colletotrichum gloeosporioides* of apple (Shimo et al. 2013). Some endophytic fungi produced antifungal compounds against *Colletotrichum gloeosporioides* and *Rhizoctonia cerealis* (Zhang et al. 2015).

Interestingly, *T. gamsii* YIM PH30019 produced many volatile compounds comprising dimethyl disulfide, dibenzofuran, methanethiol, and ketones that might be very effective in the growth inhibition of plant fungal pathogens (Chen et al. 2016a). The endophytic fungus of *Penicillium chermesinum* was showing high efficacy in effect on the growth of three plant fungal pathogens including *Fusarium oxysporum*, *F. graminearum*, and *F. culmorum* (Mishra et al. 2016). Three endophyte fungi such as *Colletotrichum karstii*, *Corynespora cassiicola*, and *Diaporthe actinidiae* were showing high efficacy to inhibit the growth of plant pathogen *Rhizoctonia solani* but showed low effectivity on two plant pathogens such as *Fusarium oxysporum* and *Sclerotinia sclerotiorum* (Liu et al. 2016). The 28 strains of endophyte fungus *Chaetomium globosum* were able to control the plant pathogens such as *Fusarium oxysporum* f. sp. *niveum* by using the bioactive antifungal compounds (Pan et al. 2016). *Trichoderma koningiopsis* QA-3 affected on eight plant fungal pathogens by producing seven antifungal compounds comprising *Ceratobasidium cornigerum*, *Fusarium oxysporum*, *F. graminearum*, *Colletotrichum gloeosporioides* Penz, *Penicillium digitatum*, *Physalospora piricola* Nose, *Valsa mali*, and *Bipolaris sorokiniana* (Shi et al. 2017). The crude extract of *Cladosporium oxysporum* isolate PH30409 and *Trichoderma koningiopsis* isolate PH30441 showed high effectivity on the growth of five plant fungal pathogens causing root rot disease including *Fusarium oxysporum*, *F. solani*, *Phoma herbarum*, *Mycocentrospora acerina*, and *Alternaria panax* (Zhang et al. 2017). *Trichoderma polysporum* secreted the antimicrobial metabolite against five plant fungal pathogens of ginseng including *Alternaria panax*, *Cylindrocarpon destructans*, *Botrytis cinerea*, *Rhizoctonia solani*, and *Pythium* sp. (Park et al. 2017).

9.4.2.2 Induced the Plant Defense

Endophyte fungi of the medicinal plants can produce some compounds or enzymes that are inducing the defense activity in the host plant against the plant pathogens. The endophytic fungi *Gilmaniella* sp. AL12 was induced the plant defense by acting a jasmonic acid as a complementary of downstream signaling molecule at H_2O_2- and NO- mediated the accumulate of volatile oil and interaction with the SA signaling pathway (Ren and Dai 2012). Endophytic fungi could act the plant defenses against plant pathogens in the medicinal plants (Gautam 2014). The Dzf17 strain of *Fusarium oxysporum* produced two oligosaccharides (EOS and WOS) that induced three defense-related enzymes including polyphenoloxidase (PPO), phenylalanine ammonia lyase (PAL), and peroxidase (POD) in the host plant *Dioscorea zingiberensis* (Li et al. 2014a, b).

9.5 The Effect of Endophyte Fungi on the Plant Chemistry

The medicinal plants are exposed to endophytic microbes such as fungi being very normal. Fungi is everywhere around the medicinal plants. Endophytic fungi are very useful for the medicinal plants. The phenolic and flavonoid compounds were changing in total contents that are different according to type and quantity of endophyte fungi (Huang et al. 2008). The 643 samples of 20 medicinal plants produced many pharmaceutical compounds that also were produced by the endophyte fungi (Khan et al. 2010). The synthesis bases of bioactive compounds in the medicinal plants from the genetic and biochemical were reflecting the biocenotic interactions between the associated endophyte fungi and the host plants (Nicoletti and Fiorentino 2015).

The endophytic fungi *Gilmaniella* sp. AL12 of a medicinal plant *Atractylodes lancea* could improve the quality of the herb medicines affecting on the characteristic metabolites through changing in the host accumulation such as enhance the content of volatile oil (Wang et al. 2012). In the extraction method of several compounds of the cultural broth of *Colletotrichum truncatum* EF10 by using hexane solvent was showed the ability to produce oil similar of the host plant *Jatropha curcas* (Kumar and Kaushik 2013). *Gilmaniella* sp. AL12 strain improved the accumulation of volatile oil, activities of total protein phosphorylation, and Ca^{2+}-dependent protein kinase in *Atractylodes lancea* (Ren and Dai 2013). The AL12 strain induced *Atractylodes lancea* to accumulate the volatile oil following defense reaction (Chen et al. 2016b). The AL12 strain could induce ethylene as an upstream signal of SA and JA and also a downstream signal of signaling pathways for H_2O_2 and NO as well as acted as an important signal mediating sesquiterpenoid biosynthesis (Yuan et al. 2016).

Indeed, the strain of *Fusarium redolens* Dzf2 of *Dioscorea zingiberensis* could induce the production of diosgenin in cell culture and seedling of host plant by producing beauvericin (Yin et al. 2010). *Fusarium oxysporum* Dzf17 secreted three oligosaccharides, DP4, DP7, and DP10, through adding individually that could induce the cell suspension culture of host plant *Dioscorea zingiberensis* to produce the diosgenin (Li et al. 2011a). The Dzf17 strain increased the diosgenin content in

Dioscorea zingiberensis by producing polysaccharides (water-extracted mycelial polysaccharide) (Li et al. 2011b). The strain of *Schizophyllum commune* 3R-2 of *Panax ginseng* could induce the changes in the content of chemical substance implicated in the secondary metabolite biosynthetic pathway including (1) ginsenoside 20Rc, (2) ginsenoside Rg2, and (3) ginsenoside Rg3 in hairy roots by modifying the gene expression (Zhai et al. 2017). *Fusarium* sp., *Aspergillus* sp., and *Ramularia* sp. of *Rumex gmelini* Turcz could enhance the production of secondary metabolites in the host plant and have the ability to produce the secondary metabolites similar to the host plant (Ding et al. 2018). *Chaetomium globosum* D38 was very efficient in increasing the accumulation of secondary metabolites of tanshinone and salvianolic acids in the hairy roots of *Salvia miltiorrhiza* (Zhai et al. 2018).

9.6 The Effect of Endophyte Fungi on the Plant Growth

Endophytic fungi affect the medicinal plants through improving the plant growth and plant vigor (Jia et al. 2016). The endophytic fungus of *Fusarium* spp. from *Euphorbia pekinensis* (E5 and E4) was able to improve the plant growth by producing indole-3-acetic acid (IAA) and gibberellin (GA) (Dai et al. 2008). *Fusarium* sp. DL26 and *Pyrenochaeta* sp. DL351 could enhance the plant growth (Chen et al. 2010). The endophyte fungus *Acremonium strictum* AL16 of Chinese medicinal herb *Atractylodes lancea* was enhanced the plant trait plasticity and tolerance the drought stress by improve the proteins, leaf soluble sugars, antioxidant and proline enzyme activity, as well as, increasing the root/shoot ratio an host's abscisic acid level but decreasing the degree of plasmalemma (Yang et al. 2014). *Alternaria alternata* Aa_27 and sterile hyphae Sh_26 enhanced plant growth and root elongation by producing IAA and ammonia (Fouda et al. 2015). The culture filter of *Phoma* sp. could increase the rice seed germination and growth 100% by producing IAA, and *Bipolaris* sp. and *Phoma* sp. were showing higher contents of phenolic and flavonoid (Khan et al. 2017a). *Alternaria alternata* RSF-6L of common herb *Solanum nigrum* Korean was improved the plant vigor that elucidation by increasing in shoot length, dry biomass, root length, chlorophyll contents, and leaf area, as well as, affected by reducing the uptake of Cd in roots (Khan et al. 2017b). *Chaetomium globosum* D38 showed its ability in improving the plant growth of the *Salvia miltiorrhiza* as biofertilizer (Zhai et al. 2018).

9.7 Dark Septate Endophytes (DSE)

Dark septate endophytes (DSE) are the most common fungi within *Ascomycota* anamorphic and can colonize the root tissue intracellularly and intercellularly of the medicinal plants and others in the different habitats without causing any effect on the host tissues (Jumpponen 2001). *Leptodontidium* sp. isolate DQ069033 improved the plant vigor comprising the stem diameter, plant height, biomass, and new root number for seedling *Dendrobium nobile* Lindl. (Hou and Guo 2009). DSE fungal could colonize the root and enhance plant growth with improvement of the fruit chemical components of three cultivars of medicinal plant *Lycium barbarum* L. (Zhang et al. 2010). DSE fungi could colonize 22 ginger species (Uma et al. 2010). DSE fungi could colonize 23 plants among of 3 lycophytes and 44 fern plants (Muthuraja et al. 2014). DSE fungi were able to colonize the root of five species for *Asparagus* (*A. aethiopicus*, *A. umbellatus*, *A. setaceus*, *A. densiflorus*, and *A. racemosus*), but the colonization of DES fungi varied according to soil factor (Thangavelu and Raji 2016). There is an interaction between DSE fungal and medicinal plant that it can detect it and appear through the growth and formation the structures of DSE fungal (Zhang et al. 2010). DSE fungal could form structures including (A) melanized microsclerotia in different shapes in the cortex cells, (B) hyaline hyphae with lipid vacuoles, (C) septate coils, and (D) hyaline hyphae with lipid body.

However, the LBF-2 isolate of *Paraphoma chrysanthemicola* from cortical cells of root *Lycium barbarum* L. improved chlorophyll fluorescence, and it could increase the total biomass by 39.2%, total chlorophyll by 22.8%, and chlorophyll by 21.3% of host plant (Zhang et al. 2012). DSE fungi may be able to enhance potential the physiology of the medicinal plant for the tolerance of heavy metal in the soils. Some DSE genera of *Phialophora*, *Cladosporium*, *Leptodontidium*, and *Exophiala* were colonizing the root of *Alnus nepalensis* (Nepal alder) very intensively (Xu et al. 2015). The DSE isolate of *Leptodontidium orchidicola* DSE8 from *Epimedium wushanense* improved the plant vigor of host plant including root length, area and number of leaf, plant height, and biomass of the shoot and root (Zhu et al. 2015). The DSE8 isolate could enhance and increase the total flavonoid and icariin content at 20.24–237.97% (Zhu et al. 2015). On the other hand, some isolates of DSE isolated from medicinal plants cause negative effect on the host plant. Zhu et al. (2015) found three isolates including DSE1, DSE7, and DSE3 that have high effect on the medicinal plant by reducing the content of bioactive and plant growth.

9.8 The Relationship Between Endophyte Fungi and the Host Medicinal Plant

Endophytic fungi could be isolated from leaf, stem, root, bark, rhizome, seed, and petiole of the medicinal plants. This state is indicating endophyte fungi are able to colonize the parts of medicinal plants for the positive relationship between the fungi and host plant. The positive relationship from the side of endophytic fungi comprising some attention is as follows:

1. Endophyte fungi may induce the dormant gene or genes. This gene may lead to enhance the trait of the medicinal plant such as increase of the interesting compounds including antifungal, antibacterial, antimicrobial, and anticancer and the different enzymes and induce the defense against plant pathogens.

2. The possible endophytic fungi are able to secrete many useful compounds in the intercellular space that increases the importance of medicinal plants to utilize in various fields.
3. Endophytic fungi provide a good service for the environment by reducing usage of the synthetic chemical as the alternative method of pesticides and chemical fertilizers.
4. In addition, endophytic fungi may be used as an alternative method of the synthetic chemicals such as synthetic pesticides or synthetic fertilizer. Endophyte fungi are more beneficial for medicinal plants such as controlling the plant diseases, conferring the protection of the infection with the plant pathogens, enhancing the plant growth, and improving the plant chemistry. The usage endophyte fungi as an alternative method of synthetic pesticides or synthetic fertilizer is more interesting because it is leading to utilize the medicinal plants by the consumer and the pharmacology industry without any residue of synthetic chemical compounds.

For a side of the host, the indication that medicinal plant is allowing for endophytic fungi to colonize the intercellular space without any barrier both of chemical or physical. This back to the special theory of the relationship between the microbes and the host plants is a gene to gene (Flor 1971) that means virulence to virulence and also avirulence to avirulence. This is a special connection being between the fungi and host plant. This interaction between them may give some attention:

(A). Endophytic fungi during colonization to the tissues of the host medicinal plant, may send a message (chemical signal) as symbiosis interaction nor parasitic, and this message may reflect on the role of the host plant by don't use the defense tools both of chemical or physical barriers against the attacker fungus.

(B). The host medicinal plant cannot confront the colonization of fungi because it doesn't have the active defenses or the fungi can degrade the host plant defenses.

9.9 Conclusion

Endophytic fungi isolated from medicinal plant showing several advantages. Medicinal plant is considered as a high source of diverse fungi either a new species or new strains or other strains of filamentous fungi. The diversity of endophyte fungi is a huge number of genera around 174, leading it to be separated into 3 levels (low, middle, and high levels). This estimate can provide some idea about the fungi genera that are able to live as nonpathogenic fungi. This may more advantage to approximation the ideas about more interesting fungi that can utilize instead of synthetic chemicals used in the manufactureof pesticides and fertilizer synthetic. Some fungi genera are more dominant among others. *Trichoderma*, *Aspergillus*, *Fusarium*, *Colletotrichum*, *Penicillium*, *Phomopsis*, *Alternaria*, *Cladosporium*, *Curvularia*, *Phoma*, *Pestalotiopsis*, *Diaporthe*, and *Chaetomium*, with DSE, are spreading into the different parts of the medicinal plants. This diversity of genera is showing 13 genera preferring the exudates from these medicinal plants and living as endophytic fungi. But this diversity is more critical because it is having two sides, one useful and another dangerous (not mention here for these points).

In addition, the results indicate these fungi genera have a special genome to confront the natural defense of the medicinal plants. Then, these fungi can colonize the tissues as nonpathogenic without causing any damage for the host but appeared to be more useful for the host. These genera are living in the host plant as nonpathogenic fungi that can be useful to the biocontrol of both the plant pathogens and insect and use them in the agriculture without the dangerous residue coming from the synthetic chemicals of pesticides in the environment. Therefore, these genera can be utilized in (1) agriculture as biopesticides and biofertilizer, (2) in the industry by extracting their natural compounds and using them in different industries, and (3) in the pharmacy by using them in the drug industry as antifungal, antibacterial, antimicrobial, and anticancer.

Finally, endophyte fungi of the medicinal plants are showing high efficacy to produce the bioactive compounds. These compounds as natural products can be utilized to inhibit the growth of different pathogenic microbes for human, animal, and plant. These compounds can be more effective in saving the environment from the synthetic chemical compounds that are causing pollution in the ecosystem. Therefore, for further future need to isolate endophytic fungi from the medicinal plants and detect the bioactive compounds to utilize instead of the synthetic chemicals to save the ecosystem, due to endophytic fungi look like astrologer of the natural compounds.

References

Abdel-Wahab FA, Bahkai AHA, El-Gorban AM, Hodhod MS (2017) Natural products of *Nothophoma multilocularis* sp. nov. an endophyte of the medicinal plant *Rhazya stricta*. Mycosphere 8(8):1185–1200. https://doi.org/10.5943/mycosphere/8/8/15

Agusta A, Ohashi K, Shibuya H (2006) Composition of the endophytic filamentous fungi isolated from the tea plant *Camellia sinensis*. J Nat Med 60:268–272

Al-Ani LKT (2006) Induce resistance against Cucumber mosaic virus by Pseudomonas fluorescens Migula. M.Sc., Department of Plant Protection, College of Agriculture, University of Baghdad, Baghdad, Iraq, p. 90

Al-Ani LKT (2017a) Potential of utilizing biological and chemical agents in the control of Fusarium wilt of banana. PhD, School of Biology Science, Universiti Sains Malaysia, Pulau Pinang, Malaysia, p 259

Al-Ani LKT (2017b) PGPR: a good step to control several of plant pathogens. In: Singh HB, Sarma BK, Keswani C (eds) Advances in PGPR research. CABI, pp 398–410

Al-Ani LKT (2018a) Trichoderma: beneficial role in sustainable agriculture by plant disease management. In: Egamberdieva D, Ahmad P (eds) Plant microbiome: stress response, microorganisms for sustainability, vol 5. Springer, Singapore, pp 105–126

Al-Ani LKT (2018b) Trichoderma from extreme environments: physiology, diversity, and antagonistic activity. In: Egamberdieva D, Birkeland N-K, Panosyan H, Li W-J (eds) Extremophiles in Eurasian ecosystems: ecology, diversity, and applications. Microorganisms for sustainability, vol 8. Springer, Singapore, pp 388–403

Al-Ani LKT (2019a) Secondary metabolites of nonpathogenic Fusarium spp.; scope in agriculture. In: Singh HB, Keswani C, Reddy MS, Royano ES, García-Estrada C (eds) Secondary metabolites of plant growth promoting rhizomicroorganisms - discovery and applications. Springer, Singapore, pp 59–76

Al-Ani LKT (2019b) Bioactive secondary metabolites of Trichoderma spp. for efficient Management of Phytopathogens. In: Singh HB, Keswani C, Reddy MS, Royano ES, García-Estrada C (eds) Secondary metabolites of plant growth promoting rhizomicroorganisms - discovery and applications. Springer, Singapore, pp 125–143

Al-Ani LKT (2019c) A patent survey on Trichoderma spp. (from 2007-2017). In: Singh HB, Keswani C, Singh SP (eds) Intellectual property issues in microbiology. Springer, Singapore, (In Press)

Al-Ani LKT (2019d) Entomopathogenic fungi in intellectual property and using in biotechnology. In: Singh HB, Keswani C, Singh SP (eds) Intellectual property issues in microbiology. Springer, Singapore, (In Press)

Al-Ani LKT (2019e) Recent patents on Endophytic Fungi and their international market. In: Singh HB, Keswani C, Singh SP (eds) Intellectual property issues in microbiology. Springer, Singapore, (In Press)

Al-Ani LKT (2019f) Mechanisms of fungi to control for plant diseases. In: Egbuna C, Sawicka B (eds) Natural remedies for pest, disease and weed control. Elsevier Science, (In press)

Al-Ani LKT (2019g) Stable coexistence of biological control agents. In: Egbuna C, Sawicka B (eds) Natural remedies for pest, disease and weed control. Elsevier Science, (In press)

Al-Ani LKT (2019h) Mycovirus and the genome of plant fungi pathogen. In: Tripathi V, Kumar P, Tripathi V, Singh AK (eds) Microbial genomics in sustainable Agroecosystems. Springer, Singapore, (In Press)

Al-Ani LKT (2019i) Mycorrhiza in plant disease control. In: Tripathi V, Kumar P, Tripathi V, Singh AK (eds) Microbial genomics in sustainable Agroecosystems. Springer, Singapore, (In Press)

Al-Ani RA, Al-Ani LKT (2011) Induced of systemic resistance in cucumber plants against cucumber mosaic virus (CMV) by *Pseudomonas fluorescens* Migula. Arab J Plant Protect 29:36–42

Al-Ani LKT, Albaayit SFA (2018a) Antagonistic of some Trichoderma against Fusarium oxysporum sp. f. Cubense tropical race 4 (FocTR4) International conference on Research in Education & Science, ICRES April 28 – May 1, Marmaris, Turkey, pp 271 (Abstract)

Al-Ani LKT, Albaayit SFA (2018b) Antagonistic of some Trichoderma against Fusarium oxysporum sp. f. Cubense tropical race 4 (FocTR4). In: The Eurasia proceedings of science, technology, Engineering & Mathematics (EPSTEM), vol 2, pp 35–38

Al-Ani LKT, Salleh B (2010) Biological control of Fusarium wilt of banana by non pathogenic Fusarium oxysporum. PPSKH colloquium, Pust Pengajian Sains Kajihayat/School of Biological Sciences, USM, June, pp 10

Al-Ani LKT, Negim E-S, Mohammed AM, Salleh B, Saleh MI (2012) Antifungal activity of novel Binary grafting polymers. 1st USM – KAZNU International Conference on: Challenges of Teaching and Chemistry Research in Institutions of Higher Learning, 11–13 July, p 44

Al-Ani LKT, Salleh B, Ghazali AHA (2013a) Biocontrol of Fusarium wilt of Banana by Trichoderma spp. 8th PPSKH colloquium, Pust Pengajian Sains Kajihayat/School of Biological Sciences, USM, June 5–6

Al-Ani LKT, Salleh B, Mohammed AM, Ghazali AHA, Al-Shahwany AW, Azuddin NF (2013b) Biocontrol of Fusarium wilt of Banana by Non-pathogenic Fusarium spp. International symposium on tropical fungi, ISTF, IPB International Convention Center, Bogor, Indonesia; 09/2013, pp 50–51

Al-Ani LKT, Yonus MI, Mahdii BA, Omer MA, Taher JK, Albaayit SFA, Al-Khoja SB (2018) First record of use Fusarium proliferatum fungi in direct treatment to control the adult of wheat flour *Tribolium confusum*, as well as, use the entomopathogenic fungi Beauveria bassiana. Ecology Environ Conservation 24(3):29–34

Alberto RN, Costa AT, Polonio JC, Santos MS, Rhoden SA, Azevedo JL, Pamphile JA (2016) Extracellular enzymatic profiles and taxonomic identification of endophytic fungi isolated from four plant species. Genet Mol Res 15(4):gmr15049016. https://doi.org/10.4238/gmr15049016

Ashiq S, Hussain M, Ahmad B (2014) Natural occurrence of mycotoxins in medicinal plants: a review. Fungal Genet Biol 66:1–10. https://doi.org/10.1016/j.fgb.2014.02.005

Attitalla IH, Mansour SE, Mohamed WS, Al-Ani LKT, Mohammed AM, Faturi MY, Balal IAA, El-Maraghy SSM (2010a). Influence of Aspergillus Flavus and Aspergillus Terreus on the protein value of the two varieties of peanut grains. International conference, International Mycotoxin Conference, MycoRed, Penang–Malaysia, 1-4 Dec., (177)

Attitalla IH, Al-Ani LKT, Nasib MA, Balal IAA, Zakaria M, El-Maraghy SSM, Karim SMR (2010b) Screening of Fungi associated with commercial grains and animal feeds in Al-Bayda governorate, Libya. World Appl Sci J 9(7):746–756

Banerjee D, Strobel G, Geary B, Sears J, Ezra D, Liarzi O, Combs J (2010) Muscodor albus strain GBA, an endophytic fungus of Ginkgo biloba from the United States of America, produces volatile antimicrobials. Mycology 1:179–186. https://doi.org/10.1080/21501203.2010.506204

Basha NS, Ogbaghebriel A, Yemane K, Zenebe M (2012) Isolation and screening of endophytic fungi from Eritrean traditional medicinal plant *Terminalia brownii* leaves for antimicrobial activity. Int J Green Pharm 6:40–44. https://doi.org/10.4103/0973-8258.97124

Bezerra JDP, Nascimento CCF, Barbosa RN, Silva DCV, Svedese VM, Silva-Nogueira EB, Gomes BS, Paiva LM, Souza-Motta CM (2015) Endophytic fungi from medicinal plant Bauhinia forficata: diversity and biotechnological potential. Braz J Microbiol 46(1):49–57. https://doi.org/10.1590/S1517-838246120130657

Bhagat J, Kaur A, Sharma M, Saxena AK, Chadha BS (2012) Molecular and functional characterization of endophytic fungi from traditional medicinal plants. World J Microbiol Biotechnol 28(3):963–971

Bhagobaty RK, Joshi SR (2012) Enzymatic activity of fungi endophytic on five medicinal plant species of the pristine sacred forests of Meghalaya, India. Biotechnol Bioprocess Eng 17:33–40. https://doi.org/10.1007/s12257-011-0453-4

Bhagobaty RK, Joshi SR, Kumar R (2010) Penicillium verruculosum RS7PF: a root fungal endophyte associated with an ethno-medicinal plant of the indigenous tribes of Eastern India. Afr J Microbiol Res 4(9):766–770

Bhagya N, Sheik S, Sharma MS, Chandrashekar KR (2011) Isolation of endophytic *Colletotrichum gloeosporioides* Penz. from *Salacia chinensis* and its antifungal sensitivity. J Phytology 3:20–22

Bhattacharyya LH, Borah G, Parkash V, Bhattacharyya PN (2017) Fungal endophytes associated with the ethnomedicinal plant *Meyna spinosa* Roxb. Curr Life Sci 3(1):1–5. https://doi.org/10.5281/zenodo.258104

Carvalho CR, Gonçalves VN, Pereira CB, Johann S, Galliza IV, Alves TMA, Rabello A, Sobral MEG, Zani CL, Rosa CA, Rosa LH (2012) The diversity, antimicrobial and anticancer activity of endophytic fungi associated with the medicinal plant *Stryphnodendron adstringens* (Mart.) Coville (Fabaceae) from the Brazilian savannah. Symbiosis 57:95–107. https://doi.org/10.1007/s13199-012-0182-2

Centko RU, Ratnaweera PB, Tysoe C, Withers SG, de Silva ED, Andersen RJ (2017) Alpha-glucosidase and alpha-amylase inhibiting thiodiketopiperazines from the endophytic fungus Setosphaeria rostrata isolated from the medicinal plant Costus speciosus in Sri Lanka. Phytochem Lett 22:76–80. https://doi.org/10.1016/j.phytol.2017.09.004

Chen XM, Dong HL, Hu KX, Sun ZR, Chen J, Huo SX (2010) Diversity and antimicrobial and plant-growth-promoting activities of endophytic fungi in *Dendrobium loddigesii* Rolfe. J Plant Growth Regul 29(3):328–337. https://doi.org/10.1007/s00344-010-9139-y

Chen J, Wang H, Guo S-X (2012) Isolation and identification of endophytic and mycorrhizal fungi from seeds and roots of *Dendrobium* (Orchidaceae). Mycorrhiza 22:297–307. https://doi.org/10.1007/s00572-011-0404-0

Chen HJ, Awakawa T, Sun JY, Wakimoto T, Abe I (2013) Epigenetic modifier-induced biosynthesis of novel fusaric acid derivatives in endophytic fungi from Datura stramonium L. Nat Prod Bioprospect 3:20–23. https://doi.org/10.1007/s13659-013-0010-2

Chen LJ, Sun SZ, Miao CP, Wu K, Chen YW, Xu LH, Guan HL, Zhao LX (2016a) Endophytic Trichoderma gamsii YIM PH 30019: a promising biocontrol agent with hyperosmolar, mycoparasitism and antagonistic activities of induced volatile organic compounds on root-rot pathogenic fungi of Panax notoginseng. J Ginseng Res 40:315–324. https://doi.org/10.1016/j.jgr.2015.09.006

Chen F, Ren CG, Zhou T, Wei YJ, Dai CC (2016b) A novel exopolysaccharide elicitor from endophytic fungus *Gilmaniella* sp. AL12 on volatile oils accumulation in *Atractylodes lancea*. Sci Rep 6:34735. https://doi.org/10.1038/srep34735

Chow YY, Ting ASY (2015) Endophytic L-asparaginase-producing fungi from plants associated with anticancer properties. J Adv Res 6:869–876. https://doi.org/10.1016/j.jare.2014.07.005

Chowdhary K, Kaushik N (2015) Fungal endophyte diversity and bioactivity in the Indian medicinal plant *Ocimum sanctum* Linn. PLoS One 10(11):e0141444. https://doi.org/10.1371/journal.pone.0141444

Chowdhary K, Kaushik N, Coloma AG, Raimundo CM (2012) Endophytic fungi and their metabolites isolated from Indian medicinal plant. Phytochem Rev 11:467–485. https://doi.org/10.1007/s11101-012-9264-2

Cosoveanu A, Hernandez M, Cabrera R (2017) Host Artemisia: what is on the phylogenetic menu of the endophytic fungi. J Hortic For Biotechnol 21(4):88–106

Cosoveanu A, Rodriguez Sabina SR, Cabrera R (2018) Fungi as endophytes in Artemisia thuscula: juxtaposed elements of diversity and phylogeny. J Fungi 4:17. https://doi.org/10.3390/jof4010017

Cui JL, Guo SX, Xiao PG (2011) Antitumor and antimicrobial activities of endophytic fungi from medicinal parts of *Aquilaria sinensis*. J Zhejiang Univ-Sci B 12(5):385–392. https://doi.org/10.1631/jzus.B1000330

Cui Y, Yi D, Bai X, Sun B, Zhao Y, Zhang Y (2012) Ginkgolide B produced endophytic fungus (*Fusarium oxysporum*) isolated from *Ginkgo biloba*. Fitoterapia 83:913–920

Dai C-C, Yu B-Y, Li X (2008) Screening of endophytic fungi that promote the growth of *Euphorbia pekinensis*. Afr J Biotechnol 7(19):3505–3510

Darah Ibrahim N, Hong LS (2018) Anti-candidal activity of Aspergillus flavus IBRL-C8, an endophytic fungus isolated from Cassia siamea Lamk leaf. J Appl Pharm Sci 8(02):083–087. https://doi.org/10.7324/JAPS.2018.8213

Das S, Narzary D (2017) Diversity study on endophytic fungi associated with Oroxylum indicum and their interactions with some phytopathogens. J Adv Plant Sci 9(2):27–41

de Siqueira VM, Conti R, Araujo JM, Souza-Motta CM (2011) Endophytic fungi from the medicinal plant Lippia sidoides Cham. and their antimicrobial activity. Symbiosis 53:89–95

Deka D, Jha DK (2018) Antimicrobial activity of endophytic fungi from leaves and barks of *Litsea cubeba* Pers., a traditionally important medicinal Plant of North East India. Jordan J Biol Sci 11(1):73–79

Desale MG (2016) Bioprospecting of endophytic fungi from certain medicinal plants. PhD thesis, Bharati Vidyapeeth Deemed University, Pune, India, pp 54:107–108

Devi NN, Singh MS (2015) Endophytic fungi associated with traditional medicinal plants of Manipur. Int J Pharm Sci Rev Res 33(1):127–132

Devi NN, Prabaka JJ, Wahab F (2012) Phytochemical analysis and enzyme analysis of endophytic fungi from *Centella asiatica*. Asian Pac J Trop Biomed 2:1280–1284

Ding C, Wang Q, Guo S, Wang Z (2018) The improvement of bioactive secondary metabolites accumulation in Rumex gmelini Turcz through co-culture with endophytic fungi. Braz J Microbiol 49(2):362–369. https://doi.org/10.1016/j.bjm.2017.04.013

Egbuna C, Saravanan K, Verla AW, Al-Ani LKT, Godwin A, Jabeen NS (2019) Homemade preparations of natural biopesticides and applications. In: Egbuna C, Sawicka B (eds) Natural remedies for pest, disease and weed control. Elsevier Science, (In press)

Eo J-K, Choi M-S, Eom A-H (2014) Diversity of endophytic fungi isolated from Korean ginseng leaves. Mycobiology 42(2):147–151. https://doi.org/10.5941/MYCO.2014.42.2.147

Eyberger AL, Dondapati R, Porter JR (2006) Endophyte fungal isolates from *Podophyllum peltatum* produce podophyllotoxin. J Nat Prod 69(8):1121–1124. https://doi.org/10.1021/np060174f

Ferreira MC, Vieira MLA, Zani CL, Alves TMA, Junior PAS, Murta SMF, Romanha AJ, Gil LHVG, Carvalho AGO, Zilli JE, Vital MJS, Rosa CA, Rosa LH (2015) Molecular phylogeny, diversity, symbiosis and discover of bioactive compounds of endophytic fungi associated with the medicinal Amazonian plant Carapa guianensis Aublet (Meliaceae). Biochem Syst Ecol 59:36e44. https://doi.org/10.1016/j.bse.2014.12.017

Flor HH (1971) Current status of gene-for-gene concept. Annu Rev Phytopathol 9:275–296

Fouda AH, Hassan SE-D, Eid AM, Ewais EE-D (2015) Biotechnological applications of fungal endophytes associated with medicinal plant *Asclepias sinaica* (Bioss.). Ann Agric Sci 60(1):95–104. https://doi.org/10.1016/j.aoas.2015.04.001

Gangadevi V, Muthumary J (2008) Isolation of *Colletotrichum gloeosporioides*, a novel endophytic taxol-producing fungus from the leaves of a medicinal plant, *Justicia gendarussa*. Mycol Balc 5:1–4

Gangadevi V, Muthumary J (2009) A novel endophytic Taxol-producing fungus *Chaetomella raphigera* isolated from a medicinal plant, *Terminalia arjuna*. Appl Biochem Biotechnol 158(3):675–684. https://doi.org/10.1007/s12010-009-8532-0

García A, Rhoden SA, Bernardi-Wenzel J, Orlandelli RC, Azevedo JL, Pamphile JA (2012) Antimicrobial activity of crude extracts of endophytic fungi isolated from medicinal plant *Sapindus saponaria* L. J App Pharm Sci 2(10):035–040. https://doi.org/10.7324/JAPS.2012.21007

Garyali S, Kumar A, Reddy MS (2014) Diversity and antimitotic activity of taxol-producing endophytic fungi isolated from Himalayan yew. Ann Microbiol 64:1413–1422. https://doi.org/10.1007/s13213-013-0786-7

Gautam AK (2014) Diversity of fungal endophytes in some medicinal plants of Himachal Pradesh, India. Arch Phytopathol Plant Protect 47(5):537–544. https://doi.org/10.1080/03235408.2013.813678

Gond SK, Mishra A, Sharma VK, Verma SK, Kumar J, Kharwar RN, Kumar A (2012) Diversity and antimicrobial activity of endophytic fungi isolated from Nyctanthes arbor-tristis, a well-known medicinal plant of India. Mycoscience 53:113–121. https://doi.org/10.1007/s10267-011-0146-z

Guo B, Dai J-R, Ng S, Huang Y, Leong C, Ong W, Carte BK (2000) Cytonic acids A and B: novel tridepside inhibitors of hcmv protease from the endophytic fungus cytonaema species. J Nat Prod 63(5):602–604

Hiruma K, Gerlach N, Sacristán S, Nakano RT, Hacquard S, Kracher B, Neumann U, Ramírez D, Bucher M, O'Connell RJ, Schulze-Lefert P (2016) Root endophyte *Colletotrichum tofieldiae* confers plant fitness benefits that are phosphate status dependent. Cell 165(2):464–474

Hou XQ, Guo SX (2009) Interaction between a dark septate endophytic isolate from *Dendrobium* sp. and roots of *D. nobile* seedlings. J Integr Plant Biol 51(4):374–381. https://doi.org/10.1111/j.1744-7909.2008.00777.x

Huang W-Y, Cai Y-Z, Xing J, Corke H, Sun M (2007) A potential antioxidant resource: endophytic fungi from medicinal plants. Econ Bot 61:14–30

Huang WY, Cai YZ, Hyde KD, Corke H, Sun M (2008) Biodiversity of endophytic fungi associated with 29 traditional Chinese medicinal plants. Fungal Divers 33:61–75

Huang Q, An H, Song H, Mao H, Shen W, Dong J (2015) Diversity and biotransformative potential of endophytic fungi associated with the medicinal plant Kadsura angustifolia. Res Microbiol 166(1):45–55. https://doi.org/10.1016/j.resmic.2014.12.004

Huang R, Jiang B-G, Li X-N, Wang Y-T, Liu S-S, Zheng K-X, He J, Wu S-H (2018) Polyoxygenated cyclohexenoids with promising α-glycosidase inhibitory activity produced by Phomopsis sp. YE3250, an endophytic fungus derived from Paeonia delavayi. J Agric Food Chem 66(5):1140–1146. https://doi.org/10.1021/acs.jafc.7b04998

Idrissi FEJ, Ouchbani T, Berrada I, Nounsi A, Douira A, El Hourch A (2016) Isolation and identification of Carpobrotus edulis endophytic fungi growing in Morocco. Biolife 4(3):441–445

Ji ZQ, Wu WJ, Wang MA, Gu AG (2005) Identification of fungicidal compounds from endophytic fungi *Fusarium proliferatum* in *Celastrus angulatus*. J North-West Sci-Tech Univ Agric For 33(5):61–64

Jia M, Chen L, Xin H-L, Zheng C-J, Rahman K, Han T, Qin L-P (2016) A friendly relationship between endophytic fungi and medicinal plants: a systematic review. Front Microbiol 7:906. https://doi.org/10.3389/fmicb.2016.00906

Jumpponen A (2001) Dark septate endophytes – are they mycorrhizal? Mycorrhiza 11:207–211. https://doi.org/10.1007/s005720100112

Kaul S, Gupta S, Ahmed M, Dhar MK (2013) Endophytic fungi from medicinal plants: a treasure hunt for bioactive metabolites. Phytochem Rev 11:487–505. https://doi.org/10.1007/s11101-012-9260-6

Khan R, Shahzad S, Choudhary MI, Khan SA, Ahmad A (2010) Communities of endophytic fungi in medicinal plant *Withania somnifera*. Pak J Bot 42(2):1281–1287

Khan AL, Gilani SA, Waqas M, Al-Hosni K, Al-Khiziri S, Kim Y-H, Ali L, Kang S-M, Asaf S, Shahzad R, Hussain J, Lee I-J, Al-Harrasi A (2017a) Endophytes from medicinal plants and their potential for producing indole acetic acid, improving seed germination and mitigating oxidative stress. J Zhejiang Univ-Sci B 18:125–137. https://doi.org/10.1631/jzus.B1500271

Khan AR, Ullah I, Waqas M, Park GS, Khan AL, Hong SJ, Ullah R, Jung BK, Park CE, Ur-Rehman S, Lee IJ (2017b) Host plant growth promotion and cadmium detoxification in Solanum nigrum, mediated by endophytic fungi. Ecotoxicol Environ Saf 136:180–188. https://doi.org/10.1016/j.ecoenv.2016.03.014

Kharwar RN, Verma SK, Mishra A, Gond SK, Sharma VK, Afreen T, Kumar A (2011) Assessment of diversity, distribution and antibacterial activity of endophytic fungi isolated from a medicinal plant Adenocalymma alliaceum Miers. Symbiosis 55:39–46. https://doi.org/10.1007/s13199-011-0142-2

Kharwar RN, Maurya AL, Verma VC, Kumar A, Gond SK, Mishra A (2012) Diversity and antimicrobial activity of endophytic fungal community isolated from medicinal plant Cinnamomum camphora. Proc Natl Acad Sci India Sect B Biol Sci 82(4):557–565. https://doi.org/10.1007/s40011-012-0063-8

Khiralla A, Mohamed I, Thomas J, Mignard B, Spina R, Yagi S, Laurain- Mattar D (2015) A pilot study of antioxidant potential of endophytic fungi from some Sudanese medicinal plants. Asian Pac J Trop Med 8:701–704. https://doi.org/10.1016/j.apjtm.2015.07.032

Kour A, Shawl AS, Rehman S, Sultan P, Qazi PH, Suden P, Khajuria RK, Verma V (2008) Isolation and identification of an endophytic strain of fusarium oxysporum producing podophyllotoxin from Juniperus recurva. World J Microbiol Biotechnol 24:1115–1121. https://doi.org/10.1007/s11274-007-9582-5

Kumar S, Kaushik N (2013) Endophytic fungi isolated from oil-seed crop Jatropha curcas produces oil and exhibit antifungal activity. PLoS One 8(2):e56202. https://doi.org/10.1371/journal.pone.0056202

Kumaran RS, Kim HJ, Hur B-K (2010) Taxol promising fungal endophyte, Pestalotiopsis species isolated from Taxus cuspidate. J Biosci Bioeng 110(5):541–546. https://doi.org/10.1016/j.jbiosc.2010.06.007

Kusari P, Kusari S, Spiteller M, Kayser O (2013) Endophytic fungi harbored in Cannabis sativa L.: diversity and potential as biocontrol agents against host plant-specific phytopathogens. Fungal Divers 60:137–151. https://doi.org/10.1007/s13225-012-0216-3

Lakshmi PJ, Selvi KV (2013) Anticancer potentials of secondary metabolites from endophytes of *Barringtonia acutangula* and its molecular characterization. Int J Curr Microbiol App Sci 2(2):44–45

Li P, Mao Z, Lou J, Li Y, Mou Y, Lu S, Peng Y, Zhou L (2011a) Enhancement of diosgenin production in Dioscorea zingiberensis cell cultures by oligosaccharides from its endophytic fungus *Fusarium oxysporum* Dzf17. Molecules 16:10631–11044. https://doi.org/10.3390/molecules161210631

Li P, Mou Y, Shan T, Xu J, Li Y, Lu S, Zhou L (2011b) Effects of polysaccharide elicitors from endophytic *Fusarium oxysporium* Dzf17 on growth and diosgenin production in cell suspension culture of *Dioscorea zingiberensis*. Molecules 16:9003–9016. https://doi.org/10.3390/molecules16119003

Li H-Q, Li X-J, Wang Y-L, Zhang Q, Zhang A-L, Gao J-M, Zhang X-C (2011c) Antifungal metabolites from *Chaetomium globosum*, an endophytic fungus in *Ginkgo biloba*. Biochem Syst Ecol 39:876–879

Li H, Xiao J, Gao YQ, Tang JJ, Zhang AL, Gao J-M (2014a) Chaetoglobosins from *Chaetomium globosum*, an endophytic fungus in *Ginkgo biloba*, and their phytotoxic and cytotoxic activities. J Agric Food Chem 62:3734–3741. https://doi.org/10.1021/jf500390h

Li P, Luo H, Meng J, Sun W, Wang X, Lu S (2014b) Effects of oligosaccharides from endophytic fusarium oxysporum Dzf17 on activities of defense related enzymes in Dioscorea zingiberensis suspension cell and seedling cultures. Electron J Biotechnol 17:156–161. https://doi.org/10.1016/j.ejbt.2014.04.012

Liu XL, Huang KH, Zhou JZ, Meng L, Wang Y, Zhang LX (2012) Identification and antibacterial characteristics of an endophytic fungus *Fusarium oxysporum* from *Lilium lancifolium*. Lett Appl Microbiol 55:399–406. https://doi.org/10.1111/j.1472-765X.2012.03306.x

Liu YH, Hu XP, Li W, Cao XY, Yang HR, Lin ST, Xu CB, Liu SX, Li CF (2016) Antimicrobial and antitumor activity and diversity of endophytic fungi from traditional Chinese medicinal plant *Cephalotaxus hainanensis* Li. Genet Mol Res 15(2):1–11. https://doi.org/10.4238/gmr.15028633

Mahdi T, Mohamed I, Yagi S (2014) Endophytic fungal communities associated with ethnomedicinal plants from Sudan and their antimicrobial and antioxidant prospective. J For Prod Ind 3(6):248–256

Manganyi MC, Regnier T, Kumar A, Bezuidenhout CC, Ateba CN (2018) Biodiversity and antibacterial screening of endophytic fungi isolated from *Pelargonium sidoides*. S Afr J Bot 116:192–199. https://doi.org/10.1016/j.sajb.2018.03.016

Marcellano JP, Collanto AS, Fuentes RG (2017) Antibacterial activity of endophytic fungi isolated from the bark of *Cinnamomum mercadoi*. Pharm J 9(3):405–409. https://doi.org/10.5530/pj.2017.3.69

Mbouobda HD, Fotso B, Muyang RF, Chiatoh TA, Denis ON (2014) Enzymes and qualitative phytochemical screening of endophytic fungi isolated from Lantana camara Linn. Leaves J Appl Biol Biotechnol 2(6):1–6. https://doi.org/10.7324/JABB.2014.2601

Michalczyk A, Cieniecka-Rosłonkiewicz A, Cholewińska M (2015) Plant endophytic fungi as a source of paclitaxel. Herb Pol 60(4):22–33. https://doi.org/10.1515/hepo-2015-0002

Mishra VK, Singh G, Passari AK, Yadav MK, Supta VK, Singh BP (2016) Distribution and antimicrobial potential of endophytic fungi associated with ethnomedicinal plant *Melastoma malabathricum* L. J Environ Biol 37:229–237

Mohammed AM, AL-Ani LKT, Bekbayeva L, Salleh B (2011) Biological control of *Fusarium oxysporum* f. sp. *cubense* by *Pseudomonas fluorescens* and BABA in vitro. World Appl Sci J 15(2):189–191

Mohammed AM, Negim E-S, Al-Ani LKT, Salleh B, Saleh MI (2012) Utilization of amino-azines polymers as antifungal activity for banana. 1st USM – KAZNU International Conference on:

Challenges of Teaching and Chemistry Research in Institutions of Higher Learning, 11–13 July, p 29

Mohammed AM, Al-Ani LKT, Salleh B (2013) Potential management of *Fusarium oxysporum* f. sp. *cubense*, the banana wilt pathogen by using Pseudomonas and beta-amino-butyric acid (BABA). International symposium on tropical fungi, ISTF, IPB International Convention Center, Bogor, Indonesia; 09/2013, p 37

Mohammed AM, Al-Ani LKT, Salleh B, Ghazali AMA (2014) Determining plant growth promoting and biocontrol factor of bacterial culture media. The 3rd conference on Pests management, Crop Protection Research Centre, Sudan, 3–4 February, p 103

Muthuraja R, Muthukumar T, Sathiyadash K, Uma E, Priyadharsini P (2014) Arbuscular mycorrhizal (AM) and dark septate endophyte (DSE) fungal association in lycophytes and ferns of the Kolli Hills Eastern Ghats, Southern India. Am Fern J 104(2):67–102. https://doi.org/10.1640/0002-8444-104.2.67

Nadeem M, Ram M, Alam P, Ahmad MM, Mohammad A, Al-Qurainy F, Khan S, Zainul Abdin M (2012) *Fusarium solani*, P1, a new endophytic podophyllotoxin-producing fungus from roots of Podophyllum hexandrum. Afr J Microbiol Res 6:2493–2499. https://doi.org/10.5897/AJMR11.1596

Nagda V, Gajbhiye A, Kumar D (2017) Isolation and characterization of endophytic fungi from *Calotropis procera* for their antioxidant activity. Asian J Pharm Clin Res 10(3):254–258. https://doi.org/10.22159/ajpcr.2017.v10i3.16125

Nalini MS, Sunayana N, Prakash HS (2014) Endophytic fungal diversity in medicinal plants of Western Ghats. Int J Biodivers 2014:1–9. https://doi.org/10.1155/2014/494213

Nath A, Chattopadhyay A, Joshi SR (2013) Biological activity of endophytic fungi of *Rauwolfia serpentina* Benth, an ethnomedicinal plant used in folk medicines in Northeast India. Proc Natl Acad Sci India Sect B Biol Sci 85(1):233–240. https://doi.org/10.1007/s40011-013-0184-8

Nath A, Pathak J, Joshi SR (2014) Bioactivity assessment of endophytic fungi associated with *Centella asiatica* and *Murraya koengii*. J Appl Biol Biotechnol 2(05):006–011. https://doi.org/10.7324/JABB.2014.2502

Nicoletti R, Fiorentino A (2015) Plant bioactive metabolites and drugs produced by endophytic fungi of Spermatophyta. Agriculture 5:918–970. https://doi.org/10.3390/agriculture5040918

Nithya K, Muthumary J (2009) Growth studies of *Colletotrichum gloeosporioides* (Penz.) Sacc. – a taxol producing endophytic fungus from *Plumeria acutifolia*. Indian J Sci Technol 2(11):16–19

Ntuba-Jua GM, Mih AM, Bechem EET (2017) Diversity and distribution of endophytic fungi in different *Prunus africana* (Hook. f.) Kalkman Provenances in Cameroon. Int J Curr Res Biosci Plant Biol 4(6):7–23. https://doi.org/10.20546/ijcrbp.2017.406.002

Pai G, Chandra M (2017) Screening of phytochemicals and isolation of endophytic fungi from medicinal plant *Helicteres isora* L. IOSR J Pharm 7(12):01–05

Pan F, Hou K, Gao F, Hu B, Chen Q, Wu W (2014) Peimisine and peiminine production by endophytic fungus *Fusarium* sp. isolated from *Fritillaria unibracteata* var. *wabensis*. Phytomedicine 21:1104–1109. https://doi.org/10.1016/j.phymed.2014.04.010

Pan F, Liu Z-Q, Chen Q, Xu Y-W, Hou K, Wu W (2016) Endophytic fungus strain 28 isolated from *Houttuynia cordata* possesses wide-spectrum antifungal activity. Braz J Microbiol 47:480–488. https://doi.org/10.1016/j.bjm.2016.01.006

Park YH, Kim YC, Park SU, Lim HS, Kim JB, Cho BK, Bae H (2012a) Age-dependent distribution of fungal endophytes in Panax ginseng roots cultivated in Korea. J Ginseng Res 36:327–333. https://doi.org/10.5142/jgr.2012.36.3.327

Park SU, Lim H-S, Park K-C, Park Y-H, Bae H (2012b) Fungal endophytes from three cultivars of *Panax ginseng* Meyer cultivated in Korea. J Ginseng Res 36(1):107–113. https://doi.org/10.5142/jgr.2012.36.1.107

Park Y-H, Lee S-G, Ahn DJ, Kwon TR, Park SU, Lim H-S, Bae H (2012c) Diversity of fungal endophytes in various tissues of *Panax ginseng* Meyer cultivated in *Korea*. J Ginseng Res 36(2):211–217

Park Y-H, Kim Y, Mishra RC, Bae H (2017) Fungal endophytes inhabiting mountain-cultivated ginseng (Panax ginseng Meyer): diversity and biocontrol activity against ginseng pathogens. Sci Rep 7:16221. https://doi.org/10.1038/s41598-017-16181-z

Pharamat T, Palaga T, Piapukiew J, Whalley AJS, Sihanonth P (2013) Antimicrobial and anticancer activities of endophytic fungi from Mitrajyna javanica Koord and Val. Afr J Microbiol Res 7(49):5565–5572. https://doi.org/10.5897/AJMR12.2352

Prabukumar S, Rajkuberan C, Ravindran K, Sivaramakrishnan S (2015) Isolation and characterization of endophytic fungi from medicinal plant *Crescentia cujete* L. and their antibacterial, antioxidant and anticancer properties. Int J Pharm Pharm Sci 7(11):316–321

Puri SC, Nazir A, Chawla R, Arora R, Riyaz-Ul-Hasan S, Amna T, Ahmed B, Verma V, Singh S, Sagar R, Sharma A, Kumar R, Sharma RK, Qazi GN (2006) The endophytic fungus *Trametes hirsuta* as a novel alternative source of podophyllotoxin and related aryl tetralin lignans. J Biotechnol 122:494–510. https://doi.org/10.1016/j.jbiotec.2005.10.015

Raviraja NS (2005) Fungal endophytes in five medicinal plant species from Kudremukh range, Western Ghats of India. J Basic Microbiol 45(3):230–235

Ren C-G, Dai C-C (2012) Jasmonic acid is involved in the signaling pathway for fungal endophyte-induced volatile oil accumulation of *Atractylodes lancea* plantlets. BMC Plant Biol 12:128. http://www.biomedcentral.com/1471-2229/12/128

Ren C-G, Dai C-C (2013) Nitric oxide and brassinosteroids mediated fungal endophyte-induced volatile oil production through protein phosphorylation pathways in *Atractylodes lancea* plantlets. J Integr Plant Biol 55:1136–1146. https://doi.org/10.1111/jipb.12087

Rhoden SA, Garcia A, Rubim Filho CJ, Azevedo JL, Pamphile JA (2012) Phylogenetic diversity of endophytic leaf fungus isolates from the medicinal tree *Trichilia elegans* (Meliaceae). Genet Mol Res 11(3):2513–2522. https://doi.org/10.4238/2012.June.15.8

Rosa LH, Tabanca N, Techen N, Wedge DE, Pan Z, Bernier UR, Becnel JJ, Agramonte NM, Walker LA, Moraes RM (2012) Diversity and biological activities of endophytic fungi associated with micropropagated medicinal plant Echinacea purpurea (L.) Moench. Am J Plant Sci 3:1105–1114. https://doi.org/10.4236/ajps.2012.38133

Ruchikachorn N (2005) Endophytic fungi of cassia fistula L. Liverpool John Moores University, Liverpool, p 245

Saxena S, Meshram V, Kapoor N (2015) Muscodor tigerii nov.-volatile antibiotic producing endophytic fungus from the Northeastern Himalayas. Ann Microbiol 65(1):47–57. https://doi.org/10.1007/s13213-014-0834-y

Selim KA, El-Beih AA, Abd El-Rahman TM, El-Diwany AI (2012) Biology of endophytic fungi. Curr Res Environ Appl Mycol 2(1):31–82. https://doi.org/10.5943/cream/2/1/3

Selvi KB, Balagengatharathilagam P (2014) Endophytic fungi from medicinal plants of Virudhunagar district for antimicrobial activity. Int J Sci Nat 25(1):147–155

Shankar Naik B, Shashikala J, Krishnamurthy YL (2008) Diversity of fungal endophytes in shrubby medicinal plants of Malnad region, Western Ghats, Southern India. Fungal Ecol 1:89–93

Shankar Naik B, Krishnappa M, Krishnamurthy YL (2014) Biodiversity of endophytic fungi from seven herbaceous medicinal plants of Malnad region, Western Ghats, southern India. J For Res 25(3):707–711

Sharma D, Pramanik A, Agrawal PK (2016) Evaluation of bioactive secondary metabolites from endophytic fungus Pestalotiopsis Neglecta BAB-5510 isolated from leaves of *Cupressus Torulosa* D.Don. 3 Biotech 6:210. https://doi.org/10.1007/s13205-016-0518-3

Shi YW, Zhang X, Lou K (2013) Isolation, characterization, and insecticidal activity of an endophyte of drunken horse grass, *Achnatherum inebrians*. J Insect Sci 13:151. http://www.insect-science.org/13.151

Shi X-S, Wang D-J, Li X-M, Li H-L, Meng L-H, Li X, Pi Y, Zhou X-W, Wang B-G (2017) Antimicrobial polyketides from *Trichoderma koningiopsis* QA-3, an endophytic fungus obtained from the medicinal plant *Artemisia argyi*. RSC Adv 7:51335–51342. https://doi.org/10.1039/c7ra11122c

Shimo H, Yu B, Haiqun C, Xiangwei W, Qingqing X, Silver O (2013) Method for preparing phytopathogen-resisting *Ginkgo biloba* L. endophytic fungus fermentation broth. CN Patent, CN103194490A

Shubha J, Srinivas C (2017) Diversity and extracellular enzymes of endophytic fungi associated with *Cymbidium aloifolium* L. Afr J Biotechnol 16(48):2248–2258. https://doi.org/10.5897/AJB2017.16261

Silva GH, Oliveira CM, Teles HL, Pauletti PM, Gamboa IC, Silva DHS, Bolzani VS, Young MCM, Costa-Neto CM, Pfenning LH, Berlinck RGS, Araujo RA (2010) Sesquiterpenes from *Xylaria* sp., an endophytic fungus associated with *Piper aduncum* (Piperaceae). Phytochem Lett 3:164–167. https://doi.org/10.1016/j.phytol.2010.07.001

Sonaimuthu V, Krishnamoorthy S, Johnpaul M (2010) Taxol producing endophytic fungus fusarium culmorum SVJM072 from medicinal plant of Tinospora cordifolia – a first report. J Biotechnol 150:425. https://doi.org/10.1016/j.jbiotec.2010.09.588

Strobel GA, Dirkse E, Sears J, Markworth C (2001) Volatile antimicrobials from Muscodor albus a novel endophytic fungus. Microbiology 147:2943e2950

Sun J, Awakawa T, Noguchi H, Abe I (2012) Induced production of mycotoxins in an endophytic fungus from the medicinal plant *Datura stramonium* L. Bioorg Med Chem Lett 22:6397–6400. https://doi.org/10.1016/j.bmcl.2012.08.063

Sunayana N, Nalini MS, Sampath Kumara KK, Prakash HS (2014) Diversity studies on the endophytic fungi of *Vitex negundo* L. Mycosphere 5(4):578–590

Taware AS, More YW, Ghag SV, Rajurkar SK (2017) Screening of endophytic fungi isolated from Azadirachta indica A. Juss. for production of enzyme. Biosci Discov 8(4):688–694

Tayung K, Barik BP, Jha DK, Deka DC (2011) Identification and characterization of antimicrobial metabolite from an endophytic fungus, *Fusarium solani* isolated from bark of Himalayan yew. Mycosphere 2(3):203–213

Teles HL, Silva GH, Castro-Gamboa I, Bolzani VS, Pereira JO, Costa-Neto CM, Haddad R, Eberlin MN, Young MCM, Araújo AR (2005) Benzopyrans from *Curvularia* sp., an endophytic fungus associated with Ocotea corymbosa (Lauraceae). Phytochemistry 66:2363–2367

Thangavelu M, Raji M (2016) Arbuscular mycorrhizal and dark septate endophyte fungal associations in *Asparagus*. Turk J Bot 40:662–675. https://doi.org/10.3906/bot-1602-11

Thongsandee W, Matsuda Y, Ito S (2012) Temporal variations in endophytic fungal assemblages of *Ginkgo biloba* L. J For Res 17:213–218. https://doi.org/10.1007/s10310-011-0292-3

Toghueo KRM, Dinkar S, Boyom FF (2016a) Stimulation of the production of new volatile and non-volatile metabolites by endophytic *Aspergillus niger* using small organic chemicals. Curr Res Environ Appl Mycol 6(4):256–267. https://doi.org/10.5943/cream/6/4/3

Toghueo KRM, Eke P, Zabalgogeazcoa Í, de Aldana RVB, Wakam NL, Fekam BF (2016b) Biocontrol and growth enhancement potential of two endophytic *Trichoderma* spp. from *Terminalia catappa* against the causative agent of common bean root rot (*Fusarium solani*). Biol Control 96:8–20. https://doi.org/10.1016/j.biocontrol.2016.01.008

Toghueo RMK, Zabalgogeazcoa I, Vázquez de Aldana BR, Boyom FF (2017) Enzymatic activity of endophytic fungi from the medicinal plants Terminalia catappa, Terminalia mantaly and Cananga odorata. S Afr J Bot 109:146–153. https://doi.org/10.1016/j.sajb.2016.12.021

Tripathy V, Basak BB, Varghese TS, Saha A (2015) Residues and contaminants in medicinal herbs – a review. Phytochem Lett 14:67–78. https://doi.org/10.1016/j.phytol.2015.09.003

Uma Maheswari N, Saranya P (2018) Isolation and identification and phytochemical screening of endophytes from medicinal plants. Int J Biol Res 3(1):16–24

Uma E, Muthukumar T, Sathiyadash K, Muniappan V (2010) Mycorrhizal and dark septate fungal associations in gingers and spiral gingers. Botany 88(5):500–511. https://doi.org/10.1139/B10-021

Uzma F, Konappa NM, Chowdappa S (2016) Diversity and extracellular enzyme activities of fungal endophytes isolated from medicinal plants of Western Ghats, Karnataka. Egypt J Basic Appl Sci 3(4):335–342. https://doi.org/10.1016/j.ejbas.2016.08.007

Vega FE, Posada F, Peterson SW, Gianfagna TJ, Chaves F (2006) *Penicillium* species endophytic in coffee plants and ochratoxin A production. Mycologia 98(1):31–42

Verma SK, Gond SK, Mishra A, Sharma VK, Kumar J, Singh DK, Kumar A, Goutam J, Kharwar RN (2014) Impact of environmental variables on the isolation, diversity and antibacterial activity of endophytic fungal communities from Madhuca indica Gmel. At different locations in India. Ann Microbiol 64(2):721–734. https://doi.org/10.1007/s13213-013-0707-9

Vieira ML, Johann S, Hughes FM, Rosa CA, Rosa LH (2014) The diversity and antimicrobial activity of endophytic fungi associated with medicinal plant *Baccharis trimera* (Asteraceae) from the Brazilian savannah. Can J Microbiol 60(12):847–856. https://doi.org/10.1139/cjm-2014-0449

Voříšková A, Jansa J, Püschel D, Krüger M, Cajthaml T, Vosátka M, Janoušková M (2017) Real-time PCR quantification of arbuscular mycorrhizal fungi: does the use of nuclear or mitochondrial markers make a difference? Mycorrhiza 27(6):577–585

Wang Y, Dai CC, Cao JL, Xu DS (2012) Comparison of the effects of fungal endophyte Gilmaniella sp. and its elicitor on *Atractylodes lancea* plantlets. World J Microbiol Biotechnol 28:575–584. https://doi.org/10.1007/s11274-011-0850-z

Win PM, Matsumura E, Fukuda K (2018) Diversity of tea endophytic fungi: cultivar- and tissue preferences. Appl Ecol Environ Res 16(1):677–695. https://doi.org/10.15666/aeer/1601_677695

Wu H, Yang HY, You XL, Li YH (2013) Diversity of endophytic fungi from roots of *Panax ginseng* and their saponins yield capacities. Springerplus 2(107):1–9. https://doi.org/10.1186/2193-1801-2-107

Xiao Y, Li H-X, Li C, Wang J-X, Li J, Wang M-H, Ye Y-H (2013) Antifungal screening of endophytic fungi from Ginkgo biloba for discovery of potent anti-phytopathogenic fungicides. FEMS Microbiol Lett 339:130–136

Xu RB, Li T, Gui HL, Wang JL, Yu X, Ding YH, Wang CJ, Yang ZL, Zhao ZW (2015) Diversity and characterization of Cd-tolerant dark septate endophytes (DSEs) associated with the roots of Nepal alder (*Alnus nepalensis*) in a metal mine tailing of southwest China. Appl Soil Ecol 93:11–18. https://doi.org/10.1016/j.apsoil.2015.03.013

Yadav V, Kumar M, Deep DK, Kumar H, Sharma R, Tripathi T, Tuteja N, Saxena AK, Johri AK (2010) A phosphate transporter from the root endophytic fungus Piriformospora indica plays a role in phosphate transport to the host plant. J Biol Chem 285:26532–26544

Yadav M, Yadav A, Yadav JP (2014) In vitro antioxidant activity and total phenolic content of endophytic fungi isolated from Eugenia jambolana Lam. Asian Pac J Trop Med 7(S1):S256–S261. https://doi.org/10.1016/S1995-7645(14)60242-X

Yang T, Ma S, Dai CC (2014) Drought degree constrains the beneficial effects of a fungal endophyte on *Atractylodes lancea*. J Appl Microbiol 117:1435–1449. https://doi.org/10.1111/jam.12615

Yao YQ, Lan F, Qiao YM, Wei JG, Huang RS, Li LB (2017) Endophytic fungi harbored in the root of *Sophora tonkinensis* Gapnep: diversity and biocontrol potential against phytopathogens. Microbiol Open 6:e437. https://doi.org/10.1002/mbo3.437

Yin C, Li P, Li H, Xu L, Zhao J, Shan T, Zhou L (2010) Enhancement of diosgenin production in *Dioscorea zingiberensis* seedling and cell cultures by beauvericin from the endophytic fungus *Fusarium redolens* Dzf2. J Med Plants Res 5(29):6550–6554. https://doi.org/10.5897/JMPR11.921

Yu H, Zhang L, Li L, Zheng C, Guo L, Li W, Sun P, Qin L (2010) Recent developments and future prospects of antimicrobial metabolites produced by endophytes. Microbiol Res 165(6):437–449. https://doi.org/10.1016/j.micres.2009.11.009

Yuan ZL, Chen YC, Yang Y (2009) Diverse non-mycorrhizal fungal endophytes inhabiting an epiphytic, medicinal orchid (*Dendrobium nobile*): estimation and characterization. World J Microbiol Biotechnol 25:295–303. https://doi.org/10.1007/s11274-008-9893-1

Yuan J, Sun K, Deng-Wang M-Y, Dai C-C (2016) The mechanism of ethylene signaling induced by endophytic fungus Gilmaniella sp. AL12 mediating Sesquiterpenoids biosynthesis in Atractylodeslancea. Front Plant Sci 7:361. https://doi.org/10.3389/fpls.2016.00361

Zhai X, Chen L, Jia M, Li CH, Shen H, Ye BZ, Qin LP, Han T (2017) A stable beneficial symbiotic relationship between endophytic fungus *Schizophyllum commune* and host plant *Panax ginseng*. bioRxiv 175885. https://doi.org/10.1101/175885

Zhai X, Luo D, Li X, Han T, Jia M, Kong Z, Ji J, Rahman K, Qin L, Zheng C (2018) Endophyte Chaetomium globosum D38 promotes bioactive constituents accumulation and root production in Salvia miltiorrhiza. Front Microbiol 8:2694. https://doi.org/10.3389/fmicb.2017.02694

Zhang HH, Tang M, Chen H, Wang YJ, Ban YH (2010) Arbuscular mycorrhizas and dark septate endophytes colonization status in medicinal plant *Lycium barbarum* L. in arid Northwestern China. Afr J Microbiol Res 4(18):1914–1920

Zhang G, Sun S, Zhu T, Lin Z, Gu J, Li D, Gu Q (2011) Antiviral isoindolone derivatives from an endophytic fungus *Emericella* sp. associated with *Aegiceras corniculatum*. Phytochemistry 72:1436–1442. https://doi.org/10.1016/j.phytochem.2011.04.014

Zhang H-H, Tang M, Chen H, Wang Y-J (2012) Effects of a dark-septate endophytic isolate LBF-2 on the medicinal plant *Lycium barbarum* L. J Microbiol 50(1):91–96. https://doi.org/10.1007/s12275-012-1159-9

Zhang GZ, Zhang YH, Qin JC, Qu XY, Liu JL, Li X, Pan HY (2013) Antifungal metabolites produced by *Chaetomium globosum* no.04, an endophytic fungus isolated from *Ginkgo biloba*. Indian J Microbiol 53(2):175–180. https://doi.org/10.1007/s12088-013-0362-7

Zhang HW, Ying C, Bai XL (2014) Advancement in endophytic microbes from medicinal plants. Int J Pharm Sci Res 5:1589–1600. https://doi.org/10.13040/IJPSR.0975-8232.5(5).1589-1600

Zhang H, Ruan C, Bai X (2015) Isolation and antimicrobial effects of endophytic fungi from Edgeworthia chrysantha. Bangladesh J Pharmacol 10:529–532. https://doi.org/10.3329/bjp.v10i3.23575

Zhang D, Tao X, Liu J, Chen R, Zhang M, Li L, Fang X, Yu L, Dai J (2016) Periconiasin G, a new cytochalasan with unprecedented 7/6/5 tricyclic ring system from the endophytic fungus *Periconia* sp. Tetrahedron Lett 57:796–799. https://doi.org/10.1016/j.tetlet.2016.01.030

Zhang L, Niaz SI, Wang Z, Zhu Y, Lin Y, Li J, Liu L (2017) α-Glucosidase inhibitory and cytotoxic botryorhodines from mangrove endophytic fungus *Trichoderma* sp. 307. Nat Prod Res. https://doi.org/10.1080/14786419.2017.1385023

Zheng YK, Miao CP, Chen HH, Huang FF, Xia YM, Chen YW, Zhao LX (2017) Endophytic fungi harbored in Panax notoginseng: diversity and potential as biological control agents against host plant pathogens of root-rot disease. J Ginseng Res 41(3):353–360. https://doi.org/10.1016/j.jgr.2016.07.005

Zhu ZB, Fan JY, Guo QS, Liu ZY, Zhu GS (2015) The growth and medicinal quality of Epimedium wushanense are improved by an isolate of dark septate fungus. Pharm Biol 53(9):1344–1351. https://doi.org/10.3109/13880209.2014.982296

Chapter 10
Medicinal Plant-Associated Microbes as a Source of Protection and Production of Crops

Osama Abdalla Abdelshafy Mohamad, Jin-Biao Ma, Yong-Hong Liu, Li Li, Shaimaa Hatab, and Wen-Jun Li

Abstract Symbiosis research has been undertaken by researchers working independently of one another and often focused on a broad range of symbiotic interactions ranging from bipartite microbial consortia to multicellular hosts and their complex microbial communities. Recent investigations in symbiosis can impact areas such as agriculture sustainability, where a basic understanding of plant-microbe symbiosis will provide foundational information on the increasingly important issue of climate change. In this respect, in this chapter, we provided comments and references to finally establish symbiosis as an overdue central discipline of biological science. The interactions between medicinal plants and beneficial microorganisms, such as some *Actinobacteria*, *Firmicutes*, etc., proved the importance of these interactions to both symbionts in terms of enhanced adaptability, survival, and fitness of plants under different environmental stresses.

O. A. A. Mohamad
Key Laboratory of Biogeography and Bioresource in Arid Land, Xinjiang Institute of Ecology and Geography, Chinese Academy of Sciences, Urumqi, China

Department of Biological, Marine Sciences, and Environmental Agriculture, Institute for Post Graduate Environmental Studies, Arish University, Al-Arish, Egypt

J.-B. Ma · Y.-H. Liu · L. Li
Key Laboratory of Biogeography and Bioresource in Arid Land, Xinjiang Institute of Ecology and Geography, Chinese Academy of Sciences, Urumqi, China

S. Hatab
Department of Food Science and Technology, College of Environmental Agricultural Sciences, Arish University, El-Arish, Egypt

W.-J. Li (✉)
Key Laboratory of Biogeography and Bioresource in Arid Land, Xinjiang Institute of Ecology and Geography, Chinese Academy of Sciences, Urumqi, China

State Key Laboratory of Biocontrol and Guangdong Provincial Key Laboratory of Plant Resources, School of Life Sciences, Sun Yat-Sen University, Guangzhou, China
e-mail: liwenjun3@mail.sysu.edu.cn

D. Egamberdieva, A. Tiezzi (eds.), *Medically Important Plant Biomes: Source of Secondary Metabolites*, Microorganisms for Sustainability 15,
https://doi.org/10.1007/978-981-13-9566-6_10

Keywords Environmental microbiology · Medicinal plants · Endophytes · Biofertilizer · Biocontrol · Bioactive compounds

10.1 Introduction

In both natural and agricultural ecosystems, plants are constantly challenged by beneficial microbes that can be beneficial to promote plant growth, health, and productivity with important ecological and economic consequences. Medicinal plants are used globally as medication for various diseases including gastrointestinal symptoms, skin disorders, asthma, respiratory and urinary problems, and hepatic diseases because traditional medicines are often effective and easily available alternatives to pure pharmaceuticals (Cushnie et al. 2014). It is well known that endophyte-herb symbiosis having an ethnobotanical history may host beneficial microbes that can produce a diverse array of industrially important metabolites under different environmental stresses that are stressful to the host plant (Strobel et al. 2004; Bafana 2013; Li et al. 2018). However, unless key plant and microbial processes associated with the beneficial outcomes are well understood, these approaches will not reach their full potential with regard to antagonistic activity against plant pathogens (Bakker et al. 2013; Bhuvaneswari et al. 2013; Egamberdieva et al. 2017a).

There is a growing interest in the use of these beneficial microbes as microbial fertilizers and biocontrol for plants under different environmental stresses (Li et al. 2018; Mohamad et al. 2018a). Beneficial microbes (endophytes) reside in the roots or in above-ground organs of all plants (Gasser et al. 2011; Malfanova et al. 2011). These symbiotic interactions contribute significantly to nutrients and water acquisition in plants, particularly in resource-limited environments (Sánchez-López et al. 2018). In addition, many recent studies have begun to probe the importance of endophytic bacteria to medicinal plants, particularly those growing in unusual or stressed environments (Strobel and Daisy 2003; Vieira et al. 2011; Liu et al. 2016, 2017; Egamberdieva et al. 2017b). In view of the importance of endophytes to plant health and traditional herbal remedies as alternatives to synthetic pharmaceuticals, this chapter will be focused on the exploitation of beneficial microbes (endophytes) associated with medicinal plants on agriculture sustainability to alleviate climatic stresses by stimulating plant growth and protection.

10.1.1 What Are the Beneficial Microbes (Endophytes)?

The word endophyte means "inside the plant" (endon in Greek = within, phyton = plant). Endophytic bacteria are distinct microbial communities living in healthy roots, stems, leaves, and other tissues (Malfanova et al. 2011; Li et al. 2018). The usage of this term is as broad as its literal definition and spectrum of potential hosts and inhabitants (Malfanova et al. 2013). Since the first decent studies about the isolation of endophytic bacteria from plants tissue (Samish and Etinger-Tulczynska

1963; Mundt and Hinkle 1976), more than 200 bacterial genera from 16 phyla have been reported as endophytes. However, the most studied endophytes belong to three major phyla (*Firmicutes*, *Proteobacteria*, and *Actinobacteria*) and include members of *Pseudomonas* (Taghavi et al. 2009), *Acetobacter* (renamed as *Gluconobacter*) (Bertalan et al. 2009), *Herbaspirillum* (Pedrosa et al. 2011), *Burkholderia* (Weilharter et al. 2011), *Enterobacter* (Taghavi et al. 2010), *Azoarcus* (Krause et al. 2011), *Serratia* (Taghavi et al. 2009), *Stenotrophomonas* (Ryan et al. 2009), *Streptomyces* (Suzuki et al. 2005), and *Bacillus* (Deng et al. 2011). These microorganisms are often hypothesized to help their hosts by producing resources that afford protection and facilitate survival of the host plant (Sanchez et al. 2017).

10.1.2 How the Beneficial Microbes (Endophytes) Colonize the Plant?

There is a number of alternative ways by which plant tissues such as rhizoplane, plant cortex, xylem, and reproductive organs can be colonized by endophytes. Bacteria generally start with their establishment in the rhizosphere. Many studies showed that attachment of bacterial cells to the root is a crucial step for subsequent endophytic establishment (Lugtenberg et al. 2001; Lugtenberg and Kamilova 2009; Huang et al. 2011). The bacterial cell entry the apical root zone with the thin-walled surface root layer such as lateral roots (Zachow et al. 2010). Moreover, endophytes could also colonize the plant cortex by crossing the exodermal barrier, and they can remain at the site of entry (Timmusk et al. 2005) or move deeper inside and occupy the intercellular space of the cortex (James et al. 1994; Roncato-Maccari et al. 2003; Compant et al. 2005; Krid et al. 2012). For colonization of the xylem, a few bacteria can penetrate the endodermal barrier and invade the xylem vessels (James et al. 2002; Roncato-Maccari et al. 2003; Compant et al. 2005; Krid et al. 2012). The colonization of endophytic bacteria in reproductive organs of plants was approved by cultivation (Samish and Etinger-Tulczynska 1963; Mundt and Hinkle 1976; Graner et al. 2003; Okunishi et al. 2005; Furnkranz et al. 2012) and by microscopic visualization (Coombs and Franco 2003; Compant et al. 2005). Although the rhizosphere is assumed to be the major source of endophytic colonization in the plant tissue, other colonization sites of entry cannot be neglected. For instance, some bacteria are able to enter a plant tissue through stomata (James et al. 2001; Suzuki et al. 2005).

10.1.3 How to Prepare Plant Samples to Isolate Beneficial Microbes (Endophytes)

The isolation of endophytes from plants should have the surface sterilization as first step. Overall, the surface sterilization of root tissue consists of the following protocol:

1. Wash thoroughly the hall plant organ (root, stem, and leaves) under tap water to remove adhering soil particles and microbial surface epiphytes.
2. Pretreatment to remove hydrophobic substances on the surface of the plant.
3. Surface sterilization to remove remaining microbial colonizers from the plant surface.
4. Several rinses by sterilized distal water under aseptic conditions.
5. Check the surface sterilization protocol to confirm complete sterilization of the plant surface by culturing aliquots of distal water from the last rinsing onto nutrient media (Ryan et al. 2008).

All steps in the sterilization procedure should be conducted with sterilized tools under a laminar flow.

A number of pretreatment procedures could be done by sterilizing agents and surfactants. Commonly used sterilizing agents are sodium hypochlorite (Gardner et al. 1982; Schulz et al. 1993; Quadt-Hallmann et al. 1997; Sieber 2002), ethanol (Dong et al. 1994; Yi et al. 2013), and hydrogen peroxide (Misaghi and Donndelinger 1990; Sieber 2002; Ryan et al. 2008). In addition, as surfactants, there are some commonly used chemicals such as Tween 20 (Mahaffee and Kloepper 1997), Tween 80 (Sturz 1995), or Triton X-100 (Misaghi and Donndelinger 1990). The combinations of agents can improve the effectiveness of surface sterilization.

10.1.4 How to Isolate Beneficial Microbes (Endophytes)

This method is commonly used for isolation of endophytic bacteria. The surface-sterilized plant tissues are aseptically cut into very small and thin segments, which are then plated, i.e., pressed directly, onto an appropriate nutrient medium (Boyle et al. 2001; Carroll 2011). This method should select for fast-growing microorganisms, therefore representing a more qualitative than a quantitative approach.

10.1.5 Media for Beneficial Microbes (Endophytes) Isolating

The choice of the growth medium is very important as it directly affects the number and type of endophytic microorganisms that can be isolated from the root tissue. For isolation of endophytic bacteria, commonly used media include tryptic soya agar

Table 10.1 Compositions of the different media used for the isolation of endophytic bacteria

Medium	Composition (g/L)
M1	Yeast 0.25 g, K_2HPO_4 0.5 g, agar 15 g
M2	Trehalose 6 g, proline 1 g, KNO_3 0.5 g, Na_2HPO_4 0.3 g, $MgSO_4{\cdot}7H_2O$ 0.2 g, $CaCl_2$ 0.5 g, agar 15 g
M3	Raffinose 5 g, L-histidine 1 g, KNO_3 1 g, NaCl 1 g, $CaCl_2$ 2 g, K_2HPO_4 1 g, $MgSO_4{\cdot}7H_2O$ 1 g, agar 15 g
M4	Sodium propionate 2 g, arginine 1 g, NH_4NO_3 0.1 g, KCl 0.1 g, $MgSO_4{\cdot}7H_2O$ 0.05 g, $FeSO_4{\cdot}7H_2O$ 0.05 g, agar 15 g
M5	Cellulose 2.5 g, sodium pyruvate 2 g, proline 1 g, KNO_3 0.25 g, $MgSO_4{\cdot}7H_2O$ 0.2 g, K_2HPO_4 0.2 g, $CaCl_2$ 0.5 g, $FeSO_4{\cdot}7H_2O$ 0.01 g, agar 15 g
M6	Glycerol 10 g, asparagine 1 g, K_2HPO_4 1 g, trace salt 1 mL, agar 15 g
M7	Sodium succinate 1 g, L-asparagine 0.2 g, KH_2PO_4 0.9 g, K_2HPO_4 0.6 g, $MgSO_4{\cdot}7H_2O$ 0.1 g, $CaCl_2{\cdot}2H_2O$ 0.2 g, KCl 0.3 g, $FeSO_4{\cdot}7H_2O$ 0.001 g, agar 15 g
M8	Dulcitol 2 g, proline 0.5 g, K_2HPO_4 0.3 g, NaCl 0.3 g, $MgSO_4{\cdot}7H_2O$ 1 g, $CaCl_2{\cdot}2H_2O$ 1 g, agar 15 g
M9	Sodium propionate 2 g, L-asparagine 1 g, NH_4NO_3 0.1 g, KCl 0.1 g, $MgSO_4{\cdot}7H_2O$ 0.05 g, $FeSO_4{\cdot}7H_2O$ 0.05 g, agar 15 g
M10	Yeast 0.3 g, casein 0.3 g, glycose 0.3 g, K_2HPO_4 2 g, agar 15 g

(TSA), which supports the growth of a broad range of bacteria (Gardner et al. 1982), R2A for bacteria requiring low levels of nutrients (Reasoner and Geldreich 1985), nutrient broth-yeast extract medium for the growth of less selective bacteria (Zinniel et al. 2002), and King's B medium for the growth of *Pseudomonas* (King et al. 1954; Misaghi and Donndelinger 1990). However, the selection of isolation media is very important and influences microbial diversity, due to the fact that the high nutrient concentration media allows fast-growing bacteria to overgrow slower-growing ones. In addition, there are some other media used for isolation of endophytic bacteria associated with medicinal plant (Liu et al. 2016, 2017; Li et al. 2018) and listed in Table 10.1.

10.2 Medicinal Plants-Associated Microorganisms

The collective communities of medicinal plant-associated microorganisms have been widely reported and are assigned as the plant microbiome. Microbial communities inhabiting various varieties of plants were analyzed by 16S rRNA gene-based techniques. The molecular approaches for the identification of bacterial endophytes associated with plant and communities have been reviewed in detail (Yang et al. 2016).

However, the diversity of bacterial species and the extent of colonization to medicinal plants may vary depending on host plant species, soil properties, growing season, climate, and environmental factors (Ashraf et al. 2004). The medicinal plant-associated microbiome consists of distinct microbial communities living in roots, stems, leaves, and other tissues considered as beneficial microbes to host plant (Li et al. 2018; Mohamad et al. 2018a). Symbioses between medicinal plants and endophytic bacteria are mutually beneficial through several ways by maintaining the soil ecosystem and biological exchange of metabolites and sharing of physiological processes (Radhakrishnan and Lee 2016; Li et al. 2018). The host plant provides endophytes with a stable habitat, while endophytic bacteria supply nutrients such as fixed nitrogen, soluble potassium, iron, phosphate, and indole acetic acid (Liu et al. 2016, 2017; Li et al. 2018). Nowadays, a number of studies have been conducted to investigate endophytic bacteria associated with medicinal plants; among them, a few examples of medicinal plants and associated endophytes are listed in Table 10.2.

10.2.1 Beneficial Properties of Medicinal Plant-Associated Microbes

For their growth and development, plants need 16 essential elements such as C, N, H, O, and P and 11 more. They are obtained from the atmosphere, soil, water, and organic matter, and endophytes play an important role in the uptake of these nutrients (Baugh and Escobar 2007).

The symbioses between plants and endophytic bacteria are beneficial through the exchange of metabolites and sharing of physiological processes inside the plant organs (Reinhold-Hurek and Hurek 2011). The plant provides endophytes with a stable habitat, while endophytes supply nutrients such as fixed nitrogen, soluble potassium, iron, phosphate, and indole acetic acid (Liu et al. 2016, 2017; Li et al. 2018). These endophytes are part of a broader strategy for eco-friendly sustainable agriculture because they may contribute in the alleviation of environmental pollution due to chemical fertilizers. Endophytic bacteria associated with medicinal plants can adopt sophisticated survival strategies to promote plant growth (Boor 2006; Daffonchio et al. 2015). The medicinal plants forming association with various microorganisms can be formulated as biofertilizer and biocontrol tools.

10.2.2 Production of Indole-3-Acetic Acid (IAA)

Indole-3-acetic acid (IAA) is the phytohormone responsible for stimulating cell enlargement, division, differentiation, and gene regulation to promote plant growth (Lewis 2005; Khan et al. 2014). Endophytic bacteria are known to exhibit a wide variety of plant growth-promoting activities. Recently, many endophytic isolates associated with the medicinal plants *Hypericum perforatum* and *Ziziphora capitata* were observed to have the highest level of IAA production belonging to *Arthrobacter* sp., *Enterobacter* sp., and *Pantoea* sp. (Egamberdieva et al. 2017a). In addition, the endophytic isolate associated with *Glycyrrhiza uralensis* F. includes members of the

Table 10.2 Medicinal plants and associated bacteria

Plant species	Microorganisms	References
Glycyrrhiza uralensis F.	*Actinobacteria, Brevibacterium, Microbacterium, Streptomyces, Kocuria, Micromonospora, Pantoea, Phyllobacterium, Stenotrophomonas, Brevundimonas, Achromobacter, Catellatospora, Dietzia, Janibacter, Methylobacterium, Mycobacterium, Nocardioides, Rhodococcus, Staphylococcus, Starkeya*	Li et al. (2018)
Ferula sinkiangensis K. M. Shen	*Actinobacteria, Proteobacteria, Firmicutes*	Liu et al. (2017)
Ferula songorica	*Actinobacteria, Proteobacteria, Firmicutes*	Liu et al. (2016)
Ziziphora capitata	*Achromobacter, Pseudomonas*	Egamberdieva et al. (2017a)
Chlorophytum borivilianum	*Bacillus pumilus, Bacillus megaterium, Bacillus subtilis, Pseudomonas mendocina*	Krid et al. (2012)
Astragalus membranaceus	*Geodermatophilus obscurus*	Jin et al. (2014)
Phytolacca acinosa	*Aspergillus fumigatus*	Kaplan et al. (2013)
Agathosma betulina	*Cryptococcus laurentii*	Qin et al. (2012)
Ocimum sanctum, Coleus forskohlii, Catharanthus roseus, Aloe vera	*Azospirillum, Azotobacter, Pseudomonas*	Egamberdieva et al. (2017b)
Plectranthus tenuiflorus	*Bacillus* sp., *Micrococcus* sp., *Pseudomonas* sp.	El-Deeb et al. (2013)
	Acinetobacter sp., *Paenibacillus* sp.	
Annona squamosa	*Bacillus, Pseudomonas, Enterobacter, Corynebacterium, Micrococcus, Serratia*	Rais et al. (2017)
Eclipta alba		
Cassia auriculata		
Fritillaria thunbergii	*Proteobacteria, Acidobacteria, Actinobacteria*	Radhakrishnan et al. (2017)
	Bacteroidetes	
Nerium indicum	*Pontibacter*	Jha and Subramanian (2014)
Ajuga bracteosa	*Pseudomonas*	Egamberdieva and da Silva (2015)
Hypericum silenoides	*Acinetobacter, Enterobacter, Pseudomonas*	Hashem et al. (2016)
	Sphingobium, Stenotrophomonas, Agrobacterium, Pantoea, Serratia	
Ginseng plants	*Actinomycetes*	Ebrahim and Saleem (2017)
Typhonium giganteum	*Kribbella flavida, K. karoonensis, K. alba*	Prasanth Reddy et al. (2014)
Origanum vulgare	*Pseudomonas, Stenotrophomonas*	Mandal and DebMandal (2016)

(continued)

Table 10.2 (continued)

Plant species	Microorganisms	References
Vochysia divergens	*Actinomadura, Streptomyces, Aeromicrobium, Microbacterium, Microbispora* sp., *Micrococcus* sp., *Sphaerisporangium* sp., *Williamsia serinedens*	Bacon et al. (2005)
Rumex patientia	*Proteobacterium, Bacteroidetes, Acidobacteria*	Ahmad et al. (2012)
	Gemmatimonadetes, Verrucomicrobia, Planctomycetes, Actinobacteria, Firmicutes, Chloroflexi	
Matricaria chamomilla Calendula officinal, Solanum distichum	*Bacillus* sp.	Leveau and Lindow (2005)
Angelica sinensis	*Terriglobus saanensis, Mucilaginibacter polysacchareus, Mucilaginibacter myungsuensis, Mucilaginibacter ximonensis*	Hameeda et al. (2008) and Lemanceau et al. (2009)
Peganum harmala L.	*Firmicutes, Proteobacteria, Actinobacteria*	Chowdhury et al. (2015b)
Ginseng	*Arthrobacter* sp., *Bacillus* sp., *Pseudomonas* sp., *Microbacterium* sp., *Rahnella* sp., *Pseudoclavibacter* sp., *Paneibacillus* sp., *Kocuria* sp., *Serratia* sp., *Pantoea* sp., *Pectobterium* sp.	Chowdhury et al. (2015a)

Source: Modified from Horikoshi (2008)

genera *Achromobacter*, *Bacillus*, *Brevibacterium*, and *Stenotrophomonas* (Li et al. 2018). Liu et al. (2017) reported that about 79.4% of the endophytic isolates associated with medicinal plant *Ferula sinkiangensis* were capable of producing IAA. Moreover, the root-associated bacteria of *Ajuga bracteosa* exhibited a wide range of plant growth-promoting activities by producing siderophores and indole acetic acid (Naragani et al. 2016). However, the effect of indole-3-acetic acid (IAA) on plants depends on plant sensitivity to indole-3-acetic acid, the amount of IAA produced from plant-associated bacteria, and the capability to induce the production of other phytohormones (Janisiewicz and Korsten 2002). Moreover, endophytes associated with medicinal plant *Capsicum annuum* and *Cynara cardunculus* were reported to produce IAA (Jošić et al. 2012; Ramyasmruthi et al. 2012).

10.2.3 Siderophore Production

In soils, iron is found predominately as ferric ions, a form that cannot be directly assimilated by microorganisms (Droby 2005; Knight et al. 2018). The production of siderophores by endophytes is one of the indirect plant growth promotion regulation and is involved in prevention of deleterious effects of phytopathogenic organisms by removing iron from the environment (Khatiwora et al. 2012; Mohamad et al. 2018b). In addition, it has been previously reported that endophytic bacteria can produce different structural types of siderophores such as catecholate, hydroxamates, and citrate-based polycarboxylates (Kavitha et al. 2009). About 23 & 57% of the endophytes associated with the medicinal plants *Glycyrrhiza uralensis* F. and *Ferula sinkiangensis* produced siderophores, respectively, and these isolates belonged to various species within *Bacillus*, *Achromobacter*, and *Janibacter* (Liu et al. 2017; Li et al. 2018). Moreover, the root-associated bacteria of *Ajuga bracteosa* exhibited a wide range of plant growth-promoting activities by producing siderophores and indole acetic acid (Naragani et al. 2016). It has been reported that two root-associated bacterial endophytes isolated from *Ralstonia solanacearum* belonging to *Pseudomonas* sp. and *Pantoea* sp. showed siderophore production (Fernandes et al. 2003). The endophytes associated with medicinal plants *Capsicum annuum*, *Launaea nudicaulis*, *Jatropha curcas*, *Arachis Hypogaea*, *Brassica oxyrrhina*, and *Brassica napus* were reported to produce siderophore (Mansoor et al. 2007; Dell'Amico et al. 2008; Ma et al. 2010; Nithya and Halami 2012; Ramyasmruthi et al. 2012; Saraf et al. 2013).

10.2.4 Phosphate Solubilization

Phosphorus (P) is one of the essential elements needed by plants. The surrounding soils may have various contents of insoluble phosphates, but the amounts of available phosphorus to plants are usually a tiny amount. The plants can absorb it in two soluble forms, the monobasic (H_2PO_4) and the diabasic (HPO_4) ions (Strobel et al. 2001). This low availability of phosphorus to plants is due to the majority of soil P found in insoluble forms. Several phosphate-solubilizing microorganisms (PSMs) are now recorded to convert the insoluble form of phosphorus into accessible forms through acidification, secretion of organic acids, or protons (Sharma et al. 2016). The most efficient phosphate-solubilizing microorganisms mainly belong to the genera *Bacillus*, *Rhizobium*, and *Pseudomonas* (El-Deeb et al. 2013). The phosphate solubilization processes by beneficial microorganism are considered an important PGP mechanism (Kavitha et al. 2009; Hsouna et al. 2011). In contrast, Li et al. (2018) reported that about 13% of isolated endophytes from the medicinal plant *Glycyrrhiza uralensis* F. were able to solubilize phosphate, and phosphate solubilization was restricted to *Bacillus* and *Microbacterium* genera (Li et al. 2018). In addition, almost 19% of the endophytic bacteria associated with the medicinal plant *Ferula sinkiangensis* can solubilize phosphate (Liu et al. 2016). Moreover, many publications demonstrated that endophytes associated with medicinal plants such as *Capsicum annuum*, *Trigonella foenum-graecum*, *Trigonella foenum*, *Vitis vinifera*, and *Lactuca sativa* were able to solubilize phosphate (Tank and Saraf 2003; Barka et al. 2006; Arkhipova et al. 2007; Ramyasmruthi et al. 2012).

10.2.5 Nitrogen Fixation

Nitrogen is an essential element for all forms of life being the most important for synthesizing nucleic acids. Nitrogen is a basic requisite for plant nutrient, and the ability to reduce and derive such appreciable amounts of nitrogen from atmosphere and soil could be done by microorganisms (Young 1992). The ability of microbes to be diazotrophs and fix nitrogen symbiotically in plant and enhance crop yield could replace the use of chemical fertilizers (Vessey 2003). Biological nitrogen fixation includes symbiotic nitrogen fixation and in the case of endophytes includes members of *Azotobacter*, *Azospirillum*, *Azoarcus*, *Acetobacter*, and *Cyanobacteria* genera; these microorganisms are able to convert unutilizable nitrogen sources to accessible forms which make nutrients available for the plant (Döbereiner 1997; Esitken et al. 2006). In fact, biological nitrogen fixation processes are considered an important plant growth promotion mechanism especially in arid lands. Endophytic bacteria ubiquitously inhabit medicinal plant species and are most likely to possess special functions. Limited studies have been done to explore the endophytes associated with medicinal plant in the past 10 years.

However, recently, some studies reported that about 88% of the strains associated with the medicinal plant *Ferula songorica* have the capacity to fix nitrogen (Liu et al. 2016). In addition, a large number of isolates (75.8%) associated with medicinal plant *Ferula sinkiangensis* showed the ability to fix nitrogen (Liu et al. 2017). It has been demonstrated that the majority of endophytic bacteria isolates (76%) from the medicinal plant *Glycyrrhiza uralensis* F. were able to fix nitrogen (Li et al. 2018).

10.2.6 Lytic Enzymes Production

Various kinds of lytic enzymes are produced by microbes such as endophytes. The antagonistic activity against different types of pathogens may also be attributed to the production of cell wall-degrading enzymes that are produced by endophytes.

Chitinase and lipase enzymes are very useful enzymes in agriculture which act as biocontrol agents against fungal phytopathogens because of their ability to hydrolyze the chitinous fungal cell wall (Suresh et al. 2010; Wahyudi et al. 2011). The endophytic bacteria belonging to *Pseudomonas* sp. associated with the medicinal plants *Launaea nudicaulis*, *Coleus forskohlii*, and *Cupressus sempervirens* were able to produce chitinolytic enzymes and to control plant pathogens such as *Macrophomina phaseolina*, *Fusarium solani*, *Seiridium cardinale*, *Fusarium chlamydosporum*, *Ralstonia solanacearum*, and *Fusarium oxysporum* (Mansoor et al. 2007; Raio et al. 2011; Singh et al. 2013). Moreover, endophytic bacteria belonging to *Bacillus* sp. and associated with the medicinal plants *Panax quinquefolius*, *Arachis hypogaea*, and *Glycyrrhiza uralensis* were able to produce chitinolytic enzymes and control plant diseases (Nautiyal et al. 2013; Song et al. 2014; Mohamad et al. 2018a). A similar work carried out by Egamberdieva et al. (2017a) reported that endophytic bacteria associated with the medicinal plant *Ziziphora capitata* were able to produce chitinolytic enzymes. In addition, endophytic bacteria isolated from the medicinal plants *Ferula songorica*, *Hypericum perforatum*, *Glycyrrhiza uralensis*, and *Ferula sinkiangensis* secreted lytic enzymes such as protease, cellulase, and lipase (Liu et al. 2016, 2017; Egamberdieva et al. 2017a; Li et al. 2018).

10.2.7 Antimicrobial Activity

Plant pathogens cause many different kinds of symptoms that include galls and overgrowths, wilts, leaf spots, specks and blights, soft rots, scabs, and cankers. Plant diseases need to be controlled in order to maintain the quality and abundance of food around the world. Different approaches may be used to control plant diseases. The use of pesticides has contributed significantly in crop productivity over the past 100 years. However, the environmental pollution caused by an excessive use of pesticides has led to considerable changes in people's attitude toward the use of chemical pesticides. Biocontrol of plant diseases involves the use of beneficial organisms to inhibit the pathogens and reduce disease incidence and severity by direct or indirect manipulation of microorganisms (Baker and Cook 1974; Cook and Baker 1983; Maloy 1993).

Various mechanisms employed by the biocontrol agents in controlling plant diseases are broadly classified into direct and indirect antagonism. Direct antagonism results from the physical contact by biocontrol agents. Indirect antagonisms result from the stimulation of plant host defense pathways by beneficial microorganisms in the natural environment and ecosystems. Microorganisms associated with medicinal plants are of interest as the producers of important bioactive compounds. Such alternative approaches like investigating plant-microbe interactions with medicinal plants and to produce enhanced levels of phytochemicals have recently been reviewed (Sekar and Kandavel 2010; Singh 2013).

Several reports have investigated bacterial endophytes as possible biocontrol agents against diverse pathogenic fungi (Lacava et al. 2007; Erdogan and Benlioglu 2010; Egamberdieva et al. 2017a, b). Recently, several studies have described bacterial endophytes with biological control activity on a number of crops as potential sources of antimicrobial metabolites. The host plants in these studies have included *Solanum trilobatum* (Bhuvaneswari et al. 2013), *Nicotiana attenuata* (Santhanam et al. 2014), *Solanum melongena*, and *Solanum torvum* (Achari and Ramesh 2014).

The bacterial isolates *Arthrobacter crystallopoietes*, *Bacillus* sp., *Pseudomonas koreensis*, *Serratia liquefaciens*, and *Stenotrophomonas* sp. associated with the medicinal plants *Hypericum perforatum* and *Ziziphora capitata* exhibited in vitro antagonistic activity against the following pathogens: *Fusarium oxysporum*, *Botrytis cinerea*, *Pythium ultimum*, *Fusarium culmorum*, *Fusarium solani*, *Gaeumannomyces graminis*, and *Alternaria alternate* (Egamberdieva et al. 2017a). In addition, Passari et al. (2015) reported that out of 42 isolates associated with medicinal plants, 22 exhibited antagonistic activity against at least two of the four tested following pathogens: *Staphylococcus aureus*, *Pseudomonas aeruginosa*, *Escherichia coli*, and *Candida albicans*.

The endophytic bacteria isolated from different organs of the medicinal plant *Plectranthus tenuiflorus* in Saudi Arabia desert showed antimicrobial activity to at least one of six common antibiotic-resistant human pathogens such as *Salmonella typhi*, *Staphylococcus aureus*, *Escherichia coli*, *Klebsiella pneumoniae*, *Streptococcus agalactiae*, *Proteus mirabilis*, and the fungal pathogen *Candida albicans* (El-Deeb et al. 2013).

One hundred and fourteen endophytic strains isolated from the wild ethnomedicinal plant *Glycyrrhiza uralensis* in Xinjiang desert were screened for their in vitro antimicrobial activities against nine fungal phytopathogens such as *Fusarium oxysporum* f. sp., *Fulvia fulva* (Cooke) Cif., *Alternaria solani* Sorauer, *Fusarium oxysporum* f. sp. vasinfectum, *Verticillium dahliae* Kleb, *Ceratocystis fimbriata*, *Colletotrichum gloeosporioides*, *Pestalotiopsis microspora*, and *Fusarium graminearum*; the results showed that the beneficial endophytes isolates belonged to *Bacillus halotolerans*, *Bacillus atrophaeus*, *Brevibacterium frigoritolerans*, *Nocardioides alkalitolerans*, and *Bacillus mojavensis* (Mohamad et al. 2018a). Moreover, of the 114 isolates examined against common antibiotic-resistant human pathogens, 56 belonged to 6 different genera, namely, *Bacillus*, *Microbacterium*, *Brevibacterium*, *Phyllobacterium*, *Pantoea*, and *Stenotrophomonas*, and resulted antagonistic to pathogens such as *Staphylococcus aureus*, *Bacillus cereus*, *Escherichia coli*, and *Salmonella enteritidis*. The inhibitory activity was ranging from 8.6 to 11.8 mm (Mohamad et al. 2018a).

10.3 Environmental and Host-Plant Factors Affecting the Endophytes

The ecological or environmental conditions, such as temperature, humidity, geographic location, soil nutrition, and seasons, were reported as important factors which significantly affected the distribution pattern of endophytic microorganisms associated with host plant and would indirectly affect the production of secondary metabolites (Song et al. 1992; Suryanarayanan et al. 2005; Zhang and Qi-Yong 2007). These findings suggest that in particular conditions such as low sunshine and high humidity, the host medicinal plants would produce more nutrients suitable for the colonization of the endophytic microorganisms during the entire life cycle of host plant (Dai et al. 2003; Wu et al. 2013).

However, it should be noted that the distribution of certain endophytic populations was only restricted to particular host plant tissues and geographic location. For example, the study of medicinal plant *Glycyrrhiza uralensis* showed that the number of distinct bacterial isolates, selected on the base of distinct colony morphology, was 54 from root tissues, 34 from leaves, and 28 from stems. Similarly, the number of genera associated with roots, stems, and leaves was 14, 9, and 6, respectively (Li et al. 2018).

A similar study on isolation of endophytes (Liu et al. 2016) reported that more isolates obtained from roots than stems and leaves tissues of the medicinal plant *Ferula songorica*, *Methylobacterium*, *Agrococcus*, *Microbacterium*, *Brevundimonas*, *Micrococcus*, and *Rhizobium* genera were associated with root tissues, whereas in leaves and stems, *Ralstonia*, *Acinetobacter*, *Streptomyces*, *Arthrobacter*, *Kocuria*, *Sphingomonas*, and *Williamsia* genera were found.

On the other side, the endophytic microorganisms associated with *Plectranthus tenuiflorus* showed in roots a CFU value of 1.5×10^2, much lower than in stems (2.4×10^3). The leaves had higher CFU values than the other organs (2.9×10^4) (El-Deeb et al. 2013).

10.4 Endophytic Bacteria: A New Source of Bioactive Compounds

The "bioactive" or "biologically active" compounds are extra-nutritional constituents present in small quantities and mostly produced by plants and microbes. Such compounds have broad pharmaceutical properties including anticancer, cardiovascular, anti-lipidemic, antihypertensive, anti-glycemic, antithrombotic, antiatherogenic, and antidiabetic (Puri et al. 2005; Chang et al. 2013; Atanasov et al. 2015; Pastor-Villaescusa et al. 2015). Nowadays, bioactive compounds are used as preferred synthetic medicines for various diseases with very few side effects (Chang et al. 2013). Beneficial microbes (endophytes) spend most of their life cycle within the plant tissues without causing any visible damage to the host plant; they also have the potential to secrete unique secondary metabolites which can be used pharmaceutically and agriculturally (Liarzi et al. 2016). Bioactive compounds synthesized by endophytes help the host plant develop systemic resistance against pathogens and are also used in the pharmaceutical industries as antibiotics, anticancer, antiviral, antidiabetic, and other bioactive compounds (Guo et al. 2008). Endophytes associated with ethnomedicinal plants serve as a potential source of natural products for application in oxidative stress and as new bioactive agents (Nongkhlaw and Joshi 2015). Therefore, there is a growing interest of researchers from all over the world in bioprospecting of endophytic microbial communities inhabiting the plants and medicinal plants from various ecosystems. In this chapter, we showed some of bioactive compounds produced by endophytic bacteria isolated from medicinal plants (Table 10.3).

Moreover, in recent years, the scientists from different countries are endeavoring to search new antibiotic compounds in order to tackle the big problem represented by pathogens. As endophytic bacteria associated with medicinal plants have been found to synthesize many novel antibiotics against pathogenic bacteria and fungi with broad-spectrum microbiocidal potential of antibiotics (Table. 10.4), new interesting research perspectives are open.

Table 10.3 Endophytic actinobacteria isolated from medicinal plants

Species of microorganism	Host plant	Bioactive compounds	References
Streptomyces sp. MSU-2110	*Monstera* sp.	Coronamycin	Ezra et al. (2004)
Streptomyces sp. TP-A0569	*Allium fistulosum*	Fistupyrone 70 -demethyl novobiocin, 500 – demethyl novobiocin, novobiocin, 6-prenylindole, Anicemycin Pteridic acids A and B	Igarashi (2004)
Streptomyces hygroscopicus TP-	*Pteridium aquilinum*	Clethramycin	Igarashi (2004)
Streptomyces hygroscopicus	ND	Pterocidin	Igarashi et al. (2006)
Micromonospora lupini	ND	Lupinacidins	Igarashi et al. (2007)
Streptomyces laceyi MS53	*Ricinus communis*	6-Alkylsalicylic acids (salaceyins A and B)	Kim et al. (2006)
Streptomyces sp.	*Cistanches deserticola*	Tyrosol (possible ligand for GPR12) phenylethylamine derivatives, cyclic dipeptides, nucleosides and their aglycones, N-acetyltryptamine, pyrrole-2-carboxylic acid	Lin et al. (2008)
Streptomyces sp. CS	*Maytenus hookeri*	24-demethyl-bafilomycin C1 (Naphthomycin A)	Lu and Shen (2003)
Streptomyces sp. CS	*Maytenus hookeri*	Naphthomycin K, A and E	Lu and Shen (2007)
Streptomyces olivochromogenes	*Tinospora crispa, Phaleria macrocarpa, Curcuma aeruginosa, Andrographis paniculata, Caesalpinia sappan, Curcuma xanthoriza, Gynura procumbens, Physalis peruviana, Hibiscus sabdariffa*	Inhibitor of alpha-glucosidase	Pujiyanto et al. (2012)
Streptomyces setonii, Streptomyces sampsonii, Streptomyces sp. Q21, *Streptomyces* sp. MaB- QuH- 8	*Maytenu saquifolia, Putterlickia retrospinosa, Putterlickia verrucosa*	Celastramycins A and B	Pullen et al. (2002)
Streptomyces aureofaciens	*Zingiber officinale*	5,7- dimethox y-4-p-methoxylphenylcoumarin, 5,7-dimethoxy-4-phenylcoumarin	Taechowisan et al. (2005, 2007)

(continued)

Table 10.3 (continued)

Species of microorganism	Host plant	Bioactive compounds	References
Streptomyces sp. neau-D50	*Soybean*	3-acetonylidene-7-prenylindolin-2-one (isoprenoids, 7-isoprenylindole-3-carboxylic acid, 3-cyanomethyl-6-prenylindole, 6- isoprenylindole-3-carboxylic acid and 7,40 -dihydroxy-5-methoxy-8-(g, g- dimethyl allyl)-flavanone)	Zhang et al. (2014)
Streptomyces sp. YIM66017	*Alpinia oxyphylla*	2,6-dimethoxy terephthalic acid, yangjinhualine A	Zhou et al. (2014)
Bacillus atrophaeus	*Glycyrrhiza uralensis*	Ethylbenzene; p-xylene; dimethyl phthalate; 1,2-benzenedicarboxylic acid; bis (2-methyl propyl) ester; 9,12-octadecadienoic acid (Z,Z)-methyl ester; 9-octadecenoic acid methyl ester, (E)-; eicosane; heptadecane; tetracosane; bis (2-ethylhexyl) phthalate; and decanedioic acid, bis (2-ethylhexyl) ester	Mohamad et al. (2018a)

Source: Modified from Golinska et al. (2015)

Table 10.4 Bioactivity of compounds from endophytic actinobacteria isolated from medicinal plants

Compound	Microorganism	References
Munumbicins A, B, C, and D from *Streptomyces* sp. NRRL 30562	*Pseudomonas aeruginosa*, *Vibrio fischeri*, *Enterococcus faecalis*, *Staphylococcus aureus*, *Acinetobacter* sp., *Neisseria gonorrhoeae*, *Streptococcus pneumoniae*, *Bacillus anthracis*, *Escherichia coli*, *Pythium ultimum*, *Rhizoctonia solani*, *Phytophthora cinnamomic*, *Geotrichum candidum*, *Sclerotinia sclerotiorum*, *Pseudomonas syringe*, *Cryptococcus neoformans*, *Candida albicans*, *Aspergillus fumigates*, *Staphylococcus aureus* ATCC 33591, *Staphylococcus aureus* MH II, *Enterococcus faecalis* ATCC 51299, *Mycobacterium tuberculosis* MDR-P, *Mycobacterium tuberculosis* H37Rv (ATCC 25618)	Castillo et al. (2002)

(continued)

Table 10.4 (continued)

Compound	Microorganism	References
Kakadumycin A from Streptomyces sp. NRRL 30566	*Bacillus anthracis* 40/BA 100, *Bacillus anthracis* 14578, *Bacillus anthracis* 28, *Bacillus anthracis* 62-8, *Staphylococcus simulans* ATCC 11631, *Enterococcus faecalis* ATCC 29212, *Enterococcus faecalis* VRE, ATCC 51299, *Enterococcus faecium* ATCC 49624, *Listeria monocytogenes* ATCC 19114, *Listeria monocytogenes* ATCC 19115, *Shigella dysenteriae* ATCC 11835, *Staphylococcus epidermidis* ATCC 12228, *Staphylococcus aureus* ATCC 29213, *Staphylococcus aureus* MRSA, ATCC 33591, *Staphylococcus aureus* GISA, ATCC 700787, *Staphylococcus aureus* ATCC 27734, *Streptococcus pneumoniae* ATCC 49619, *Streptococcus pneumoniae* ATCC 70674, *Streptococcus pneumoniae* ATCC 70676, Inhibitor of human breast cancer cell line BT20	Castillo et al. (2003)
Munumbicins E-4 and E-5 from *Streptomyces* sp. NRRL 30562	*Burkholderia thailandensis, Escherichia coli, Staphylococcus aureus* ATCC 29213, *Staphylococcus aureus* 43,000 (MRSA), *Staphylococcus aureus, Pythium ultimum, Bacillus subtilis, Rhizoctonia solani,* Cytotoxic activity against *Plasmodium falciparum*	Castillo et al. (2006)
Saadamycin/5,7-Dimethoxy-4-p-methoxylphenyl coumarin from *Streptomyces* sp. Hedaya48	*Trichophyton rubrum, Trichophyton mentagrophytes, Microsporum gypseum, Epidermophyton floccosum, Aspergillus niger, Aspergillus fumigates, Fusarium oxysporum, Candida albicans, Cryptococcus humicolus*	El-Gendy and EL-Bondkly (2010)

(continued)

Table 10.4 (continued)

Compound	Microorganism	References
Coronamycin from *Streptomyces* sp. MSU-2110	*Pythium ultimum, Phytophthora cinnamomic, Aphanomyces cochlioides, Geotrichum candidum, Aspergillus fumigates, Aspergillus ochraceus, Fusarium solani, Rhizoctonia solani, Cryptococcus neoformans* (ATCC 32045), *Candida* parapsilosis (ATCC 90018), *Candida albicans* (ATCC 90028), *Saccharomyces cerevisiae* (ATCC 9763), *Candida parapsilosis* (ATCC 22019), *Candida albicans* (ATCC 24433), *Candida krusei* (ATCC 6258), *Candida tropicalis* (ATCC 750)	Ezra et al. (2004)
Pterocidin from *Streptomyces hygroscopicus* TP-A0451	Cytotoxicity against human cancer cell lines NCI-H522, OVCAR-3, SF539, and LOX-IMVI	Igarashi et al. (2006)
Lupinacidins A and B from *Micromonospora*	Inhibitor of in vitro invasion of colon 26-L5 cells	Igarashi et al. (2007)
6-Alkalysalicyclic acids (salaceyins A and B) from *Streptomyces laceyi* MS53	Cytotoxicity against human breast cancer cell line SKBR3	Kim et al. (2006)
Naphthomycin K from *Streptomyces* sp. CS	*Penicillium avellaneum* UC-4376, *Staphylococcus aureus, Mycobacterium tuberculosis,* cytotoxicity against P388 and A-549 human tumor cells	Lu and Shen (2003) and Lu and Shen (2007)
Celastramycins A/B from *Streptomyces* MaB-QuH-8	*Staphylococcus aureus* MRSA 134/93, *Staphylococcus aureus* MR 994/93, *Enterococcus faecalis* V-r 1528, *Mycobacterium smegmatis* SG 987, *Mycobacterium aurum* SB 66, *Mycobacterium vaccae* IMET 10670, *Mycobacterium fortuitum, Bacillus subtilis* ATCC 6633	Pullen et al. (2002)
5,7-dimethoxy-4-pmethoxylphenylcoumarin; 5,7-dimethoxy-4-phenylcoumarin from *Streptomyces aureofaciens* CMUAc130	*Colletorichum musae*	Taechowisan et al. (2005)
Actinomycin D from *Streptomyces* sp. Tc022	*Colletotrichum musae, Candida albicans*	Taechowisan et al. (2006)
5,7-Dimethoxy-4-pmethoxylphenylcoumarin; 5,7-dimethoxy-4-phenylcoumarin from *Streptomyces aureofaciens* CMUAc130	Antitumor activity	Taechowisan et al. (2007)

(continued)

Table 10.4 (continued)

Compound	Microorganism	References
6-Prenylindole from *Streptomyces* sp. TP-A0595, Fistupyrone from *Streptomyces* sp. TP-A0569, Clethramycin from *Streptomyces hygroscopicus* TP-A0326, Cedarmycin from *Streptomyces* sp. TP-A0456, Anicemycin from *Streptomyces thermoviolaceus* TP-A0648 Pterocidin from *Streptomyces ygroscopicus* TP-A0451	*Alternaria brassicola*, suppressing spore germination of *Alternaria brassicicola*, *Candida albicans*, *Cryptococcus neoformans*, Candida *glabrata*, cytocidal activity against tumor cell lines cytotoxicity against human cancer cell lines NCI- H522, OVCAR-3, SF539, and LOX-IMVI	Igarashi (2004)
Perlolyrine, 1-hydroxy-b-carboline, lumichrome, 1H-indole-3-carboxaldehyde from *Jishengella endophytica* 161111	Antiviral activity	Wang et al. (2014)
3-Acetonylidene-7-prenylindolin-2-one (isoprenoids, 7-isoprenylindole-3-carboxylic acid, 3-cyanomethyl-6-prenylindole, 6-isoprenylindole-3-carboxylic acid and 7,40 – dihydroxy-5-methoxy-8-(g,g-dimethylallyl)- flavanone) from *Streptomyces* sp. neau-D50	Cytotoxic activity against human lung adenocarcinoma cell line A549, *Colletotrichum orbiculare*, *Phytophthora capsici*, *Corynespora cassiicola*, *Fusarium oxysporum*	Zhang et al. (2014)

Source: Modified from Golinska et al. (2015)

References

Achari GA, Ramesh R (2014) Diversity, biocontrol, and plant growth promoting abilities of xylem residing bacteria from *solanaceous* crops. Int J Microbiol 2014:296521

Ahmad P, Kumar A, Ashraf M, Akram NA (2012) Salt-induced changes in photosynthetic activity and oxidative defense system of three cultivars of mustard (*Brassica juncea* L.). Afr J Biotechnol 11:2694–2703

Arkhipova T, Prinsen E, Veselov S, Martinenko E, Melentiev A, Kudoyarova G (2007) Cytokinin producing bacteria enhance plant growth in drying soil. Plant Soil 292:305–315

Ashraf M, Hasnain S, Berge O, Mahmood T (2004) Inoculating wheat seedlings with exopolysaccharide-producing bacteria restricts sodium uptake and stimulates plant growth under salt stress. Biol Fertil Soils 40:157–162

Atanasov AG et al (2015) Discovery and resupply of pharmacologically active plant-derived natural products: a review. Biotechnol Adv 33:1582–1614

Bacon C, Hinton D, Snook M (2005) Tentative identification of *Bacillus mojavensis* antifungal inhibitor. Phytopathology 95:S5

Bafana A (2013) Diversity and metabolic potential of culturable root-associated bacteria from *Origanum vulgare* in sub-Himalayan region. World J Microbiol Biotechnol 29:63–74

Baker K, Cook RJ (1974) Biological control of plant pathogens. WH Freeman, San Francisco, p 433

Bakker PA, Berendsen RL, Doornbos RF, Wintermans PC, Pieterse CM (2013) The rhizosphere revisited: root microbiomics. Front Plant Sci 4:165

Barka EA, Nowak J, Clément C (2006) Enhancement of chilling resistance of inoculated grapevine plantlets with a plant growth-promoting rhizobacterium, Burkholderia phytofirmans strain PsJN. Appl Environ Microbiol 72:7246–7252

Baugh CL, Escobar B (2007) The genus Bacillus and genus Trichoderma for agricultural bioaugmentation. Rice Farm Mag 1:1–4

Bertalan M et al (2009) Complete genome sequence of the sugarcane nitrogen-fixing endophyte Gluconacetobacter diazotrophicus Pal5. BMC Genomics 10:450. https://doi.org/10.1186/1471-2164-10-450

Bhuvaneswari S, Madhavan S, Panneerselvam A (2013) Enumeration of endophytic bacteria from *Solanum trilobatum* L. World J Pharm Res 3:2270–2279

Boor KJ (2006) Bacterial stress responses: what doesn't kill them can make them stronger. PLoS Biol 4:e23

Boyle C, Götz M, Dammann-Tugend U, Schulz B (2001) Endophyte-host interactions III. Local vs. systemic colonization. Symbiosis 31:259–281

Carroll G (2011) Forest endophytes: pattern and process. Can J Bot 73:1316–1324

Castillo UF et al (2002) Munumbicins, wide-spectrum antibiotics produced by *Streptomyces* NRRL 30562, endophytic on *Kennedia nigriscansa*. Microbiology 148:2675–2685

Castillo U et al (2003) Kakadumycins, novel antibiotics from *Streptomyces* sp. NRRL 30566, an endophyte of *Grevillea pteridifolia*. FEMS Microbiol Lett 224:183–190

Castillo UF et al (2006) Munumbicins E-4 and E-5: novel broad-spectrum antibiotics from *Streptomyces* NRRL 3052 FEMS. Microbiol Lett 255:296–300

Chang CL, Lin Y, Bartolome AP, Chen Y-C, Chiu S-C, Yang W-C (2013) Herbal therapies for type 2 diabetes mellitus: chemistry, biology, and potential application of selected plants and compounds. Evid-Based Complement Alternat Med 2013:378657

Chowdhury SP, Hartmann A, Gao X, Borriss R (2015a) Biocontrol mechanism by root-associated *Bacillus amyloliquefaciens* FZB42 – a review. Front Microbiol 6:780

Chowdhury SP et al (2015b) Cyclic lipopeptides of Bacillus amyloliquefaciens subsp. plantarum colonizing the lettuce rhizosphere enhance plant defense responses toward the bottom rot pathogen *Rhizoctonia solani*. Mol Plant-Microbe Interact 28:984–995

Compant S, Reiter B, Sessitsch A, Nowak J, Clement C, Ait Barka E (2005) Endophytic colonization of *Vitis vinifera* L. by plant growth-promoting bacterium *Burkholderia* sp. strain PsJN. Appl Environ Microbiol 71:1685–1693. https://doi.org/10.1128/aem.71.4.1685-1693.2005

Cook RJ, Baker KF (1983) The nature and practice of biological control of plant pathogens. Am Phytopathol Soc

Coombs JT, Franco CMM (2003) Visualization of an endophytic *Streptomyces* species in wheat seed. Appl Environ Microbiol 69:4260–4262

Cushnie TT, Cushnie B, Lamb AJ (2014) Alkaloids: an overview of their antibacterial, antibiotic-enhancing and antivirulence activities. Int J Antimicrob Agents 44:377–386

Daffonchio D, Hirt H, Berg G (2015) Plant-microbe interactions and water management in arid and saline soils. In: Principles of plant-microbe interactions. Springer, Cham, pp 265–276

Dai C, Yu B, Xu Z, Yuan S (2003) Effect of environmental factors on the growth and fatty acid composition of five endophytic fungi from *Sapium sebiferum*. Ying Yong Sheng Tai Xue Bao 14:1525–1528

Dell'Amico E, Cavalca L, Andreoni V (2008) Improvement of Brassica napus growth under cadmium stress by cadmium-resistant rhizobacteria. Soil Biol Biochem 40:74–84

Deng Y et al (2011) Complete genome sequence of *Bacillus subtilis* BSn5, an endophytic bacterium of Amorphophallus konjac with antimicrobial activity for the plant pathogen *Erwinia carotovora* subsp. Carotovora J Bacteriol 193:2070–2071

Döbereiner J (1997) Biological nitrogen fixation in the tropics: social and economic contributions. Soil Biol Biochem 29:771–774

Dong Z, Canny MJ, McCully ME, Roboredo MR, Cabadilla CF, Ortega E, Rodes R (1994) A nitrogen-fixing endophyte of sugarcane stems (a new role for the apoplast). Plant Physiol 105:1139–1147

Droby S (2005) Improving quality and safety of fresh fruits and vegetables after harvest by the use of biocontrol agents and natural materials. In: I international symposium on natural preservatives in food systems 709, p 45–52

Ebrahim MK, Saleem A-R (2017) Alleviating salt stress in tomato inoculated with mycorrhizae: photosynthetic performance and enzymatic antioxidants. J Taibah Univ Sci 11:850–860

Egamberdieva D, da Silva JAT (2015) Medicinal plants and PGPR: a new frontier for phytochemicals. In: Plant-growth-promoting rhizobacteria (PGPR) and medicinal plants. Springer, Cham, pp 287–303

Egamberdieva D, Wirth S, Behrendt U, Ahmad P, Berg G (2017a) Antimicrobial activity of medicinal plants correlates with the proportion of antagonistic endophytes. Front Microbiol 8:199

Egamberdieva D, Wirth SJ, Shurigin VV, Hashem A, Abd_Allah EF (2017b) Endophytic bacteria improve plant growth, symbiotic performance of chickpea (*Cicer arietinum* L.) and induce suppression of root rot caused by *Fusarium solani* under salt stress. Front Microbiol 8:1887

El-Deeb B, Fayez K, Gherbawy Y (2013) Isolation and characterization of endophytic bacteria from *Plectranthus tenuiflorus* medicinal plant in Saudi Arabia desert and their antimicrobial activities. J Plant Interact 8:56–64

El-Gendy MM, EL-Bondkly AM (2010) Production and genetic improvement of a novel antimycotic agent, saadamycin, against dermatophytes and other clinical fungi from endophytic *Streptomyces* sp. Hedaya48. J Ind Microbiol Biotechnol 37:831–841

Erdogan O, Benlioglu K (2010) Biological control of *Verticillium* wilt on cotton by the use of fluorescent *Pseudomonas* spp. under field conditions. Biol Control 53:39–45

Esitken A, Pirlak L, Turan M, Sahin F (2006) Effects of floral and foliar application of plant growth promoting rhizobacteria (PGPR) on yield, growth and nutrition of sweet cherry. Sci Hortic 110:324–327

Ezra D et al (2004) Coronamycins, peptide antibiotics produced by a verticillate Streptomyces sp.(MSU-2110) endophytic on Monstera sp. Microbiology 150:785–793

Fernandes P, Ferreira BS, Cabral JMS (2003) Solvent tolerance in bacteria: role of efflux pumps and cross-resistance with antibiotics. Int J Antimicrob Agents 22:211–216

Furnkranz M, Lukesch B, Muller H, Huss H, Grube M, Berg G (2012) Microbial diversity inside pumpkins: microhabitat-specific communities display a high antagonistic potential against phytopathogens. Microb Ecol 63:418–428

Gardner JM, Feldman AW, Zablotowicz RM (1982) Identity and behavior of xylem-residing bacteria in rough lemon roots of Florida citrus trees. Appl Environ Microbiol 43:1335–1342

Gasser I et al (2011) Analysis of the endophytic lifestyle and plant growth promotion of *Burkholderia terricola* ZR2-12. Plant Soil 347:125

Golinska P, Wypij M, Agarkar G, Rathod D, Dahm H, Rai M (2015) Endophytic actinobacteria of medicinal plants: diversity and bioactivity. Antonie Van Leeuwenhoek 108:267–289

Graner G, Persson P, Meijer J, Alstrom S (2003) A study on microbial diversity in different cultivars of *Brassica napus* in relation to its wilt pathogen, *Verticillium longisporum*. FEMS Microbiol Lett 224:269–276

Guo B, Wang Y, Sun X, Tang K (2008) Bioactive natural products from endophytes: a review. Appl Biochem Microbiol 44:136–142

Hameeda B, Harini G, Rupela O, Wani S, Reddy G (2008) Growth promotion of maize by phosphate-solubilizing bacteria isolated from composts and macrofauna. Microbiol Res 163:234–242

Hashem A, Abd_Allah E, Alqarawi A, Al-Huqail A, Shah M (2016) Induction of osmoregulation and modulation of salt stress in *Acacia gerrardii* benth. by arbuscular mycorrhizal fungi and *Bacillus subtilis* (BERA 71). Biomed Res Int 2016:1–11

Horikoshi K (2008) Past, present and future of extremophiles. Extremophiles 12:1–2

Hsouna AB, Trigui M, Mansour RB, Jarraya RM, Damak M, Jaoua S (2011) Chemical composition, cytotoxicity effect and antimicrobial activity of *Ceratonia siliqua* essential oil with preservative effects against Listeria inoculated in minced beef meat. Int J Food Microbiol 148:66–72

Huang B, Lv C, Zhuang P, Zhang H, Fan L (2011) Endophytic colonisation of *Bacillus subtilis* in the roots of *Robinia pseudoacacia* L. Plant Biol 13:925–931

Igarashi Y (2004) Screening of novel bioactive compounds from plant-associated actinomycetes. Actinomycetologica 18:63–66

Igarashi Y, Miura S-S, Fujita T, Furumai T (2006) Pterocidin, a cytotoxic compound from the endophytic *Streptomyces hygroscopicus*. J Antibiot 59:193

Igarashi Y et al (2007) Antitumor anthraquinones from an endophytic actinomycete *Micromonospora lupini* sp. Nov Bioorg Med Chem Lett 17:3702–3705

James EK, Reis VM, Olivares FL, Baldani JI, Döbereiner J (1994) Infection of sugar cane by the nitrogen-fixing bacterium *Acetobacter diazotrophicus*. J Exp Bot 45:757–766

James EK, Olivares FL, de Oliveira AL, dos Reis FB Jr, da Silva LG, Reis VM (2001) Further observations on the interaction between sugar cane and *Gluconacetobacter diazotrophicus* under laboratory and greenhouse conditions. J Exp Bot 52:747–760

James EK et al (2002) Infection and colonization of rice seedlings by the plant growth-promoting bacterium *Herbaspirillum seropedicae* Z67. Mol Plant-Microbe Interact 15:894–906

Janisiewicz WJ, Korsten L (2002) Biological control of postharvest diseases of fruits. Annu Rev Phytopathol 40:411–441

Jha Y, Subramanian R (2014) PGPR regulate caspase-like activity, programmed cell death, and antioxidant enzyme activity in paddy under salinity. Physiol Mol Biol Plants 20:201–207

Jin H et al (2014) Characterization of rhizosphere and endophytic bacterial communities from leaves, stems and roots of medicinal *Stellera chamaejasme* L. Syst Appl Microbiol 37:376–385

Jošić D, Protolipac K, Starović M, Stojanović S, Pavlović S, Miladinović M, Radović S (2012) Phenazines producing *Pseudomonas* isolates decrease *Alternaria tenuissima* growth, pathogenicity and disease incidence on cardoon. Arch Biol Sci 64:1495–1503

Kaplan D et al (2013) A survey of the microbial community in the rhizosphere of two dominant shrubs of the Negev Desert highlands, *Zygophyllum dumosum* (Zygophyllaceae) and *Atriplex halimus* (Amaranthaceae), using cultivation-dependent and cultivation-independent methods. Am J Bot 100:1713–1725

Kavitha A, Prabhakar P, Vijayalakshmi M, Venkateswarlu Y (2009) Production of bioactive metabolites by *Nocardia levis* MK-VL_113. Lett Appl Microbiol 49:484–490

Khan AL et al (2014) Bacterial endophyte *Sphingomonas* sp. LK11 produces gibberellins and IAA and promotes tomato plant growth. J Microbiol 52:689–695

Khatiwora E, Adsul VB, Kulkarni M, Deshpande N, Kashalkar R (2012) Antibacterial activity of dibutyl phthalate: a secondary metabolite isolated from *Ipomoea carnea* stem. J Pharm Res 5:150–152

Kim N, Shin JC, Kim W, Hwang BY, Kim BS, Hong Y-S, Lee D (2006) Cytotoxic 6-alkylsalicylic acids from the endophytic *Streptomyces laceyi*. J Antibiot 59:797

King EO, Ward MK, Raney DE (1954) Two simple media for the demonstration of pyocyanin and fluorescin. J Lab Clin Med 44:301–307

Knight C, Bowman MJ, Frederick L, Day A, Lee C, Dunlap CA (2018) The first report of antifungal lipopeptide production by a *Bacillus subtilis* subsp. inaquosorum strain. Microbiol Res 216:40–46

Krause A, Bischoff B, Miche L, Battistoni F, Reinhold-Hurek B (2011) Exploring the function of alcohol dehydrogenases during the endophytic life of *Azoarcus* Sp. strain BH72. Mol Plant-Microbe Interact 24:1325–1332

Krid S, Triki MA, Gargouri A, Rhouma A (2012) Biocontrol of olive knot disease by *Bacillus subtilis* isolated from olive leaves. Ann Microbiol 62:149–154

Lacava PT, Li W, Araújo WL, Azevedo JLC, Hartung JS (2007) The endophyte *Curtobacterium flaccumfaciens* reduces symptoms caused by *Xylella fastidiosa* in *Catharanthus roseus*. J Microbiol 45:388–393

Lemanceau P, Bauer P, Kraemer S, Briat J-F (2009) Iron dynamics in the rhizosphere as a case study for analyzing interactions between soils, plants and microbes. Plant Soil 321:513–535

Leveau JH, Lindow SE (2005) Utilization of the plant hormone indole-3-acetic acid for growth by *Pseudomonas putida* strain 1290. Appl Environ Microbiol 71:2365–2371

Lewis GP (2005) Legumes of the world. Royal Botanic Gardens, Kew

Li L et al. (2018) Synergistic plant–microbe interactions between endophytic bacterial communities and the medicinal plant *Glycyrrhiza uralensis* F. Antonie Van Leeuwenhoek 111(10): 1735–1748
Liarzi O, Bucki P, Miyara SB, Ezra D (2016) Bioactive volatiles from an endophytic *Daldinia cf. concentrica* isolate affect the viability of the plant parasitic nematode *Meloidogyne javanica*. PLoS One 11:e0168437
Lin Z-J, Lu X-M, Zhu T-J, Fang Y-C, Gu Q-Q, Zhu W (2008) GPR12 selections of the metabolites from an endophytic *Streptomyces* sp. asociated with *Cistanches deserticola*. Arch Pharm Res 31:1108
Liu Y-H et al (2016) Culturable endophytic bacteria associated with medicinal plant *Ferula songorica*: molecular phylogeny, distribution and screening for industrially important traits 3. Biotech 6:209
Liu Y et al (2017) Endophytic bacteria associated with endangered plant *Ferula sinkiangensis* KM Shen in an arid land: diversity and plant growth-promoting traits. J Arid Land 9:432–445
Lu C, Shen Y (2003) A new macrolide antibiotic with antitumor activity produced by Streptomyces sp. CS, a commensal microbe of *Maytenus hookeri*. J Antibiot 56:415–418
Lu C, Shen Y (2007) A novel ansamycin, naphthomycin K from *Streptomyces* sp. J Antibiot 60:649
Lugtenberg B, Kamilova F (2009) Plant-growth-promoting rhizobacteria. Annu Rev Microbiol 63:541–556
Lugtenberg BJ, Dekkers L, Bloemberg GV (2001) Molecular determinants of rhizosphere colonization by *Pseudomonas*. Annu Rev Phytopathol 39:461–490
Ma W et al (2010) The NS segment of an H5N1 highly pathogenic avian influenza virus (HPAIV) is sufficient to alter replication efficiency, cell tropism, and host range of an H7N1 HPAIV. J Virol 84:2122–2133
Mahaffee WF, Kloepper JW (1997) Temporal changes in the bacterial communities of soil, rhizosphere, and endorhiza associated with field-grown cucumber (*Cucumis sativus* L.). Microb Ecol 34:210–223
Malfanova N, Kamilova F, Validov S, Shcherbakov A, Chebotar V, Tikhonovich I, Lugtenberg B (2011) Characterization of *Bacillus subtilis* HC8, a novel plant-beneficial endophytic strain from giant hogweed. Microb Biotechnol 4:523–532
Malfanova N, Lugtenberg BJJ, Berg G (2013) Bacterial endophytes: who and where, and what are they doing there? In: de Bruijn FJ (ed) Molecular microbial ecology of the rhizosphere, Wiley, pp 391–403
Maloy OC (1993) Plant disease control: principles and practice. Wiley, New York
Mandal S, DebMandal M (2016) Chapter 94 – Thyme (*Thymus vulgaris* L.) oils. In: Preedy VR (ed) Essential oils in food preservation, flavor and safety. Academic, San Diego, pp 825–834
Mansoor F, Sultana V, Ehteshamul-Haque S (2007) Enhancement of biocontrol potential of *Pseudomonas aeruginosa* and *Paecilomyces lilacinus* against root rot of mungbean by a medicinal plant *Launaea nudicaulis* L. Pak J Bot 39:2113–2119
Misaghi IJ, Donndelinger CR (1990) Endophytic bacteria in symptom-free cotton plants. Phytopathology 80:808–811
Mohamad OA et al (2018a) Evaluation of the antimicrobial activity of endophytic bacterial populations from Chinese traditional medicinal plant licorice and characterization of the bioactive secondary metabolites produced by *Bacillus atrophaeus* against *Verticillium dahliae*. Front Microbiol 9:924
Mohamad OAA et al (2018b) Halophilic actinobacteria biological activity and potential applications. In: Extremophiles in Eurasian ecosystems: ecology, diversity, and applications. Springer, Cham, pp 333–364
Mundt JO, Hinkle NF (1976) Bacteria within ovules and seeds. Appl Environ Microbiol 32:694–698
Naragani K, Mangamuri U, Muvva V, Poda S, Munaganti R (2016) Antimicrobial potential of *Streptomyces cheonanensis* VUK-a from mangrove origin. Int J Pharm Pharm Sci 8:53–57
Nautiyal J, Christian M, Parker MG (2013) Distinct functions for RIP140 in development, inflammation, and metabolism. Trends Endocrinol Metab 24:451–459

Nithya V, Halami PM (2012) Novel whole-cell reporter assay for stress-based classification of antibacterial compounds produced by locally isolated *Bacillus* spp. Indian J Microbiol 52:180–184
Nongkhlaw FM, Joshi SR (2015) Investigation on the bioactivity of culturable endophytic and epiphytic bacteria associated with ethnomedicinal plants. J Infect Dev Countries 9:954–961
Okunishi S, Sako K, Mano H, Imamura A, Morisaki H (2005) Bacterial flora of endophytes in the maturing seed of cultivated rice (*Oryza sativa*). Microbes Environ 20:168–177
Passari AK, Mishra VK, Saikia R, Gupta VK, Singh BP (2015) Isolation, abundance and phylogenetic affiliation of endophytic actinomycetes associated with medicinal plants and screening for their in vitro antimicrobial biosynthetic potential. Front Microbiol 6:273
Pastor-Villaescusa B, Rangel-Huerta OD, Aguilera CM, Gil A (2015) A systematic review of the efficacy of bioactive compounds in cardiovascular disease: carbohydrates, active lipids and nitrogen compounds. Ann Nutr Metab 66:168–181
Pedrosa FO et al (2011) Genome of *Herbaspirillum seropedicae* strain SmR1, a specialized diazotrophic endophyte of tropical grasses. PLoS Genet 7:e1002064
Prasanth Reddy V, Ravi Vital K, Varsha P, Satyam S (2014) Review on *Thymus vulgaris* traditional uses and pharmacological properties. Med Aromat Plants 3:164
Pujiyanto S, Lestari Y, Suwanto A, Budiarti S, Darusman LK (2012) Alpha-glucosidase inhibitor activity and characterization of endophytic actinomycetes isolated from some Indonesian diabetic medicinal plants. Int J Pharm Pharm Sci 4:327–333
Pullen C et al (2002) New and bioactive compounds from Streptomyces strains residing in the wood of *Celastraceae*. Planta 216:162–167
Puri SC, Verma V, Amna T, Qazi GN, Spiteller M (2005) An endophytic fungus from *Nothapodytes foetida* that produces Camptothecin. J Nat Prod 68:1717–1719
Qin S et al (2012) Abundant and diverse endophytic actinobacteria associated with medicinal plant *Maytenus austroyunnanensis* in Xishuangbanna tropical rainforest revealed by culture-dependent and culture-independent methods. Environ Microbiol Rep 4:522–531
Quadt-Hallmann A, Hallmann J, Kloepper JW (1997) Bacterial endophytes in cotton: location and interaction with other plant-associated bacteria. Can J Microbiol 43:254–259
Radhakrishnan R, Lee I-J (2016) Gibberellins producing *Bacillus methylotrophicus* KE2 supports plant growth and enhances nutritional metabolites and food values of lettuce. Plant Physiol Biochem 109:181–189
Radhakrishnan R, Hashem A, Abd Allah EF (2017) Bacillus: a biological tool for crop improvement through bio-molecular changes in adverse environments. Front Physiol 8:667
Raio A, Puopolo G, Cimmino A, Danti R, Della Rocca G, Evidente A (2011) Biocontrol of cypress canker by the phenazine producer *Pseudomonas chlororaphis* subsp. aureofaciens strain M71. Biol Control 58:133–138
Rais A, Jabeen Z, Shair F, Hafeez FY, Hassan MN (2017) Bacillus spp., a bio-control agent enhances the activity of antioxidant defense enzymes in rice against *Pyricularia oryzae*. PLoS One 12:e0187412
Ramyasmruthi S, Pallavi O, Pallavi S, Tilak K, Srividya S (2012) Chitinolytic and secondary metabolite producing *Pseudomonas fluorescens* isolated from Solanaceae rhizosphere effective against broad spectrum fungal phytopathogens. Asian J Plant Sci Res 2:16–24
Reasoner DJ, Geldreich EE (1985) A new medium for the enumeration and subculture of bacteria from potable water. Appl Environ Microbiol 49:1–7
Reinhold-Hurek B, Hurek T (2011) Living inside plants: bacterial endophytes. Curr Opin Plant Biol 14:435–443
Roncato-Maccari LD et al (2003) Endophytic *Herbaspirillum seropedicae* expresses nif genes in gramineous plants. FEMS Microbiol Ecol 45:39–47
Ryan RP, Germaine K, Franks A, Ryan DJ, Dowling DN (2008) Bacterial endophytes: recent developments and applications. FEMS Microbiol Lett 278:1–9
Ryan RP et al (2009) The versatility and adaptation of bacteria from the genus. Nat Rev Microbiol 7:514–525

Samish Z, Etinger-Tulczynska R (1963) Distribution of bacteria within the tissue of healthy tomatoes. Appl Microbiol 11:7–10

Sanchez A, Thijs S, Beckers B, Gonzalez Chavez MDC, Weyens N, Carrillo R, Vangronsveld J (2017) Community structure and diversity of endophytic bacteria in seeds of three consecutive generations of *Crotalaria pumila* growing on metal mine residues. 422(1–2):51–66

Sánchez-López AS, Thijs S, Beckers B, González-Chávez MC, Weyens N, Carrillo-González R, Vangronsveld J (2018) Community structure and diversity of endophytic bacteria in seeds of three consecutive generations of *Crotalaria pumila* growing on metal mine residues. Plant Soil 422:51–66

Santhanam R, Groten K, Meldau DG, Baldwin IT (2014) Analysis of plant-bacteria interactions in their native habitat: bacterial communities associated with wild tobacco are independent of endogenous jasmonic acid levels and developmental stages. PLoS One 9:e94710

Saraf M, Pandya U, Thakkar A, Patel P (2013) Evaluation of rhizobacterial isolates for their biocontrol potential of seed borne fungal pathogens of *Jatropha curcas* L. Int J Innov Res Sci Eng Technol 2:7560–7566

Schulz B, Wanke U, Draeger S, Aust HJ (1993) Endophytes from herbaceous plants and shrubs: effectiveness of surface sterilization methods. Mycol Res 97:1447–1450

Sekar S, Kandavel D (2010) Interaction of plant growth promoting rhizobacteria (PGPR) and endophytes with medicinal plants "new avenues for phytochemicals." J Phytol 2(7):91–100

Sharma D, Pramanik A, Agrawal PK (2016) Evaluation of bioactive secondary metabolites from endophytic fungus *Pestalotiopsis neglecta* BAB-5510 isolated from leaves of Cupressus torulosa D. Don. 3 Biotech 6:210

Sieber TN (2002) Fungal root endophytes. In: Waisel Y, Eshel A, Kafkafi U (eds) Plant roots: the hidden half third ed. Marcel Dekker, New York/Basel, pp 887–917

Singh JS (2013) Plant growth promoting rhizobacteria. Resonance 18:275–281

Singh R, Soni SK, Kalra A (2013) Synergy between *Glomus fasciculatum* and a beneficial Pseudomonas in reducing root diseases and improving yield and forskolin content in *Coleus forskohlii* Briq. under organic field conditions. Mycorrhiza 23:35–44

Song S-Q, Otkur M, Zhang Z-D, Tang Q-Y (1992) Isolation and characterization of endophytic microorganisms in *Glaycyrrhiza inflat* Bat. from Xinjiang. Microbiology

Song M, Yun HY, Kim YH (2014) Antagonistic Bacillus species as a biological control of ginseng root rot caused by *Fusarium cf. incarnatum*. J Ginseng Res 38:136–145

Strobel G, Daisy B (2003) Bioprospecting for microbial endophytes and their natural products. Microbiol Mol Biol Rev 67:491–502

Strobel GA, Dirkse E, Sears J, Markworth C (2001) Volatile antimicrobials from *Muscodor albus*, a novel endophytic fungus. Microbiology 147:2943–2950

Strobel G, Daisy B, Castillo U, Harper J (2004) Natural products from endophytic microorganisms. J Nat Prod 67:257–268

Sturz AV (1995) The role of endophytic bacteria during seed piece decay and potato tuberization. Plant Soil 175:257–263

Suresh A, Pallavi P, Srinivas P, Kumar VP, Reddy SR (2010) Plant growth promoting activities of *fluorescent pseudomonads* associated with some crop plants. Afr J Microbiol Res 4:1491–1494

Suryanarayanan TS, Wittlinger SK, Faeth SH (2005) Endophytic fungi associated with cacti in Arizona. Mycol Res 109:635–639

Suzuki T, Shimizu M, Meguro A, Hasegawa S, Nishimura T, Kunoh H (2005) Visualization of infection of an endophytic actinomycete *Streptomyces galbus* in leaves of tissue-cultured. Rhododendron 19:7–12

Taechowisan T, Lu C, Shen Y, Lumyong S (2005) Secondary metabolites from endophytic *Streptomyces aureofaciens* CMUAc130 and their antifungal activity. Microbiology 151:1691–1695

Taechowisan T, Wanbanjob A, Tuntiwachwuttikul P, Taylor WC (2006) Identification of *Streptomyces* sp. Tc022, an endophyte in *Alpinia galanga*, and the isolation of actinomycin D. Ann Microbiol 56:113–117

Taechowisan T, Lu C, Shen Y, Lumyong S (2007) Antitumor activity of 4-arylcoumarins from endophytic *Streptomyces aureofaciens* CMUAc130. J Cancer Res Ther 3:86

Taghavi S et al (2009) Genome survey and characterization of endophytic bacteria exhibiting a beneficial effect on growth and development of poplar trees. Appl Environ Microbiol 75:748–757

Taghavi S et al (2010) Genome sequence of the plant growth promoting endophytic bacterium *Enterobacter* sp. 638. PLoS Genet 6:e1000943

Tank N, Saraf M (2003) Phosphate solubilization, exopolysaccharide production and indole acetic acid secretion by rhizobacteria isolated from *Trigonella foenum-graecum*. Indian J Microbiol 43:37–40

Timmusk S, Grantcharova N, Wagner EGH (2005) *Paenibacillus polymyxa* invades plant roots and forms biofilms. Appl Environ Microbiol 71:7292–7300

Vessey JK (2003) Plant growth promoting rhizobacteria as biofertilizers. Plant Soil 255:571–586

Vieira ML et al (2011) Diversity and antimicrobial activities of the fungal endophyte community associated with the traditional Brazilian medicinal plant *Solanum cernuum* Vell.(Solanaceae). Can J Microbiol 58:54–66

Wahyudi AT, Astuti RP, Widyawati A, Mery A, Nawangsih AA (2011) Characterization of *Bacillus* sp. strains isolated from rhizosphere of soybean plants for their use as potential plant growth for promoting rhizobacteria. J Microbiol Antimicrob 3:34–40

Wang P, Kong F, Wei J, Wang Y, Wang W, Hong K, Zhu W (2014) Alkaloids from the mangrove-derived actinomycete *Jishengella endophytica* 161111. Mar Drugs 12:477–490

Weilharter A, Mitter B, Shin MV, Chain PSG, Nowak J, Sessitsch A (2011) Complete genome sequence of the plant growth-promoting endophyte *Burkholderia phytofirmans* strain PsJN. J Bacteriol 193:3383–3384

Wu L, Han T, Li W, Jia M, Xue L, Rahman K, Qin L (2013) Geographic and tissue influences on endophytic fungal communities of *Taxus chinensis* var. mairei in China. Curr Microbiol 66:40–48

Yang A, Akhtar SS, Iqbal S, Amjad M, Naveed M, Zahir ZA, Jacobsen S-E (2016) Enhancing salt tolerance in quinoa by halotolerant bacterial inoculation. Funct Plant Biol 43:632–642

Yi H-S, Yang JW, Ryu C-M (2013) ISR meets SAR outside: additive action of the endophyte *Bacillus pumilus* INR7 and the chemical inducer, benzothiadiazole, on induced resistance against bacterial spot in field-grown pepper. Front Plant Sci 4:122

Young J (1992) Phylogenetic classification of nitrogen-fixing organisms. Biol Nitrogen Fixation 1544:43–86

Zachow C, Fatehi J, Cardinale M, Tilcher R, Berg G (2010) Strain-specific colonization pattern of *Rhizoctonia antagonists* in the root system of sugar beet. FEMS Microbiol Ecol 74:124–135

Zhang S, Qi-Yong Z (2007) Isolation and characterization of endophytic microorganisms in *Glaycyrrhiza inflat* bat. from Xinjiang. J Microbiol 5:014

Zhang J, Wang J-D, Liu C-X, Yuan J-H, Wang X-J, Xiang W-S (2014) A new prenylated indole derivative from endophytic actinobacteria *Streptomyces* sp. neau-D50. Nat Prod Res 28:431–437

Zhou H, Yang Y, Peng T, Li W, Zhao L, Xu L, Ding Z (2014) Metabolites of *Streptomyces* sp., an endophytic actinomycete from *Alpinia oxyphylla*. Nat Prod Res 28:265–267

Zinniel DK et al (2002) Isolation and characterization of endophytic colonizing bacteria from agronomic crops and prairie plants. Appl Environ Microbiol 68:2198–2208

Chapter 11
Endophytic Microbial Diversity in the Halophytic Medicinal Plant *Ferula* and Their Bioapplicable Traits

Nimaichand Salam, Mipeshwaree Devi Asem, Yong-Hong Liu, Min Xiao, and Wen-Jun Li

Abstract *Ferula* spp. are halophytic plants grown in arid region and are known for their traditional medicinal values. Large-scale excavation for extraction of medicinal compounds and other man-made destructions have severely reduced the number of wild *Ferula* and even extinction of few plant species. Alternate way to explore the bioapplicability of such plants is to isolate the microbial communities from these plant tissues and screen for their beneficial effects. Analysis of microbial diversity in two *Ferula* spp. indicated that members of phylum *Actinobacteria* are quite prominent besides members of the phyla *Proteobacteria* and *Firmicutes*. Many constituent members of these phyla are known to produce chemically active biomolecules.

Keywords Endophyte · *Ferula* spp. · *Actinobacteria* · Plant growth promotion · Biocontrol activity

N. Salam · M. D. Asem · M. Xiao
State Key Laboratory of Biocontrol and Guangdong Provincial Key Laboratory of Plant Resources, School of Life Sciences, Sun Yat-Sen University, Guangzhou, People's Republic of China

Y.-H. Liu
Key Laboratory of Biogeography and Bioresource in Arid Land, Xinjiang Institute of Ecology and Geography, Chinese Academy of Sciences, Ürümqi, People's Republic of China

W.-J. Li (✉)
State Key Laboratory of Biocontrol and Guangdong Provincial Key Laboratory of Plant Resources, School of Life Sciences, Sun Yat-Sen University, Guangzhou, People's Republic of China

Key Laboratory of Biogeography and Bioresource in Arid Land, Xinjiang Institute of Ecology and Geography, Chinese Academy of Sciences, Ürümqi, People's Republic of China
e-mail: liwenjun3@mail.sysu.edu.cn

D. Egamberdieva, A. Tiezzi (eds.), *Medically Important Plant Biomes: Source of Secondary Metabolites*, Microorganisms for Sustainability 15,
https://doi.org/10.1007/978-981-13-9566-6_11

11.1 Introduction

Endophytic bacteria are endosymbionts that live within a plant for at least a part of its life cycle without causing apparent disease (Hallmann et al. 1997). The earliest report of existence of endophytic microbes dates back to 1904 (Tan and Zou 2001). The most effective study with the endophytes started with the discovery of several new bioactive molecules (Strobel and Daisy 2003). Besides, the presence of endophytes is known to be advantageous to the host plants as they can enhance growth promotion and protect them against external microbial infections (Wang 2015). The necessity for search of beneficial plant-associated microbial communities among the diverse plant varieties is considerably enormous (Matsumoto and Takahashi 2017), and the number of plants studied so far is still limited.

Ferula spp. of the family Umbelliferae are native to Mediterranean, central Asia, and its adjacent areas (Pimenov and Leanov 2004). Of the 180 species reported so far, 26 are found along in the arid region of Xinjiang Uyghur Autonomous Region, China. Many of these species are widely excavated for extraction of biomedical compounds particularly in traditional medicines such as treatments of digestive disorders, rheumatism, headache, dizziness, toothache, etc. (Sun et al. 2013). Developmental activities of modern cities further escalate the destruction of the natural ecosystems, which in turn is another cause for loss of habitats for such medicinal plants. Though the natural values of the plants can never be substituted, an alternate way to evaluate the natural compounds from the plants is to isolate the associated microbial communities from such plants rather than utilizing the plant themselves. Besides the endophytes from plant from special habitats such as *Ferula* spp. from arid Xinjiang desert are also likely to harbor special metabolites.

11.2 Isolation of Endophytic Bacteria

Isolation of endophytic bacteria depends on the effectiveness of sterilization procedure. In most cases, surface sterilization of plant tissues to be used for bacterial isolation is done by using one or more of chemical treatments (0.1% Tween 20; 5% NaClO; 2.5% $Na_2S_2O_3$; 75% ethanol; 10% $NaHCO_3$) followed by washing with sterile water (Qin et al. 2009; Li et al. 2012). Initial sterilization of the plants is usually done by washing under tap water (Liu et al. 2016) or using ultrasonic water bath (Qin et al. 2009) to remove any adhering soil and subsequently drying under normal conditions. Another important factor that needs to be taken care of is the selection of disease symptoms-free tissues and segregation of the plant part into different tissues which in most cases are leaves, stems, and roots (Liu et al. 2016, 2017). During the isolation of bacteria from two species of *Ferula* collected from Xinjiang province in P.R. China, a three-step sterilization procedure was employed (Liu et al. 2016, 2017). The first step of this sterilization includes washing in 75% ethanol for 1 min. The sterilized tissues are then treated with sodium hypochlorite solution for

8 min and are then washed five times using sterile water. In earlier report for isolation of endophytic actinobacteria from *Artemisia annua* (Li et al. 2012), prior to the last sterilization step, an additional step was added to remove the possible residual chlorine by rinsing with 2.5% sodium thiosulfate for 10 min (Miche and Balandreau 2001). Checking the effectiveness of the surface sterilization can be carried out either by imprinting the surface-sterilized tissue onto the isolation media or by inoculating the sterile distilled water used in the final rinse on the isolation media. The sterilization is considered to be effective if no microbial growth is observed.

The next step toward isolation of endophytic bacteria involved suspension of the sterilized tissues onto the isolation media. In the two reported cases for isolation from *Ferula* spp., the surface-sterilized tissues were air-dried for 2 days at normal temperature and aseptically blended using sterilized commercial blender (Liu et al. 2016, 2017). The homogenates are then placed directly onto the bacterial isolation media or made into tissue suspensions and inoculate to the media. Other methods may involve chemical pretreatment of the tissue homogenates before inoculation, or the tissue might be placed directly on the media and check for external outgrowth from within the tissue onto the media.

11.3 Endophytic Microbial Diversity in *Ferula songorica* vs *Ferula sinkiangensis*

Cultivable endophytic diversity in the two species of species is reported to be quite enormous (Liu et al. 2016, 2017). The bacteria isolates obtained from three different types of tissues of *Ferula songorica* represented 58 taxa distributed along three bacterial phyla (Liu et al. 2016). As reported with many other endophytic bacterial diversity (Chen et al. 2012; Ma et al. 2013), root harbors the maximum diversity of the isolates accounting 52% of the total isolates. Distribution pattern of the strains indicated that the maximum diversity is related with the members of the phylum *Actinobacteria*. This diversity of *Actinobacteria* within plant tissues might be related with the distinct properties of *Actinobacteria* to produce different cell wall-degrading enzymes and their ability to regulate the root exudates (Nimaichand et al. 2016). The distribution of the different bacterial strains isolated from *Ferula songorica* is listed in Table 11.1.

Maximum numbers of strains among the cultivable strains from the tissues of *Ferula songorica* are, however, related with the proteobacterial genera *Brevundiomonas* and *Sphingomonas* and the spore forming firmicutes *Bacillus* (Liu et al. 2016). A representative dendrogram of selective strains is presented in Fig. 11.1. The isolation of these genera is biotechnologically and ecologically significant. First, members of the *Brevundimonas* are particularly associated with soil, and their abundance in root tissues might be influenced by the soil bacterial communities colonizing the rhizosphere (Long et al. 2010). Similarly, the isolation of *Sphingomonas* and *Bacillus* could be related with the differential biocontrol and

Table 11.1 Distribution of strains isolated from *Ferula songorica*

Phyla	Orders	Families	Genera	Species	Strains
Actinobacteria	*Micrococcales*	*Microbacteriaceae*	*Microbacterium*	3	9
			Curtobacterium	1	1
			Agrococcus	1	3
		Micrococcaceae	*Arthrobacter*	4	5
			Kocuria	3	11
			Micrococcus	3	5
		Brevibacteriaceae	*Brevibacterium*	1	1
		Promicromonosporaceae	*Promicromonospora*	2	5
	Streptosporangiales	*Nocardiopsaceae*	*Nocardiopsis*	3	8
	Corynebacteriales	*Nocardiaceae*	*Williamsia*	1	4
			Rhodococcus	1	1
			Nocardia	1	1
	Streptomycetaceae	*Streptomycetales*	*Streptomyces*	5	7
	Pseudonocardiales	*Pseudonocardiaceae*	*Pseudonocardia*	2	2
			Saccharopolyspora	1	1
	Propionibacteriales	*Nocardioidaceae*	*Nocardioides*	1	1
	Corynebacteriales	*Dietziaceae*	*Dietzia*	2	2
Proteobacteria	*Sphingomonadales*	*Sphingomonadaceae*	*Sphingomonas*	4	22
	Pseudomonadales	*Moraxellaceae*	*Acinetobacter*	2	7
	Gammaproteobacteria	*Xanthomonadales*	*Pseudomonas*	1	1
	Rhizobiales	*Methylobacteriaceae*	*Methylobacterium*	2	7
		Rhizobiaceae	*Rhizobium*	1	14
	Rhodobacterales	*Rhodobacteraceae*	*Paracoccus*	1	1
	Burkholderiales	*Burkholderiaceae*	*Ralstonia*	1	5
	Caulobacterales	*Caulobacteraceae*	*Brevundimonas*	1	23
Firmicutes	*Bacillales*	*Paenibacillaceae*	*Paenibacillus*	1	1
		Bacillaceae	*Bacillus*	9	22

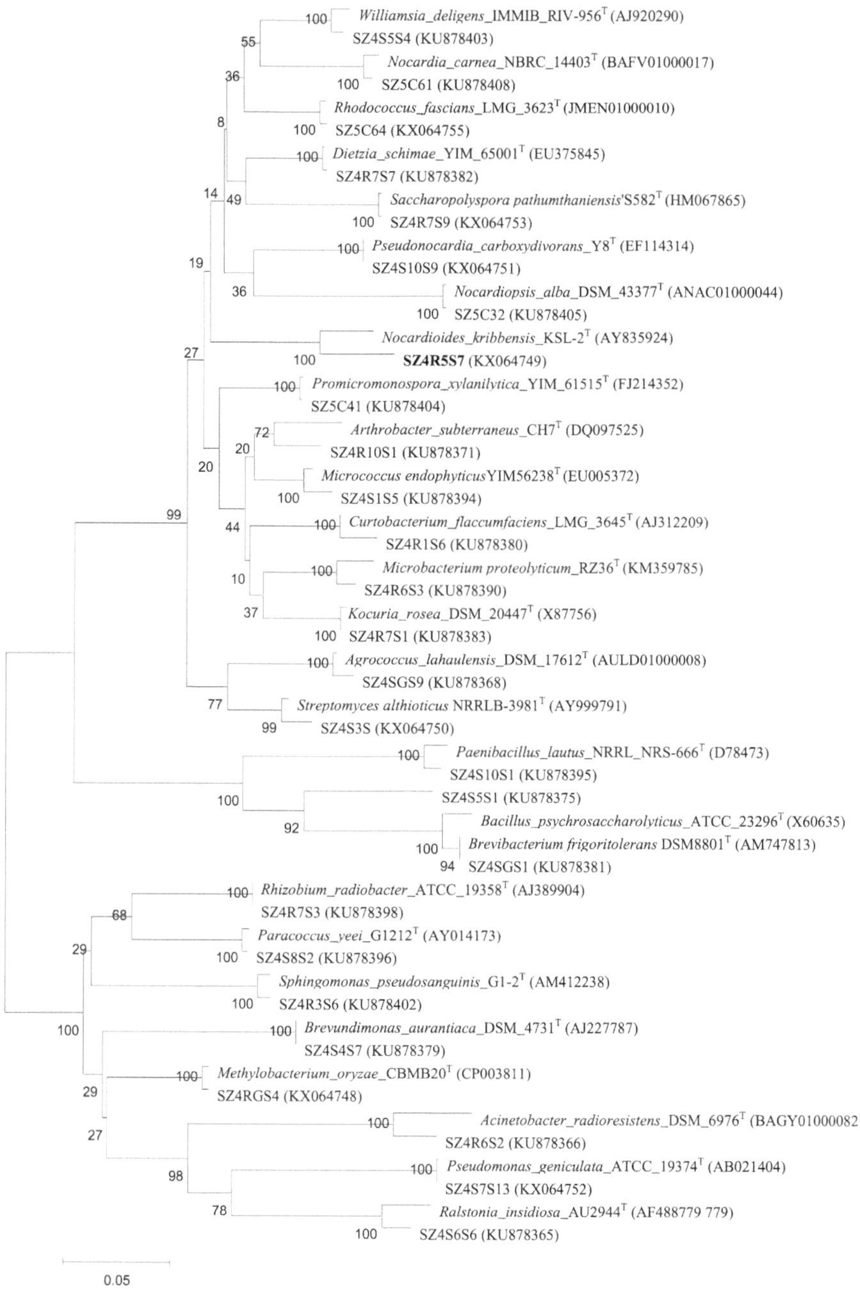

Fig. 11.1 Maximum likelihood tree based on 16S rRNA gene sequence representing few isolated endophytic bacteria from the tissues of *Ferula songorica*

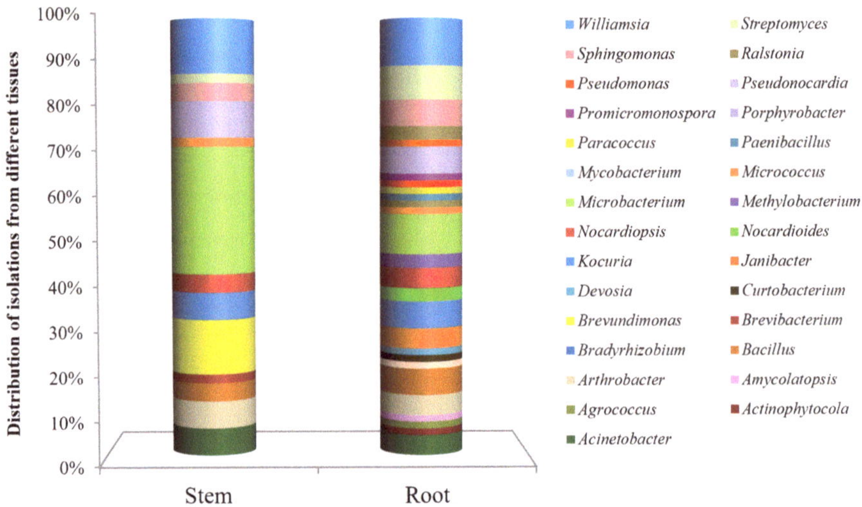

Fig. 11.2 Barplots indicating the distribution of endophytes isolated from tissues of *Ferula sinkiangensis*

plant growth-promoting ability they exert on the host plants, as reported with many other cereal crops (Kumar et al. 2012; Chen et al. 2014; Khan et al. 2014; Yang et al. 2014; Halo et al. 2015).

Similar to the above findings, endophytes isolated from the tissues of *Ferula sinkiangensis* are also relates with the phyla *Actinobacteria*, *Proteobacteria*, and *Firmicutes*, based on the 16S rRNA gene sequencing profile. They are classified into 13 orders, 23 families, and 29 genera, with maximum diversity among the phylum *Actinobacteria* (Fig. 11.2). Maximum abundance are related to families *Microbacteriaceae*, *Micrococcaceae*, and *Nocardiaceae* and are mostly isolated from root tissues, indicating these groups might be laterally transferred from rhizosphere (Santoyo et al. 2016).

11.4 Bioapplicable Traits

Naturally, the interactions between plant and endophyte range from commensalism to mutualism to pathogenicity, but these interactions may well be influenced by the external environmental conditions (Hardoim et al. 2015). The ecological significance of the symbiosis of endophytes is still to be properly decipher, but they are known to implicate considerable biological traits in the host plants, such as helping them in stress amelioration, enhancing plant fitness, etc. (Aschehoug et al. 2014; Karpinets et al. 2014; Nimaichand et al. 2016), while some have no apparent role on plant performance (Hardoim et al. 2015). The most widely studied beneficial effects of endophytes are the plant growth promotion activity and protection against biotic and abiotic stresses including iron homeostasis, production of secondary metabolites, induce systemic resistance, and nitrogen fixation (Hardoim et al. 2015). Besides, endophytes may secrete additional metabolites which are directly necessary for the growth and metabolism of the host plant or used indirectly to regulate the exudates concentration (Compant et al. 2010). However, the effect of endophytes on the host needs not to be necessarily effected by all the constituent microbiomes but is likely strain specific (Newcombe et al. 2009). The endophytes from the *Ferula* spp. are analyzed with direct growth-promoting traits (nitrogen fixation, phosphate solubilization, phytohormone, and siderophore production) as well as indirect traits such as antagonistic assays.

The plant growth potential traits of endophytes isolated from *Ferula sinkiangensis* (Liu et al. 2017) are represented in Table 11.2. Interestingly most of the strains that showed positive activities for phosphate solubilization are limited to members of the genera *Acinetobacter*, *Microbacterium*, and *Ralstonia*. These strains are also reported with other plant growth-promoting traits in other plant as well (Qin et al. 2014). The nitrogen-fixing abilities are relatively expressed by at least 75% of the endophyte isolated from *Ferula sinkiangensis* (Liu et al. 2017), indicating that biological nitrogen fixation is an important means to access the un-utilizable nitrogen from atmosphere by *Ferula* spp. found in Xinjiang. Production of phytohromones such as indole acetic acid in soil and siderophores by endophytes is another means for improving the fitness of the host plants by increasing mineral, nutrients, and iron uptake (Nimaichand et al. 2016). In the study of *Ferula* spp., 79% of the isolates, most prominently the *Bacillus* strains, are reported with indole acetic acid capacity, while 57% are capable of siderophore formation. These functions are also associated with increasing yield with cereal group as well (Tamreihao et al. 2018).

For the antagonistic assay against the causative agents of *Verticullium* wilt and leaf blight, most of the Firmicutes particularly the *Bacillus* strains showed strong inhibitory action against the pathogenic fungi *Alternaria alternata* and *Verticillium dahlia* (Table 11.2). Besides the production of bioactive secondary metabolites, production of siderophores by beneficial microbes was also reported to be an alternative route of biocontrol activity (Schippers et al. 1987).

11.5 Conclusion

It is assumed that endophyte exhibits different characteristics inside the plant depending on the physiological conditions of the host during the period of colonization. Applicability of the beneficial effects under in vivo condition may therefore be species specific and the environmental condition. To extract any beneficial traits from endophytic bacteria in plant with important biotechnological effects such as *Ferula* spp. which possesses medicinal values, besides having its adaptive capability to grow under arid condition, a concerted study involving modern technologies, especially the "-omics" technologies along with environmental parameters, may be necessary.

Table 11.2 Direct and indirect plant growth-promoting traits of the endophytes from *Ferula sinkiangensis*

Strain	Closely related homolog	Tissue	Growth-promoting activities				Biocontrol potential		
			Phosphate solubilization	Nitrogen fixation	IAA	Siderophore production	*Alternaria alternata.*	*Verticillium dahlia* 991	*Verticillium dahlia* 7
S5S6	*Acinetobacter* spp.	Stem	++	++	+	–	+	–	–
S10S4	*Acinetobacter* spp.	Stem	+++	+	+	–	+	–	++
R4S5	*Acinetobacter* spp.	Root	+++	+	+	+	+	–	+
R4S7	*Acinetobacter* spp.	Root	++	+	+	+	+	–	–
3A6	*Bacillus* spp.	Root	–	+	+++	++	++	+++	++
3A8	*Bacillus* spp.	Root	–	++	+++	+	+++	+++	+++
4B5	*Bacillus* spp.	Root	–	–	++	–	+	+	+
S3S4	*Bacillus* spp.	Stem	–	–	++	+	+	–	–
S4S19	*Bacillus* spp.	Stem	–	–	+++	++	+++	+++	+++
S5S2	*Bacillus* spp.	Stem	–	+	++	–	–	–	–
R5S1	*Bacillus* spp.	Root	–	+	++	+	+	–	–
R10S11	*Bacillus* spp.	Root	–	+	+++	+++	+++	++	+++
S1S8	*Brevibacterium* spp.	Stem	–	–	+	++	+	++	–
S10S2	*Brevibacterium* spp.	Stem	–	+	++	++	–	–	+
S3S6	*Brevibacterium* spp.	Stem	–	+	+	+	–	+	–
S3S7	*Brevibacterium* spp.	Stem	–	–	+	+	+	–	–
R4S4	*Janibacter* spp.	Root	–	+	++	+	–	–	–
S3S2	*Kocuria* spp.	Stem	–	+	/	/	+	–	–
S5S11	*Kocuria* spp.	Stem	–	+	++	–	+	–	+++
R7S1	*Kocuria* spp.	Root	–	+++	++	–	–	–	–
S3S1	*Microbacterium* spp.	Stem	–	++	+	+	+++	–	–
S3S9	*Microbacterium* spp.	Stem	+	++	+	–	+	–	–

S3S10	*Microbacterium* spp.	Stem	−	++	++	+	+	−	−
S3S12	*Microbacterium* spp.	Stem	−	++	+	−	−	+	+
S4S13	*Microbacterium* spp.	Stem	−	−	++	+	+	−	−
S4S17	*Microbacterium* spp.	Stem	−	+++	+	++	+	−	−
S5S4	*Microbacterium* spp.	Stem	−	−	++	+	+	−	−
S6S5	*Microbacterium* spp.	Stem	−	+	++	+	−	++	−
S6S8	*Microbacterium* spp.	Stem	−	+++	−	+	−	−	−
R3S6	*Microbacterium* spp.	Root	−	+	+	+	−	+	+
R6S4	*Microbacterium* spp.	Root	−	+++	+	+	−	+	−
S10S3	*Micrococcus* spp.	Stem	−	+	+	−	−	−	++
RGS8	*Porphyrobacter* spp.	Root	/	/	++	−	+	−	−
R1S10	*Ralstonia* spp.	Root	++	−	+	+	−	−	−
S4S2	*Sphingomonas* spp.	Stem	−	+	/	/	+	−	−
S7S1	*Sphingomonas* spp.	Stem	−	++	+	+	/	/	/
S1S2	*Williamsia* spp.	Stem	−	+	−	+	−	−	−

(continued)

Table 11.2 (continued)

Strain	Closely related homolog	Tissue	Growth-promoting activities				Biocontrol potential		
			Phosphate solubilization	Nitrogen fixation	IAA	Siderophore production	*Alternaria alternata.*	*Verticillium dahlia* 991	*Verticillium dahlia* 7
S4S1	*Williamsia* spp.	Stem	–	+	–	+	–	–	–
S6S10	*Williamsia* spp.	Stem	–	++	–	++	–	–	+
R3S10	*Williamsia* spp.	Root	–	++	–	–	++	–	–
R3S13	*Williamsia* spp.	Root	–	++	++	–	–	+	–
R6S11	*Williamsia* spp.	Root	–	++	–	+	–	–	–

The ability of growth promotion potential were represented by "+", "-", or "/". "+++", strongly positive; "++", moderately positive; "+", weakly positive; "-", negative; and "/", not tested

References

Aschehoug ET, Callaway RM, Newcombe G, Tharayil N, Chen S (2014) Fungal endophyte increases the allelopathic effects of an invasive forb. Oecologia 175:285–291

Chen WM, Tang YQ, Mori K, Wu XL (2012) Distribution of culturable endophytic bacteria in aquatic plants and their potential for bioremediation in polluted waters. Aquat Biol 15:99–110

Chen B, Zhang Y, Rafiq MT, Khan KY, Pan F, Yang X, Feng Y (2014) Improvement of cadmium uptake and accumulation in *Sedum alfredii* by endophytic bacteria *Spingomonas* SaMR12: effects on plant growth and root exudates. Chemosphere 117:367–373

Compant S, Clement C, Sessitsch A (2010) Plant growth-promoting bacteria in the rhizo- and endosphere of plants: their role, colonization, mechanisms involved and prospects for utilization. Soil Biol Biochem 42:669–678

Hallmann J, Quadt-Hallmann A, Mahaffee WF, Kloepper JW (1997) Bacterial endophytes in agricultural crops. Can J Microbiol 43:895–914

Halo BA, Khan AL, Waqas M, Al-Harrasi A, Hussain J, Ali L, Adnan M, Lee IJ (2015) Endophytic bacteria (*Sphingomonas* sp. LK11) and gibberellin can improve *Solanum lycopersicum* growth and oxidative stress under salinity. J Plant Interact 10:117–125

Hardoim PR, van Overbeek LS, Berg G, Pirttila AM, Compant S, Campisano A, Döring M, Sessitsch A (2015) The hidden world within plants: ecological and evolutionary considerations for defining functioning of microbial endophytes. Microbiol Mol Biol Rev 79:293–320

Karpinets TV, Park BH, Syed MH, Klotz MG, Uberbacher EC (2014) Metabolic environments and genomic features associated with pathogenic and mutualistic interactions between bacteria and plants. Mol Plant-Microbe Interact 27:664–677

Khan AL, Waqas M, Kang SM, Al-Harrasi A, Hussain J, Al-Rawahi A, Al-Khiziri S, Ullah I, Ali L, Jung HY, Lee IJ (2014) Bacterial endophyte *Sphingomonas* sp. LK11 produces gibberellins and IAA and promotes tomato plant growth. J Microbiol 52:689–695

Kumar P, Khare S, Dubey RC (2012) Diversity of bacilli from disease suppressive soil and their role in plant growth promotion and yield enhancement. New York Sci J 5:90–111

Li J, Zhao GZ, Huang HY, Qin S, Zhu WY, Zhao LX, Xu LH, Zhang S, Li WJ, Strobel G (2012) Isolation and characterization of culturable endophytic actinobacteria associated with *Artemisia annua* L. Antonie Van Leeuwenhoek 101:515–527

Liu YH, Guo JW, Salam N, Li L, Zhang YG, Han J, Mohamad OA, Li WJ (2016) Culturable endophytic bacteria associated with medicinal plant *Ferula songorica*: molecular phylogeny, distribution and screening for industrially important traits. 3 Biotech 6:209. https://doi.org/10.1007/s13205-016-0522-7

Liu Y, Guo J, Li L, Asem MD, Zhang Y, Mohamad OA, Salam N, Li W (2017) Endophytic bacteria associated with endangered plant *Ferula sinkiangensis* K.M. Shen in an arid land: diversity and plant growth-promoting traits. J Arid Land 9:432–445

Long HH, Sonntag DG, Schmidt DD, Baldwin IT (2010) The structure of the culturable root bacterial endophyte community of *Nicotiana attenuate* is organized by soil composition and host plant ethylene production and perception. New Phytol 185:554–567

Ma B, Lv X, Warren A, Gong J (2013) Shifts in diversity and community structure of endophytic bacteria and archaea across root, stem and leaf tissues in the common reed, *Phragmites australis*, along a salinity gradient in a marine tidal wetland of northern China. Antonie Van Leeuwenhoek 104:759–768

Matsumoto A, Takahashi Y (2017) Endophytic actinomycetes: promising source of novel bioactive compounds. J Antibiot 70:514–519

Miche L, Balandreau J (2001) Effects of rice seed surface sterilization with hypochlorite on inoculated *Burkholderia vietnamiensis*. Appl Environ Microbiol 67:3046–3052

Newcombe G, Shipunov A, Eigenbrode S, Raghavendra AK, Ding H, Anderson CL, Menjivar R, Crawford M, Schwarzlander M (2009) Endophytes influence protection and growth of an invasive plant. Commun Integr Biol 2:29–31

Nimaichand S, Devi AM, Li WJ (2016) Direct plant growth-promoting ability of *Actinobacteria* in grain legumes. In: Subramanium G, Arumugam S, Rajendran V (eds) Plant growth promoting Actinobacteria. Springer, Singapore, pp 1–16

Pimenov MG, Leonov MV (2004) The Asian *Umbelliferae* biodiversity database (ASIUM) with particular reference to South-West Asian taxa. Turk J Bot 28:139–145

Qin S, Li J, Chen HH, Zhao GZ, Zhu WY, Jiang CL, Xu LH, Li WJ (2009) Isolation, diversity, and antimicrobial activity of rare actinobacteria from medicinal plants of tropical rain forests in Xishuangbanna, China. Appl Environ Microbiol 75:6176–6186

Qin S, Zhang YJ, Yuan B, Xu PY, Xing K, Wang J, Jiang JH (2014) Isolation of ACC deaminase-producing habitat-adapted symbiotic bacteria associated with halophyte *Limonium sinense* (Girard) Kuntze and evaluating their plant growth-promoting activity under salt stress. Plant Soil 374:753–766

Santoyo G, Moreno-Hagelsieb G, del Carment Orozco-Mosqueda M, Glick BR (2016) Plant growth-promoting bacterial endophytes. Microbiol Res 183:92–99

Schippers B, Bakker AW, Bakker PAHM (1987) Interactions of deleterious and beneficial rhizosphere microorganisms and the effect of cropping practices. Annu Rev Phytopathol 25:339–358

Strobel G, Daisy B (2003) Bioprospecting for microbial endophytes and their natural products. Microbiol Mol Biol Rev 67:491–502

Sun L, Shi SB, Zhu J, Li XJ (2013) Traditional use of modern research review of *Ferula* in China. Mod Chinese Med 15:620–626

Tamreihao K, Nimaichand S, Ningthoujam DS (2018) Antagonistic and plant growth promoting *Bacillus* sp. MBRL 576 enhances the growth and yield of rice (*Oryza sativa* L.). Oryza 55:202–213

Tan RX, Zou WX (2001) Endophytes: a rich source of functional metabolites. Nat Prod Rep 18:448–459

Wang HF (2015) Study of biodiversity of endophytic bacteria isolated from four *Chenopodiaceae* halophytes in Xinjiang and evaluation of their growth-promoting function and salt-tolerance ability. PhD dissertation. Xinjiang Institute of Ecology and Geography, Chinese Academy of Sciences, Ürümqi (in Chinese)

Yang S, Zhang X, Cao Z, Zhao K, Wang S, Chen M, Hu X (2014) Growth-promoting *Sphingomonas paucimobilis* ZJSH1 associated with *Dendrobium officinale* through phytohormone production and nitrogen fixation. Microb Biotechnol 7:611–620

Chapter 12
Endophytic Bacteria-Mediated Regulation of Secondary Metabolites for the Growth Induction in *Hyptis suaveolens* Under Stress

Yachana Jha

Abstract Two endophytic bacteria *Pseudomonas pseudoalcaligenes* and *Pseudomonas aeruginosa* have been isolated from the root of paddy and *Suaeda nudiflora* wild mosque plant to evaluate their growth promoting ability in one of the important medicinal plant *Hyptis suaveolens* under stress. The endophytic isolates have been inoculated either alone or in combination in *Hyptis suaveolens* and have enhanced the growth and essential oil content as well as yield of the *Hyptis suaveolens* under normal condition and under stress. It reduces the negative effects of stress and shows enhanced plant growth compared to control, while the essential oil content and yield decrease due to inoculation of endophytic bacteria compared to control. But due to enhanced vegetative growth, overall yield increases. Inoculation with endophytic bacteria increases the phenolic content under normal condition and decreases under stress. Medicinal plants have bioactive compounds which are used for curing of various human diseases and also play an important role in healing of different diseases. So inoculation with such beneficial bacteria serves as a better option for utilization of degraded wastelands for cultivation of medicinal and aromatic plants.

Keywords Endophytic bacteria · *Hyptis suaveolens* · Secondary metabolite · Essential oil · Stress

---

Y. Jha (✉)
N. V. Patel College of Pure and Applied Sciences, S. P. University, Anand, Gujarat, India

D. Egamberdieva, A. Tiezzi (eds.), *Medically Important Plant Biomes: Source of Secondary Metabolites*, Microorganisms for Sustainability 15,
https://doi.org/10.1007/978-981-13-9566-6_12

12.1 Introduction

Nature has been a source of medicinal agents for thousands of years, and an impressive number of modern drugs have been isolated from natural sources. The cultivation of medicinal and aromatic plants has gained more importance, and the trade in products of these plants is estimated to be over US $ 3000 million per annum. *Hyptis suaveolens* (L.) Poit plant commonly known as Wilayati tulsi belongs to the family *Lamiaceae* and is an ethnobotanically important medicinal plant. It is a coarse, erect, branched, and hairy annual herb commonly called wild spikenard. The plant is bitter, minty, and aromatic. The plant has been considered as an obnoxious weed, distributed throughout the tropics and subtropics. Almost all parts of this plant are being used in traditional medicine to treat various diseases, as its medicinal constituents include the strong essential oils, tannin, flavonoids, etc. The phytochemical screening of leaf extracts shows the presence of secondary metabolites and essential oils having antibacterial activity. Essential oils and secondary metabolites are the major reason for medicinal properties of *Hyptis suaveolens* (L.) Poit plant (Moreira et al. 2010). The leaves have been utilized as a stimulant, carminative, sudorific, galatogogue, and to cure parasitic cutaneous diseases (Mandal et al. 2007). It has also been used as analgesic, decongestant, and antipyretic and stimulates blood circulation. Decoction of the herb as tea is effective for fever associated with cold and flatulence. *Hyptis suaveolens* (L.) Poit is a medicinal plant used in traditional medicines. In higher plants a wide variety of secondary metabolites are synthesized from primary metabolites and are needed in plant defense against stress like protection against pathogens or environmental stresses. Plant secondary metabolites are unique sources for food additives, flavors, and pharmaceuticals, and such chemicals include calcium, abscisic acid (ABA), salicylic acid (SA), polyamines, jasmonates (JA), and nitric oxide which are involved in stress responses in plants (Tuteja and Sopory 2008). The accumulation of such natural products strongly depends on the growing conditions, such as the temperature, the light regime, and the nutrient supply (Ballhorn et al. 2011). The levels and composition of secondary metabolites in plants vary according to genotype, climate factors such as seasonal variation, light intensity, relative humidity and temperature, environment stimuli, and agronomical practices. Cultivation factors such as soil type, compost, mulching, and fertilization can also affect the plant secondary metabolites. In addition, more severe environmental influences, such as various stress conditions, will also impact on the metabolic pathways responsible for the accumulation of secondary plant products. Plant secondary metabolites are often referred to as compounds that have no fundamental role in the maintenance of life processes in the plants, but they are important for the plant to interact with its environment for adaptation and defense. The generation of ROS in plants is triggered by different kinds of environmental stresses, such as high light, high or low temperature, salinity, drought, nutrient deficiency, and pathogen attack. ROS also act as signaling molecules involved in growth and developmental processes, pathogen defense responses such as hypersensitive reaction and systemic acquired resistance, stress hormone production, acclimation, and programmed cell death (Apel and Hirt 2004). It is well-known that free radical and other reactive oxygen species formed in the living cells play an important role in metabolism. Natural products from medicinal plants are known to be chemically balanced, effective, and least injurious with none or much reduced side effects as compared to synthetic medicines. The objective of this study is to investigate the effect of endophytic bacteria, inoculated on the growth, secondary metabolite production, and oil yield of *Hyptis suaveolens* plants. The collaboration of endophytic bacteria and their effect on the biological growth response of plants under stress is complex. The inoculation of endophytic bacteria in plant, alone or in groups, can confer tolerance to plant against adverse environmental condition, improves nutrient availability, and helps the plant to overcome stress by regulating secondary metabolite production.

12.2 Secondary Metabolites of Plant

Secondary metabolites are organic molecules that are not involved in the normal growth and development of an organism. These compounds are an extremely diverse group of natural products synthesized by plants, fungi, bacteria, algae, and animals. Most of secondary metabolites, such as terpenes, phenolic, and alkaloids, are classified based on their biosynthetic origin. Different classes of these compounds are often associated to a narrow set of species within a phylogenetic group and constitute the bioactive compound in several medicinal, aromatic, colorant, and spice plants.

Secondary metabolites are frequently produced at highest levels during transition from active growth to stationary phase. The producer organism can grow in the absence of their synthesis, suggesting that secondary metabolism is not essential, at least for short-term survival. The genes involved in secondary metabolism provide a "genetic playing field" that allows mutation and natural selection to fix new beneficial traits via evolution. The secondary metabolism as an integral part of cellular metabolism relies on primary metabolism to supply the required enzymes, energy, substrates, and cellular machinery and contributes to the long-term survival of the plant (Roze et al. 2011). A simple classification of secondary metabolites includes three main groups: terpenes (such as plant volatiles, cardiac glycosides, carotenoids, and sterols), phenolics (such as phenolic acids, coumarins, lignans, stilbenes, flavonoids, tannins, and lignin), and nitrogen-containing compounds (such as alkaloids and glucosinolates). Nowadays, medicinal and aromatic plants have undergone a transition from unknown or minor agricultural plantings to major crops that farmers may consider as alternatives to traditional food or feed crops. The steadily increasing agricultural role is driven by consumer interest in these plants for culinary, medicinal, and other anthropogenic applications. The use of plants, foods, and herbal products is increasing due to consumer awareness of their various health benefits. Due to the ever-increasing population, the pressure on arable lands for cultivation of food crops has amplified; therefore, utilization of degraded wastelands is a viable option for cultivation of medicinal and aromatic plants. So, for the enhanced production of secondary metabolite, endophytic bacteria-inoculated *Hyptis suaveolens* can be cultivated at normal as well as stressed condition for human welfare.

12.3 Isolation and Inoculation of Endophytic Bacteria

Plant-bacterial associations have been studied for many decades. However, a complete understanding of the mechanisms utilized by plant endophytic bacteria had remained somewhat indefinable, often making it difficult to take full advantage of these complex relationships to reproducibly improve the growth of plants. The bacterial endophytes can offer several benefits to the host plant, particularly growth promotion and protection from pathogens and diverse environmental conditions. Under adverse condition bacterial endophytes are able to communicate and interact with the plant more efficiently (Coutinho et al. 2015).

The autotrophic plants need minerals for life cycle, and an adequate supply of mineral nutrients is necessary for optimum plant growth. However, when adequate amounts of essential nutrients are present in soil, plants may still show deficiencies due to the nonavailability of these mineral nutrients. Microorganisms can help plants to grow by providing soluble mineral nutrients converted by acidification from insoluble mineral or via mobilization of essential nutrients that can also help in plants' growth improvement. To increase plant nutrient status by the endophytic bacteria by associative nitrogen fixation, phosphorus and potassium solubilization, and siderophores production, changing the absorptivity is a good option for sustainable plant growth. Bacterial genera such as *Bacillus*, *Pseudomonas*, and *Brevibacillus* are well-known to promote growth and yield in different nonleguminous plant (Karlidag et al. 2007). Two bacterial strains are isolated from the root of paddy and *Suaeda nudiflora* wild mosque plant from Khambhat near the seashore of Gujarat as a previously published method (Jha et al. 2011). And the soil sample is tested in SICART (Sophisticated Instrumentation Centre for Applied Research and Testing) laboratory by extracted water sample method. The soil possesses the following physiochemical properties: pH 6.58, electrical conductivity 1480 μS/cm, salinity 8.6%, nitrate 112.5 mg kg^{-1}, chloride 128 mg kg^{-1}, sulfate 155 mg kg^{-1}, ammonia nitrogen 23.3 mg kg^{-}1, CEC 3 cmol, and organic carbon 5500 mg kg^{-1}. Molecular identification of bacterial isolates has been done by isolation of total genomic DNA and PCR amplification with 16S rDNA specific primers 16S F: 5′AGAGTTTGATCCTGGCTCAG3′ and 16S R: 5′AGGTTACCTTGTTACGACTT3′ followed by sequencing as our published method (Jha and Subramanian 2012). PCR amplicons of 16S rDNA of about 1500 bp are obtained for both the isolates as discrete bands in agarose gel (Fig. 12.1). The phylogenetic trees have been constructed using BLAST software by the comparison of the 16S rDNA sequence of isolates and related genera from a database using the neighbor-joining (NJ) algorithm and maximum likelihood (ML) method. The two isolates are identified by molecular analysis by nucleotides homology and phylogenetic analysis as *Pseudomonas pseudoalcaligenes* (GenBank Accession Number: EU921258) and *Pseudomonas aeruginosa* (GenBank Accession Number: JQ790515).

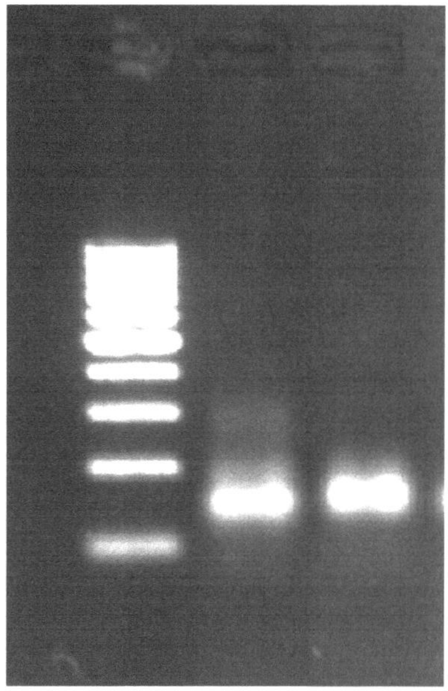

Fig. 12.1 Agarose gel showing the amplified 16S rDNA of isolates, where M = 1 KB marker, L1 = *Pseudomonas aeruginosa*, and L2 = *Pseudomonas pseudoalcaligenes*

Seeds of *Hyptis suaveolens* have been washed thoroughly with distilled water followed by surface sterilization with 0.1% $HgCl_2$ solution for 4 min and 70% ethanol for 10 min. The washed seeds are kept in a shaker for 1 h in autoclaved distilled water on a rotary shaker and then transferred to Petri dishes containing tryptone glucose yeast extract agar medium to test for possible contamination at 30 °C. The germinated seedlings devoid of any contamination are used for inoculation experiments. To study the effect of the isolated bacteria on the physiological and biochemical parameters, 4 days old germinated seedlings devoid of any contamination are transferred to culture tubes containing 400 μl Hoagland's nutrient medium, 400 μl micronutrients, and 1% agar in 40 ml distilled water. Before the transfer, bacterial inoculums of the isolated bacteria *Pseudomonas pseudoalcaligenes* and *Pseudomonas aeruginosa* have been added with the medium at a concentration of 6×10^8 cfu ml^{-1}. To obtain a mixture of both bacterial cultures, an equal volume of both the cultures is mixed in the medium to give a concentration of 6×10^8 cfu ml^{-1}. The tubes are incubated at 27 °C in a 12 h light–dark cycle in a growth chamber. Seven days old plants are carefully removed from different test tubes inoculated with the strain of bacterium and planted in a pot. Similarly the control plants (uninoculated) are also transferred to a fresh pot. Seedlings are planted at the rate of four plants per pot and watered at the time of transplantation of the seedlings. The present study indicates that inoculation with endophytic bacteria, viz., *P. pseudoalcaligenes* and *P. aeruginosa*, both alone or in combination significantly enhanced all the growth parameter and led to recovery of the plants from the saline stress (Table 12.1). The results obtained clearly demonstrated that the stress adversely affects the growth of the plants. However, when the plants were inoculated with the endophytic bacteria, the extent of growth suppression decreased suggesting participation of the endophytic bacteria in alleviation of some of the debilitating effects of stress.

Table 12.1 Effect of endophytic bacteria on growth parameters of *Hyptis suaveolens* under stress

Treatments	Germination %	Survival %	Plant height (m)	Dry weight (kg)	RGR relative growth rate
Normal					
Control	71.1^{d}	85.7^{d}	1.223^{fg}	0.423^{cde}	31.28^{fg}
Control + *P. aeruginosa*	75.4^{bc}	89.1^{bc}	1.431^{c}	0.398^{fg}	33.63^{c}
Control + *P. pseudoalcaligenes*	76.8^{b}	91.3^{b}	1.489^{a}	0.479^{ab}	34.32^{b}
Control + *P. aeruginosa* + *P. pseudoalcaligenes*	82.2^{a}	93.2^{a}	1.470^{ab}	0.497^{a}	35.97^{a}
Stressed					
Control	22.3^{d}	32.2^{cd}	1.021^{h}	0.378^{gh}	29.26^{h}
Control + *P. aeruginosa*	24.1^{bc}	32.6^{bc}	1.276^{def}	0.414^{def}	31.92^{ef}
Control + *P. pseudoalcaligenes*	25.7^{b}	33.1^{ab}	1.312^{cd}	0.442^{bc}	32.54^{de}
Control + *P. aeruginosa* + *P. pseudoalcaligenes*	27.2^{a}	33.8^{a}	1.297^{de}	0.431^{bcd}	33.32^{cd}

Values are the means of replicates. Values with different letters are significantly different at $P < 0.05$ (Duncan's Test). Values in columns followed by the same letter are not significantly different at ($P \leq 0.05$)

12.4 Endophytic Bacteria-Mediated Regulation of Nutrients for Secondary Metabolites Production

When plants are stressed, secondary metabolite production may increase, and growth is often inhibited, because the carbon fixed is predominantly allocated to secondary metabolites. According to the theory of functional balance (Hendrik et al. 2012), plants increase the allocation of biomass to shoots if carbon gain is affected by limited resources above ground, such as light and CO_2. Similarly, plants increase biomass allocation to roots in the presence of low levels of below-ground resources, such as water and nutrients. Under prolonged stress, the nutrients allocated to secondary metabolites are predominantly reverted back to maintain cell osmotic balance to maintain metabolic activity of plant. Stress adversely affects plant nutrient acquisition, especially in the root, resulting in a significant decrease in shoots dry biomass.

The collaboration of endophytic bacteria and their effect on the biological growth response of plants under stress is complex. In our study, the foliar contents of N, P, K, Na, and Ca in endophytic bacteria-inoculated plant are estimated by taking 1 g of plant material digested in tri-acid mixture in the ratio of 9:3:1 by using specific filter on digital flame photometry. The foliar Na concentration is higher in the non-inoculated control plants, while P concentration is higher in the plants inoculated with the endophytic bacteria under stress. The plants inoculated with endophytic bacteria alone and in combination show higher levels of foliar K. Potassium is an osmotically active solute that contributes to water absorption at the cell and whole

Table 12.2 Effect of endophytic bacteria on nutrient minerals concentration of *Hyptis suaveolens* under stress

Treatments	N (mg kg^{-1})	P (mg kg^{-1})	K (mg kg^{-1})	Na (mg kg^{-1})	Ca (mg kg^{-1})
Normal					
Control	0.926^{d}	0.671^{d}	0.647^{cd}	0.823^{ef}	0.924^{cd}
Control + *P. aeruginosa*	1.093^{c}	0.845^{c}	0.754^{bc}	0.718^{cd}	0.835^{ab}
Control + *P. pseudoalcaligenes*	1.321^{b}	0.894^{ab}	0.837^{ab}	0.644^{bc}	0.746^{bc}
Control + *P. aeruginosa* + *P. pseudoalcaligenes*	1.471^{a}	0.973^{a}	0.915^{a}	0.528^{b}	0.651^{a}
Stressed					
Control	0.782^{h}	0.516^{h}	0.516^{gh}	0.953^{a}	0.998^{de}
Control + *P. aeruginosa*	0.897^{fg}	0.677^{f}	0.672^{ef}	0.825^{de}	0.839^{f}
Control + *P. pseudoalcaligenes*	0.928^{ef}	0.766^{fg}	0.716^{fg}	0.723^{gf}	0.771^{fg}
Control + *P. aeruginosa* + *P. pseudoalcaligenes*	1.021^{e}	0.811^{e}	0.854^{e}	0.642^{h}	0.732^{h}

Values are the means of replicates. Values with different letters are significantly different at $P < 0.05$ (Duncan's Test). Values in columns followed by the same letter are not significantly different at ($P \leq 0.05$)

plant level (Table 12.2) and helps stressed plant in maintaining central metabolic activity for its survival. In our study, inoculations of plants with endophytic bacteria always have higher N_2 and carbon concentration under normal and stress conditions. Deficiencies of important nutrient like N, P, K, and S usually cause a greater concentration of secondary metabolite like phenolic compounds, and abundant N generally reduces phenolic accumulation in plant (Gershenzon 1983), which is easily regulated and maintained by endophytic bacteria. The levels of phenolic compounds are directly related to secondary metabolism and show the sensitivity of plant response to nutrient deficiency. The inoculation of endophytic bacteria in plant, alone or in groups, can confer tolerance to plant against adverse environmental condition and also improves other nutrient availability and helps the plant to overcome stress by regulating secondary metabolite production.

12.5 Endophytic Bacteria-Mediated Regulation of Concentration of Photosynthetic Pigments for Secondary Metabolites Production

The plant growth is controlled by a multitude of physiological, biochemical, and molecular processes. However, stressful environments considerably hamper the plant growth by altering the ultrastructure of the organelles and concentration of various photosynthetic pigments (Wu and Kubota 2008).Chlorophyll is vital for photosynthesis, which allows plants to absorb energy from light. Fresh leaves are used for chlorophyll measurements, by placing fresh leaf samples (0.5 g) in a shaker

Table 12.3 Effect of endophytic bacteria on the photosynthetic pigment and oil yield of *Hyptis suaveolens* under stress

Treatments	Chl a (mg g-1FW)	Chl b (mg g-1FW)	Carotenoid (mg g-1FW)	Oil Content %	Oil yield (ml Pot^{-1})
Normal					
Control	0.876^{d}	0.481^{d}	0.427^{cd}	0.763^{ef}	0.832^{cd}
Control + *P. aeruginosa*	1.075^{c}	0.672^{c}	0.564^{bc}	0.828^{cd}	0.975^{ab}
Control + *P. pseudoalcaligenes*	1.082^{b}	0.612^{ab}	0.517^{ab}	0.784^{bc}	0.924^{bc}
Control + *P. aeruginosa* + *P. pseudoalcaligenes*	1.541^{a}	0.873^{a}	0.725^{a}	0.958^{b}	1.141^{a}
Stressed					
Control	0.562^{h}	0.356^{h}	0.296^{gh}	0.853^{a}	0.868^{de}
Control + *P. aeruginosa*	0.683^{g}	0.437^{f}	0.412^{ef}	0.815^{de}	0.779^{f}
Control + *P. pseudoalcaligenes*	0.678^{ef}	0.376^{fg}	0.366^{fg}	0.762^{gf}	0.751^{fg}
Control + *P. aeruginosa* + *P. pseudoalcaligenes*	0.861^{e}	0.541^{e}	0.484^{e}	0.744^{h}	0.732^{h}

Values are the means of replicates. Values with different letters are significantly different at $P < 0.05$ (Duncan's Test). Values in columns followed by the same letter are not significantly different at ($P \leq 0.05$)

with 80% acetone until the leaves are completely bleached. The extract is centrifuged at 13,000 rpm for 10 min, and the supernatant is used to measure chlorophyll a (Chl a), chlorophyll b (Chl b), and carotenoid by taking absorbance at 663, 645, and 470 nm, respectively, using spectrophotometer. In the present study, endophytic bacteria-inoculated plants show increased chlorophyll a, b, and carotenoids in comparison to control plants (Table 12.3).Chlorophyll content is influenced by the nitrogen concentration; as the levels of N_2 increased, chlorophyll a, b, and total chlorophyll are also enhanced. Endophytic bacteria having nitrogen fixing ability must enhance concentration of N_2 in plant and result in high level of chlorophylls and carotenoids also. The increase in chlorophyll content with increasing nitrogen has been reported by Suza and Valio (2003).

Carotenoids are necessary for photoprotection of photosynthesis, and they play an important role as a signaling precursor during the plant development under stress. They have significant potential to enhance nutritional quality and plant yield (Jha and Subramanian 2013). The chlorophyll a, b, and total chlorophyll are negatively related with secondary metabolites. Competition between secondary metabolites and chlorophyll contents fits well with the prediction of protein competition model, where secondary metabolites content is controlled by the competition between protein and secondary metabolites biosynthesis pathway and its regulation. The negative relationship between secondary metabolites and chlorophyll content is a sign of gradual switch of investment from protein to polyphenolics production (Meyer et al. 2006). Results of present study indicate that the production of chlorophyll content competes with the production of secondary metabolites due to endophytic bacteria inoculation under stress. Affendy et al. (2010) have reported an increase in the production of secondary metabolites of *O. stimaneus* under low irradiance due to increase in availability of phenylalanine, a precursor for secondary metabolites and protein production. The production of secondary metabolites is more prioritized under low nitrogen levels due to the restriction of protein production as exhibited by reduced chlorophyll production.

12.6 Endophytic Bacteria-Mediated Regulation of Essential Oil

Essential oils are secondary metabolites whose production is associated with primary metabolism and with availability of soil nutrients (Shulka et al. 1992). Plants synthesize essential oils for a variety of purposes, including protection of the plant against fungi and bacteria, allelopathic activity, defense against insects (terpenoids), attraction of pollinators, and dispersal agents to favor the dispersion of seeds and pollens. The major activities of essential oils are antimicrobial, sedative, anti-inflammatory, bactericidal, antiviral, antifungal (fungicidal), and preservative for foods. The essential oils found in the genus *Hyptis* have a great importance as a source of bioactive constituents, especially due to their biological properties such as antimicrobial, cytotoxic, and insecticide (Kuhnt et al. 1995).

Essential oils are volatile liquids that can be synthesized by all plant organs and are stored in secretory cells, cavities, canals, and epidermic cells. In the present study, essential oil (EO) extraction from the fresh herbage is performed by hydrodistillation in Clevenger's apparatus for 1 h and 30 min. Oil content (w/v) and total oil yield (ml) are estimated. The essential oil has been analyzed on an Agilent 4890D gas chromatograph fitted with a column (30 m × 0.25 mm, film thickness 0.25 μm, Supelco Wax-10).The leaves being the source of the essential oil are thus the most economically viable parts of the *Hyptis suaveolens*, and the oil yield is hence directly proportional to the number of leaves. In the present study, the leaf-stem ratio and thus the oil yield in endophytic bacteria-inoculated plants are higher at normal condition. But under stress inoculation with endophytic bacteria shows no significant effect on oil yield as shown in Table 12.3. Inoculation with endophytic bacteria under stress reduces the oil yield and has also been reported by Arvini et al. (2012), but due to enhanced plant growth, the overall yield per plant increased. The decreased oil yield due to inoculation with endophytic bacteria may be due to reduced stress which ultimately reduces the oil content. In the present study, the gas chromatographic (GC) analysis of the *Hyptis suaveolens* essential oil enabled us to compare 11 major compounds, viz., α-pinene, β-pinene, sabinene, myrcene, limonene, 1, 8-cineole, menthone, isomenthone, menthyl acetate, neomenthol, and menthol (Table 12.4). The percentage concentration of menthol, the characteristic constituent of the *Hyptis suaveolens* essential oil, varied with the treatments and stress. The effect of stress is more profound on menthol content as it decreased under stress irrespective of the microbiological application. The oil content in a plant is largely dependent on the physiological state of the plant. Enhanced production of secondary metabolites including essential oil may depend directly on improved nutritional status and on primary metabolism of

Table 12.4 Effect of endophytic bacteria on the essential oil contents of *Hyptis suaveolens*

Treatments	α-pinene	β-pinene	Sabinene	Myrcene	Limonene	1,8-Cineole	Menthone	Iso-menthone	Menthyl acetate	Neo-menthol	Menthol
	1.478[a]	2.214[a]	2.572[a]	3.312[a]	3.752[a]	4.211[a]	12.232[a]	12.891[a]	13.543[a]	14.342[a]	16.348[a]
Normal											
Control	0.54^{gh}	0.55^{gh}	0.13^{fg}	0.23^{fg}	0.71^{ef}	0.113^{ef}	0.895^{f}	2.436^{ef}	3.423^{e}	1.641^{hi}	81.21^{ef}
Control + *P. aeruginosa*	0.74^{e}	0.68^{d}	0.317^{d}	0.33^{cd}	0.846^{d}	0.119^{cd}	0.964^{d}	2.655^{d}	2.861^{h}	1.786^{cd}	83.11^{cd}
Control + *P. pseudoalcaligenes*	0.831^{c}	0.876^{c}	0.336^{c}	0.388^{c}	0.984^{c}	0.129^{c}	1.237^{c}	2.982^{bc}	3.142^{f}	1.987^{c}	81.87^{e}
Control + *P. aeruginosa* + *P. pseudoalcaligenes*	0.954^{b}	1.129^{b}	0.386^{b}	0.473^{b}	1.181^{b}	0.191^{b}	1.541^{b}	3.028^{b}	3.342^{c}	2.182^{b}	83.19^{c}
Stressed											
Control	0.723^{d}	0.614^{e}	0.17^{e}	0.28^{de}	0.753^{e}	0.121^{de}	0.904^{de}	2.541^{de}	3.553^{d}	1.845^{de}	84.36^{b}
Control + *P. aeruginosa*	0.53^{ef}	064^{fg}	0.14^{ef}	0.19^{ef}	0.665^{g}	0.118^{fg}	0.791^{fg}	2.347^{fg}	3.424^{b}	1.744^{fg}	76.82^{g}
Control + *P. pseudoalcaligenes*	0.54^{fg}	0.59^{ef}	0.13^{gh}	0.17^{gh}	0.572^{h}	0.089^{gh}	0.725^{h}	1.981^{gh}	3.123^{fg}	1.792^{ef}	71.37^{gh}
Control + *P. aeruginosa* + *P. pseudoalcaligenes*	0.511^{i}	0.489^{i}	0.08^{i}	0.143^{i}	0.391^{i}	0.067^{i}	0.611^{i}	1.649^{i}	2.971^{hi}	1.659^{gh}	68.21^{i}

Values are the means of replicates. Values with different letters are significantly different at $P < 0.05$ (Duncan's Test)
[a]Retention time (min)

plants following endophytic bacterial inoculation (Farag et al. 2006; Jha and Subramanian 2016). Plants may be considered as a famous chemical factory for biosynthesis of a huge array of secondary metabolites. Many of these chemicals are utilized as medicine, scent, dyes, and pesticides and are of commercial importance. Secondary metabolites are those compounds produced by plant which are not essential for plant growth and development. Environmental factors including biotic and abiotic stimuli, carbon-nutrition balance, genotype, and ontogenesis usually control and regulate the biosynthesis of secondary metabolites in plants (Mary Ann Lila 2006). With regard to plant-microbe interactions, coevolution between plants and their microbial partners are mediated via plant chemical defense.

12.7 Endophytic Bacteria-Mediated Regulation of Secondary Metabolites Production Under Biotic Stress

Plants are facing numerous biotic stress and adverse environmental conditions. They respond to such stress through several morphological, biochemical, molecular mechanisms and by their interactions among respective signaling pathways (Nejat and Mantri 2017). Biotic stresses in plants are caused by pests, parasites, and pathogens, primarily responsible for plant diseases. Prone to attacks by pathogens and pests, plants utilize complicated chemical defense mechanisms consisting of metabolic adaptations. However, many plant pathogens can manipulate the host metabolism to induce favorable nutritional condition and to counteract defense responses. The plants employ a highly intricate defense system that is capable of protecting themselves from the majority of attackers. Plant immunity is multilayered and consists of preformed, constitutive as well as inducible defense mechanisms (Pieterse et al. 2009). Besides physical preformed barriers such as the cell wall, plants also possess highly effective preformed chemical defenses called secondary metabolite (González-Lamothe et al. 2009). These are constitutively present products of secondary plant metabolism and represent first-line defense, which are released and activated as antimicrobial compounds upon pathogen entry. Biocontrol using endophytic bacteria may be an alternative method for controlling plant diseases to maintain plant growth (Jha 2018). Systemic plant resistance induced by endophytic bacteria as elicitor represents ISR and can protect plants against a wide range of pathogens through the activation of secondary plant metabolites, i.e., PAL, β-1, 3-glucanases, and phenolic (Jha et al. 2011). The result of the present study showed that plant inoculated with endophytic bacteria has better induction of PR proteins like PAL, β-1, 3-glucanases, and phenolic compared to non-inoculated *Hyptis suaveolens* plant, which is due to its elicitation effect prior to infection (Table 12.5). Elicitation is the induced or enhanced biosynthesis of metabolites due to addition of trace amounts of elicitors. In the same way, metabolic profiling of two rice cultivars inoculated with two different endophytic bacteria under stress conditions showed modified profiles of secondary metabolites with phenolic compounds such as flavonoids and hydroxyl cinnamic derivatives (Chamam et al. 2013). However inoculation with

Table 12.5 Effect of endophytic bacteria PAL, β-1, 3-glucanases, phenolic of maize under stress ($n = 5$)

Treatment	Proline mMol min^{-1} g^{-1}	Glycine betaine mMol min^{-1} g^{-1}	PAL (nmol of trans cinnamic acid min^{-1} g^{-1})	β-1, 3-glucanases (nmol of Glucose min^{-1} g^{-1})	Phenolic (mgg^{-1} of the gallic acid equivalent)
Normal					
Control	1.32[cd]	0.8[d]	0.25[d]	0.221[d]	0.74[cd]
Control + *P. aeruginosa*	1.49[c]	1.02[c]	0.29b[c]	0.229[c]	0.82[bc]
Control + *P. pseudoalcaligenes*	1.53[b]	1.23[b]	0.32[b]	0.237[b]	0.87[ab]
Control + *P. pseudoalcaligenes* + *P. aeruginosa*	1.59[a]	1.32[a]	0.38[a]	0.244a	0.99[a]
Stressed					
Control	1.41[d]	1.09[cd]	0.23[d]	0.219[d]	0.95[cd]
Control + *P. aeruginosa*	1.56[bc]	1.11[c]	0.28[ab]	0.235[b]	1.08[c]
Control + *P. pseudoalcaligenes*	1.58[b]	1.29[ab]	0.26[a]	0.241[a]	1.13[b]
Control + *P. pseudoalcaligenes* + *P. aeruginosa*	1.61[a]	1.31[a]	0.31[c]	0.2547[bc]	1.24[a]

For each parameter, values in columns followed by the same letter are not significantly different at ($P \leq 0.05$)

endophytic bacteria affects the composition of secondary metabolites in shoots, pointing toward systemic effects, and helps plant to survive under stress. Therefore, understanding the relationship between growth and the production of secondary metabolites under different stress environments is significant in managing plants' growth conditions to acquire the maximal yield of biomass and phyto-medicinal compounds (Jha et al. 2014a, b). Plant secondary metabolites are often referred to as compounds that have no fundamental role in the maintenance of life processes in the plants, but they are important for the plant to interact with its environment for adaptation and defense. Secondary metabolites have significant practical applications in medicinal, nutritive, and cosmetic purposes, besides, importance in plant stress physiology for adaptation.

12.8 Endophytic Bacteria-Mediated Regulation of Secondary Metabolites Production Under Abiotic Stress

The vast metabolic diversity observed in plants is the direct result of continuous evolutionary processes. Environmental factors significantly affect plant growth and biosynthesis of secondary metabolites. Plant growth and productivity are negatively affected by abiotic stress. Plant secondary metabolites are compounds that play an essential part in the interaction of plants with abiotic stress. Abiotic stress causes reduction in plant growth with a resulting increase in production of secondary metabolites like phenolic compounds as a defense mechanism (Arbona et al. 2013). Phenolics are compounds possessing one or more aromatic rings with one or more hydroxyl groups. They are broadly distributed in the plant kingdom and are the most abundant secondary metabolites of plants. In the present study, the leaf extract is used for the determination of the total phenol content using gallic acid for standard by spectrophotometer (absorbance 735 nm). This study showed that inoculation of endophytic bacteria alone is sufficient to increase the phenolic content in *Hyptis suaveolens* plant in normal state and stress don't caused any further increase in it (Table 12.5). The synthesis of phenolics in ginger can be increased and affected under abiotic stress which has also been reported by Ghasemzadeh et al. (2010). Plants produce diverse secondary plant products that are triggered by a wide range of abiotic factors, to cope with environmental changes, but when inoculated with endophytic bacteria, it will regulate and modulate both primary and secondary metabolites of the plant for the better survival of plant, especially under stress (Jha 2018). Global changes in environmental conditions due to human activities appear to influence endogenous plant metabolites for adaptation. Moreover, plants have adapted to produce several metabolites that are species-specific and dependent on environmental factors. Various plant metabolites, such as polyamines, flavonoids, jasmonic acid, methyl jasmonate, glycine betaine, and so on, have a protective role during abiotic stress. These phytochemical derivatives of secondary metabolism confer a multitude of adaptive and evolutionary advantages to the producing plants (Bilgin et al. 2010).

The secondary metabolite compound is measured as glycine betaine (GB) equivalents and proline as a major osmoprotectant. In our study, accumulation of glycine betaine-like quaternary compounds and proline is significantly higher in the plants' leaves inoculated with both *P. pseudoalcaligenes* and *P. aeruginosa* (Table 12.5). Accumulation of QACs enhanced in the *Hyptis suaveolens* leaves inoculated with both *P. pseudoalcaligenes* and *P. aeruginosa* compared to *Hyptis suaveolens* plant treated with either of the *P. pseudoalcaligenes* and *P. aeruginosa* alone under stress. Many plant species naturally accumulate QACs and proline as major organic osmolytes when subjected to different abiotic stresses. These compounds are thought to play an adaptive role in mediating osmotic adjustment and protecting subcellular structures in stressed plants (Asharf and Foolad 2007). When the environment is adverse and plant growth is affected, metabolism is profoundly involved in signaling, physiological regulation, and defense responses. At the same time, in feedback, abiotic stresses affect the biosynthesis, concentration, transport, and storage of primary and secondary metabolites.

12.9 Endophytic Bacteria-Mediated Regulation of Gene Expression for Secondary Metabolites Production

During evolution, plants have developed a wide variety of highly complicated and competent mechanisms to sense, respond, and adapt to a wide range of environmental changes. Under adverse or limiting growth conditions, plants respond by activating tolerance mechanisms at multiple levels of organization (molecular, tissue, anatomical, and morphological), by adjusting the membrane system and the cell wall architecture, by altering the cell cycle and rate of cell division, and by metabolic tuning. Plants perceive stress signals through receptors that trigger molecular cascades to transmit the signals to regulatory systems via ion channels, signaling proteins, and secondary messengers (Choudhary et al. 2012). At a molecular level, many genes are induced or repressed by stress, involving a precise regulation of extensive stress-gene networks (Müller and Stelling 2009). Plants have several mechanisms to overcome such stress by stimulating production of secondary metabolite or induction of defensive gene. The regulatory system is composed of various components, including phytohormones, transcription factors (TFs), mitogen-activated protein kinases, and phosphatases that regulate the expression of various stress-responsive genes (Osakabe et al. 2014). In order to establish a favorable energy balance for defense, the upregulation of defense-related pathways is compensated by the downregulation of genes involved in other metabolic pathways. Plant secondary metabolites are often referred to as compounds that play important role for the plant to interact with its environment for adaptation and defense. The biochemical change involved in plant stress responses is by the synthesis of new proteins and hormone that have direct or indirect action on the course of stress. It is therefore surprising that induction of stress-related gene can be induced prior to stress, i.e., merely by inoculation with endophytic bacteria (Jha et al. 2014a, b), but till date there is no report on the induction or expression of gene in *Hyptis suaveolens* under stress. Endophytic bacteria induced different small protein molecules in plant under stress as well as control conditions to establish itself in the host plant and to protect the plant under stress. Mechanisms of endophytic bacteria-mediated phytostimulation would help us to find more capable strains having the ability to function efficiently for sustainable production of important secondary metabolites under different agro-ecological conditions (Jha and Subramanian 2018a, b).

12.10 Conclusion

Secondary metabolism comprises a coordinate series of coupled enzymatic conversions that utilizes limited products of primary metabolism as substrates. Secondary metabolism uses highly organized systematic mechanisms that integrate into developmental, morphological, and biochemical regulatory patterns of the entire plant metabolic network. To improve the production of secondary metabolites, one of the main problems is the lack of basic knowledge of the biosynthetic routes and mechanisms responsible for the production of plant metabolites. The productivity of the desired metabolites is limited by the lack of particular precursors, and biotransformation using an exogenous supply of biosynthetic precursors may improve the accumulation of such compounds. Elicitation or compounds produced by the endophytic bacteria or stress also responsible for triggering the formation of secondary metabolites. The interdisciplinary intensive research efforts are required for identification of genes and enzymes involved in plant secondary metabolism of metabolic pathways leading to the biosynthesis of secondary metabolites. While coordinated induction of all genes is not easy, endophytic bacteria have the ability to modulate the primary product for the production of desired secondary metabolite for the survival of the host plant in normal as well as under stress condition.

References

Affendy H, Aminuddin M, Azmy M, Amini MA, Assis K, Tamir AT (2010) Effects of light intensity on *Orthosiphon stamineus* Benth treated with different organic fertilizers. Int J Agric Res 5:201–207

Ann Lila M (2006) The nature-versus-nurture debate on bioactive phytochemicals: the genome versus terroir. J Sci Food Agric 86:2510–2515

Apel K, Hirt H (2004) Reactive oxygen species: metabolism, oxidative stress, and signal transduction. Annu Rev Plant Biol 55:373–399

Arbona V, Manzi M, Ollas CD, Gómez-Cadenas A (2013) Metabolomics as a tool to investigate abiotic stress tolerance in plants. Int J Mol Sci 14:4885–4911

Arvini P, Vafabakhsh J, Mazaheri D, Noormohamadi G, Azizi M (2012) Study of drought stress and plant growth promoting rhizobacteria (PGPR) on yield, yield components and seed oil content of different cultivars and species of brassica oilseed rape. Ann Biol Res 3(9):4444–4451

Ashraf M, Foolad MR (2007) Roles of glycine betaine and proline in improving plant abiotic stress resistance. Environ Exp Bot 59:206–216

Ballhorn DJ, Kautz S, Jensen M, Schmitt S, Heil M, Hegeman AD (2011) Genetic and environmental interactions determine plant defences against herbivores. J Ecol 99:313–326

Bilgin DD, Zavala JA, Zhu J, Clough SJ, Ort DR, De Lucia EH (2010) Biotic stress globally downregulates photosynthesis genes. Plant Cell Environ 33:1597–1613

Chamam A, Sanguin H, Bellvert F, Meiffren G, Comte G, Wisniewski-Dyé F (2013) Plant secondary metabolite profiling evidences strain-dependent effect in the *Azospirillum Oryza sativa* association. Phytochem Lett 87:65–77

Choudhary P, Yu JQ, Yamaguchi-Shinozaki K, Shinozaki K, Tran LS (2012) Benefits of brassino steroid crosstalk. Trends Plant Sci 17:594–605

Coutinho BG, Licastro D, Mendonça-Previato L, Cámara M, Venturi V (2015) Plant-influenced gene expression in the rice endophyte *Burkholderia kururiensis* M130. Mole Plant-Microbe Interact 28:10–21

Farag MA, Ryu CM, Sumner LW (2006) GC-MS SPME profiling of rhizobacterial volatiles reveals prospective inducers of growth promotion and induced systemic resistance in plants. Phytochemistry 67:2262–2268

Gershenzon J (1983) Changes in levels of plant secondary metabolites under water and nutrient stress. Recent Adv Phytochem 18:273–320

Ghasemzadeh A, Jaafar HZE, Rahmat A (2010) Elevated carbon dioxide increases contents of flavonoids and phenolic compounds, and antioxidant activities in Malaysian young ginger (*Zingiber officinale* Roscoe.) varieties. Molecules 15:7907–7922

González-Lamothe R, Gabriel M, Mariza G, Moussa SD, Malouin F, Bouarab K (2009) Plant Antimicrobial Agents and Their Effects on Plant and Human Pathogens. IJMS https://doi.org/10.3390/ijms10083400

Hendrik P, Karl JN, Peter B, Reich J, Oleksyn P, Poot Liesje M (2012) Biomass allocation to leaves, stems and roots: meta-analyses of inter specific variation and environmental control. New Phytol 193:30–50

Jha Y (2018) Induction of anatomical, enzymatic, and molecular events in maize by PGPR under biotic stress. In: Meena V (ed) Role of rhizospheric microbes in soil. Springer, Singapore

Jha Y, Subramanian RB (2012) Isolation of root associated bacteria from the local variety of rice GJ-17. World Res J Geoinformatics 1:21–26

Jha Y, Subramanian RB (2013) Paddy physiology and enzymes level is regulated by rhizobacteria under saline stress. J Appl Bot Food Qual 85:168–173

Jha Y, Subramanian RB (2016) Rhizobacteria enhance oil content and physiological status of *Hyptis suaveolens* under salinity stress. Rhizosphere 1:33–35

Jha Y, Subramanian RB (2018a) From interaction to gene induction: an eco-friendly mechanism of PGPR-mediated stress management in the plant. In: Egamberdieva D, Ahmad P (eds) Plant microbiome: stress response. Microorganisms for sustainability, vol 5. Springer, Singapore

Jha Y, Subramanian RB (2018b) Chapter 12: Effect of root-associated bacteria on soluble sugar metabolism in plant under environmental stress. In: Ahmad P, Ahanger MA, Singh VP, Tripathi DK, Alam P, Alyemeni MN (eds) Plant metabolites and regulation under environmental stress. Academic, pp 231–240

Jha Y, Subramanian RB, Patel S (2011) Endophytic bacteria induced enzymes against *M. grisea* in *O. sativa* under biotic stress. Afr J Basic Appl Sci 3:136–146

Jha Y, Sablok G, Subbarao N, Sudhakar R, Fazil MHU, Subramanian RB, Squartini A, Kumar S (2014a) Bacterial-induced expression of RAB18 protein in Oryza sativa salinity stress and insights into molecular interaction with GTP ligand. J Mol Recognit 27:521–527

Jha Y, Subramanian RB, Sahoo S (2014b) Antifungal potential of fenugreek coriander, mint, spinach herbs extracts against Aspergillus niger and *Pseudomonas aeruginosa* phyto-pathogenic fungi. Allelopath J 34:325–334

Karlidag H, Esitken A, Turan M, Sahin F (2007) Effects of root inoculation of plant growth promoting rhizobacteria (PGPR) on yield, growth and nutrient element contents of leaves of apple. Sci Hortic 114:16–20

Kuhnt M, Probstle A, Rimpler H, Bauer R, Heinrich M (1995) Biological and pharmacological activity and further constituents of *Hyptis verticillata*, planta. Medica 61:227–232

Mandal SM, Mandal KC, Day S, Pati BR (2007) Antimicrobial activity of the leaf extracts of *Hyptis suaveolens*. Indian J Pharm Sc 69:568–569

Meyer S, Cerovic ZG, Goulas Y, Montpied P, Demotes-Mainard S, Bidel LPR, Moya I, Dreyer E (2006) Relationships between optically assessed polyphenols and chlorophyll contents, and leaf mass per area ratio in woody plants: a signature of the carbon–nitrogen balance within leaves? Plant Cell Envirn 29:1338–1348

Moreira ACP, Lima EO, Wanderley PA, Carmo ES, Souza EL (2010) Chemical composition and antifungal activity of *Hyptis suaveolens* (L.) Poit leaves essential oil against *Aspergillus* species. Braz J Microbiol 41:28–33

Müller D, Stelling J (2009) Precise regulation of gene expression dynamics favors complex promoter architectures. PLoS Comput Biol 5(1):e1000279

Nejat N, Mantri N (2017) Plant immune system: crosstalk between responses to biotic and abiotic stresses the missing link in understanding plant defense. Curr Issues Mol Biol 23:1–16

Osakabe Y, Yamaguchi-Shinozaki K, Shinozaki K, Tran LS (2014) ABA control of plant macroelement membrane transport systems in response to water deficit and high salinity. New Phytol 202:35–49

Pieterse CMJ, León-Reyes A, Van der Ent S, Van Wees SCM (2009) Networking by small-molecule hormones in plant immunity. Nat Chem Biol 5:308–316

Roze LV, Chanda A, Linz JE (2011) Compartmentalization and molecular traffic in secondary metabolism: a new understanding of established cellular processes. Fungal Genet Biol 48:35–48

Shulka A, Abad Farooqi AH, Shukla YN, Sharma S (1992) Effect of triacontanol and chlormequat on growth, plant hormones and artemisinin yield in Artemisia annuaL. Plant Growth Regul 11:165–171

Suza R, Valio IFM (2003) Leaf optical properties as affected by shade in samplings of six tropical tree species differing in successional status. Braz J Plant Physiol 15:49–54

Tuteja N, Sopory SK (2008) Chemical signaling under abiotic stress environment in plants. Plant Signal Behav 3:525–536

Wu M, Kubota C (2008) Effects of high electrical conductivity of nutrient solution and its application timing on lycopene, chlorophyll and sugar concentrations of hydroponic tomatoes during ripening. Sci Hortic 116:122–129

Chapter 13
Bioactive Potentials of Novel Molecules from the Endophytes of Medicinal Plants

Monnanda S. Nalini, Harischandra S. Prakash, and Mysore V. Tejesvi

Abstract Microbial endophytes have a long-standing association with numerous plant species. A closer look into their diversity indicates the existence of novel species from biologically diverse regions on the earth, especially the tropics. The novelty is related to their ability to produce diverse chemical structures with reliable bioactive potentials, which has resulted in the addition of new compounds to the unending list of natural products. Hyphenated techniques have fastened the cumbersome screening of crude extracts with reliable bioassays resulting in the elucidation of novel compounds or molecules of interest. Biotechnological approaches are of added advantage in the production of such compounds with remarkable bioactivities. This chapter highlights the fungi and actinomycetes as endophytes from the medicinal and pharmaceutical plants of relevance, host-related metabolites, novel bioactive metabolites of endophytes, and approaches for the augmentation of metabolites and a special mention of the metabolites by *Pestalotiopsis* species. Therefore, endophytes are microbial chemical factories more suited for the production of novel metabolites with therapeutic potentials.

Keywords Endophytic microbes · Diversity · Bioprospecting · Novel structures · Medicinal plants

M. S. Nalini (✉)
Department of Studies in Botany, University of Mysore, Mysore, Karnataka, India

H. S. Prakash
Department of Studies in Biotechnology, University of Mysore, Mysore, Karnataka, India

M. V. Tejesvi
Department of Ecology & Genetics, University of Oulu, Oulu, Finland

Chain Antimicrobials Ltd., Oulu, Finland

D. Egamberdieva, A. Tiezzi (eds.), *Medically Important Plant Biomes: Source of Secondary Metabolites*, Microorganisms for Sustainability 15,
https://doi.org/10.1007/978-981-13-9566-6_13

13.1 Introduction

Medicinal plants play a pivotal role in the health-care systems, and traditional systems of medicine are in existence due to their healing properties. The traditional use of herbal medicines is in vogue due to their historic and cultural significance. One of the most practiced medicinal systems is the Traditional Chinese Medicines often used to treat 5000 remedies (Li 2000), followed by the Japanese and the Indian traditional medicinal systems. The popularity of these medicinal systems has led to the inclusion of plant-derived medicines in the pharmacopoeias. Of the 10,000–11,500 species documented to be of medicinal use in China (Huang et al. 2015; Pei and Huai 2015), ~563 species are cited in the Chinese Pharmacopoeia (http://www.kew.org/mpns). The Japanese traditional medicine, which is practiced in the Japanese society for more than a thousand years, may be divided into folk medicine and Chinese medicine (Kampo medicine). The popularity of this medicine has augmented the per capita consumption of herbal medicine and is viewed as an alternate to modern medicines due to its safe and efficacious use (Saito 2000).

In the Indian system of medicine, 7000–7500 plants are listed as cure for treating diseases, and Ayurveda, "the science of life," utilizes 2000 medicinal plants in plant-based formulations (Pandey et al. 2013). A survey conducted by Medicinal Plant Names Services (MPNS), The Royal Botanic Gardens, Kew, lists the use of 28,187 species as medicinal (Willis 2017). Regardless of their natural distributions across the continents, they are preferred sources of pharmaceutical drugs for treating diseases. Exploitation of medicinal plants from their original habitat for pharmacological benefits has posed serious threat to biodiversity. Therefore, as an alternative, microbes living inside plants – "the endophytes" – are capable of producing bioactive compounds similar to those of host plants. One of the phenomenal success stories first reported in science is the production of Taxol®, the million dollar drug by the novel endophytic fungus, *Taxomyces andreanae*, isolated from the inner bark of the medicinal tree, *Taxus brevifolia* (Stierle et al. 1993). It is estimated that 10,000 kg of bark from *Taxus* is needed to ensure the yield of 2 g of Taxol for the treatment of various types of cancers (Strobel et al. 1996). Microbial endophytes can be viewed as the most reliable and potential sources of bioactive molecules with immense applications in pharmaceutics and agriculture.

13.2 Endophytes: Terminology and Estimates

The term "endophyte" is derived from the Greek words *endon* = inside and *phyton* = plants and was first introduced by Heinrich Friedrich Link, a German biologist in 1809. It is applied to organisms causing asymptomatic infections within plant tissues (Carroll 1986), as latent pathogens (Clay 1988), inhabiting plant organs without causing harm to hosts (Petrini 1991), and microbes that colonize living, internal tissues of plants without causing any immediate, overt negative effects (Bacon and

White 2000) or residing within the tissues of apparently healthy host plants (Schulz and Boyle 2006). The host range for microbial endophytes include bacteria, archaea, fungi inclusive of yeasts, and unicellular eukaryotes such as algae (Trémouillaux-Guiller et al. 2002) and amoeba (Müller and Döring 2009). Endophytes range from mutualists, symbionts, plant growth enhancers, and biocontrol agents.

Fungi are taxonomically diverse group of organisms holding important roles in the cycling of nutrients and modulation of plant growth in the environment (Taylor and Sinsabaugh 2014). The richness of fungal species globally is estimated at millions, yet <2% of the species are described. Fungal diversity has earlier been estimated conservatively at 1.5 million species (Hawksworth 2001), while Schmit and Mueller (2007) estimated the diversity of fungal species to be 712,000. Reliable approaches such as pyrosequencing and Illumina platforms are often useful to estimate the fungal diversity (Buée et al. 2009; Schmidt et al. 2013). The endophytic fungal estimate includes one million species (Dreyfuss and Chapela 1994).

Actinomycetes are Gram-positive bacteria with high G+C content. They are ubiquitous in field-grown crop plants, which are related to their abundance in rhizosphere soils, and are important component of microbial diversity. They are being recovered from a range of habitats and unusual environments. The diversity estimates of these organisms are a matter of debate as soil samples from various depths and types tend to vary with the actinomycete populations. Vieira and Nahas (2005) provided the total counts for actinomycete populations in three soil types (crop field, tree, and forest) ranging from 79.6 to 88.0×10^6 colony-forming units (CFU) g^{-1}. Unlike fungi, such estimates of diversity have not been predicted for the actinomycetes despite their omnipresence in soil.

The diversity, bioactivity, novel molecules, and applications of endophytic fungi (Strobel and Daisy 2003; Kaul et al. 2012) and actinomycetes (Qin et al. 2011; Golinska et al. 2015; Gao et al. 2018) with biotechnological and pharmaceutical relevance are documented. Therefore, this chapter highlights the importance of medicinal plants as sources of drugs, definitions of endophytes and their isolations, endophyte diversity in biodiverse areas, and bioprospection for novel molecules and biological activities.

13.3 Biodiversity Hotspots and the Isolation of Endophytes

The world's plant wealth is harbored in the biodiverse regions known for plant diversity; the epicenters are known as "hotspots," the term coined by Myers (1988), a conservation biologist. In order to qualify for the hotspot status, the area needs to contain endemic vascular plants (1500, 0.5% of the world total of 300,000) whose populations have declined by 70%. At present, 34 hotspot locations are documented by Conservation International (Mittermeier et al. 2004).

A perusal of literature indicates that among the 34 hotspots of plant diversity, a handful of them have been subjected to the isolations of endophytic microbes. The rationale for the selection of plants is based on the ethnobotanical uses, location in unique environments, and undisturbed habitats (Strobel et al. 2004), and these considerations have led to the listing of the tropical Andes as the lead hotspot location, which boasts of harboring ~6.7% of the global (44%) endemic plant species (Myers et al. 2000).

Inspired by the works of late Monroe Wall and Mansuhk Wani, natural product chemists from the Research Triangle Institute, Gary Strobel, Professor of Plant Sciences from the Montana State University, worked with endophytic microbes from the success achieved by the isolation of Taxol of microbial origin. Deriving impetus from the natural product research, he travelled to tropical wilderness, rain forests, and undisturbed places to collect plant species of ethnobotanical significance, their location in unique environments resulting in the isolation and characterization of a number of endophytes and their products with potential bioactivities (Strobel and Daisy 2003; Strobel et al. 2004).

The Kinshasa reserve in Congo yielded an insulin-mimetic small molecule from the endophytic *Pseudomassaria* sp. (Zhang et al. 1999), and from the Venezuela-Guyana border in the Tepuis range, a unique *Siematoantelerium tepuiense* with Taxol-producing potentials *was* documented (Strobel et al. 1999). A unique *Streptomyces* MSU-2110 producing the peptide antibiotics was isolated from a vine, *Monstera* sp., from the Manu region of the upper Amazon (Ezra et al. 2004). Three hundred medicinally rich plants, sampled from Lake Sandoval area in the Northern Territory, Australia region, yielded 14 strains with antimicrobial potentials (Bascom-Slack et al. 2009). The Northern Territory of Australia is the abode of the aboriginal community, who use stem pieces of snake vine plant (*Kennedia nigriscans*) to treat wounds and infections. The plating of stem fragments resulted in the isolation of a *Streptomyces* sp., producing a newly described class of broad-spectrum peptide antibiotics, munumbicins (Castillo et al. 2002). The nomenclature of this antibiotic was dedicated to Mr. Reggie Munumbi Miller of the Manyallaluk community. The antibiotic kakadumycin was isolated from *Streptomyces* sp., as endophyte of the fern-leaved tree *Grevillea pteridifolia* from this region (Castillo et al. 2003).

The mountainous Southwest China representing the tropical rain forests of Xishuangbanna in the Yunnan Province and Panxi plateau of the Sichuan Province and nature conservation areas of Fujian Province, Southeast China, are treasure houses of pharmaceutically important medicinal plants owing to the unique geographical conditions and abundant rainfall. Plants sampled from these locations have had a long-standing use in TCM, and unique actinomycetes have been reported (Li et al. 2008; Qin et al. 2009a, 2012; Yuan et al. 2008; Zhao et al. 2011). Few medicinal plants of the Malayan peninsula and two hotspots in India, representing the Himalayas and the Western Ghats, regions with endemic species, were sampled for the isolation of endophytic actinomycetes (Zin et al. 2007; Passari et al. 2015; Akshatha et al. 2014).

Papua New Guinea (PNG), Solomon, and Mborokua islands adjacent in the archipelago are the relics of tropical wilderness of biodiversity. Unique endophytic actinomycetes were isolated from the tropical plants of this region (Janso and Carter 2010). PNG is covered with undisturbed rain forests receiving 80–100 cm of rainfall, and antimycotic compounds were isolated from the endophyte, *Pestalotiopsis jesteri* (Li and Strobel 2001). The Hawaiian Islands in the mid of the Pacific Ocean are separated from the mainland with natural organisms and endophytes identified with bioactive potentials (Li et al. 2016a, b). Endemism, as defined by the high richness of species, is distinct in oceanic islands than the mainland (Mutke et al. 2011). There is tremendous scope for the isolation of newer and more novel endophyte taxa from these centers of plant diversity. The tropical rain forests represent one of the biologically rich and diverse regions on the earth.

The Xishuangbanna tropical rain forest in Southwest China contains plant diversity with 3000 endemic species. Streptomycete endophytes with potential bioactivities were isolated from this region (Li et al. 2008). Many novel species of endophytic actinomycetes reported from the medicinal plants of China have been documented (Nalini and Prakash 2017). It is worthy to note that the tropical rain forest plants are a treasure house of novel endophytic genera and many novel species. A single medicinal tree *Maytenus austroyunnanensis* yielded many novel endophytic actinomycete species, which indeed supports the fact that southwest Chinese tropical rain forest harbors rare and diverse actinomycetes. The actinomycete abundance was related to the geographical conditions in the tropical rain forests at the time of sampling (Qin et al. 2012). Owing to these factors, the chances of finding novel endophytic microbes are high.

13.4 Bioprospecting of Microbial Endophytes for Novel Biologically Active Metabolites

Natural products continue to play a significant role in the discovery of drugs. Microbial natural products represent an important path to the discovery of novel chemicals as therapeutic agents. Due to the onset of new emerging diseases, drug resistance among the pathogenic strains has often resulted in the search for biologically active compounds. Hence, there is a need to pursue microbial sources of novel drugs as therapeutic agents.

13.4.1 Host-Specific Novel Metabolites Produced by Microbial Endophytes

Plant-based therapy in the traditional medicinal practices has often cured many diseases and has led to the search for efficient antimalarial drugs quinine and artemisinin. Health-care practitioners in the TCM and Ayurvedic medicinal systems utilize drugs prepared from the traditional medicinal plants or their analogues to treat ailments (Newman and Cragg 2016), notably malarial fever (artemisinin; *Artemisia annua*), hypertension (reserpine; *Rauwolfia densiflora*), asthma (ephedrine; *Ephedra sinica*), and cancer (vincristine, vinblastine; *Catharanthus roseus*). The continuous harvesting of medicinal plants from their natural populations is already a threat to biodiversity and, therefore, an alternate strategy to provide inexhaustible supply of drugs needs immediate attention.

Microorganisms have contributed phenomenally through the production of secondary metabolites that are potential sources of drugs. The success achieved in the production of the anticancer drug Taxol® by the endophytic fungus *Taxomyces andreanae* (Stierle et al. 1994) has opened new vistas related to the bioprospecting of microbial endophytes for novel bioactive molecules. In view of this concept, a number of endophytes produce bioactive metabolites akin to the novel host plant metabolites (Table 13.1). The following are the examples for host-specific novel metabolites produced by endophytes.

13.4.1.1 Paclitaxel (Taxol®)

One of the most exciting discoveries in science is the isolation of the anticancer drug Taxol® from the bark of Pacific yew tree *Taxus brevifolia* by natural product chemist group Wani et al. (1971). The uniqueness of the approved anticancer drug lies in its ability to prevent the depolymerization of tubulin during cell division and is effective against ovarian and breast cancer cells. Treating a patient requires 2 g of the drug, which otherwise would mean felling of 12 large yew trees (Hartzell 1991). One of the milestones achieved in the natural product leads is the identification of the Taxol-producing fungus *Taxomyces andreanae* from the bark of the Pacific yew (Stierle et al. 1993). The yield of Taxol produced by this fungus is low, and hence, a promising source of Taxol-producing fungus was later identified as *Pestalotiopsis microspora* from the bark of Himalayan yew tree, *Taxus wallichiana* (Strobel et al. 1996). The Taxol production from *P. microspora* is 60–70 μg L^{-1}. Taxol-producing endophytes are reported from various plant sources (Kaul et al. 2012).

Endophytic actinomycetes are being bioprospected for their ability to produce anticancer metabolites. Actinomycetes isolated from the lignified woody tissues and herbaceous tissues of yew from different locations in Italy were identified as taxane producers (Caruso et al. 2000). The genera *Streptomyces*, *Micromonospora*, and *Kitasatospora* were identified as taxane producers (50–100 ng L^{-}). This is the first report of endophytic actinomycetes as taxane producers from the woody and herbaceous yew tissues.

13.4.1.2 Camptothecin

Camptothecin (CPT) is an antitumor chemotherapeutic agent isolated from the bark of the Chinese medicinal tree *Camptotheca acuminata* Decne. (Icacinaceae) (Wall et al. 1966) and was first approved by FDA for the treatment of colon cancer, and reports indicated its efficacy for treating various types of cancers. The drug has severe side effects and hence was thoroughly discouraged for clinical use in 1972,

Table 13.1 Host-related novel bioactive metabolites produced by the fungal endophytes of medicinal plants

Fungal endophyte	Plant species	Host-related metabolites	Chemical structure	Chemical group	Bioactivity	References
Taxomyces andreanae	*Taxus brevifolia* Nutt.	Taxol	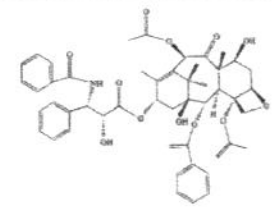	Diterpenoid	Anticancer	Stierle et al. (1993)
Pestalotiopsis microspora	*Taxus wallichiana* Zucc.	Taxane			Antileukemia; antitumor	Strobel et al. (1996)
Mycelia sterilia 97 CY-3	*Catharanthus roseus* L.	Vincristine	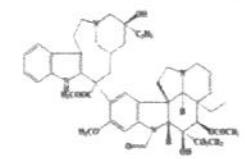	Monoterpenoid indole alkaloids	NA	Yang et al. (2004)
Fusarium oxysporum		Vincristine and vinblastine			Antiproliferative/ apoptosis	Kumar et al. (2013)
Talaromyces radicus-Cr-P20		Vincristine and vinblastine	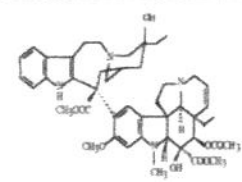		Cytotoxic (MDA-MB-231)	Palem et al. (2015)
Nigrospora sphaerica		Vinblastine				Ayob et al. (2017)
Chaetomium globosum		Vinblastine				Zafari et al. (2018)
Phialocephala fortinii	*Podophyllum peltatum* L.	Podophyllotoxin	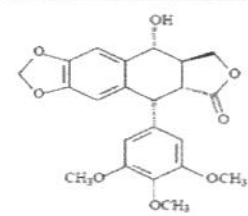	Aryl tetralin lignan	Antineoplastic	Eyberger et al. (2006)
Trametes hirsuta						Puri et al. (2006)
Entrophosphora infrequens	*Nothapodytes foetida* (Wight) Sluemer	Camptothecin	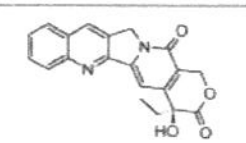	Alkaloid	Antineoplastic; anticancer	Puri et al. (2005), Rehman et al. (2008), and Kusari et al. (2009)
Neurospora sp.						
Fusarium solani						
Trichoderma atroviride LY357	*Camptotheca acuminata* Decne					Pu et al. (2013)

(continued)

Table 13.1 (continued)

Fungal endophyte	Plant species	Host-related metabolites	Chemical structure	Chemical group	Bioactivity	References
C. globosum	*Hypericum perforatum* L.	Hypericin	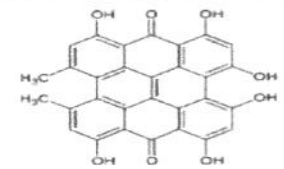	Naphthodianthrone derivative	Antiviral; cytotoxic	Kusari et al. (2008)
Eupenicillium parvum	*Azadirachta indica* A. Juss. (Meliaceae)	Azadirachtin A-B	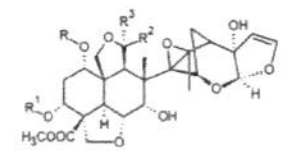	Oxygenated tetranortriterpenoid	Antifeedant; insect-growth regulating	Kusari et al. (2012)
F. solani (ERP-07)	*Cajanus cajan* L. (Millsp.) (Fabaceae)	Cajaninstilbene acid	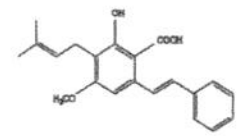	Stilbene carboxylic acid	Antioxidative	Zhao et al. (2012)
F. oxysporum (ERP-10)						
Fusarium proliferatum (ERP-13)						
T. atroviride D16	*Salvia miltiorrhiza* Bunge (Lamiaceae)	Tanshinones	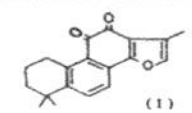	Diterpenoid quinones	NA	Ming et al. (2012)
Phoma glomerata D14		Salvianolic acids	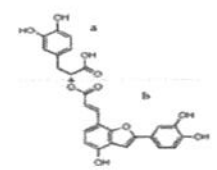	Phenolic acids		Li et al. (2016d)

NA not available

and in the 1980s the drug gained attention as it acts as a poison to the DNA topoisomerase I (TOP I) and as a cellular target (Jaxel et al. 1989). CPT is isolated from non-related plant families, and several structural analogues have been chemically synthesized.

Camptothecin, an alkaloid and a potent antineoplastic agent, was isolated from an endophytic fungus *Entrophospora infrequens* residing in *Nothapodytes foetida* (Puri et al. 2005). In vitro cytotoxicity assay of the compound against human cancer cell lines (A-549/lung cancer, HEP-2/liver cancer, OVCAR-5 for ovarian cancer) was comparable with the standard (Puri et al. 2005). Camptothecin and its analogues 9-methoxycamptothecin and 10-hydroxycamptothecin were extracted from *Fusarium solani* inhabiting *Camptotheca acuminata* (Kusari et al. 2009) collected from Yunnan Province of China. The latter two are precursors for the synthesis of anticancer drugs topotecan and irinotecan. Standardized conditions for the production of CPT from endophytic *Trichoderma atroviride* LY357 at 197.82 μg ml^{-1} was developed by Pu et al. (2013). A 50- to 75-fold increase of CPT yield was obtained when the optimized fermentation conditions, elicitor, and adsorbent resin were combined and applied to the culture of the seventh and eighth generations.

13.4.1.3 Hypericin

Hypericin is obtained from *Hypericum perforatum* (Clusiaceae), the perennial medicinal herb commonly known as St. John's wort, and has long been associated with wound healing, diuretic, antibiotic, and antiviral properties (Bombardelli and Morazzoni 1995). It acts as an antidepressant due to monoamine oxidase inhibiting property in comparison to the standard drug imipramine. Kusari et al. (2008) identified a fungal endophyte, *Chaetomium globosum*, from the surface-sterilized stem fragments of *H. perforatum* collected from wild high altitudinal populations of Jammu and Kashmir, India. The fungus produced hypericin and its precursor emodin intracellularly, at 35 ± 2 μg/100 g and 113 ± 1 μg/100 g dry weight of fungal mycelia under shake flask conditions after 6–7 days of incubation.

13.4.1.4 Podophyllotoxin

Podophyllotoxin (PTOX) was first isolated from the North American plant *Podophyllum peltatum* L., commonly known as the American mandrake in 1880. This natural product has been also isolated from the Indian podophyllum, *Podophyllum emodi* (Ramos et al. 2001). PTOX is the most abundant lignan in podophyllin, a resin produced by species of the genus *Podophyllum, and the derivatives etoposide and teneposide are used in the treatment of cancers and venereal warts.* PTOX prevents cell growth via polymerization of tubulin, leading to the arrest in cell cycle and suppression of the formation of the mitotic spindles and microtubules (Ardalani et al. 2017). The endophytic fungi *Trametes hirsute* and *Phialocephala fortinii* isolated from *Podophyllum peltatum* produce podophyllotoxin (0.5–189 μg L^{-1}) and other related aryl tetralin lignans with anticancer potential (Puri et al. 2006; Eyberger et al. 2006).

13.4.1.5 Azadirachtin

Azadirachta indica A. Juss., commonly known as the Indian neem or Indian lilac, is one of the most used medicinal plants growing abundantly in India. Traditionally, neem-based formulations have been used to cure fever, pain, leprosy, and malaria in Ayurvedic and Unani medical treatments, but the most striking property of neem tree reported to date is its insect-repellent property (Veitch et al. 2008) due to azadirachtin that acts as antifeedant. Azadirachtins A and B were characterized from the secondary metabolites of the fungal endophyte *Eupenicillium parvum* isolated from the healthy plant parts of neem tree from northern India (Kusari et al. 2012). Quantification of azadirachtins indicated that A was measured at 0.4 μg 100 g^{-1} dry weight of fungal mycelia and 43 μg L^{-1} from the spent broth, whereas B contained 0.05 μg 100 g^{-1} dry weight of fungal mycelia and 11 μg L^{-1} from the spent broth, respectively.

13.4.1.6 Vincristine

The endemic medicinal plant of Madagascar *Catharanthus roseus* known as Madagascar periwinkle is used in the treatment of solid tumors and leukemia. The vinca alkaloids are monoterpenoid indole alkaloids with high therapeutic value. Vincristine and vinblastine were discovered in the 1950s by Eli Lilly Pharmaceutical Company in Indianapolis, USA, and Noble Research Group in Toronto (Duge de Bernonville et al. 2015). Due to the importance of vinca alkaloids in cancer therapy, numerous research groups have identified fungal sources of alkaloid production. *Fusarium oxysporum* (Kumar et al. 2013) produced vinca alkaloids vincristine (670 μg L^{-1}) and vinblastine (70 μg L^{-1}). In *Talaromyces radicus*, different culture media induced the production of vincristine (67 μg L^{-1}) and vinblastine (76 μg L^{-1}) (Palem et al. 2015). The compounds showed antiproliferative activity tested against HeLa cell line with IC_{50} value of 20 μg ml^{-1}. The apoptosis-inducing activity of fungus-derived vincristine was proven through cell cycle analysis, loss of mitochondrial membrane potential, and DNA fragmentation patterns. Ayob et al. (2017) reported the intracellular quantitation of vinblastine produced (0.868 μg ml^{-1}) from the mycelia of endophytic *Nigrospora sphaerica*. Cytotoxicity studies with the human breast cancer cell line MDA-MB 231 with various concentrations (6.35–400 μg ml^{-1}) showed positive results with IC_{50} value of >32 and 350 μg ml^{-1} for vinblastine that was purified from the crude fungal and leaf extracts, respectively. Recently, vinblastine (78 μg L^{-1}) characterized from the culture filtrate of *C. globosum* was evaluated for the cytotoxic bioactivity against conidial germination of the rice blast fungus, *Pyricularia oryzae*. The compound exhibited cytotoxicity at IC_{50} value of 5 μg ml^{-1} (Zafari et al. 2018).

13.4.1.7 Cajaninstilbene Acid

Cajaninstilbene acid (CSA), a major stilbene, is found as a phytoconstituent in the legume crop, *Cajanus cajan* (L.) Millsp. and has traditional medicinal use. The pharmacological effects of CSA are well known, and the antioxidant activity is on par with the natural antioxidant resveratrol (Wu et al. 2011). Three CSA-producing fungi were isolated from *C. cajan* plants growing in China and identified as *Fusarium solani* (ERP-07), *Fusarium oxysporum* (ERP-10), and *Fusarium proliferatum* (ERP-13), respectively (Zhao et al. 2012). ERP-13 produced CSA in the culture medium at 504.8 ± 20.1 μg/L. In DPPH radical scavenging assay, the inhibition percentage was on par to that of standard CSA. This study for the first time reported the characterization of natural antioxidant CSA from the endophytic fungi *F. solani* and *F. proliferatum* from pigeon pea.

13.4.1.8 Tanshinones and Salvianolic Acids

The Chinese herb *Salvia miltiorrhiza* also known as "danshen" is a traditional medicinal plant. Two main compounds from this plant are salvianolic acids and tanshinones, the novel phenolic acids and diterpenoid quinones, respectively. The quinone compounds are known to have anticarcinogenic, antihypertensive, antiatherogenic properties, while the phenolic acids promote cardio- and cerebrovascular health (Ming et al. 2012; Chun-Yan et al. 2015).

S. miltiorrhiza plants collected from the Shanxi Province of China harbored the fungal endophytes *Trichoderma atroviride* D16 and *Phoma glomerata* D14, and the secondary metabolites of the endophytes contained the host-specific compounds, tanshinones (Ming et al. 2012) and salvianolic acids (Li et al. 2016d). The compounds were detected in the mycelia as well as the culture broth. The root tissue-colonized endophyte *C. globosum* D38 was able to increase the production of tanshinones in cocultivation with hairy root cultures of *S. miltiorrhiza* (Zhai et al. 2018).

A number of traditionally useful medicinal plants have yielded endophytes producing novel molecules with bioactive potentials (Venieraki et al. 2017). Many endophytes are potential producers of host-plant metabolites, with promising yield under the influence of elicitors or by altering the fermentation conditions. Modern methods to augment the yield can be exploited for pharmaceutical benefits.

13.4.2 Bioactive Potentials of Novel Microbial Metabolites

Metabolites from soil microorganisms especially fungi and actinomycetes are reliable platforms for drug discovery. Novel antibiotics from microbial strains, especially soil actinomycetes, have contributed to the development of life-saving drugs. Microbial endophytes from plant sources are by far proven to be reliable sources of

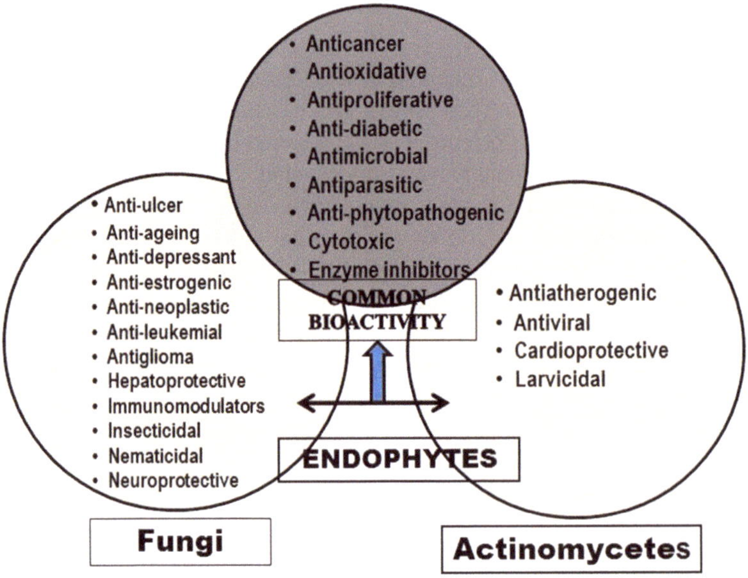

Fig. 13.1 Illustration of biological activities of endophytic fungi and actinomycetes from the medicinal plants

new metabolites with potent biological activities. Over the past 25 years, metabolites from endophytes have been bioprospected for various bioactivities, and some of them are common to both fungal and actinomycete groups (Fig. 13.1). Based on a detailed literature review, the distinct biological activities of novel molecules, the chemical class of compounds, and their sources for both fungal (Table 13.2) and the actinomycete endophytes (Table 13.3) are presented in the following sections.

13.4.2.1 Antimicrobial Compounds

Antimicrobial agents are the low-molecular-weight organic natural substances produced by microorganisms that are active at low concentrations against test microorganisms (Guo et al. 2000).

Novel Antimicrobial Metabolites from Fungal Endophytes

Phomopsichalasin, a cytochalasin-type compound, was the first antimicrobial agent produced by endophytic *Phomopsis* sp. It markedly differs from other cytochalasans in possessing a 13-membered tricyclic system that replaces the macrolide ring (Horn et al. 1995). The antimicrobial activity of the compound was tested positive against the test bacteria and yeast at 4 μg/disk in the disk diffusion assay. The potent antimycotics characterized from the cultures of the endophytic fungus *C. quercina*

Table 13.2 Novel metabolites produced by the fungal endophytes of medicinal plants and their bioactivities

Fungal endophyte	Plant species	Place of collection	Metabolites	Chemical group	Bioactivity	References
Anti-inflammatory						
Phomopsis sp.	*Erythrina crista-galli*	Argentina	Phomol	Polyketide lactone	Anti-inflammatory	Weber et al. (2004)
			Mevinic acid			
Periconia sp.	*Annona muricata*		Periconianone A	Sesquiterpenoids	Neural anti-inflammatory	Zhang et al. (2014a)
T. atroviride	*Lycoris radiata*	Hubei Province, China	Atrichodermone A	3-amino-5-hydroxy-5-vinyl-2-cyclo- penten-1-one dimer	Anti-inflammatory	Zhou et al. (2017)
			Atrichodermone B	Cyclopentenone		
			Atrichodermone C	Sesquiterpene		
Aspergillus terreus	–		Asperimides A–D	Butanolides	Anti-inflammatory	Liao et al. (2018)
Aspergillus sp.			Terrusnolides A–D	Butanolides	Anti-inflammatory	Qi et al. (2018)
Fusarium tricinctum SYPF 7082	*Panax notoginseng*	Yunnan Province, China	Rigidiusculamide E; [–(a-oxyisohexanoyl-N-methyl-leucyl)2-]	Alkaloids	Anti-inflammatory	Sun et al. (2018)
Fusarium sp.	*Mentha longifolia*		Fusaristerols B–D	Ergosterols	Anti-inflammatory	Khayat et al. (2019)
Immunosuppressants						
Fusarium subglutinans	*T. wilfordii*		Subglutinols A–B	α-Pyrone	Immunosuppressants	Lee et al. (1995a, b)
Colletotrichum dematium		Tropical rain forest, Costa Rica	Colutellin A	Peptide	Immunosuppressants	Ren et al. (2008)
Antimicrobial						
Phomopsis sp.	*Salix gracilistyla* var. *melanostachys*	Wakehurst, UK	Phomopsichalasin	Cytochalasin (3-ring system)	Antifungal; antibacterial	Horn et al. (1995)
Cryptosporiopsis cf. *quercina*	*T. wilfordii*	–	Cryptocandin A	Peptide antibiotic	Antimycotic	Strobel et al. (1999)
			Cryptocin	Tetramic acid	Antifungal	Li et al. (2000)

(continued)

Table 13.2 (continued)

Fungal endophyte	Plant species	Place of collection	Metabolites	Chemical group	Bioactivity	References
Unidentified fungus	*Daphnopsis americana*	Guanacaste National Park, Costa Rica	Guanacastepene A	Diterpenoids	Antibacterial	Brady et al. (2000c)
Cytospora sp.	*Conocarpus erectus*	Guanacaste National Park, Costa Rica	Cytoskyrins A–B	Bisanthraquinones	Antibacterial; biochemical induction assay (BIA)	Singh et al. (2007), Brady et al. (2000b)
			Cytosporones A–E	Octaketide antibiotics		
Colletotrichum sp.	*Artemisia annua*	Nanjing, China	Isoprenylindole-3-carboxylic acid; 3b,5a-dihydroxy-6b-acetoxy-ergosta-7,22-diene; 3b,5a-dihydroxy-6b-phenylacetyloxyergosta-7,22-diene	Ergosterol derivatives	Antifungal (fungistatic)	Lu et al. (2000)
Colletotrichum gloeosporioides	*Artemisia mongolica*	Nanjing, China	Colletrotic acid		Antibacterial; antifungal	Zou et al. (2000)
Periconia sp. OBW-15	*Taxus cuspidata*	Kangwon region, Korea	Periconicins A–B	Fusicocane diterpenes	Antibacterial	Kim et al. (2004)
Phomopsis sp.	*E. crista -galli*	Argentina	Phomol	Polyketide lactone	Antifungal; antibacterial	Weber et al. (2004)
Non-sporulating sp.	*Knightia excelsa*	Valley forests of New Zealand	Spiro-mamakone	Spirobisnaphthalene	Antibacterial	Van der Sar et al. (2006)
Cephalosporium acremonium	*Trachelospermum jasminoides*		Cephalosol	Unprecedented carbon skeleton	Antimicrobial	Zhang et al. (2008a, b)

Blennoria sp.	*Carpobrotus edulis*	Canary Islands	Blennolides A–G	Benzopyran polyketides (chromanone subunits/γ-lactone moiety)	Antifungal	Zhang et al. (2008a)
Chalara sp.	*Artemisia vulgaris*		Isofusidienols A–D	Benzopyran with chromone oxepine moiety	Antibacterial	Lösgen et al. (2008)
Microsphaeropsis sp.	*Lycium intricatum*		Microsphaeropsones A–C	Benzopyran with oxepino[2,3-b] chromen-6-one		Krohn et al. (2009)
Fusidium sp.	*Mentha arvensis*	Lower Saxony, Germany	Fusidilactones A–C	Polycyclic lactones with bicyclic and oxoadamantane skeleton	Antifungal; weak antibacterial	Krohn et al. (2002)
Cryptosporiopsis sp.	*Viburnum tinus*	Gomera	Viburspiran	8-membered maleic anhydride	Antifungal	Saleem et al. (2011)
			Cryptosporioptide	Polyketide (functionalized benzopyrone)	Antibacterial	Saleem et al. (2013)
Phomopsis CMU-LMA	*Alpinia malaccensis*		Benquinone	Lactone (14 membered)	Antibacterial	Adelin et al. (2011)
Microsphaeropsis arundinis PSU-G-18	*Garcinia hombroniana*	Songkhla Province, Thailand	Microsphaerodiolin	Modiolin	Moderate antifungal activity	Sommart et al. (2012)
			Microsphaerophthalides A–G	Phthalides		
Chaetomium sp.	*Zanthoxylum leprieurii*	Cameroon	Chaetosidone A	Depsidone (orsellic acid derivative)	Antibacterial	Talontsi et al. (2013)
Penicillium raciborskii	*Rhododendron tomentosum*	Oulu, Finland	Outovirins A–C	Epipolythiodiketopiperazines	Antifungal	Kajula et al. (2014)
Penicillium namyslowskii	*Rhododendron tomentosum*	Oulu, Finland	Dechlorodehydrogriseofulvin	Polyketide	Antifungal	Wubshet et al. (2013)
Aspergillus sp.	*M. azedarach*		Aspertryptanthrins A–C	Diketopiperazine alkaloids		Lhamo et al. (2015)

(continued)

Table 13.2 (continued)

Fungal endophyte	Plant species	Place of collection	Metabolites	Chemical group	Bioactivity	References
Trichoderma sp.	*Myoporum bontioides*	China	Dichlorodiaportinolide	Isocoumarin	Antifungal	Li et al. (2016c)
Simplicillium sp. PSU-H41	*Hevea brasiliensis*	Songkhla province, Thailand	Simplicidones A–I	Depsidones	Weak antibacterial and antifungal	Saetang et al. (2017)
			Simplicilopyrone	α-Pyrone		
Dendrothyrium variisporum	*Globularia alypum*	Algeria	(5S)-cis-gregatin B	Furanone derivatives	Antifungal	Teponno et al. (2017)
			Graminin D	Anthranilic acid derivatives		
			2-phenylethyl 3-hydroxyanthranilate;			
			phenylmethyl anthranilate; 3-hydroxy-3-methylbutyl anthranilate			
Anticancer, antitumor, antiproliferative, and cytotoxic compounds						
T. andreanae	*T. brevifolia*	Montana, USA	Taxol; taxane	Diterpenoids	Anticancer	Stierle et al. (1993)
Rhinocladiella sp.	*Tripterygium wilfordii*		Cytochalasins H-J	Cytochalasins	Antitumor	Wagenaar et al. (2000)
			Epoxycytochalasin H			
Phomopsis longicolla	*Mentha* sp.		Dicerandols	Dimer	Cytotoxic	Wagenaar and Clardy (2001)
Chaetomium sp.	*Adenophora axilliflora*		Chaetominine	Tripeptide alkaloid	Cytotoxic	Jiao et al. (2006)
Non-sporulating sp.	*Knightia excelsa*	Valley forests of New Zealand	Spiro-mamakone	Spirobisnaphthalene	Anticancer (leukemia)	Van der Sar et al. (2006)

Chaetomium globosum IFB-E019	*Imperata cylindrica*	Jiangsu Province, China	Chaetoglobosin U	Cytochalasan alkaloid	Cytotoxic	Ding et al. (2006)
Hypoxylon truncatum	*A. annua*	Nanjing, China	Daldinones C–D	Benzo[j]fluoranthene	Cytotoxic	Gu et al. (2007)
C. globosum	*Imperata cylindrica*		Chaetoglobins A–B	Heterocyclic azaphilone alkaloids	Cytotoxic	Ge et al. (2008)
Edenia sp.	*Petrea volubilis*	Coiba National Park, Panama	Palmarumycin CP_{17}–CP_{18}	1,8-dihydroxynaphthalene derived spiroketal unit linked to a second oxidized naphthalene unit	Weak cytotoxic	Martinez-Luis et al. (2008)
Alternaria sp.	*Polygonum senegalense*	Egypt	Alternariol 5-*O*-sulfate	Sulfated derivatives of alternariol and its monomethyl ethers	Cytotoxic	Aly et al. (2008)
			Alternariol 5-*O*-methyl			
			Ether-4′-*O*-sulfate			
			3′-Hydroxyalternariol			
			5-*O*-methyl ether, desmethylaltenusin, alterlactone			
			Alternaric acid			
Alternaria sp.	*Carex aridula*		(−) Alternarlactam	Polyketide with cyclopentenone and isoquinolinone pharmacophores	Antitumor	Zhang et al. (2010)

(continued)

Table 13.2 (continued)

Fungal endophyte	Plant species	Place of collection	Metabolites	Chemical group	Bioactivity	References
Phomopsis CMU-LMA	*Alpinia malaccensis*		Benquinone	Lactone (14 membered)	Cytotoxic	Adelin et al. (2011)
Fusarium sp. LN-10	*M. azedarach*	China	Fusariumine	Isocoumarin	Cytotoxic	Yang et al. (2012)
Fusarium sp. BCC14842	*Bambusa* sp.	Nam Nao National Park, Thailand	4-Hydroxydihydronorjavanicin; dihydronaphthalenone; diastereomer	Dihydronaphthalenones	Weak to moderate cytotoxic	Kornsakulkarn et al. (2011)
			3, 5-hydroxydihydrofusarubins A, B, and D; methyl ether derivatives			
Stemphylium globuliferum	*Mentha pulegium*	Morocco	Tetrahydroanthraquinone	Anthracene derivatives	Antitumor/cytotoxic	Debbab et al. (2009)
			Tetrahydroanthraquinone dimers			
Penicillium sp.	*C. roseus*		Citreoviripyrone A	α-Pyrone polyketide	Cytotoxic	Asai et al. (2013)
Chaetomium sp.	*Zanthoxylum leprieurii*	Cameroon	Chaetosidone A	Depsidone	Weak cytotoxicity	Talontsi et al. (2013)
Trichoderma atroviride	*Taxus baccata*	France	Harzianes 1–4	Tetracyclic diterpene	Weak cytotoxic	Adelin et al. (2014)
Penicillium manginii	*P. notoginseng*	China	Duclauxamide A1	Heptacyclic oligophenalenone dimer	Cytotoxic	Cao et al. (2015)
Trichoderma gamsii			Trichoderones A–B	Cytochalasans	Weak cytotoxic	Ding et al. (2012, 2013)
			Trichodermone			
Aspergillus versicolor	*Paris polyphylla* var. *yunnanensis*	China	Aspergillines A–E	Oxygenated cyclopiazonic acid-derived alkaloids	Anti-tobacco mosaic virus (TMV); cytotoxic	Zhou et al. (2014)
Alternaria phragmospora	*Vinca rosea*	–		α-Pyrone	Antileukemic	Metwaly et al. (2014b)
Fusarium sp. PDB51F5.	NA	NA	Fusaraisochromenone; fusaraisochromanone	Isochromenone; isochromanone	Weak cytotoxicity	Boonyaketgoson et al. (2015)

Peyronellaea coffeae-arabicae	*Pritchardia lowreyana*	Hawaii	Peyronellins A–C	Polyketide sesquiterpenes	Anticancer	Li et al. (2016b)
Phomopsis sp.			Phomopchalasins A–B	Cytochalasans	Anticancer (antimigrative)	Yan et al. (2016)
T. atroviride	*Lycoris radiata*	Hubei Province, China	Atrichodermone A	3-amino-5-hydroxy-5-vinyl-2-cyclopenten-1-one dimer	Cytotoxic	Zhou et al. (2017)
			Atrichodermone B	Cyclopentenone		
			Atrichodermone C	Sesquiterpene		
Phoma sp. YN02-P-3	–	–	Phomones	α-Pyrone	Cytotoxic	Sang et al. (2017)
Nigrospora BCC47789	–	–	Hydroanthraquinone	Anthraquinones	Cytotoxic	Kornsakulkarn et al. (2018)
			Nigrosporones A–B			
Aspergillus sp.	*Paeonia ostii*		Aspergillates A–E	Globoscinic acid derivatives	Cytotoxic	Wang et al. (2018)
Dendrothyrium variisporum	*Globularia alypum*	Algeria	2-Phenylethyl 3-hydroxyanthranilate	Anthranilic acid derivative	Moderate cytotoxicity	Teponno et al. (2017)
Aspergillus sp.	*Pinellia ternata*		Seco-cytochalasins A–F	Cytochalasans	Cytotoxic	Xin et al. (2019)
			Asperlactones G–H			
Phoma bellidis	*Tricyrtis maculata*		Bellidisins A–D	Polyketides	Cytotoxic	Wang et al. (2019)
T. wortmannii	*T. wilfordii*		Wortmannines F–G	Pyranones	PI3Kα inhibition (cancer therapeutics)	Zhao et al. (2019)
Antidiabetic compounds						
Pseudomassaria sp.	–	Kinshasa Republic, Congo	Demethylasterriquinone	Nonpeptide	Insulin-mimetic	Zhang et al. (1999)

(continued)

Table 13.2 (continued)

Fungal endophyte	Plant species	Place of collection	Metabolites	Chemical group	Bioactivity	References
Nigrospora sphaerica	*Oxya chinensis*	Guangzhou, China	Nigrosporamide A	Pyrrolidinone derivative	α-Amylase inhibition	Zhu et al. (2018)
			4-Prenyloxyclavatol	Acetophenone derivative		
Antiviral						
Cytonaema sp.	*Quercus* sp.	UK	Cytonic acids A–B	*p*-tridepsides	hCMV protease inhibitors	Guo et al. (2000)
Aspergillus versicolor	*Paris polyphylla* var. *yunnanensis*	China	Aspergillines A–E	Oxygenated cyclopiazonic acid-derived alkaloids	Anti-tobacco mosaic virus (TMV)	Zhou et al. (2014)
Periconia sp. F-31	*Annona muricata*	China	Periconiasin G	Cytochalasan with 7/6/5 tricyclic ring system	Anti-HIV	Zhang et al. (2016)
Nigrospora sp. YE3033	*Aconitum carmichaelii*		6-*O*-demethyl-4-dehydroxyaltersolanol A;8,11-didehydrochermesinone B;(7*S*)-7-hydroxy-3,7-dimethyl-isochromene-6,8-dione	Hydroanthroquinone	Anti-H1N1	Zhang et al. (2016)
				Azaphilones		
Phoma sp.	*Aconitum vilmoriniana*	–	Phomanolide	Sesquiterpene (14-nordrimane type)	Anti-H1N1	Liu et al. (2019)
Anti-parasitic						
Phomopsis sp.	*Tectona grandis*	Thailand	Phomoxanthones A–B	Xanthone dimers	Antimalarial	Isaka et al. (2001)
Exserohilum rostratum	*Stemona* sp.	Thailand	11-hydroxymonocerin	Monocerin derivative	Antimalarial	Sappapan et al. (2008)
Edenia sp.	*Petrea volubilis*	Coiba National park, Panama	Palmarumycin CP_{17}–CP_{18}	1,8-Dihydroxynaphthalene-derived spiroketal unit linked to a second oxidized naphthalene unit	Anti-parasitic (leishmanial)	Martinez-Luis et al. (2008)

Codinaeopsis gonytrichoides	*Vochysia guatemalensis*	Costa Rica	Codinaeopsin	Tryptophan polyketide	Anti-parasitic (plasmodial)	Kontnik and Clardy (2008)
Chalara alabamensis	*Asterogyne martiana*	Costa Rica	Asterogynins A–B	Steroids (isoprenoids)	Anti-parasitic	Cao et al. (2010)
Delitzchia winteri	–	Costa Rica	Delitzchianones A–B;	Naphthaquinones	Moderate anti-plasmodial activity	Cao and Clardy (2011)
Phomatospora bellawinteri			8-Acetoxy pestalopyrone	δ-lactone		
Insecticidal						
Unidentified sp.	*Gaultheria procumbens*		5-Hydroxy-2-(1′hydroxy-5′-methyl4′-hexenyl)benzofuran; 5-Hydroxy-2-(1′oxo-5′-methyl4′-hexenyl) benzofuran	Benzofurans	Anti-insect	Findlay et al. (1997)
Nodulisporium sp.	*Bontia daphnoides*		Nodulisporic acids	Indole diterpenes	Anti-insect	Demain (2000)
Novel chemical structures with unusual bioactivities						
Fusarium pallidoroseum		Merck	Apicidins A–C	Cyclic tetrapeptides	Anti-protozoal	DarkinRattray et al. (1996)
Fusidium sp.	*Mentha arvensis*	Lower Saxony, Germany	Fusidilactones A–C	Polycyclic lactones with bicyclic and oxoadamantane skeleton	Antialgal	Krohn et al. (2002)
			Fusidilactones D–E	γ-Lactones		Qin et al. (2009a)
Penicillium chrysogenum	*Cistanche deserticola*	Northwest China	Chrysogenamide A	Alkaloid	Neuroprotective	Lin et al. (2008)

(continued)

Table 13.2 (continued)

Fungal endophyte	Plant species	Place of collection	Metabolites	Chemical group	Bioactivity	References
A. terreus	*A. annua*	Zijin Mountain, China	16-α-hydroxy-5*N*-acetyllaardeemin	Alkaloid	Anti-acetylcholinesterase	Ge et al. (2010)
Penicillium dangeardii	*Lysidice rhodostegia*	China	Penicillactones A–C	Spirocyclic anhydride moiety	Inhibiting β-glucoronidase from leucocytes	Liu et al. (2013)
Phoma sp.	*Garcinia* sp.	Thailand	Phomoxanthones A–B	Lactones	Antitubercular	Isaka et al. (2001)
Phomopsis sp.	*Garcinia dulcis*	Thailand	Phomoenamide	Amides	Antitubercular	Rukachaisirikul et al. (2008)
			Phomonitroester	Ester		
Diaporthe sp.	*Pandanus amaryllifolius*		Diaporthenone	Benzopyranones	Antitubercular	Bungihan et al. (2011)
T. wortmannii	*T. wilfordii*	–	Secovironolide	Furanosteroid	Antidepressant (monoamine oxidase inhibitory)	Ding et al. (2015)
			Epoxyvirone			
F. subglutinans	*Tripterygium wilfordii*		Subglutinols A–B	α-Pyrone	Antiestrogenic	Lim et al. (2015)
Novel chemical structures with no bioactivities						
Penicillium janthinellum	*M. azedarach*	Brazil	Janthinone	Lactone	No antimalarial activity detected	Marinho et al. (2005)
Eupenicillium sp.	*Glochidion ferdinandi*	Toohey Forest, QLD, Australia	Phomoxins B–C	Polyketide with cyclic carbonate moiety	No antimicrobial activity	Davis et al. (2005)

Fusarium sp. LN-12	*M. azedarach*	China	Fusarimine	Isoquinoline alkaloid	Inactive to phytotoxicity and cytotoxicity	Yang et al. (2012)
Xylaria sp. PSU-H182	*H. brasiliensis*	Trang Province, Thailand	Xylaromanones A – B;	Dimeric chromanones; cyclohexenone; benzamide	No activity in antimalarial, antibacterial, and cytotoxic assays	Maha et al. (2016)
			(R)-4-Hydroxy-2-ethyl-2-cyclohexen-1-one;			
			2,3-Dihydroxy-N-methoxy-6-propylbenzamide			
T. wortmannii LGT-4	*T. wilfordii*		Wortmannines A–C	Wortmannin derivative with an unusual five-membered B ring	No cytotoxicity	Fu et al. (2016)
Chaetoconis sp. FT087	*Osmoxylon novoguineensis*	Waimea Valley, Hawaii, USA	Chaetopenoids A–F	Sesquiterpene derivatives (eremophilane type)	No antibacterial and anti-proliferative activities	Li et al. (2016a)
N. sphaerica	*V. rosea*	Egypt	Nigrosphaerin A	Isochromene derivatives	No activity in antileishmanial, antimicrobial, and cytotoxic assays	Metwaly et al. (2014a)
Simplicillium sp. PSU-H41	*H. brasiliensis*	Songkhla Province, Thailand	Simplicidones A, C	Depsidones	No cytotoxicity	Saetang et al. (2017)

'–' not available

Table 13.3 Novel bioactive metabolites produced by the actinomycete endophytes of medicinal plants

Actinomycetes	Plant species	Sampling site	Metabolites	Chemical group	Bioactivity	Reference
Streptomyces NRRL 30562	*Kennedia nigriscans*	Northern Territory, Australia	Munumbicins A–D	Peptide antibiotics	Antibacterial; antimalarial	Castillo et al. (2002)
Streptomyces NRRL 3052			Munumbicins E-4 and E-5			Castillo et al. (2006)
Streptomyces NRRL 30566	*Grevillea pteridifolia*	Northern Territory, Australia	Kakadumycin A	Peptide antibiotics	Antibacterial; antimalarial	Castillo et al. (2003)
Streptomyces (MSU-2110)	*Monstera* sp.	Manu region, Upper Amazon, Peru	Coronamycin	Peptide antibiotics	Antifungal; antimalarial	Ezra et al. (2004)
Nonomuraea sp.	*Artemisia vulgaris*	Sao Paulo, Brazil	Brartemicin	Trehalose-derived antibiotic	Anti-metastatic/ anti-invasive	Igarashi et al. (2009)
Micromonospora sp.	*Abrus pulchellus* subsp. *pulchellus*	Thailand	Maklamicin	Polyketide	Antibacterial	Igarashi et al. (2011)
Microbispora sp. GMKU 363	*Clinacanthus siamensis*	Thailand	Linfuranone A	Polyketide	Antidiabetic; antiatherogenic	Indananda et al. (2013)
Streptomyces sp. YIM 65408	*T. wilfordii*	China	1″-*O*-methyl-8-hydroxymethyl-daidzein	Isoflavone	Antioxidative	Yang et al. (2013)
Streptomyces sp. YIM 66017	*Alpinia oxyphylla*	China	2–6-Dimethoxy, terephthalic acid	Benzenedicarboxylic acid	Antimicrobial; antioxidative	Zhou et al. (2014)
			Flavensomycinoic acid	Alkaloid	Cytotoxic	Zhou et al. (2013)

Streptomyces sp. BT01	*Boesenbergia rotunda*	Thailand	7-Methoxy-3, 3′,4′,6-tetrahydroxy flavones; 2′,7-dihydroxy-4′,5′-dimethoxyisoflavone	Phenolics (flavonoids)	Antibacterial	Taechowisan et al. (2014)
Streptomyces sp. YIM 66142	–	–	Medilamine C	ʊ-Hydroxy alkylamine derivative	Reduced cytotoxicity	Zhang et al. (2014b)
Streptomyces sp.	*Camellia sinensis*	China TCM	Rubrolone B	Tropolone alkaloid	Cardioprotection	Yan et al. (2016)
Streptomyces kebangsaanensis	*Portulaca oleracea*	Nensai reserve forest, Pahang, Malaysia	6-(2-Hydroxy-4-methoxyphenoxy carbonyl)	Phenazine compound	–	Remali et al. (2017)
			Phenazine-1-carboxylic acid			

'–' not available

are cryptocandin, a peptide with unique amino acid 3-hydroxy-4-hydroxy methyl proline, and cryptocin, a tetramic acid (Strobel et al. 1999; Li et al. 2000). Both compounds were effective in inhibiting selective phytopathogens *Sclerotinia sclerotiorum*, *Botrytis cinerea*, and *P. oryzae* (MIC 0.39 μg ml^{-1}).

Terpenoids are the major group of compounds produced by fungal endophytes and possess antimicrobial activity. Guanacastepene, a diterpenoid from the unidentified Costa Rican endophytic fungus, showed potent antibacterial activity against *Enterococcus faecalis* (MRSA and VREF strain) (Brady et al. 2000c, 2001). Three new ergosterol derivatives, isoprenylindole-3-carboxylic acid; 3b,5a-dihydroxy-6b-acetoxy-ergosta-7,22-diene; and 3b,5a-dihydroxy-6b-phenylacetyloxyergosta-7,22-diene, from *Colletotrichum* sp. showed antibacterial and fungistatic effects (200 μg ml^{-1}, Lu et al. 2000). Colletotric acid inhibited the growth of *Bacillus subtilis*, *Staphylococcus aureus*, and *Sarcina lutea* with MICs of 25, 50, and 50 μg ml^{-1}, respectively, and the crop pathogenic fungus *Helminthosporium sativum* (MIC: 50 μg ml^{-1}) (Zou et al. 2000). Periconicins A and B are fusicoccane diterpenes, i.e., diterpenes with a glycosylated isoprene unit resembling the carbon skeleton as fusicoccin, but differ in possessing a trans stereochemistry with C-1 methyl and C-3 hydrogen. They are novel compounds isolated from *Periconia* sp., from the inner bark of *Taxus cuspidata*, from Korea (Kim et al. 2004). Periconicin A exhibited significant antibacterial activity against *B. subtilis*, *S. aureus*, *Klebsiella pneumoniae*, and *Salmonella typhimurium* with minimum inhibitory concentrations in the range of 3.12–12.5 g ml^{-1} with the antibiotic gentamicin. Periconicin B exhibited modest antibacterial activity against the same strains of bacteria with MICs in the range of 25–50 μg ml^{-1}. Therefore, these compounds could be suggested as lead compounds for the development of antibacterial agents for many bacterial strains.

The polyketide groups of compounds are prevalent among the fungal secondary metabolites and are identified bearing alternating carbonyl and methylene groups and have antimicrobial and immunosuppressive properties. Phomol, a polyketide lactone, was identified from the endophyte of the Argentinian medicinal tree *E. crista-galli* and exhibited antifungal activity (Weber et al. 2004). Benzopyran and polyketides consisting of rare chromanones, blennolides A–G, with highly substituted γ–lactone moiety and two unusual chromanone units were isolated from endophytic *Blennaria* sp. from Canary Islands (Zhang et al. 2008a), which displayed moderate antifungal activity against *Microbotryum violaceum* (50 μg/disk). New benzopyrans with chromone oxepine moiety (isofusidienols A-D) and oxepino [2,3-b] chromen-6-one (microsphaeropsones A-C) were characterized from endophytic *Chalara* sp. (Lösgen et al. 2008) and *Microsphaeropsis* sp. (Krohn et al. 2009). Isofusidienols A and B exhibited strong antibacterial activity against *B. subtilis* (15 μg/disk).

Novel fusidilactones A and B and a rare fusidilactone C with an oxoadamantane skeleton, a spiro acetal structure, and two ether-bridged hemiacetals were isolated from the fungal endophyte *Fusidium* sp. (Krohn et al. 2002) and exhibited antifungal and weak antibacterial activity. Benquinone, a polyketide with 14-membered lactone formed due to the cyclization of benquinol, is a novel antibacterial compound found to inhibit *B. subtilis* (Adelin et al. 2011). New bioactive polyketide, cryptosporioptide, a functionalized benzopyrone was obtained from *Cryptosporiopsis* sp., with antifungal activity against *Bacillus megaterium* (Saleem et al. 2013).

Structurally unique metabolites such as depsidone with 2,4-dihydroxy-benzoic acid linked with ether and ester bonds such as chaetosidone A with orsellic acid derivatives were isolated from the Cameroonian endophyte, *Chaetomium* sp. (Talontsi et al. 2013), which displayed antibacterial activity against *B. cereus* and *S. aureus* at 40 μg/disk.

Epipolythiodiketopiperazine (ETP) alkaloids constitute a large and diverse family of biologically active secondary metabolites produced by a number of filamentous fungi including the genus *Penicillium* (Boyer et al. 2013). These small-molecule natural products are characterized by the incorporation of an intramolecular polysulfide bridge at the α,α'-positions of a *cyclo*-dipeptide (or diketopiperazine – DKP). Three novel α-β bridged ETP alkaloids, outovirins A–C, were characterized from the cultures of *P. raciborskii*, the endophytic fungus of *R. tomentosum* (Kajula et al. 2014). Outovirin C exhibited active antifungal assay against *B. cinerea* at low concentration (0.38 mM).

Indole alkaloids, with one or more indole/indoline moieties, are one of the largest classes of nitrogen-containing secondary metabolites. Indole diketopiperazine alkaloids are a special subclass of indole alkaloids. They are biogenetically derived from tryptophan, and commonly isolated from fungi of the genus *Penicillium* and *Aspergillus* (Wen et al. 2018). Aspertryptathrins A–C, new indole diketopiperazine alkaloids from *Aspergillus* sp. from *M. azedarach*, possess a 6/5/6/6 trypanthrin framework formed from tryptophan unit and an anthranilate residue. An unusual 16-membered ring skeleton is characteristic of C (Lhamo et al. 2015).

Viburspiran, a new antifungal class of maleic anhydride, belongs to octadride, having an eight-membered ring with two maleic anhydride units. It was isolated from endophytic *Cryptosporiopsis* sp. and exhibited antifungal activity against *B. cinerea* and *M. violaceum* (Saleem et al. 2011). Two new bisanthraquinones, cytoskyrins A and B, and five new related octaketides, cytosporones A–E, were isolated from the endophyte *Cytospora* sp. from Guanacaste National Park, Costa Rica (Singh et al. 2007). Cytoskyrin A exhibited potent in vitro antibacterial (MICs 0.03–0.25 μg/mL) and DNA-damaging activities (10 ng/spot), whereas cytoskyrin B was inactive in these assays. Among the cytosporones, only D and E exhibited Gram-positive activity, but they were inactive in the biochemical induction assay. Novel furanone and anthranilic acid derivatives were isolated from *Dendrothyrium variisporum*; the compound 2-phenylethyl 3-hydroxyanthranilate showed antimicrobial activity against a panel of test organisms (Teponno et al. 2017).

Two novel compounds from *Nigrospora sphaerica*, nigrosporamide A and 4-prenyloxyclavatol, were tested for their antifungal potentials (Zhu et al. 2018). Nigrosporamide exhibited higher antifungal activity against *C. gloeosporioides* with a MIC of 25.14 μM than triadimefon (MIC 272.39 μM), the positive control, which showed weak activity to *F. oxysporum* (MIC 401.62 μM) and *C. musae* (MIC 803.23 μM), respectively. The antifungal activities of 4-prenyloxyclavatol were weak, with MIC values of 402.71–805.41 μM.

Novel Antimicrobial Metabolites from Actinomycetes as Pharmaceutical and Agricultural Agents

Antibiotics are important drugs for health care and are preferred due to their potent therapeutic applications and have desired pharmacokinetic properties for the clinical use (Farner and Zazopoulos 2005). The actinomycetes are prolific producers of antibiotics. Plant-associated endophytic actinomycetes have produced a wide range of antibiotics with novel chemical structures (Matsumoto and Takahashi 2017). The genera *Streptomyces* and *Micromonospora* are the potential producers of antibiotics. Munumbicins are novel peptide antibiotics produced by the endophytic *Streptomyces* spp., from the ethnomedicinal plants of the Upper Amazon and Northern Territory of Australia, and were effective against Gram-positive bacteria *Bacillus anthracis* and *Mycobacterium tuberculosis* (Castillo et al. 2002, 2006). Kakadumycins produced by *Streptomyces* sp. 30,566 depicted impressive activity against *B. anthracis* (MIC0.2 to 0.3 μg ml^{-1}) (Castillo et al. 2003), and the antimycotic coronamycin produced by *Streptomyces* NRRL 30562 at 2 μg ml^{-1} (MIC) is effective against pythiaceous fungi and the human pathogen *Cryptococcus neoformans* (MIC 4 μg ml^{-1}) (Ezra et al. 2004). The endophyte was tested against agriculturally important plant pathogens along with *S. griseoviridis* formulation (Mycostop). The former produced inhibition zones twice that of the latter, to be considered for the product development as a potential agricultural agent.

Maklamycin, an antibacterial polyketide from *Micromonospora* isolated from the Thai medicinal plant Maklam phueak (*Abrus pulchellus*), has shown activity against Gram-positive bacteria at 0.2–13 μg ml^{-1} (Igarashi et al. 2011). Recently, novel flavonoids, benzamide, and lactone-producing *Streptomyces* spp. from Thai, Chinese, and Vietnamese medicinal species were identified to have antibacterial and/or antifungal activities (Taechowisan et al. 2014; Yang et al. 2015a, b; Vu et al. 2018). There is a need to discover newer antimicrobial agents from the endophytic actinomycetes for antibiotics.

13.4.2.2 Novel Anticancer and Cytotoxic Compounds

Natural compounds derived from plants and microorganisms are being screened for the anticancer properties and cytotoxicity studies. Recent mechanisms in the ontogeny of tumor cells and their conversion into metastasis have led to the classification and therapeutic applications of anticancer compounds. Antibiotics are used in cancer therapy.

Novel Anticancer/Antitumor Metabolites from Endophytic Fungi

An insight into novel anticancer and cytotoxic agents from plant-derived drugs had tremendous impact on the drug industry with the identification of paclitaxel as the world's first anticancer drug. Many type genera of noted plant families have yielded cytotoxic agents and are discussed in Sect. 13.4.1 of the chapter. However, the path to the discovery of anticancer agents by endophytic microbes was the identification of the Taxol-/Paclitaxel-producing fungus *T. andreanae*, from the inner bark of *T. brevifolia* (Stierle et al. 1993). It is not an exaggeration to document the exceptional enthusiasm of natural product researchers generated by this outcome over the past 25 years in the quest to isolate, identify, and characterize the bioactive compounds and their analogues from endophytic microbes.

Novel cytochalasin compounds cytochalasins H–J from *Rhinocladiella* sp. (Wagenaar et al. 2000) and a cytochalasan alkaloid chaetoglobosin U from a fungal endophyte *C. globosum* IFB-E019 (Ding et al. 2006) exhibited cytotoxic activity against nasopharyngeal epidermoid tumor cell line. New sulfated derivatives of alternariol and its monomethyl ethers along with four new compounds were isolated from *Alternaria* sp., an endophyte of *P. senegalense* (Aly et al. 2008). Alternariol 5-*O*-sulfate and desmethylaltenusin exhibited cytotoxicity against L5178Y lymphoma cells with EC_{50} values of 4.5 and 6.2 μg ml^{-1}, respectively.

Six new seco-cytochalasins A–F and two new asperlactones G–H were isolated from the solid cultures of an endophytic fungus *Aspergillus* sp. from the tubers of *P. ternata* (Xin et al. 2019). Compounds E and F were rare seco-cytochalasins possessing an α, β-unsaturated furanone structure in their side chains. These isolates exhibited cytotoxicity against human lung cancer A-549 cell line with IC_{50} values ranging from 7.8 to 70.2 μM. D and exerted a three-fold enhancement of doxorubicin susceptibility on doxorubicin-resistant human breast cancer (MCF-7/DOX) cell line (16 μM).

Four new polyketides bellidisins A–D were isolated from the rice fermentation extract of endophytic *Phoma bellidis* (Wang et al. 2019). The cytotoxicity was evaluated against human cancer cell lines HL-60, A549, SMMC-7721, MCF-7, and SW480. Compound D showed significant cytotoxicity on all five cell lines with IC_{50} value ranged between 3.40 to 15.25 μM, stronger than cisplatin (4.86–27.70 μM). Benquoine, a polyketide with 14-membered lactone from *Phomopsis* sp., exhibited weak cytotoxicity against the colonic epithelial cancer cell line HCT-116 with IC_{50} value of 210 nM (Adelin et al. 2011).

Three new anthracene derivatives, tetrahydroanthraquinone and two tetrahydroanthraquinone hetero dimers, were isolated from *Stemphylium globuliferum* (Debbab et al. 2012). Nigrosporone A, a new hydroanthraquinone, and a new naturally occurring nigrosporone B together were isolated from an endophytic fungus, *Nigrospora* sp. BCC 47789. Nigrosporone A showed cytotoxic activity (Kornsakulkarn et al. 2018). The compounds were analyzed for their cytotoxic activities. Tetrahydroanthraquinone showed cytotoxicity against murine cancer cell line L5178Y. Five novel globoscinic acid derivatives, aspergillates A–E, were isolated from the endophytic fung-us, *Aspergillus* sp. derived from *Paeonia ostii* (Wang et al. 2018). Cytotoxic activities against five selected tested tumor cell lines were evaluated.

Few novel metabolites from endophytic fungi do exhibit weak cytotoxicity as in chaetosidone A (Talontsi et al. 2013), fusaraisochromenone and fusaraisochromanone (Boonyaketgoson et al. 2015), palmarumycins (Martinez-Luis et al. 2008), trichoderones A–B (Ding et al. 2012, 2013), and harzianes (Adelin et al. 2014). Cytoskyrin A, a bisanthraquinone from *Cytospora* sp., exhibited poor cytotoxicity against tumor cell lines (IC_{50} > 5 μg/mL) compared to known antitumor agents (Singh et al. 2007).

The phosphatidylinositol 3-kinase (PI3K) pathway is frequently activated in human cancers. Therefore, PI3K has become an important anticancer drug target, and currently, pharmaceutical developments of PI3K inhibitors are in pipeline (Zhao et al. 2010). Two new pyranone compounds, wortmannines F and G, were characterized from *T. wortmannii*, the endophytic fungus of the medicinal plant, *T. wilfordii* (Zhao et al. 2019). The compounds were tested for the phosphatidylinositol 3-kinase inhibition. Both compounds showed inhibitory activity with IC_{50} value of 25 and 5 μM, respectively.

Novel metabolites from endophytes inactive in cytotoxicity assays are documented for janthinone, phomoxins, fusarimines, xylaromanones, wortmannines A–C, chaetopenoids, nigrosphaerin, and simplicillins A and C (Table 13.2).

Novel Anticancer/Cytotoxic Metabolites from Endophytic Actinomycetes

Endophytes do produce beneficial compounds with medicinal applications, as demonstrated by Taxol, the anticancer drug produced by *Taxomyces andreanae* (Strobel and Daisy 2003). With impetus received from the endophytic fungal success on Taxol, Caruso et al. (2000) isolated the first taxane-producing *Streptomyces*, *Micromonospora*, and *Kitasatospora* spp., from the bark tissue of *Taxus baccata* and *T. brevifolia.* Anticancer compounds are being isolated and tested for their efficacy in various cell lines. The peptide antibiotic coronamycin from *Streptomyces* sp. (MSU-2110) showed cytotoxic potentials on par with Taxol by inhibiting the HMEC and BT20 cell lines (IC_{50} 5–10 μg ml^{-1}) (Ezra et al. 2004).

Brartemicin, a trehalose-derived antibiotic and a novel inhibitor of metastasis produced by *Nonomuraea* sp., was isolated from the Brazilian medicinal plant, *Artemisia vulgaris* (Igarashi et al. 2009). This compound indicated anti-invasive property of murine colon carcinoma cells (IC_{50} 0.39 μM) with no toxicity. Flavensomycinoic acid, a cytotoxic alkaloid from *Streptomyces* sp. YIM66017 (Zhou et al. 2013), exhibited potent cytotoxicity to MCF-7, the breast cancer cell line (IC_{50} 17.0 μM). Medilamine C, a novel ʊ-hydroxy alkylamine derivative from *Streptomyces* sp. YIM 66142, showed reduced cytotoxicity (Zhang et al. 2014b). Tumor metastasis involves a cascade of events leading to high mortality rates, and therefore, the discovery of tumor metastasis inhibitory or anti-invasive compounds from the actinomycetes holds great promise.

13.4.2.3 Antioxidants

Antioxidants are defined as "any substance when present in low concentrations compared to those of an oxidisable substrate significantly delays or prevents the oxidation of that substance" (Halliwell and Gutteridge 1989). Antioxidants are derived from food sources and have been proven to have numerous health benefits. Microbial endophytes have provided few antioxidant molecules with unique structures, but their clinical applications are limited.

Novel Antioxidant Compounds from Endophytic Fungi

The antioxidant compounds pestacin and isopestacin were isolated from the cultures of *Pestalotiopsis microspora* residing in *Terminalia morobensis* near the Sepik River drainage of Papua New Guinea (Harper et al. 2003). 4,6-Dihydroxy-5-methoxy-7-methylphthalide, a new isobenzofuranone derivative obtained from *Cephalosporium* sp. AL031 in *Sinarundinaria nitida* sampled from Yunnan Province of China, exhibited EC_{50} value of 10 μM in 1,1-diphenyl-2-picryhydrazyl (DPPH) radical scavenging assay (Huang et al. 2012). Cajaninstilbene acid (CSA), 3-hydroxy-4-prenyl-5-methoxystilbene-2-carboxylic acid, has been reported from *Fusarium* spp., an endophyte of pigeon pea *Cajanus cajan* (Zhao et al. 2012) with medicinal benefits. The antioxidative potential was tested in DPPH radical scavenging assay, with CSA exhibiting 80% scavenging potentials on par with the standard at a concentration of 500 μg ml^{-1}.

Novel Antioxidant Compounds from Endophytic Actinomycetes

A new daidzein derivative, 1″-*O*-methyl-8-hydroxymethyl-daidzein, was isolated from *Streptomyces* sp. YIM 65408, an endophyte of *Tripterygium wilfordii* cultivated in soybean powder containing medium (Yang et al. 2013). In the radical scavenging assay by 2,2-diphenyl-1-picrylhydrazyl method, the compound showed an IC_{50} value at 0.60 mmol/L.

2,6-Dimethoxy terephthalic acid, a new natural product isolated from the culture filtrate of *Streptomyces* sp. YIM66017, from *Alpinia oxyphylla* exhibited significant antioxidant activity in anti-radical assay with IC_{50} values of 4.61 and 57.12 μg ml^{-1}, respectively (Zhou et al. 2014). The actinomycetes are undoubtedly the largest producers of bioactive substances and are a challenge to microbiologists and natural product chemists to examine the potentiality of the products for therapeutic applications.

13.4.2.4 Novel Antidiabetic Compounds from Endophytes

Diabetes is a serious medical condition resulting in the early breakdown of starch into sugar resulting in hypo- or hyperglycemia. Therapeutic targets to counter diabetes by employing antidiabetic agents, through the inhibitory action of carbohydrate hydrolyzing enzymes such as α-glucosidase and α-amylase enzymes, glucose transporters, insulin receptor substrates, etc.

A fungal nonpeptidal metabolite was isolated from the endophytic fungus *Pseudomassaria* sp., collected from the rain forest plant, near Kinshasa, Republic of Congo (Zhang et al. 1999). The nonpeptidyl, small molecule, insulin-mimetic compound (demethylasterriquinone B-1, DMAQ-B1) was isolated from a mixture of metabolites. Oral administration of DMAQ-B1 resulted in significant lowering of glucose in two mouse models of diabetes (Salituro et al. 2001). DMAQ-B1 represents the first orally active insulin-mimetic agent and as a novel therapeutic target in diabetic patients.

Nigrosporamide A and 4-prenyloxyclavatol, two novel compounds isolated from *N. sphaerica* (Zhu et al. 2018), were tested for their in vitro α-glucosidase inhibitory activity with the positive control acarbose currently used clinically in combination with antidiabetic agents to control the blood glucose level of patients. Nigrosporamide A exhibited high α-glucosidase inhibitory activity (IC_{50} value 120.3 μM), which was three times more potent than acarbose (IC_{50}, 446.7 μM). 4-prenyloxyclavatol was inactive with $IC_{50} > 520.34$ μM.

The well-proven inhibitors induce side effects, and therefore the need is to screen newer inhibitors and to identify potential α-amylase and glucosidase inhibitors from natural sources to minimize the side effects. Linfuranone A, a novel polyketide isolated from the actinomycete endophyte *Microbispora* sp. from Thai medicinal plant *Clinacanthus siamensis*, displayed antidiabetic potential in mouse cell lines (Indananda et al. 2013). Recent findings indicate the potential of endophytic actinomycetes as glucosidase and amylase inhibitors (Pujianto et al. 2012; Akshatha et al. 2014).

13.4.2.5 Anti-inflammatory Molecules

Inflammation is a protective attempt by the organism to remove the injurious stimuli and to initiate the healing process. An effective anti-inflammatory drug should be able to inhibit the development of inflammation without interfering in normal homeostasis. Current approaches to overcome the inflammation include the use of immune selective anti-inflammatory derivatives, selective glucocorticoid receptor agonist, resolvins and protectins, and TNF inhibitors (Dhingra et al. 2015). A number of herbal drugs have been identified in the past that can target inflammatory cytokines. Therefore, a safe and efficient drug molecule to confer protection against inflammation is urgently needed.

Novel anti-inflammatory class of compounds such as lactones, sesquiterpenes (oids), butanolides, and ergosterols are characterized from the endophytic fungi. Phomol and mevinic acid are the first anti-inflammatory novel compounds isolated from *Phoma* sp. Periconianones A from the endophytic *Periconia* sp. are novel sesquiterpenes and exhibited anti-inflammatory activity in mouse microglia BV2 cells (Zhang et al. 2014a), the immune cells of the nervous system. Asperimides A–D isolated from the tropical *A. terreus* are novel aromatic butanolides consisting of maleimide core (Liao et al. 2018). The compounds were tested for their inhibitory effects in lipopolysaccharide-mediated RAW 264.7 cells, and inhibitory effects were noted in A and C with IC_{50} value of 1.26 and 0.78 and μM, respectively.

The endophytic fungus *T. atroviride* from the bulb of *Lycoris radiata* produced a novel 3-amino-5-hydroxy-5-vinyl-2-cyclopenten-1-one dimer, atrichodermone A; a new cyclopentenone derivative, atrichodermone B; and a new sesquiterpene, atrichodermone C, together with three known cyclopentenone derivatives (Zhou et al. 2017). Compounds were evaluated for their cytotoxicity against HL60 and U937 cell lines, as well as anti-inflammatory effect against the production of the pro-inflammatory cytokines TNF-α and IL-1β.

Novel butanolides, terrusnolides A–D from *A. terreus* with anti-inflammatory potential, were isolated from an endophyte of the Chinese medicinal plant *T. wilfordii* (Qi et al. 2018). The anti-inflammatory effects of compounds were evaluated in vitro in lipopolysaccharide (LPS)-stimulated RAW 264.7 macrophages. The compounds exhibited excellent inhibitory effects on the production of interleukin-1β (IL-1β), tumor necrosis factor-α (TNF-α), and nitric oxide (NO) in LPS-induced macrophages, comparable with the positive control (indomethacin). Results indicate terrusnolides A–D as new natural compounds for the treatment of inflammation.

Two new alkaloids, rigidiusculamide E and [−α-oxyisohexanoyl-N-methyl-leucyl)$_2^-$, were characterized from the endophytic *Fusarium tricinctum* from the roots of the Chinese medicinal plant *P. notoginseng* (Sun et al. 2018). The anti-inflammatory effects of compounds were evaluated in vitro in RAW 264.7 macrophage cell line for the inhibition of NO production. [−α-Oxyisohexanoyl-N-methyl-leucyl)$_2^-$ exhibited excellent inhibitory effects on the production of NO with the IC_{50} value of 18.1 μM.

Three new ergosterol derivatives, namely, fusaristerols B [(22*E*,24*R*)-3-palmitoyl-19(10→6)-abeo-ergosta-5,7,9,22-tetraen-3β-ol] (1), C [(22*E*,24*R*)-ergosta-7,22-diene-3*β*,6*β*,9*α*-triol] (3), and D [(22*E*,24*R*)-ergosta-7,22-diene-3β,5α,6β,9*α*-tetraol 6-acetate] (4), were characterized from the endophytic *Fusarium* sp. isolated from *Mentha longifolia* L. roots (Khayat et al. 2019). The metabolites were assessed for 5-lipoxygenase (5-LOX) inhibitory potential. Compound **1** possessed 5-LOX inhibitory potential with an IC_{50}s of 3.61 μM, compared to that of indomethacin (IC_{50} 1.17 μM)

13.4.2.6 Insecticidal Compounds

Insects often damage the standing crop produced by their feeding behavior and are a menace. Synthetic insecticides are used to control the damages caused by the pests and are considered harmful. Alternate methods to control insects involve the use of biopesticides and integrated approaches. One of the earliest insecticidal agents to control moths is naphthalene, the insect repellent. An endophytic fungus *Muscodor vitigenus* from a liana produces the organic compound naphthalene (Daisy et al. 2002). Two novel benzofurans, namely, 5-hydroxy-2-(1′hydroxy-5′-methyl-4′-hexenyl) benzofuran and 5-hydroxy-2-(1′oxo-5′-methyl-4′-hexenyl) benzofuran, were isolated from an unidentified endophyte from *G. procumbens* with toxic effects on the larvae of spruce bud worm (Findlay et al. 1997). Newer sources of novel insecticidal agents need to be investigated with endophytic microbes.

13.4.2.7 Anti-parasitic Compounds

Mosquito-borne parasitic diseases such as malaria and lymphatic filariasis are prevalent worldwide. The parasitic infections caused by the protozoans are a major concern, as the causal agent *Plasmodium falciparum* is responsible for the cerebral fever, which can be of great concern. Antimalarial drugs are efficacious; however over a long period of use, the parasite often tends to develop resistance, and there is a need to search for newer and safer anti-parasitic drugs. A number of plant and microbial-based insecticides are preferred due to their selective toxicity and safe handling. Bioprospecting of endophytic strains reveal their larvicidal potentials.

Novel Compounds from Fungal Endophytes

Anti-parasitic compounds such as novel xanthone dimers, phomoxanthones A and B (Isaka et al. 2001), 11-hydroxymonocerin (Sappapan et al. 2008), and two *Xylaria* benzoquinone metabolites, xylariaquinones and 2-chloro-5-methoxy-3-methylcyclohexa-2,5-diene-1,4-dione, have shown inhibition to *P. falciparum*. Novel antileishmanial metabolites, viz., CP17 and CP18, were identified from the secondary metabolites of *Edenia* sp., the endophytic fungus from Panamanian medicinal plant, *Petrea volubilis* (Martinez-Luis et al. 2008). The metabolites exhibited selective toxicity against *Leishmania donovani* with EC_{50} values of 1.34 and 0.62 μM, whereas in the positive control, amphotericin B, the EC_{50} value was 0.09 μM. The fungal endophytes from the Costa Rican medicinal plants yielded anti-plasmodial novel metabolites such as codinaeopsin, a polyketide (Kontnik and Clardy 2008), and the isoprenoid asterogynins (Cao et al. 2010).

Phomanolide from endophytic *Phoma* sp. having 14-nordrimane structure was identified with antiviral activity against influenza A virus (A/Puerto Rico/8/34, H1N1) with IC_{50} values of 2.96 ± 0.64 μg ml^{-1} (Liu et al. 2019). A novel hydroanthraquinone derivative, 6-*O*-demethyl-4-dehydroxyaltersolanol A, and two new azaphilones, 8,11-didehydrochermesinone B and (7*S*)-7-hydroxy-3,7-dimethyl-isochromene-6,8-dione, were isolated from *Nigrospora* sp. YE3033, an endophytic fungus of *Aconitum carmichaelii* (Zhang et al. 2016). The hydroanthroquinone exhibited the inhibitory effects on influenza viral strain of A/Puerto Rico/8/34 (H1N1) with the IC_{50} values of 2.59 μg/ml^{-1}, while the low cytotoxicity of 8,11-didehydrochermesinone B (CC_{50} value of 184.75 μg ml^{-1}) holds promising potential in the development of anti-influenza A virus drugs.

Novel Compounds from Endophytic Actinomycetes

A pioneering study by Castillo et al. (2002, 2006) revealed the anti-plasmodial potential of the newly described peptide antibiotics, munumbicins C and D with low toxicity and coronamycins produced by the endophytic *Streptomyces* NRRL 30562 and *Streptomyces* sp. (MSU-2110) from *K. nigriscans* the snake vine plant and the follow-me vine, *Monstera* sp. The antibiotics showed remarkable activity with IC_{50} of 0.5 and 0.87 μg ml^{-1} and 9.0 ng ml^{-1} against the parasite, *P. falciparum.*

13.4.2.8 Immunosuppressive Compounds

Immunosuppressive drugs are used to prevent allograft rejection in transplant patients and to treat autoimmune diseases such as rheumatoid arthritis and insulin-dependent diabetes. Approved immunosuppressive agents such as cyclosporin A and FK506 possess some undesirable side effects, and there is a need for better immunosuppressive agents.

The first endophytic fungal novel immunosuppressant, subglutinols A and B, was characterized from the liquid culture of *Fusarium subglutinans* residing in the medicinal plant *T. wilfordii* (Lee et al. 1995a). IC_{50} value of subglutinols in the mixed lymphocyte reaction and thymocyte proliferation assays indicated 0.1 μM and was equipotent to that of the standard drug cyclosporin in one of the assays. Therefore, the nontoxic nature of compound is a criterion to be considered as a potential drug. Endophytes produce immunomodulatory compounds.

Pestalotiopsis leucothes, isolated from the Chinese medicinal plant *T. wilfordii*, produces immunomodulatory compounds (Kumar et al. 2005). One compound BS significantly inhibited the production of cytokines such as interleukin (IL)-1β, IL-2, interferon (IFN)-γ, and tumor necrosis factor (TNF)-α, by peripheral blood mononuclear cells (PBMNC) and soluble IL-2 receptor expression at concentrations greater than 1 μg ml^{-1}. The inhibition of PHA stimulated PBMNC proliferation, and IL-2 and sIL-2R production by BS indicates that it is a T cell-specific immunosuppressant. The endophytic fungus *Colletotrichum dematium* isolated from the tropical rain forest Costa Rican medicinal plant *Pteromischum* sp. produces a novel immunosuppressive compound colutellin A (Ren et al. 2008). It inhibited CD4 T cell activation of IL-2 production with an IC_{50} of 167.3 potential immunosuppressive activity of this compound. In cytotoxicity experiments, the compound exhibited no toxicity and has potential to be developed as a novel immunosuppressive drug.

13.4.2.9 Other Biological Activities of Novel Compounds from Endophytes

The secondary metabolites or novel compounds derived from the endophytes have therapeutic applications as algicidal, antidepressant, neuroprotective, and antitubercular molecules.

Algicidal Compounds

Algicides are chemicals which check or kill the growth caused by the algal genera. Filamentous chlorophyceae, blue green algae, and dinoflagellates cause excessive growth by sporadic blooms, which adversely affect the aquatic systems. Therefore, algicides are used to check the growth of harmful algal populations. Novel compounds, fusidilactones A–E, were isolated from *Fusidium* sp. and were evaluated for the algicidal activity by the in vitro growth inhibition assay. Algicidal activity against *Chlorella fusca* was detected by measuring the zone of inhibition (Krohn et al. 2002; Qin et al. 2009a).

Antitubercular Compounds

Tuberculosis is a chronic respiration disease caused by *Mycobacterium tuberculosis*. It results in persistent cough and blood-tinged sputum. Antitubercular compounds are effective against the pathogen, but some strains develop resistance to the drugs, and this necessitates the search for novel antitubercular compounds. The products from endophytes are sources of novel antitubercular compounds.

Novel antitubercular compounds have been isolated from *Phoma* spp. as endophytes of *Garcinia* spp. from Thailand. The compounds phomoxanthones, phomoenamide, and phomonitroester exhibited significant activity against *M. tuberculosis* (Isaka et al. 2001; Rukachaisirikul et al. 2008). A virulent strain of *M. tuberculosis* was inhibited by diaporthenones A and B and benzopyranones from *Diaporthe* sp. (Bungihan et al. 2011).

Antidepressant Compounds

Major depressive disorder (MDD) is a chronic, recurring, and debilitating mental illness that is the most common mood disorder in several countries. Several monoamine-based pharmacological drug classes have been developed and approved for the treatment of MDD; however, remission rates are less than 60%, and there is a delayed onset before remission of depressive symptoms is achieved (Hillhouse and Porter 2015). Drugs derived from natural products are used to treat depression but are with serious side effects. So, lookout for new sources of drugs from microbial endophytes is in progress. Secovironolide, the first example of a furanosteroid scaffold bearing a five-membered B ring from *Talaromyces wortmannii*, was tested for monoamine oxidase (MAO) inhibitory activity and showed weak antidepressant activity (Ding et al. 2015).

Anti-acetylcholinesterase Compounds

Alzheimer's disease (AD) is the most prevalent neurodegenerative disease, with symptomology that typically includes confusion, memory loss, impaired cognitive and emotional function, and dementia (Alzheimer's Disease International 2015) Hence, drugs that mimic acetylcholine activity (cholinomimetics) or drugs that limit acetylcholine breakdown (AChE inhibitors) have provided a therapeutic strategy to augment cholinergic signalling in AD patients. 16-α-hydroxy-5*N*-acetyllaardeemin is a novel alkaloid from endophytic *A. terreus* in *A. annua* (Ge et al. 2010). The inhibitory activity of the compound was investigated in the in vitro *assay.* The compound showed inhibitory activity with EC_{50} value of 58.3 μM against the positive control, tacrine (37.9 μM).

Neuroprotective Compounds

Alzheimer's disease (AD) is an age-related neurodegenerative disorder characterized by progressive cognitive and memory impairment and neuronal cell death (Cummings 2004). Drugs approved to treat AD are acetylcholinesterase inhibitors or the receptor agonists; both have profound side effects. Therefore, alternate drugs are essential to treat AD. Microbial endophytes have tremendous applications in agriculture and therapeutics.

A novel alkaloid of the macfortine group, chrysogenamide, was isolated from root endophytic *P. chrysogenum* in the Chinese medicinal plant, *Cistanche deserticola* (Lin et al. 2008). The compound exerted neuroprotective effects in human neuroblastoma SH-SY5Y cell lines. The effect of chrysogenamide A on neurocytes was evaluated using oxidative stress-induced cell death by MTT assay. The oxidative stress by hydrogen peroxide resulted in a decrease in the cell viability by 43% as compared with control group. Chrysogenamide A inhibited cell death induced by hydrogen peroxide by improving cell viability by 59.6% at concentration of 1×10^4 μM.

Antiestrogenic Compounds

Subglutinol A, the immunosuppressive compound isolated from *F. subglutinans*, was tested for the antiestrogenic potentials (Lim et al. 2015). The compound blocked the 17β-induced estradiol activated receptor plasmids and estrogen-response target genes.

Cardioprotective Compounds

Rubralone, a novel compound from *Streptomyces* sp., was tested against rat cardiomyocytes (Yan et al. 2016). The compound had no toxicity effects at a concentration of 100 μM but showed an increase in the viability of the H_2O_2-induced injury to cardiomyocytes, suggesting protective ability.

13.5 Novel Bioactive Molecules from *Pestalotiopsis* Species

The genus *Pestalotiopsis* is distributed worldwide and is a common inhabitant of a range of substrata. Its occurrence was first documented by Raj (1993) as a rain forest species. *Pestalotiopsis* species may reside as endophytes in the bark of tree trunk, twigs, and leaves and are genetically diverse (Tejesvi et al. 2007). Since the first isolation of *Pestalotiopsis microspora* as an endophyte, producing the anticancer drug paclitaxel, the genus has contributed significantly towards understanding the concept of horizontal gene transfer from the "host to endophyte" and to drug discovery. *Pestalotiopsis* spp. are often associated as endophytes of the rain forest plant species and produce metabolites that virtually govern all bioactivities ranging from antioxidant and antimicrobial to anticancer activities (Table 13.4). They are readily identified on agar plate by the white to cream mycelia often with conidiomata, the acervulus erumpent with black ooze (Fig. 13.2a) and fusiform conidia consisting of four cells with hyaline terminal and basal cells (Fig. 13.2b). *Pestalotiopsis* spp. are distinctly recognized by the appendages in the terminal cell that are unbranched.

13.5.1 Antioxidant Compounds

Two novel antioxidant compounds, pestacin and isopestacin, were isolated from *Terminalia morobensis* endophyte, *P. microspora* in the Riverine Sepik drainage in Papua New Guinea (Harper et al. 2003). Pestacin, a 1, 3-dihydro isobenzofuran, naturally occurs as a racemic mixture and functions by cleaving a reactive C–H bond and through O–H abstraction; the antioxidant activity is one order higher than that of trolox, a vitamin E derivative. The presence of a unique doubly benzylic carbon is necessary for a strong antioxidant activity and racemization. Isopestacin, an isobenzofuranone, has structural similarities to flavonoids and attributes to its antioxidant effect by scavenging both superoxide and hydroxyl free radicals (Strobel et al. 2002). It differs from other isobenzofuranones in possessing a substituted benzene ring attached at the C-3 position of the furanone ring.

Table 13.4 Novel bioactive metabolites produced by *Pestalotiopsis* endophytes of medicinal plants

Endophyte	Plant species	Sampling site	Metabolites	Chemical group	Bioactivity	Reference
Pestalotiopsis microspora	*Torreya taxifolia*	Ravine slopes of Apalachicola River, Florida	Pestaloside, torreyanic acid	B-glucoside	Antifungal; cytotoxic	Lee et al. (1995a, b)
			Pestalopyrone; hydroxypestalopyrone	Quinone dimer	Phytotoxic	Lee et al. (1996)
				Pyrones (lactones)		
P. microspora	*T. wallichiana*	Foot hills of Himalayas, Nepal	Paclitaxel	Diterpenoid	Anticancer	Strobel et al. (1996)
P. microspora	*Taxodium distichum*	Swamp forest, central coast, South Carolina	Taxol	Diterpenoid	Anticancer	Li et al. (1996)
Pestalotiopsis spp.	*Taxus brevifolia*	Bozeman, Montana, USA	Pestalotiopsins A–B (caryophyllene type); C-methylated acetogenins; 2-hydroxydimeninol; humulane	Sesquiterpenoid	Immunosuppressant	Pulici et al. (1996a, b, c)
Pestalotiopsis guepini	*Wollemia nobilis*	Wollemi National Park, Sydney, Australia	Paclitaxel	Diterpenoid	Anticancer	Strobel et al. (1997)
P. microspora	*T. taxifolia*		Ambuic acid	Cyclohexenenone	Antifungal	Li et al. (2001)
Monochaetia sp.						
Pestalotiopsis jesteri	*Fragraea bodenii*		Jesterone	Cyclohexenenone epoxides	Antimycotic/ antioomycete	Li and Strobel (2001)
			Hydroxyjesterone			
P. microspora	*T. morobensis*	Sepik River, Papua New Guinea	Isopestacin	Isobenzofuranone	Antimicrobial	Strobel et al. (2002)
			Pestacin	1,3-Dihydro-isobenzofuran	Antioxidant	Harper et al. (2003)

(continued)

Table 13.4 (continued)

Endophyte	Plant species	Sampling site	Metabolites	Chemical group	Bioactivity	Reference
Pestalotiopsis sp.	*Pinus taeda*	–	Pestalotiopsolide A; taedolidol; 6-epitaedolidol	Sesquiterpenes	–	Magnani et al. (2003)
Pestalotiopsis foedan	Unidentified tree	Hainan Province, PRC	Pestaphthalides A–B	Isobenzofuranones	Antifungal	Ding et al. (2008)
			Pestafolide	Spiro azaphilone derivative		
Pestalotiopsis theae	Unidentified tree	Jianfeng Mountain, Hainan Province, China	Pestalotheols A–D	Isoprenylated chromenone	Anti-HIV	Liu et al. (2008a, b)
Pestalotiopsis fici			Pestaloficiols A–L	Isoprenylated chromenone	Anti-HIV	Liu et al. (2008a)
			Chloropupukeananin	Functionalized tricycle-decane skeleton	Anti-HIV	Liu et al. (2008b)
			Pestalofones A–E	Cyclohexanones	Anti-HIV	Liu et al. (2009a, b)
Pestalotiopsis terminaliae	*Terminalia arjuna*	Herbal Science Center, Chennai	Taxol	Diterpenoid	Anticancer (apoptosis	Gangadevi and Muthumary (2009)
Pestalotiopsis sp.	*Melaleuca quinquenervia* (Myrtaceae)	Toohey Forest, Queensland, Australia	Pestalactams A–C	Alkylated caprolactams	Anti-parasitic; cytotoxic	Davis et al. (2010)

Pestalotiopsis virgatula	*Terminalia chebula*	Gopalswami Hills, Mysore, India	9-Hydroxybenzo [c]oxepin-3[1] H-one;	Benzo[c]oxepin	–	Kesting et al. (2009, 2011)
			Pestalospiranes A and B	Skeleton		Li et al. (2011)
			Virgatolides A–C	Benzo[c]oxepin motif		
				Benzanulated 6,6- spiroketal class with 2 γ-lactone units		
Pestalotiopsis mangiferae	*Hyptis dilalata*	Central Panama	Mangiferaelactone	Polyhydroxylated macrolide	Antibacterial	Ortega et al. (2014)
Pestalotiopsis crassiuscula	*Fragaria chiloensis*	NITE Biological Resource	4,6-Dihydroxy	Coumarin		Yang et al. (2014)
		Center, PRC	7-Hydroxymethyl-3-Methoxymethylcoumarin			
P. microspora	*Taxus chinensis*	Hubei province, China	Pestalopyrone A	α-Pyrone	No antimicrobial activity	Li et al. (2015)
Pestalotiopsis versicolor		NITE Biological Resource	4,6-Dihydroxy-7-formyl-3-methylcoumarin;	Coumarin	No antifungal activity	Yang et al. (2015a, b)
		Center, PRC	6-[(7S,8R)-8-Propyloxiran-1-yl]-4-methoxy-pyran-2-one	α-Pyrone		
Pestalotiopsis sp. M-23	*Leucosceptrum canum* (Lamiaceae)	Kunming Botanic Gardens, PRC	2a,8adihydroxy-6,7-en-isodrimeninol;	Drimane derivatives	Antibacterial	Kuang et al. (2016)
			2ahydroxy-7a,8a-epoxy-isodrimeninol;	Isochromone derivative		
			11-Dehydro-3a-hydroxyisodrimeninol;			
			4,10-Dihydroxy-gamahorin			

(continued)

Table 13.4 (continued)

Endophyte	Plant species	Sampling site	Metabolites	Chemical group	Bioactivity	Reference
Pestalotiopsis adusta	Unidentified tree sp.	Hainan Province, Peoples Republic of China (PRC)	Pestalachlorides A–C	Chlorinated benzophenone alkaloid	Antifungal	Li et al. (2008)
	Clerodendrum canescens	Zhejiang Province, PRC	(10*S*)-12,16-epoxy-17(15→16)-abeo-3,5,8,12,15-abietapentaen-2,7,11,14-tetraone;	Diterpenoid	Cytotoxic	Xu et al. (2016)
	Sinopodophyllum hexandrum	Qinling Mountain, China	Pestalotiopsins D–G	Sesquiterpenes	Cytotoxic	Xiao et al. (2017)
			Pestalotiophols A–B			
			Pestalotiopsin H			
P. microspora	*Taxodium mucronatum*	Ootacamund, South India	7-Epi-10-deacetyltaxol	Diterpenoid	Anticancer (apoptosis)	Subban et al. (2017)
Pestalotiopsis sp. FT172	*Myrsine sandwicensis* A. DC (Myrsinaceae)	Mokuleia Forest Reserve, Hawaii	Pestallic acids A–E	Ambuic acid derivatives	Anti-proliferative	Li et al. (2017)
			Pestalotiotones A–B	Polyketides		Li et al. (2018)
Pestalotiopsis palmarum	*Sinomenium acutum* (Thunb.) Rehd et Wils.	–	Sinopestalotiollides A–D	Diphenyl ether derivatives	Cytotoxic	Xiao et al. (2018)

'–' not available

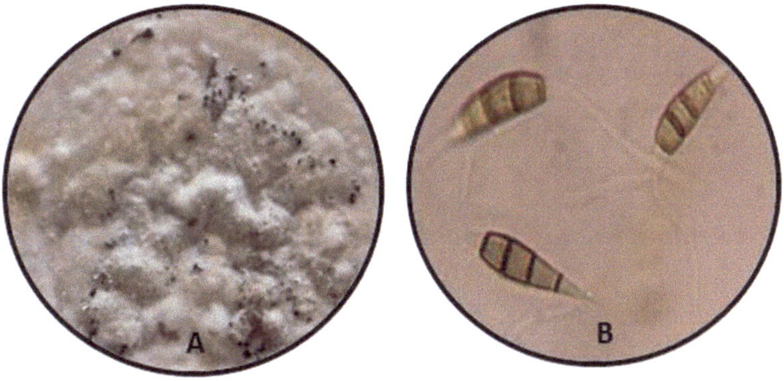

Fig. 13.2 *Pestalotiopsis* sp. isolated as an endophyte of the medicinal plant from Western Ghats, *Phyllanthus amarus* (Schum. & Thonn.) surface-sterilized stem fragments. (**a**) *Pestalotiopsis* colony on potato dextrose agar plate with pinhead-like acervulus and black conidial ooze. (**b**) Fusiform conidia with terminal unbranched appendages in light microscopy (10× magnification)

13.5.2 Anticancer and Cytotoxic Compounds

One of the first anticancer compound produced by *P. microspora*, isolated from the Himalayan yew bark, is paclitaxel, a diterpenoid (Strobel et al. 1996), confirmed by monoclonal antibody and spectroscopy techniques. The yield of the drug was 60–70 $\mu g\ l^{-1}$. This wonder molecule consists of β-phenylalanine unit which is used in the treatment of ovarian and breast cancers as it inhibits the depolymerization of tubulin molecules as a prelude to cell division (Schiff and Horowitz 1980). *P. microspora* is known to produce 7-epi-10-deacetyltaxol, a derivative of Taxol (Subban et al. 2017). *P. guepini* is an endophytic Taxol-producing fungus isolated from the ancient relic, Wollemi pine from the Wollemi Pine National Park of southwest Australia (Strobel et al. 1997).

A cytotoxic quinone dimer, torreyanic acid, isolated from *P. microspora* endophytic on the endangered tree Florida torreya is a more potent cytotoxic agent to several cancer cell lines, inducing cell death by apoptosis with values ranging from 5.1 to 65.0 μM. It is known to induce the G1 arrest of G0 synchronized cells (Lee et al. 1995a, b). Two lactone derivatives, pestalopyrones and hydroxypestalopyrones, with phytotoxic properties were derived from *P. microspora* (Lee et al. 1996). Pestaloficiols (A–L) are isoprenylated chromenone type of metabolites from *P. fici* (Liu et al. 2008b), and L showed cytotoxicity against MCF-7 and HeLa cell lines (IC_{50} 8.7 and 17.4 μM, respectively). Novel ambuic acid derivatives, pestallic acids A–E, were characterized from the Hawaiian *Pestalotiopsis* sp. FT187 with pestallic acid E exhibiting antiproliferative effect (IC_{50} 3.3 μM) in A2780 and cisplantin-resistant A2780 cell lines (Li et al. 2017). New diphenyl ether derivatives named sinopestalotiollides A–D with strong cytotoxic potentials against HeLa, HCT116, and A549 cell lines were characterized from *P. palmarum* (Xiao et al. 2018).

Three novel caprolactams, pestalactams A–C, were identified from the endophyte of Australian forest plant. Compounds A–B exhibited modest cytotoxicity in the mammalian cell lines, MCF-7 and NFF, with 12–64% inhibition at 100 μM concentration (Davis et al. 2010).

13.5.3 Antimicrobial Compounds

Pestaloside, a novel β-glucoside, is an antifungal compound inhibiting the growth of plant pathogenic fungi such as *Cladosporium* sp., *Rhizoctonia solani*, and *Geotrichum candidum* (Lee et al. 1995b). Ambuic acid, a novel organic acid with a functionalized cyclohexanone moiety found in the antibiotic tetracycline, was isolated from *Pestalotiopsis* sp. (Li et al. 2001). It was tested against a number of phytopathogens and found to be effective against *P. ultimum* at 25 μg ml^{-1}. Three caryophyllene-rich sesquiterpenes, pestalotiopsins A–C and humulane and drimane derivatives, were characterized (Pulici et al. 1996a, b, c). Two novel cyclohexanone epoxides, viz., jesterone and hydroxyjesterone, are a rare class of oxygenated cyclohexanoids and are antioomycete compounds from a novel endophytic fungus from Papua New Guinea, *P. jesteri* (Li and Strobel 2001). Noticeable activity of jesterone was against the oomycete pathogens *Phytophthora cinnamomi* and *Aphanomyces* sp. with MICs of 6.5 μg ml-^{1}.

The antioomycete activity of pestacin was determined against the root-invading pathogen *P. ultimum* with the MIC of 10 μg ml-^{1} (Harper et al. 2003). A new reduced spiro azaphilone derivative pestafolide A and pestaphthalides A and B, two new isobenzofuranones, were isolated from solid cultures of *P. foedan*. Pestafolide A displayed antifungal activity against *A. fumigatus* with a zone of inhibition of 10 mm at100 μg/disk. Pestaphthalide A showed activity against *Candida albicans*, and pestaphthalide B (43) showed activity against *G. candidum* with 11 mm zone of inhibition (fluconazole, 18–28 mm zones of inhibition for *C. albicans*, *A. fumigatus*, and *G. candidum* at 100 lg/disk) (Ding et al. 2008).

Pestalachlorides (A–C) are chlorinated benzophenones characterized from the extracts of *P. adusta*. Pestalachloride A displayed potent antifungal activity against *Fusarium culmorum* with IC_{50} value of 0.89 μM, while pestalachloride B exhibited remarkable activity against *Gibberella zeae* with an IC_{50} value of 1.1 μM. Pestalachloride C did not show antifungal activity against the test plant pathogens (Li et al. 2008). Pestalofones (C–D) are cyclohexanone derivatives from *P. fici* and exhibited inhibitory effects against *Aspergillus fumigatus* with IC_{50} values of 1.1 and 0.90 μM (Liu et al. 2009a).

Mangiferaelactone, a new polyhydroxylated macrolide, was isolated from the solid cultures of *P. mangiferae* from the Panamanian medicinal species, *H. dilatata* (Ortega et al. 2014). The compound showed antibacterial activity against *B. cereus* (MIC 0.55 μg ml^{-1}) and *Listeria monocytogenes* (MIC 1.68 μg ml^{-1}).

13.5.4 Antiviral Compounds

The isoprenylated chromenone type of metabolites, pestalofíciols (A–L), from *P. fici* displayed antiviral activity. Compounds A, B, D, G, H, J, and K inhibited the HIV-1 replication in C8166 cells (Liu et al. 2008b). Pestalotheols (A–D) from *P. theae* are derived from two isoprene units and a polyketide and displayed inhibitory activity against HIV-1 replication in C8166 cells (Liu et al. 2008b). Pestalofones (A–E) are cyclohexanone derivatives from *P. fici*. Compounds A, B, and D exhibited inhibitory effects against HIV-1 replication in C8166 cells with EC_{50} values of 90.4, 64, and 93.7 μM (Liu et al. 2009a).

13.5.5 Anti-parasitic Compounds

Pestalactams A–C are novel caprolactams identified from the endophyte of Australian plant *Melaleuca quinquenervia* (Myrtaceae). Compounds A–B exhibited modest cytotoxicity against malarial parasite cell lines, Dd2 (chloroquinone resistant) and 3D7 (chloroquinone sensitive), with 16–41% inhibition at 25 μM concentration (Davis et al. 2010).

Pestalotiopsis spp. produce the most reliable source of bioactive metabolites and are often associated with the tropical plant species. Worldwide >200 spp. are documented as pathogens and endophytes. Their nomenclature as the criterion for the evaluation of species phylogenetically is based on the relationship with the host plant (Jeewon et al. 2004), and novel species are documented. Among them, *P. microspora* is a distinct endophytic organism with metabolic diversity. Li et al. (2001) in their publication report that among the *P. microspora* strains collected from various rain forests, the strain from highland area of Papua New Guinea was of interest, as it coincided with the discovery of the perfect stage of *P. microspora* and *Pestalosphaeria hansensii*, thus providing evidence for the completion of life cycle of the endophyte. Yet, another seminal contribution from this strain is the isolation of, a novel antifungal compound, ambuic acid. The term “ambua” in Huli language of the highlands refers to the yellow color of clay, which is used as a source of facial decoration during tribal celebrations. The irony here is that the methylene chloride extract of the strain was also “yellow in color!”

13.6 Isolation, Identification, and Augmentation of Bioactive Compounds Using Worldwide Unique Technology Platform

13.6.1 *Hyphenated Techniques*

A technique where in a separation technique is coupled with an online spectroscopic detection technology is known as hyphenated technique, e.g., gas chromatography-mass spectrometry (GC-MS), liquid chromatography-mass spectrometry (LC-MS), liquid chromatography-Fourier-transform infrared (LC-FTIR) spectrometry, and liquid chromatography-nuclear magnetic resonance-mass spectrometry (LC-NMR-MS). Recent advances in hyphenated analytical techniques have remarkably widened their applications to the analysis of complex biomaterials, especially natural products (Sarker and Nahar 2012). The extracts of endophytic organisms are complex mixtures and require high-resolution bioassays and the state-of-the-art hyphenated techniques to overcome the characterization of metabolites by the usual characterization of individual metabolites (Kesting et al. 2009, 2011). The liquid cultures of bioactive culturable endophytes are upscaled to 500–2000 ml. Fungi will be grown to stationary growth phase and the extracts tested for antimicrobial activity. Extracts with activity are subjected to analysis based on simultaneous chemical and antimicrobial profiling with the aim of assessing the antimicrobial activity at the individual molecular level. This will be performed using a new worldwide unique technology platform (Schmidt et al. 2012), based on a hyphenated HPLC-SPE-NMR system further coupled with high-resolution bioassay, i.e., HPLC-MS-SPE-NMR/HR bioassay. The principles of this approach and proof-of-concept studies with a series of assays have recently been published (Grosso et al. 2013; Agnolet et al. 2012).

The simultaneous chemical and pharmacological profiling is based on the use of NMR hyphenation, i.e., an integrated and computer-managed process where components of a mixture are separated by HPLC on analytical scale (mg amounts of extracts containing μg or ng amounts of individual compounds), the individual peaks separated from HPLC mobile phase by solid-phase extraction (SPE), and the compounds submitted to structure determination by NMR (supported by MS). The use of the HPLC-SPE-NMR technology platform in natural products research has been developed and optimized by Staerk et al. (2009) and has demonstrated that the technology is applicable for the identification of a broad range of complex natural products present in crude mixtures (Staerk et al. 2009; Kesting et al. 2011). The hyphenated techniques have unravelled the bioassay-guided characterization of metabolites from *P. namyslowskii*, *P. raciborskii*, and *P. virgatula* endophytic extracts of medicinal plants (Kesting et al. 2011; Wubshet et al. 2013; Kajula et al. 2014). The significance of the technology is that the structure elucidation can target antimicrobial analytes only and performed very quickly and rigorously with very small amounts of extracts.

13.6.2 Augmentation of Bioactive Metabolite Production by Biotransformation

Endophytic microbes are bestowed with the ability to produce bioactive metabolites which is strongly influenced by a number of factors such as the host plant location, ethnomedicinal uses, and the ecosystem types. Several research groups have succeeded in elucidating the potential benefits of these microbes and their bioactive products through biotechnological processes. Bioactive metabolites from endophytes have promising potentials; the safety and concerns for human health and demands necessitate the search for synthetic drugs. Alternative methods to obtain bioactive compounds through the biotransformation method by microbial transformation have been successful for the aromatic compounds (Borges et al. 2009). Biotransformation of tetrahydrofuran lignan by endophytic *Phomopsis* sp. has led to the formation of a new compound 3,4-dimethyl-2-(4′hydroxy-3′5-dimethoxy phenyl)-5-methoxy-tetrahydrofuran with trypanocidal activity on par with the natural precursor compound against the Chagas disease parasite, *Trypanosoma cruzi* (Verza et al. 2009). The use of endophytes in the biotransformation of terpenes through the enzymatic reactions for the production of novel flavor compounds is reported (Pimentel et al. 2011).

The production of camptothecin by *F. oxysporum kolhapuriensis* isolated from the endangered medicinal plant *N. foetida* was further augmented by using whey from the dairy waste as the medium. The optimized medium yielded 283 ± 0.27 mg l^{-1} of CPT (Bhalkar et al. 2015) through the response surface methodology, which proved cost-effective. Therefore, biotechnological approaches offer tremendous potential in the production of bioactive metabolites.

13.7 Biosynthetic Potential of Endophytes

The diversity of microbes in unique environments has led to the isolation and discovery of compounds with novel chemical structures. Targeting a compound for a particular biological activity involves the screening of a number of strains against wide targets with the resulting positives designated as "leads." Deciphering the mechanisms in the biosynthesis of secondary metabolites has proven useful in knowing the metabolite-producing potential of a strain. The genomes of fungi and actinomycetes encode for the synthesis of molecules through the polyketide synthase (PKS) and non-ribosomal peptide synthetase (NPKS) genes. Using known primers, the ability of strains to produce the secondary metabolites through the detection of these genes is reported (Qin et al. 2009b; Janso and Carter 2010; Zhao et al. 2011). More so, the biosynthetic potential of culturable rhizhospheric microbes with that of Chinese medicinal plants in the Panxi Plateau yielded reliable results in terms of antimicrobial potentials and culture-independent methods (Zhao et al. 2012). This technique curtails the need to screen a number of fermentation products of the strains for bioactivities. The biosynthetic potentials of a rare strain can be exploited for the metabolite-producing capacity, but a rare actinomycete may not produce any natural products as in the case of *Planotetraspora*, a rare taxonomic group (Janso and Carter 2010).

13.8 Conclusion

The endophytic microorganisms have generated unusual interests in the bioprospection of their chemical products. Their diversity is associated with medicinal plant species, around the globe, and ensures them to be the most extensively investigated group of microbes. With the breakthrough of the anti-cancer compound to the most recently described neuroprotective and antidiabetic compounds, endophytes have diversified their metabolite-producing potentials suiting newer bioactivities. It is intriguing to understand as to why the endophytic microbes behave differently in their chemical ability to produce novel chemical structures as bioactive compounds. Interestingly, the biosynthetic pathway plays the key role in the formation of new structures, in diverse chemical class of compounds. Ultimately, the endophyte research still continues to benefit both the researcher and the society as long as the organisms tend to produce "novel" compounds of pharmacological interest.

Acknowledgments The authors wish to thank Dr. Dilfuza Egamberdieva, Institute of Landscape Biogeochemistry, Germany, for the invite to compile the book chapter. The timely help of research students is gratefully acknowledged.

Conflict of Interest The authors declare that there is no conflict of interest involved in this study.

References

Adelin E, Servy C, Cortial S et al (2011) Isolation, structure elucidation and biological activity of metabolites from Sch-642305-producing endophytic fungus *Phomopsis* sp. CMU-LMA Phytochem 72:2406–2412

Adelin E, Servy C, Martin M et al (2014) Bicyclic and tetracyclic diterpenes from a *Trichoderma* symbiont of *Taxus baccata*. Phytochemistry 97:55–61

Agnolet S, Wiese S, Verpoorte R (2012) Comprehensive analysis of commercial willow bark extracts by new technology platform: combined use of metabolomics, high-performance liquid chromatography-solid-phase extraction-nuclear magnetic resonance spectroscopy and high-resolution radical scavenging assay. J Chromatogr A 1262:130–137

Akshatha VJ, Nalini MS, D'Souza C et al (2014) *Streptomycete* endophytes from anti-diabetic medicinal plants of the Western Ghats inhibit alpha-amylase and promote glucose uptake. Lett Appl Microbiol 58:433–439

Aly AH, Edrada-Ebel R, Indriani ID et al (2008) Cytotoxic metabolites from the fungal endophyte *Alternaria* sp. and their subsequent detection in its host plant *Polygonum senegalense*. J Nat Prod 71:972–980

Alzheimer's Disease International (2015) World Alzheimer Report 2015: The Global Impact of Dementia. http://www.alz.co.uk/research/world-report-2015
Ardalani H, Avan A, Ghayour-Mobarhan M (2017) Podophyllotoxin: a novel potential natural anticancer agent. Avicenna J Phytomed 7(4):285–294
Asai T, Luo D, Yamashita K (2013) Structures and biomimetic synthesis of novel alpha-pyrone polyketides of an endophytic *Penicillium* sp. in *Catharanthus roseus*. Org Lett 15:1020–1023
Ayob FW, Simarani K, Abidin NZ et al (2017) First report on a novel *Nigrospora sphaerica* isolated from *Catharanthus roseus* plant with anticarcinogenic properties. Microbial Biotech 10(4):926–932
Bacon CW, White JF (2000) Microbial endophytes. Marcel Dekker, New York, pp 341–388
Bascom-Slack CA, Ma C, Moore E, Babbs B et al (2009) Multiple, novel biologically active endophytic actinomycetes isolated from upper Amazonian rainforests. Microb Ecol 58(2):374–383
Bhalkar BN, Bedekar PA, Patil SM et al (2015) Production of camptothecin using whey by an endophytic fungus: standardization using response surface methodology. RSC Adv 5:62828–62835
Bombardelli E, Morazzoni P (1995) Hypericum perforatum. Fitoterapia 66:43–68
Boonyaketgoson S, Trisuwan K, Bussaban B et al (2015) Isochromanone derivatives from the endophytic fungus *Fusarium* sp. PDB51F5. Tetrahedron Lett 56(36–2):5076–5078
Borges KB, Borges WDS, Dur'an-Patr R et al (2009) Stereoselective biotransformations using fungi as biocatalysts. Tetrahedron Asymmetry 20(4):385–397
Boyer N, Kim J, Morrison KC et al (2013) Synthesis and anticancer activity of epipolythiodiketopiperazine alkaloids. Chem Sci 4(4):1646–1637
Brady SF, Wagenaar MM, Singh MP et al (2000a) The cytosporones, new octaketide antibiotics isolated from an endophytic fungus. Org Lett 2(25):4043–4046
Brady SF, Singh MP, Janso JE (2000b) Cytoskyrins A and B, new BIA active bisanthraquinones isolated from an endophytic fungus. Org Lett 2:4047–4049
Brady SF, Singh MP, Janso JE et al (2000c) Guanacastepene, a fungal-derived diterpene antibiotic with a new carbon skeleton. J Am Chem Soc 122:2116–2117
Brady SF, Bondi SM, Clardy J (2001) The Guanacastepenes: a highly diverse family of secondary metabolites produced by an endophytic fungus. J Am ChemSoc 123:9900–9901
Buée M, Reich M, Murat C et al (2009) 454 pyrosequencing analyses of forest soils reveal an unexpectedly high fungal diversity. New Phytol 184:449–456
Bungihan ME, Tan MA, Kitajima M et al (2011) Bioactive metabolites of *Diaporthe* sp. P133, an endophytic fungus isolated from *Pandanus amaryllifolius*. J Nat Med 65:606–609
Cao S, Clardy J (2011) New naphthoquinones and a new d-lactone produced by endophytic fungi from Costa Rica. Tetrahedron Lett 52:2206–2208
Cao S, Ross L, Tamayo G et al (2010) Asterogynins: secondary metabolites from a Costa Rican endophytic fungus. Org Lett 12(20):4661–4663
Cao P, Yang J, Miao CP et al (2015) New duclauxamide from *Penicillium manginii* YIM PH30375 and structure revision of the duclauxin family. Org Lett 17(5):1146–1149
Carroll GC (1986) The biology of endophytism in plants with particular reference to woody perennials. In: Fokkema NJ, van den Heuvel J (eds) Microbiology of the phyllosphere. Cambridge University Press, Cambridge, pp 205–222
Caruso M, Colombo AL, Fideli L et al (2000) Isolation of endophytic fungi and actinomycetes taxane producers. Ann Microbiol 50:3–13
Castillo UF, Strobel GA, Ford EJ et al (2002) Munumbicins, wide-spectrum antibiotics produced by *Streptomyces* NRRL 30562, endophytic on *Kennedia nigriscans*. Microbiology 148:2675–2685
Castillo UF, Harper JK, Strobel GA et al (2003) Kakadumycins, novel antibiotics from *Streptomyces* sp. NRRL 30566, an endophyte of *Grevillea pteridifolia*. FEMS Microbiol Lett 234:183–190
Castillo UF, Strobel GA, Mullenberg K et al (2006) Munumbicins E-4 and E-5: novel broad-spectrumantibiotics from *Streptomyces* NRRL 3052. FEMS Microbiol Lett 255:296–300
Chun-Yan SU, Ming QL, Rahman K et al (2015) *Salvia miltiorrhiza*: traditional medicinal uses, Chemistry and Pharmacology. Chin J Nat Med 13:163–182

Clay K (1988) Clavicipitaceous fungal endophytes of grasses: coevolution and the change from parasitism to mutualism. In: Pirozynski KA, Hawksworth DL (eds) Coevolution of fungi with plants and animals. Academic, London, pp 79–105
Cummings JL (2004) Alzheimer's disease. N Engl J Med 351:56–67
Daisy B, Strobel GA, Castillo U et al (2002) Naphthalene, an insect repellent, is produced by *Muscodor vitigenus*, a novel endophytic fungus. Microbiology 148:3737–3741
DarkinRattray SJ, Gurnet AM, Myers RW et al (1996) Apicidin: a novel antiprotozoal agent that inhibits parasite histone deacetylase. Proc Natl Acad Sci U S A 93:13143–13147
Davis RA, Andjic V, Kotiw M et al (2005) Phomoxins B and C: polyketides from an endophytic fungus of the genus *Eupenicillium*. Phytochemistry 66:2771–2775
Davis RA, Carroll AR, Andrews KT et al. (2010) Pestalactams A-C: Novel caprolactams from the endophytic fungus *Pestalotiopsis* sp. Org Biomol Chem 8:1785-1790
Debbab A, Aly AH, Edrada-Ebel RA et al (2009) Bioactive metabolites from the endophytic fungus *Stemphylium globuliferum* isolated from *Mentha pulegium*. J Nat Prod 72:626–631
Debbab A, Aly AH, Edrada-Ebel R et al (2012) New anthracene derivatives – structure elucidation and antimicrobial activity. Eur J Org Chem 7:1351–1359
Demain AL (2000) Microbial natural products: a past with a future. In: Wrigley SK, Hayes MA, Thomas R et al (eds) Biodiversity: new leads for pharmaceutical and agrochemical industries. The Royal Society of Chemistry, Cambridge, pp 3–16
Dhingra AK, Chopra B, Dass R et al (2015) Antiinflammatory and antiallergy. Agents Med Chem 14(2):81–97
Ding G, Song YC, Chen JR et al (2006) Chaetoglobosin U, a cytochalasan alkaloid from endophytic *Chaetomium globosum* IFB-E019. J Nat Prod 69(2):302–304
Ding G, Liu S, Guo L et al (2008) Antifungal metabolites from the plant endophytic fungus *Pestalotiopsis foedan*. J Nat Prod 71(4):615–618. https://doi.org/10.1021/np070590f
Ding G, Wang H, Li L, Chen AJ, Lin C, Chen H, Zhang H, Liu X, Zou Z (2012) Trichoderones a and B: two Pentacyclic Cytochalasans from the plant Endophytic fungus Trichoderma gamsii. Eur J Org Chem 13:2516–2519
Ding G, Wang H, Li L, Song B, Chen H, Zhang H, Liu X, Zou Z (2013) Trichodermone, a Spirocytochalasan with a tetracyclic nucleus (7/5/6/5) skeleton from the plant Endophytic fungus. J Nat Prod 77(1):164–167
Ding H, Yang Z, Sheng L (2015) Secovironolide, a novel furanosteroid scaffold with a five-membered B ring from the endophytic fungus *Talaromyces wortmannii* LGT-4. Tetrahedron Lett 56:6754–6757
Dreyfuss MM, Chapela IH (1994) Potential of fungi in the discovery of novel, low-molecular weight pharmaceuticals. In: Gullo VP (ed) The discovery of natural products with therapeutic potential. Butterworth-Heinemann, London, pp 49–80
Duge de Bernonville T, Clastre M, Besseau S et al (2015) Phytochemical genomics of the Madagascar periwinkle: Unravelling the last twists of the alkaloid engine. Phytochemistry 113:9–23
Eyberger L, Dondapati R, Porter JR (2006) Endophyte fungal isolates from *Podophyllum peltatum* produces podophyllotoxin. J Nat Prod 69:1121–1124
Ezra D, Castillo UF, Strobel GA et al (2004) Coronamycins, peptide antibiotics produced by a verticillate *Streptomyces* sp. (MSU-2110) endophytic on *Monstera* sp. Microbiology 150:785–793
Farner CM, Zazopoulos E (2005) Improving the discovery from microorganisms. In: Zhang L, Demain DL (eds) Natural products: drug discovery and therapeutic medicine. Humana, Totowa, p 95
Findlay JA, Buthelezi S, Li G et al (1997) Insect toxins from an endophytic fungus from wintergreen. J Nat Prod 60:1214–1215
Fu GC, Yang ZD, Zhou SYG et al (2016) Wortmannines A–C, three novel wortmannin derivatives with an unusual five-membered B ring from the endophytic fungus *Talaromyces wortmannii* LGT-4. Tetrahedron Lett 57:4608–4611
Gangadevi V, Muthumary J (2009) Taxol production by *Pestalotiopsis terminaliae*, an endophytic fungus of *Terminalia arjuna* (arjun tree). Biotechnol Appl Biochem 52:9–15. https://doi.org/10.1042/BA20070243

Gao H, Li G, Lou H (2018) Structural diversity and biological activities of novel secondary metabolites from endophytes. Molecules 23:646. https://doi.org/10.3390/molecules23030646
Ge HM, Peng H, Guo ZK et al (2010) Bioactive alkaloids from the plant endophytic fungus *Aspergillus terreus*. Planta Med 76:822–824
Ge HM, Zhang WY, Ding G, Saparpakorn P, Song YC, Hannongbua S, Tan RX (2008) Chaetoglobins A and B, two unusual alkaloids from endophytic Chaetomium globosum culture. Chem Commun 45:5978
Golinska P, Wypij M, Agarkar G et al (2015) Endophytic actinobacteria of medicinal plants: diversity and bioactivity. Antonie Leewenhoek 108:267–289
Grosso C, Jäger AK, Staerk D (2013) Coupling of a high-resolution monoamine oxidase-A inhibitor assay and HPLC-SPE-NMR for advanced bioactivity profiling of plant extracts. Phytochem Anal 24:141–147
Gu W, Ge HM, Song YC et al (2007) Cytotoxic Benzo[j]fluoranthene metabolites from *Hypoxylon truncatum* IFB-18, an endophyte of *Artemisia annua*. J Nat Prod 70:114–117
Guo B, Dai J, Ng S et al (2000) Cytonic acids A and B: novel tridepside inhibitors of hCMV protease from the endophytic fungus *Cytonaema* species. J Nat Prod 63:602–604
Halliwell B, Gutteridge JMC (1989) Free radicals in biology and medicine, 2nd edn. Clarendon, Oxford
Harper JK, Arif AM, Ford EJ et al (2003) Pestacin: a 1,3- dihydroisobenzofuran from *Pestalotiopsis microspora* possessing antioxidant and antimycotics activities. Tetrahedron 59:2471–2476
Hartzell H (1991) The yew tree. Hogolosi, Orlando
Hawksworth DL (2001) The magnitude of fungal diversity: the 1.5 million species estimate revisited. Mycol Res 108:1422–1432
Hillhouse TM, Porter JH (2015) A brief history of the development of antidepressant drugs: from monoamines to glutamate. Exp Clin Psychopharmacol 23(1):1–21. https://doi.org/10.1037/a0038550
Horn WS, Simmonds MSJ, Schwartz RE et al (1995) Phomopsichalasin, a novel antimicrobial agent from an endophytic *Phomopsis* sp. Tetrahedron 14:3969–3978
Huang XZ, Zhu Y, Guan XL et al (2012) A novel antioxidant isobenzofuranone derivative from fungus *Cephalosporium* sp. AL031. Molecules 17:4219–4224
Huang HW, Oldfield S, Qian H (2015) Global significance of plant diversity in China. In: Hong DY, Blackmore S (eds) Plantsof China–a companion to the Flora of China. Science Press, Beijing, pp 7–34
Igarashi Y, Mogi T, Yanase S et al (2009) Brartemicin, an inhibitor of tumor cell invasion from the actinomycete *Nonomuraea* sp. J Nat Prod 72:980–982
Igarashi Y, Ogura H, Furihata K et al (2011) Maklamicin, an antibacterial polyketide from endophytic *Micromonospora* sp. J Nat Prod 74:670–674. https://doi.org/10.1021/np100727h
Indananda C, Igarashi Y, Ikeda M et al (2013) Linfuranone A, a new polyketide from plant-derived *Microbispora* sp. GMKU 363. J Antibiot 66:675–677
Isaka M, Jaturapat A, Rukseree K et al (2001) Phomoxanthones A and B, novel xanthone dimers from the endophytic fungus *Phomopsis* species. J Nat Prod 64(8):1015–1018
Janso JE, Carter GT (2010) Biosynthetic potential of phylogenetically unique endophytic actinomycetes from tropical plants. Appl Environ Microbiol 76:4377–4386
Jaxel C, Kohn KW, Wani MC et al (1989) Structure-activity study of the actions of camptothecin derivatives on mammalian topoisomerase I: evidence for site specific receptor site and a relation to antitumor activity. Cancer Res 49:1465–1469
Jeewon R, Liew ECY, Hyde KD (2004) Phylogenetic evaluation of species nomenclature of *Pestalotiopsis* in relation to host association. Fungal Divers 17:39–55
Jiao RH, Xu S, Liu JY et al (2006) Chaetominine, a cytotoxic alkaloid produced by endophytic *Chaetomium* sp. IFB-E015. Org Lett 8(25):5709–5712
Kajula M, Ward JM, Turpeinen A et al (2014) Bridged Epipolythiodiketopiperazines from *Penicillium raciborskii*, an endophytic fungus of *Rhododendron tomentosum* Harmaja. J Nat Prod 79(4):685–690

Kaul S, Gupta S, Ahmed M et al (2012) Endophytic fungi from medicinal plants: a treasure hunt for bioactive metabolites. Phytochem Rev. (2012) 11:487–505. https://doi.org/10.1007/s11101-012-9260-6

Kesting JR, Staerk D, Tejesvi MV et al (2009) HPLC-SPE-NMR identification of a novel metabolite containing the benzo[c]oxepin skeleton from the endophytic fungus *Pestalotiopsis virgatula* culture. Planta Med 75(10):1104–1106. https://doi.org/10.1055/s-0029-1185951

Kesting JR, Olsen L, Staerk D et al (2011) Production of unusual dispiro metabolites in *Pestalotiopsis virgatula* endophyte cultures: HPLC-SPE-NMR, electronic circular dichroism, and time-dependent density-functional computation study. J Nat Prod 74(10):2206–2215. https://doi.org/10.1021/np2005665

Khayat TM, Ibrahim SRM, Mohamed GA et al (2019) Anti-inflammatory metabolites from endophytic fungus *Fusarium* sp. Phytochem Lett 29:104–109

Kim S, Shin D, Lee T et al (2004) Periconicins, two new fusicoccane diterpenes produced by an endophytic fungus *Periconia* sp. with antibacterial activity. J Nat Prod 67:448–450

Kontnik R, Clardy J (2008) Codinaeopsin, an antimalarial fungal polyketide. Org Lett 10(18):4149–4151. https://doi.org/10.1021/ol801726k

Kornsakulkarn J, Dolsophon K, Boonyuen N et al (2011) Dihydronaphthalenones from endophytic fungus *Fusarium* sp. BCC14842. Tetrahedron 67:7540–7547

Kornsakulkarn J, Choowong W, Rachtawee P et al (2018) Bioactive hydroanthraquinones from endophytic fungus Nigrospora BCC 47789. Phytochem Lett 24:46–50

Krohn K, Biele C, Drogies KH (2002) Fusidilactones, a new group of polycyclic lactones from an endophyte, *Fusidium* sp. Eur J Org Chem 14:2331–2336

Krohn K, Kouam SF, Kuigoua GM et al (2009) Xanthones and oxepino[2, 3-b]chromones from three endophytic fungi. Chemistry 15(44):12121–12132

Kuang C, Jing S-X, Liu Y, Luo S-H, Li S-H (2016) Drimane Sesquiterpenoids and Isochromone derivative from the Endophytic fungus Pestalotiopsis sp. M-23. Nat Prod Bioprospect 6(3):155–160

Kumar DS, Lau CS, Wan JM et al (2005) Immunomodulatory compounds from *Pestalotiopsis leucothës*, an endophytic fungus from *Tripterygium wilfordii*. Life Sci 78(2):147–156

Kumar A, Patil D, Rajamohanan PR (2013) Isolation, purification and characterization of vinblastine and vincristine from endophytic fungus *Fusarium oxysporum* isolated from *Catharanthus roseus*. PLoS One 8(9):e71805. https://doi.org/10.1371/journal.pone.0071805

Kusari S, Lamshöft M, Zühlke S et al (2008) An endophytic fungus from *Hypericum perforatum* that produces hypericin. J Nat Prod 71:159–162

Kusari S, Zühlke S, Spiteller M (2009) An endophytic fungus from *Camptotheca acuminata* that produces Camptothecin and analogues. J Nat Prod 72:2–7

Kusari S, Verma VC, Lamshoeft M et al (2012) Anendophytic fungus from *Azadirachta indica* A. Juss. that produces azadirachtin. World J Microbiol Biotechnol 28:1287–1294

Lee JC, Lobkovsky E, Pliam NB et al (1995a) Subglutinols A and B: immunosuppressive compounds from the endophytic fungus *Fusarium subglutinans*. Org Chem 60:7076–7077

Lee JC, Yang X, Schwartz M et al (1995b) The relationship between an endangered North American tree and an endophytic fungus. Chem Biol 2:721–727

Lee JC, Strobel GA, Lobkovsky E et al (1996) Torreyanic acid: a selectively cytotoxic quinone dimer from the endophytic fungus *Pestalotiopsis microspora*. J Organomet Chem 61:3232–3233

Lhamo S, Wang XB, Li TX (2015) Three unusual indole diketopiperazine alkaloids from a terrestrial-derived endophytic fungus, *Aspergillus* sp. Tetrahedron Lett 56:2823–2826

Li L (2000) Opportunity and challenge of Traditional Chinese Medicine in face of the entrance to WTO (World Trade Organization). Chin J Inf Tradit Chin Med 7:7–8

Li JY, Strobel GA (2001) Jesterone and hydroxy-jesterone antioomycete cyclohexenone epoxides from the endophytic fungus *Pestalotiopsis jesteri*. Phytochemistry 57:261–265

Li JY, Strobel GA, Sidhu R et al (1996) Endophytic taxol producing fungi from bald cypress *Taxodium distichum*. Microbiology 142:2223–2226

Li JY, Strobel GA, Harper JK et al (2000) Cryptocin, a potent tetramic acid antimycotic from the endophytic fungus *Cryptosporiopsis cf. quercina*. Org Lett 2:767–770

Li JY, Harper JK, Grant DM et al (2001) Ambuic acid, a highly functionalized cyclohexenone with antifungal activity from *Pestalotiopsis* spp. and *Monochaetia* sp. Phytochemistry 56:463–468

Li J, Zhao GZ, Chen HH et al (2008) Antitumour and antimicrobial activities of endophytic streptomycetes from pharmaceutical plants in rainforest. Lett Appl Microbiol 47:574–580

Li J, Li L, Si YK et al (2011) Virgatolides A–C, benzannulated spiroketals from the plant endophytic fungus *Pestalotiopsis virgatula*. Org Lett 13:2670–2673

Liu Y et al (2013) Structures and absolute configurations of penicillactones A-C from an endophytic microorganism, *Penicillium dangeardii Pitt*. Org Lett 15(20):5206–5209

Li X, Guo Z, Deng Z et al (2015) A new α-pyrone derivative from endophytic fungus *Pestalotiopsis microspora*. Rec Nat Prod 9(4):503–508

Li CS, Ding Y, Yang B et al (2016a) Eremophilane sesquiterpenes from Hawaiian endophytic fungus *Chaetoconis* sp. FT087. Phytochemistry 126:41–46

Li CS, Ren G, Yang BJ et al (2016b) Meroterpenoids with antiproliferative activity from a Hawaiian-plant associated fungus *Peyronellaea coffeae-arabicae* FT238. Org Lett 18:2335–2338

Li W, Xu J, Li F et al (2016c) A new antifungal isocoumarin from the endophytic fungus *Trichoderma* sp. 09 of *Myoporum bontioides* A. Gray. Pharmacogn Mag 12(48):259–261. https://doi.org/10.4103/0973-1296.192204

Li X, Zhai X, Shu Z et al (2016d) *Phoma glomerata* D14: an endophytic fungus from *Salvia miltiorrhiza*. Curr Microbiol 73:31–37

Li C, Yang B, Turkson J et al (2017) Anti-proliferative ambuic acid derivatives from Hawaiian endophytic fungus *Pestalotiopsis* sp. FT172. Phytochemistry 140:77–82. https://doi.org/10.1016/j.phyochem.2017.04.017

Li C, Sarotti AM, Yoshida W et al (2018) Two new polyketides from Hawaiian endophytic fungus *Pestalotiopsis* sp. FT172 Pestalotiotones A-B. Tetrahedron Lett 59(1):42–45

Liao G, Wu P, Xue J et al (2018) Asperimides A–D, anti-inflammatory aromatic butenolides from a tropical endophytic fungus *Aspergillus terreus*. Fitoterapia 131:50–54

Lim W, Park J, Lee YH et al (2015) Subglutinol A, an immunosuppressive α-pyrone diterpenoid from *Fusarium subglutinans*, acts as an estrogen receptor antagonist. Biochem Biophys Res Commun 461(3):507–512

Lin Z, Wen J, Zhu T et al (2008) Chrysogenamide A from an endophytic fungus associated with *Cistanche deserticola* and its neuroprotective effect on SH-SY5Y cells. J Antibiot 61:81–85

Liu L, Liu SC, Jiang LH et al (2008a) Chloropupukeananin, the first chlorinated pupukeanane derivative, and its precursors from *Pestalotiopsis fici*. Org Lett 10:1397–1400

Liu L, Tian RR, Liu SC et al (2008b) Pestalofìciols A–E, bioactive cyclopropane derivatives from the plant endophytic fungus *Pestalotiopsis fici*. Bioorg Med Chem 16:6021–6026

Liu L, Liu S, Chen X et al (2009a) Pestalofones A–E, bioactive cyclohexanone derivatives from the plant endophytic fungus *Pestalotiopsis fici*. Bioorg Med Chem 17:606–613

Liu L, Liu S, Niu S et al (2009b) Isoprenylated chromone derivatives from the plant endophytic fungus *Pestalotiopsis fici*. J Nat Prod 72:1482–1486

Liu S, Jiang J, Huang R et al (2019) A new antiviral 14-nordrimane sesquiterpenoid from an endophytic fungus *Phoma* sp. Phytochem Lett 29:75–78

Lösgen S, Magull J, Schulz B et al (2008) Isofusidienols: novel chromone-3-oxepines produced by the endophytic fungus *Chalara* sp. Eur J Org Chem 4:698–703

Lu H, Zou WX, Meng JC et al (2000) New bioactive metabolites produced by *Colletotrichum* sp., an endophytic fungus in *Artemisia annua*. Plant Sci 151:67–73

Magnani RF, Rodrigues-Fo E, Daolio C et al (2003) Three highly oxygenated caryophyllene sesquiterpenes from *Pestalotiopsis* sp., a fungus isolated from bark of *Pinus taeda*. J Biosci (Z Naturforsch) 58c:319–324

Maha A, Rukachaisirikul V, Phongpaichit S et al (2016) Dimeric chromanone, cyclohexenone and benzamide derivatives from the endophytic fungus *Xylaria* sp. PSU-H182. Tetrahedron 72:2874–2879

Marinho AMR, Rodrigues Filho E et al (2005) Biologically active polyketides produced by *Penicillium janthinellum* isolated as an endophytic fungus from fruits of *Melia azedarach*. J Braz Chem Soc 16:280–283

Martinez-Luis S, Togna GD, Coley PD et al (2008) Antileishmanial constituents of the Panamanian endophytic fungus *Edenia* sp. J Nat Prod 71(12):2011–2014. https://doi.org/10.1021/np800472q
Matsumoto A, Takahashi Y (2017) Endophytic actinomycetes: promising source of novel bioactive compounds. J Antibiot 70:514–519. https://doi.org/10.1038/ja.2017.20
Metwaly AM, Kadry HA, El-Hela AA et al (2014a) Nigrosphaerin A a new isochromene derivative from the endophytic fungus *Nigrospora sphaerica*. Phytochem Lett 7:1–5
Metwaly AM, Fronczek FR, Maa G et al (2014b) Antileukemic a-pyrone derivatives from the endophytic fungus *Alternaria phragmospora*. Tetrahedron Lett 55:3478–3481
Ming QL, Han T, Li WC et al (2012) Tanshinone IIA and tanshinone I production by *Trichoderma atroviride* D16, an endophytic fungus in *Salvia miltiorrhiza*. Phytomedicine 19:330–333
Mittermeier RA, Myers N, Gill PC et al (2004) Hotspots: Earth's richest and most endangered terrestrial ecoregions. CEMEXe, Mexico City
Müller P, Döring M (2009) Isothermal DNA amplification facilitates the identification of a broad spectrum of bacteria, fungi and protozoa in *Eleutherococcus* sp. plant tissue cultures. Plant Cell Tissue Organ Cult 98:35–45. https://doi.org/10.1007/s11240-009-9536-8
Mutke J, Sommer JH, Kreft H et al (2011) Vascular plant diversity in a changing world: global centers and biome-specific patterns. In: Zachoa FE, Habel JC (eds) Biodiversity hotspots. Springer, Hiedelberg, pp 83–96
Myers N (1988) Threatened biotas: "hotspots" in tropical forests. Environmentalist 8:1–20
Myers N, Mittermeier RA, Mittermeier CG, da Fonseca GAB, Kent J (2000) Biodiversity hotspots for conservation priorities. Nature 403(6772):853–858
Nalini MS, Prakash HS (2017) Diversity and bioprospecting of actinomycete endophytes from the medicinal plants. Lett Appl Microbiol 64:261–270
Newman DJ, Cragg GM (2016) Natural products as sources of new drugs from 1981 to 2014. J Nat Prod 79:629–661
Ortega HE, Shen YY, Dyke KT (2014) Polyhydroxylated macrolide isolated from the endophytic fungus *Pestalotiopsis mangiferae*. Tetrahedron Lett 55:2642–2645
Palem PPC, Kuriakose GC, Jayabaskaran C (2015) An endophytic fungus, *Talaromyces radicus*, isolated from *Catharanthus roseus*, produces vincristine and vinblastine, which induce apoptotic cell death. PLoS One 10(12):e0144476
Pandey MM, Rastogi S, Rawat AKS (2013) Indian traditional Ayurvedic system of medicine and nutritional supplementation. Evid Based Complement Alternat Med. Article ID 376327. https://doi.org/10.1155/2013/376327
Passari AK, Mishra VK, Saikia R et al (2015) Isolation, abundance and phylogenetic affiliation of endophytic actinomycetes associated with medicinal plants and screening for their in vitro antimicrobial biosynthetic potential. Front Microbiol 6:273. https://doi.org/10.3389/fmicb.2015.00273
Pei SJ, Huai HY (2015) Medicinal plants. In: Hong DY, Blackmore S (eds) Plants of China, a companion to the flora of China. Science Press, Beijing
Petrini O (1991) Fungal endophytes of tree leaves. In: Andrews H, Hirano SS (eds) Microbial ecology of the leaves. Springer, New York
Pimentel MR, Molina G, Dion'ısio AP et al (2011) The use of endophytes to obtain bioactive compounds and their application in biotransformation process. Biotechnol Res Int. Article ID 576286. https://doi.org/10.4061/2011/576286
Pu X, Qu X, Chen F et al (2013) Camptothecin-producing endophytic fungus *Trichoderma atroviride* LY357: isolation, identification, and fermentation conditions optimization for camptothecin production. Appl Microbiol Biotechnol 97:9365–9375. https://doi.org/10.1007/s00253-013-5163-8
Pujianto S, Lestari Y, Suwanto A et al (2012) Alpha-glucosidase inhibitor activity and characterization of endophytic actinomycetes isolated from some Indonesian diabetic medicinal plants. Int J Pharm Pharm Sci 4:327–333
Pulici M, Sugawara F, Koshino H et al (1996a) Metabolites of endophytic fungi of *Taxus brevifolia*-the first highly functionalized humulane of fungal origin. J Chem Res N 8:378–379

Pulici M, Sugawara F, Koshino H et al (1996b) A new isodrimeninol from *Pestalotiopsis* sp. J Nat Prod 59:47–48

Pulici M, Sugawara F, Koshino H et al (1996c) Pestalotiopsin-A and pestalotiopsin-B: new caryophyllenes from an endophytic fungus of *Taxus brevifolia*. J Organomet Chem 61:2122–2124

Puri SC, Verma V, Amna T et al (2005) An endophytic fungus from *Nothapodytes foetida* that produces camptothecin. J Nat Prod 68:1717–1719

Puri SC, Nazir A, Chawla R et al (2006) The endophytic fungus *Trametes hirsuta* as a novel alternative source of podophyllotoxin and related aryl tetralin lignans. J Biotechnol 122:494–510

Qi C, Gao W, Wang J et al (2018) Terrusnolides A–D, new butenolides with anti-inflammatory activities from an endophytic *Aspergillus* from *Tripterygium wilfordii*. Fitoterapia 130:134–139

Qin S, Krohn K, Flörke U et al (2009a) Two new Fusidilactones from the fungal endophyte *Fusidium* sp. Eur J Org Chem:3279–3284. https://doi.org/10.1002/ejoc.200900152

Qin S, Li J, Chen HH et al (2009b) Isolation, diversity, and antimicrobial activity of rare actinobacteria from medicinal plants of tropical rain forests in Xishuangbanna, China. Appl Environ Microbiol 75:6176–6186

Qin S, Xing K, Jiang J-H, Xu L-H, Li W-J (2011) Biodiversity, bioactive natural products and biotechnological potential of plant-associated endophytic actinobacteria. Appl Microbiol Biotechnol 89(3):457–473

Qin S, Chen HH, Zhao G et al (2012) Abundant and diverse endophytic actinobacteria associated with medicinal plant *Maytenus austroyunnanensis* in Xishuangbanna tropical rainforest revealed by culture-dependent and culture-independent methods. Environ Microbiol Rep 4:522–531

Raj N (1993) Coelomycetous anamorphs with appendage bearing conidia. Edward Brothers, Ann Harbor

Ramos AC, Peláez R, López JL et al (2001) Heterolignanolides Furo-and thieno-analogues of podophyllotoxin and thuriferic acid. Tetrahedron 57:3963–3977

Rehman S, Shawl AS, Kour A et al (2008) An endophytic *Neurospora* sp. from *Nothapodytes foetida* producing camptothecin. Appl Biochem Microbiol 44:203–209. https://doi.org/10.1134/s0003683808020130

Remali J, Sarmin NIM, Ng CL et al (2017) Genomic characterization of a new endophytic *Streptomyces kebangsaanensis* identifies biosynthetic pathway gene clusters for novel phenazine antibiotic production. Peer J 5:e-3738. https://doi.org/10.7717/peerj.3738

Ren Y, Strobel GA, Graff JC et al (2008) Colutellin A, an immunosuppressive peptide from *Colletotrichum dematium*. Microbiology 154:1973–1979

Rukachaisirikul V, Sommart U, Phongpaichit S et al (2008) Metabolites from the endophytic fungus *Phomopsis* sp. PSU-D15. Phytochemistry 69:783–787

Saetang P, Rukachaisirikul V, Phongpaichit S et al (2017) Depsidones and an α-pyrone derivative from *Simplicillium* sp. PSU-H41, an endophytic fungus from *Hevea brasiliensis* leaf. Phytochemistry 143:115–123

Saito H (2000) Regulation of herbal medicines in Japan. Pharmacol Regul 41:515–519

Saleem M, Hussain H, Ahmed I et al (2011) Viburspiran, an antifungal member of the octadride class of maleic anhydride natural products. Eur J Org Chem:808–812. https://doi.org/10.1002/ejoc.201001324

Saleem M, Riaz N, Tousif MI et al (2013) Cryptosporioptide: a bioactive polyketide produced by an endophytic fungus *Cryptosporiopsis* sp. Phytochemistry 93:199–202. https://doi.org/10.1016/j.phytochem.2013.03.018

Salituro GM, Pelaez F, Zhang BB (2001) Discovery of a small molecule insulin receptor activator. Recent Prog Horm Res 56:107–126

Sang XN, Chen SF, Tang MX et al (2017) α-Pyrone derivatives with cytotoxic activities, from the endophytic fungus *Phoma* sp. YN02-P-3. Bioorg Med Chem Lett 27(16):3723–3725. https://doi.org/10.1016/j.bmcl.2017.06.079

Sappapan R, Sommit D, Ngamrojanavanich N et al (2008) 11-Hyroxymonocerin from the plant endophytic fungus *Exserohilum rostratum*. J Nat Prod 71(9):1657–1659

Sarker SD, Nahar L (2012) Hyphenated techniques and their applications in natural products analysis. In: Sarker SD, Nahar L (eds) Natural products isolations: methods and protocols. Humana, pp 301–340

Schiff PB, Horowitz SB (1980) Taxol stabilizes microtubules in mouse fibroblast cells. Proc Natl Acad Sci U S A 77:1561–1565

Schmidt JS, Lauridsen MB, Dragsted LO et al (2012) Development of a bioassay-coupled HPLC-SPE-ttNMR platform for identification of α-glucosidase inhibitors in apple peel (*Malus* × *domestica* Borkh.). Food Chem 135:1692–1699

Schmidt P-A, Bálint M, Greshake B et al (2013) Illumina metabarcoding of a soil fungal community. Soil Biol Biochem 65:128–132

Schmit JP, Mueller GM (2007) An estimate of the lower limit of global fungal diversity. Biodivers Conserv 16:99–111

Schulz B, Boyle C (2006) What are endophytes? In: Schulz B, Boyle C, Sieber TN (eds) Microbial root endophytes. Springer, Heidelberg, pp 1–13

Singh MP, Janso JE, Brady SF (2007) Cytoskyrins and cytosporones produced by *Cytospora* sp. CR200: taxonomy, fermentation and biological activities. Mar Drugs 5(3):71–84

Sommart U, Rukachaisirikul V, Tadpetch K et al (2012) Modiolin and phthalide derivatives from the endophytic fungus *Microsphaeropsis arundinis* PSU-G18. Tetrahedron 68:10005–10010

Staerk D, Kesting J, Sairafianpour M (2009) Accelerated dereplication of crude extracts using HPLC-PDA-MS-SPE-NMR: quinolone alkaloids of *Haplophyllum acutifolium*. Phytochemistry 70:1055–1061

Stierle A, Strobel GA, Stierle D (1993) Taxol and taxane production by *Taxomyces andreanae*, an endophytic fungus of Pacific yew. Science 260:214–216

Stierle A, Stierle D, Strobel G et al (1994) Bioactive metabolites of the endophytic fungi of Pacific yew, Taxus brevfolia. In: Georg GI, Chen TT, Ojima I et al (eds) Taxane anticancer agents, basic science Czlrrent Statzts, American Chemical Society symposium series, no. 583, Washington, DC, pp 81–97

Strobel GA, Daisy B (2003) Bioprospecting for microbial endophytes and their natural products. Microbiol Mol Biol Rev 67:491–502

Strobel G, Yang X, Sears J et al (1996) Taxol from *Pestalotiopsis microspora*, an endophytic fungus of *Taxus wallichiana*. Microbiology 142:435–440. https://doi.org/10.1099/13500872-142-2-435

Strobel GA, Hess WM, Li JY et al (1997) *Pestalotiopsis guepinii*, a taxol producing endophyte of the Wollemi pine, *Wollemia nobilis*. Aust J Bot 45:1073–1082. https://doi.org/10.1071/bt96094

Strobel GA, Miller RV, Miller C et al (1999) Cryptocandin, a potent antimycotic from the endophytic fungus *Cryptosporiopsis* cf. *quercina*. Microbiology 145:1919–1926

Strobel G, Ford E, Worapong J et al (2002) Isopestacin, an isobenzofuranone from *Pestalotiopsis microspora*, possessing antifungal and antioxidant activities. Phytochemistry 60(2):179–183

Strobel G, Daisy B, Castillo U (2004) Natural products from endophytic microorganisms. J Nat Prod 67:257–268. https://doi.org/10.1021/np030397v

Subban K, Singh S, Subramani R et al (2017) Fungal 7-epi-10-deacetyltaxol produced by an endophytic *Pestalotiopsis microspora* induces apoptosis in human hepatocellular carcinoma cell line (HepG2). BMC Complement Altern Med 17:504. https://doi.org/10.1186/s12906-017-1993-8

Sun W, Zhu H, Zhang T et al (2018) Two new alkaloids from *Fusarium tricinctum* SYPF 7082, an endophyte from the root of *Panax notoginseng*. Nat Prod Bioprosp 8(5):391–396

Taechowisan T, Chanaphat S, Ruensamran W et al (2014) Antibacterial activity of new flavonoids from *Streptomyces* sp. BT01; an endophyte in *Boesenbergia rotunda* (L.). Mansf J App Pharm Sci 4(4):008–013

Talontsi FM, Melib CD, Laatsch H et al (2013) Depsidones from an endophytic fungus *Chaetomium* sp. associated with *Zanthoxylum leprieurii*. Z Naturforsch 68b:1259–1264. https://doi.org/10.5560/znb.2013-3168

Taylor DL, Sinsabaugh RL (2014) The soil fungi: occurrence phylogeny and ecology. In: Paul EA (ed) Soil microbiology ecology and biochemistry, 4th edn. Academic, Burlington, p 77

Tejesvi MV, Kini KR, Prakash HS et al (2007) Genetic diversity and antifungal activity of species of *Pestalotiopsis* isolated as endophytes from medicinal plants. Fungal Divers 24:37–54

Teponno RB, Noumeur SR, Helaly SE et al (2017) Furanones and anthranilic acid derivatives from the endophytic fungus *Dendrothyrium variisporum*. Molecules 9 22(10):E1674. https://doi.org/10.3390/molecules22101674

Trémouillaux-Guiller J, Rohr T, Rohr R et al (2002) Discovery of an endophytic alga in *Ginkgo biloba*. Am J Bot 89:727–733

Van der Sar SA, Blunt JW, Munro MH (2006) Spiro-mamakone A: a unique relative of the spirobisnaphthalene class of compounds. Org Lett 8:2059–2061

Veitch GE, Pinto A, Boyer A et al (2008) Synthesis of natural products from the Indian neem tree *Azadirachta indica*. Org Lett 10(4):569–572

Venieraki A, Dimou M, Katinakis P (2017) Endophytic fungi residing in medicinal plants have the ability to produce the same or similar pharmacologically active secondary metabolites as their hosts. Helln Plant Prot J 10:51–66

Verza M, Arakawa NS, Lopes NP et al (2009) Biotransformation of a tetrahydrofuran lignan by the endophytic fungus *Phomopsis* sp. J Braz Chem Soc 20(1):195–200

Vieira FCS, Nahas E (2005) Comparison of microbial numbers in soils by using various culture media and temperatures. Microbiol Res 160:197–202

Vu HT, Nguyen DT, Nguyen HQ et al (2018) Antimicrobial and cytotoxic properties of bioactive metabolites produced by *Streptomyces cavourensis* YBQ59 isolated from *Cinnamomum cassia* Prels in Yen Bai province of Vietnam. Curr Microbiol 75:1247–1255 https://doi.org/10.1007/s00284-018-1517-x

Wagenaar MM, Clardy J (2001) Dicerandrols, new antibiotic and cytotoxic dimers produced by the fungus *Phomopsis longicolla* isolated from an endangered mint. J Nat Prod 64(8):1006–1009

Wagenaar MM, Corwin J, Strobel G et al (2000) Three new cytochalasins produced by an endophytic fungus in the genus *Rhinocladiella*. J Nat Prod 63(12):1692–1695

Wall ME, Wani MC, Cook CE et al (1966) Plant antitumor agents I. The isolation and structure of Camptothecin, a novel alkaloidal leukemia and tumor inhibitor from *Camptotheca acuminata*. J Am Chem Soc 88(16):3888–3890. https://doi.org/10.1021/ja00968a057

Wang JT, Zhang PL, Liu JS (2018) Aspergilates A to E, second metabolites from *Aspergillus* sp. isolated from *Paeonia ostii*. Fitoterapia 131:204–208

Wang W, Zheng M, Tao F et al (2019) Cytotoxic polyketides from endophytic fungus *Phoma bellidis* harbored in *Tricyrtis maculata*. Phytochem Lett 29:41–46

Wani MC, Taylor HL, Wall ME et al (1971) Plant antitumor agents VI. The isolation and structure of taxol, a novel antileukemic and antitumor agent from *Taxus brevifolia*. J Am Chem Soc 93:2325–2327

Weber D, Sterner O, Anke T et al (2004) Phomol, new antiinflammatory metabolite from an endophyte of the medicinal plant, *Erythrina crista-galli*. J Antibiot 57:559–563

Wen H, Liu X, Zhang Q et al (2018) Three new indole diketopiperazine alkaloids from *Aspergillus ochraceus*. Chem Biodivers 15(4):e1700550. https://doi.org/10.1002/cbdv.201700550

Willis KJ (ed) (2017) State of the world's plants 2017. Report. Royal Botanic Gardens, Kew, London

Wu N, Ong Y, Fu YJ et al (2011) In vitro antioxidant properties, DNA damage protective activity, and xanthine oxidase inhibitory effect of cajaninstilbene acid, a stilbene compound derived from pigeon pea [*Cajanus cajan* (L.) Millsp.] leaves. J Agric Food Chem 59:437–443

Wubshet SG, Nyberg NT, Tejesvi MV et al (2013) Targeting high-performance liquid chromatography-high-resolution mass spectrometry-solid- phase extraction-nuclear magnetic resonance analysis with high resolution radical scavenging profiles-bioactive secondary metabolites from the endophytic fungus *Penicillium namyslowskii*. J Chromatogr A 1302:34–39

Xiao J, Lin L, Hu J et al (2017) Highly oxygenated caryophyllene-type and drimane-type sesquiterpenes from *Pestalotiopsis adusta*, an endophytic fungus of *Sinopodophyllum hexandrum*. RSC Adv 7(46):29071–29079

Xiao J, Hu J, Sun H et al (2018) Sinopestalotiollides A–D, cytotoxic diphenyl ether derivatives from plant endophytic fungus *Pestalotiopsis palmarum*. Bioorg Med Chem Lett 28(3):515–518

Xin X, Chen Y, Zhang H et al (2019) Cytotoxic seco-cytochalasins from an endophytic Aspergillus sp. harbored in *Pinellia ternata* tubers. Fitoterapia 132:53–59

Xu MF, Jia OY, Wang SJ et al (2016) A new bioactive diterpenoid from P*estalotiopsis adusta*, an endophytic fungus from *Clerodendrum canescens*. Nat Prod Res 30(23):2642–2647. https://doi.org/10.1080/14786419.2016.1138297

Yan Y, Ma Y, Yang J et al (2016) Tropolone ring construction in the biosynthesis of Rubrolone B, a cationic tropolone alkaloid from endophytic *Streptomyces*. Org Lett 18:1254–1257

Yang X, Zhang L, Guo B et al (2004) Preliminary study of vincristine-producing endophytic fungus isolated from leaves of *Catharanthus roseus*. China Trade Herb Drugs 35:79–81

Yang SX, Xiao J, Laatsch H (2012) Fusarimine, a novel polyketide isoquinoline alkaloid, from the endophytic fungus *Fusarium* sp. LN12, isolated from *Melia azedarach*. Tetrahedron Lett 53:6372–6375

Yang X, Yang Y, Peng T et al (2013) A new cyclopeptide from endophytic *Streptomyces* sp. YIM 64018. Nat Prod Commun 8:1753–1754

Yang Y, Yang X, Zhang Y et al (2013) A new daidzein derivative from endophytic *Streptomyces* sp. YIM 65408. Nat Prod Res 27(19):1727–1731

Yang X, Huang L, Ruan X (2014) Epigenetic modifiers alter the secondary metabolite composition of a plant endophytic fungus, *Pestalotiopsis crassiuscula* obtained from the leaves of *Fragaria chiloensis*. J Asian Nat Prod Res 16(4):412–417. https://doi.org/10.1080/10286020.2014.881356

Yang X, Huang L, Li H et al (2015a) Two new compounds from the plant endophytic fungus *Pestalotiopsis versicolor*. J Asian Nat Prod Res 17(4):333–337. https://doi.org/10.1080/10286020.2014.961918

Yang X, Peng T, Yang Y et al (2015b) Antimicrobial and antioxidant activities of a new benzamide from endophytic *Streptomyces* sp. YIM 67086. Nat Prod Res 29:331–335

Yuan H, Zhang X, Zhao K, Zhong K, Gu Y, Lindström K (2008) Genetic characterisation of endophytic actinobacteria isolated from the medicinal plants in Sichuan. Ann Microbiol 58(4):597–604

Zafari D, Leylaie S, Tajick MA (2018) Isolation and identification of vinblastine from the fungus of *Chaetomium globosum* Cr95 isolated from *Catharanthus roseus* plant. http://bjm.ui.ac.ir/article_22831_8bba73f5d4d0d402036d4698140485b7.pdf

Zhai X, Luo D, Li X et al (2018) Endophyte *Chaetomium globosum* D38 promotes bioactive constituents accumulation and root production in *Salvia miltiorrhiza*. Front Microbiol 8:2694

Zhang B, Salituro G, Szalkowski D et al (1999) Discovery of small molecule insulin mimetic with antidiabetic activity in mice. Science 284:974–981

Zhang W, Krohn K, Zia U et al (2008a) New mono- and dimeric members of the secalonic acid family: blennolides A–G isolated from the fungus *Blennoria* sp. Chem Eur J 14:4913–4923

Zhang HW, Huang WY, Chen JR et al (2008b) Cephalosol: an antimicrobial metabolite with an unprecedented skeleton from endophytic *Cephalosporium acremonium* IFB-E007. Chem Eur J 14:10670–10674

Zhang AH, Jiang N, Gu W et al (2010) Characterization, synthesis and self-aggregation of (−)-alternarlactam: a new fungal cytotoxin with cyclopentenone and isoquinolinone scaffolds. Chem Eur J 16:14479–14485

Zhang D, Ge H, Zou JH et al (2014a) Periconianone A, a new 6/6/6 carbocyclic sesquiterpenoid from endophytic fungus *Periconia* sp. with neural anti-inflammatory activity. Org Lett 16:1410–1413

Zhang JC, Yang YB, Zhou H et al (2014b) Medelamine C, a new omega-hydroxy alkylamine derivative from endophytic Streptomyces sp. YIM 66142. Nat Prod Commun 9(1):99–100

Zhang D, Liu J, Tao X et al (2016) Periconiasin G, a new cytochalasan with unprecedented 7/6/5 tricyclic ring system from the endophytic fungus *Periconia* sp. Tetrahedron Lett 57(7). https://doi.org/10.1016/j.tetlet.2016.01.030

Zhang SP, Huang R, Li FF et al (2016) Antiviral anthraquinones and azaphilones produced by an endophytic fungus *Nigrospora* sp. from *Aconitum carmichaeli*. Fitoterapia 112:85–89. https://doi.org/10.1016/j.fitote.2016.05.013

Zhao K, Penttinen P, Guan T et al (2011) The diversity and anti-microbial activity of endophytic actinomycetes isolated from medicinal plants in Panxi plateau, China. Curr Microbiol 62:182–190. https://doi.org/10.1007/s00284-010-9685-3

Zhao JT, Fu YJ, Luo M et al (2012) Endophytic fungi from pigeon pea [*Cajanus cajan* (L.) Millsp.] produce antioxidant Cajaninstilbene Acid. J Agric Food Chem 60:4314–4319. https://doi.org/10.1021/jf205097y

Zhao J, Yang Z, Zhou S et al (2019) Wortmannine F and G, two new pyranones from *Talaromyces wortmannii* LGT-4, the endophytic fungus of *Tripterygium wilfordii*. Phytochem Lett 29:115–118

Zhou H, Yang Y, Zhang J et al (2013) Alkaloids from an endophytic *Streptomyces* sp. YIM66017. Nat Prod Commun 8:1393–1396

Zhou H, Yang Y, Peng T et al (2014) Metabolites of *Streptomyces* sp., an endophytic actinomycete from *Alpinia oxyphylla*. Nat Prod Res 28:265–267. https://doi.org/10.1080/14786419.2013.830219

Zhou M, Miao MM, Du G et al (2014) Aspergillines A–E, highly oxygenated hexacyclic indole-tetrahydrofuran-tetramic acid derivatives from *Aspergillus versicolor*. Org Lett 16(19):5016–5019. https://doi.org/10.1021/ol502307u

Zhou P, Wu Z, Tan D et al (2017) Atrichodermones A–C, three new secondary metabolites from the solid culture of an endophytic fungal strain, *Trichoderma atroviride*. Fitoterapia 123:18–22

Zhu X, Chen J, Zhu S et al (2018) Two new compounds from *Nigrospora sphaerica* ZMT05, a fungus derivated from *Oxya chinensis* Thunber. Nat Prod Res 32(20):2375–2381. https://doi.org/10.1080/14786419.2017.1413566

Zin NM, Sarmin NIM, Ghadin N, Basri DF, Sidik NM, Hess WM, Strobel GA (2007) Bioactive endophytic streptomycetes from the Malay Peninsula. FEMS Microbiol Lett 274:83–88

Zou WX, Meng JC, Lu H et al (2000) Metabolites of *Colletotrichum gloeosporioides*, an endophytic fungus in *Artemisia mongolica*. J Nat Prod 63:1529–1530

MIX
Papier aus verantwortungsvollen Quellen
Paper from responsible sources
FSC® C105338

If you have any concerns about our products,
you can contact us on
ProductSafety@springernature.com

In case Publisher is established outside the EU,
the EU authorized representative is:
Springer Nature Customer Service Center GmbH
Europaplatz 3, 69115 Heidelberg, Germany

Printed by Libri Plureos GmbH
in Hamburg, Germany